Modernization
of Agriculture
in Developing
Countries

ENVIRONMENTAL MONOGRAPHS AND SYMPOSIA

A Series in the Environmental Sciences

Convener and General Editor
NICHOLAS POLUNIN, CBE
Geneva, Switzerland

MODERNIZATION OF AGRICULTURE IN DEVELOPING COUNTRIES: Resources, Potentials and Problems

I. ARNON, *Hebrew University, Jerusalem, Agricultural Research Service, Bet Dagan, and Settlement Study Centre, Rehovot, Israel*

STRESS EFFECTS ON NATURAL ECOSYSTEMS
Edited by
G. W. BARRETT, *Institute of Environmental Sciences and Department of Zoology, Miami University, Oxford, Ohio, USA*
and
R. ROSENBERG, *Fishery Board of Sweden, Institute of Marine Research, Lysekil, Sweden*

Modernization of Agriculture in Developing Countries

RESOURCES, POTENTIALS AND PROBLEMS

I. ARNON
Hebrew University, Jerusalem, Agricultural Research Service, Bet Dagan, and Settlement Study Centre, Rehovot, Israel

A Wiley–Interscience Publication

JOHN WILEY & SONS

Chichester · New York · Brisbane · Toronto

British Library Cataloguing in Publication Data:

Arnon, Isaac
Modernization of agriculture in developing countries. — (Environmental monographs and symposia).

1. Underdeveloped areas—Agriculture—Economic aspects
I. Title II. Series
338.1′09172′4 HD1417 80-41588

ISBN 0 471 27928 5

Typeset by Preface Ltd, Salisbury, Wilts. and printed in the United States of America by Vail-Ballou Press, Inc., Binghamton, N.Y.

To my wife, Hilda,
who for fifty years has shared with me the challenges, hardships and rewards of work in a developing country.

Acknowledgements

This book is to a large extent a review of numerous studies and case histories on the problems involved in the transition from traditional to modern agriculture, that have been published in recent years in the wake of the Green Revolution. The sources of the material presented here, are recognized in the usual way in the lists of references at the end of each chapter. However, because so many disciplines are involved in the modernization process—social, economic, political, and technological, the author has drawn on the works of a number of specialists in various fields to an extent to which a simple citation of a reference cannot do adequate justice. I wish therefore to acknowledge my debt in particular to the following: Cynthia Hewlitt de Alcantera (Mexican agriculture); S. L. Barraclough (rural development strategy); D. Benor and J. Q. Harrison (extension methods); R. Chambers and J. Moris (settlement schemes in Tropical Africa); FAO Study Mission (agriculture in China); G. W. Giles (mechanization of agriculture); B. F. Johnston and associates (strategies of agricultural development); Ingrid Palmer (case studies on the Green Revolution in different countries); A. Pearse (social and economic implications of the Green Revolution); H. Ruthenberg (his definitive book on tropical farming systems); T. W. Schultz (transforming traditional agriculture); J. S. Steinhart and C. E. Steinhart (energy and agriculture); R. D. Stevens (the problems of small farms); J. C. de Wilde and P. E. M. McLoughlin (tropical agriculture).

I would like to thank the authors and publishers who have generously allowed me to use copyright material (tables and figures) from their publications; the courtesy extended is recognized in each individual case. In particular, the publications of the Food and Agriculture Organization of the United Nations have been an unfailing source of important data and information.

I wish to express my appreciation to Professor Nicholas Polunin for the privilege of being one of the first authors in the new series *Environmental Monographs and Symposia* to be published by John Wiley & Sons, of which he is the Chief Editor, and for the time and effort he has personally devoted to the preliminary editing of the manuscript. I am also most grateful for the competent work of the editors and particularly for their patience with the numerous additions and changes made in the text in order that the final script be as up-to-date and complete as possible.

Finally, I would like to thank Mrs Esther Hammermann, Mrs Erna Philip, and my wife Hilda who shared in the typing of the manuscript.

I. ARNON

Rehovot, Israel

Contents

Series Preface

For civilization to survive in anything like its present form, the world's human population will require the discipline of increasing and ever-widening knowledge about the environment, and this knowledge will need to be closely followed by concomitant action to safeguard it. The *increase* in knowledge and awareness must come about through observation, research, and applicational testing, its *widening* through environmental education, and the necessary concerted *action* through duly organized application of the knowledge that has been thus acquired and disseminated.

The environmental movement has long been an undefined but often effective vehicle for widening appreciation of the vital nature and fundamental importance of our environment, and it is hoped that the forthcoming 'World Decade of the Biosphere, 1982–92' will focus attention on the fragility of, and vital need to foster in every possible way throughout the world, this 'peripheral envelope of Earth together with its surrounding atmosphere in which living things exist'.

To help encourage such ideals and guide appropriate actions which in many cases are imperatives for Man and Nature, as well as to distil and widen knowledge in component fields of scientific and other environmental endeavours, we are founding and fostering an open-ended series of *Environmental Monographs and Symposia*. This emanates from an invitation by the publishers, John Wiley & Sons, and will consist of authoritative volumes of two kinds—monographs in the full sense of separate treatments of particular subjects by from one to three leading specialists, and contributed volumes (including symposia) by more than three specialist authors covering a particular subject between them under the guidance and editorship of a suitable specialist or up to three specialists (whether a volume results in part or wholly from an actual 'live' symposium, or consists entirely of 'contributed' papers conforming to an agreed plan).

Of possibilities for our series there seem virtually no end; we keep on getting or being given new ideas and now have several dozens of them to think about and, in chosen cases, to work on. Moreover, we would hope to complement the existing *SCOPE Reports*, emanating from what in a sense is the world's environmental 'summit'.

Themes in future volumes of *Environmental Monographs and Symposia* are expected to include: 'Ecological Commentary' (Ed. Knox, President of INTECOL), 'Air Pollution and Plant Life' (Ed. Treshow), and 'Environmental Education (Ed. Hughes-Evans & Aldrich).

Whether or not this series will in time come to cover, in a general way, the entire realm of environmental scientific and other endeavours must remain to be seen, though this was the gist of the publisher's original invitation and poses a challenge that we can scarcely forget. Yet it reminds us of a major environmental encyclopaedia which we had gone some way towards planning under the same auspices nearly a decade ago—until it was agreed that such an undertaking would be too large even for then and them (or, presumably, for anyone else who would insist on maintaining the highest possible standards throughout). We believe we have decided on a constructive compromise with this propitious and promising new series.

NICHOLAS POLUNIN

(Convener and General Editor of the Series)
Geneva, Switzerland

Preface

In most developing countries of the Third World, agriculture is still primitive, being characterized by ignorance of modern techniques and resultant low productivity of land and labour. The isolation of the undeveloped rural regions from outside influences, the numerous constraints preventing modernization—economic, social, political, and institutional—tend to perpetuate the existing situation, despite the tremendous advances in agricultural technology that have been made in the course of the twentieth century.

Until quite recently, agriculture has been largely neglected by policy-makers, economists, and planners, in many developing countries, on the assumption that all, or anyway most, resources should be devoted to industry. Only recently has awareness grown that the preponderance of the agricultural sector in the economy of developing countries signifies that, without progress in agriculture, stagnation will continue, *per caput* income will remain low, and the vicious circle of poverty and low productivity will be perpetuated.

The 1960s have been called the 'decade of development'. Actually, since the early 1960s, in most Third World countries the problems encountered have become more numerous and complex than formerly. The uncontrolled population increase has compounded these difficulties.

The great hopes pinned on the 'green revolution' have only been very partially realized and achievements are still confined to a small fraction of the areas in need of development. On the other hand, where the 'green revolution' has been successful, a number of problems have arisen in its wake—some being anticipated but others not—of which potentially the most explosive is the increasing gap between various sectors in the rural population and between regions. With the natural tendency of rural populations to increase—even explosively—with success in food production, the predictable outcome is liable to be one of boom *and* bust, as populations crash when there is no longer the food to support their increases.

The one lesson that has been learned is that, for those concerned with agricultural change—whatever their role or professional competence—an understanding of the complex of economic, social, and cultural, factors involved in agricultural development is a prerequisite to fruitful endeavour in each field of professional activity.

It is, however, a fact that the great majority of agricultural scientists and technologists are generally ignorant of the social and human implications of agricultural change, whilst social scientists and economists have little understanding of the potentials and limitations of the new technologies.

The Author has been active in agricultural research for a period extending over four decades in a country that, during this period, has passed through all the stages of development from biblical agriculture to that of the twentieth century. As the rural sector of Israel consists of a number of different communities that have evolved along different pathways and at different paces, he has been able to study at close range a number of different case-histories of modernization of agriculture in his own country. He has also been privileged to serve as a consultant on agricultural research organization and programming in a number of developing countries in Asia, Africa, and Latin America, on behalf of several UN agencies.

The lack of relevance of much of the research effort to the actual practice of agriculture in many of these countries, has caused the Author much frustration and has led him to concern himself in recent years almost exclusively with the problems involved in the adoption of modern technology by traditional farmers.

The Author's overall objective is to give an up-to-date, comprehensive, yet concise, review of present knowledge on the constraints encountered in modernizing agriculture in developing countries and the problems that arise after new technologies have been adopted.

A vast amount of literature on the technological, economic, social, political, and institutional aspects of agricultural development has been perused in the preparation of the text, and a synthesis of these studies was undertaken in the light of the Author's own experience in the field.

It is the Author's sincere hope that the information provided in this book will be found useful by those responsible for the development of agriculture at all levels of policy-making and implementation—including planners, agronomists, investigators, extension workers, and farmers—who live and work in the developing countries of the world and who must cope with the increasing complexity of the problems with which they are faced.

I. ARNON

Settlement Study Centre, Rehovot, Israel
15 June 1980

List of Figures

List of Tables

CHAPTER I

The Necessity to Transform Traditional Agriculture

THE CHARACTERISTICS OF SUBSISTENCE AGRICULTURE

Production Objectives and Methods

Agriculture is the main component of the economic sector of the low-income countries of Asia, Africa, and Latin America.

In these countries, most of the farmers are still in the subsistence class, notwithstanding the tremendous advances in agricultural technology that have been made elsewhere in the course of the twentieth century. Even in Mexico, a country that has pioneered the 'Green Revolution', about 70% of the farmers still produce at subsistence level, and in most other Third World countries the situation is still worse. On a world scale, subsistence farms cover some 40% of the cultivated land and support some 50 to 60% of mankind (Wharton, 1969).

These subsistence farms produce barely enough for the basic requirements of the family and the few inputs required for further production—such as seeds, manures, animal feeds, and home-made tools. A small surplus may be sold or bartered in order to purchase those necessities that the farm cannot produce, or to obtain cash for taxes, repayment of debts, and certain social obligations such as festivities, dowries, etc.

Because of the small—if any—surplus of production over own consumption, there is very little trade between the agricultural sector and other economic sectors of the country. Subsistence agriculture is also traditional in the constancy of the methods of production used and of the commodities produced. Agricultural techniques have developed over the centuries, and are the result of the accumulated experience of generations of farmers. This does not imply that farmers are traditionalists by choice, but that they cannot normally adopt technological innovations unless the circumstances in which they operate are first changed.

The 'Fertile Crescent' of the Middle East can serve as an example. The main commodities produced have not changed since biblical times: wheat (or barley in the drier regions) for bread; olives for fruits and oil; some sesame, cucurbits, and fruits, to vary the diet; vines for wine on the slopes; small quantities of meat, milk, and wool, produced from cattle, sheep, and goats, grazing on natural pastures and on the stubble of crops. Production methods have remained unchanged for millennia: times and methods of sowing of the crops are the same as those mentioned in the Gezer calendar (late tenth century BC) which was discovered by archeologists some years ago; the implements and tools used for tilling, harvesting, threshing, and winnowing, are identical with those described in the Bible.

Occasionally, new crops are introduced into subsistence agriculture, as for example maize and groundnuts which were introduced into African countries through the slave trade to the Americas. These, however, very soon became 'traditional' crops, with the same characteristically low yields.

The meagre 'capital' resources of subsistence agriculture are mainly the result of labour inputs of the family—sometimes over generations—such as land improvement (terracing, levelling, etc.), irrigation systems, hand-made tools, domestic animals, etc.

Stages of Transition

In most subsistence economies, it is possible to find different stages of transition from subsistence to commercial farming. For example, many subsistence farmers in West Africa produce export crops, such as cocoa, palm-oil, and rubber, on small permanent areas, while at the same time practising shifting cultivation for the production of their own staple foods, such as yams, cassava, maize, and rice (Eicher, 1969).

Where attempts have been made to replace the production of food-crops by cash-crops, without changing the methods of agricultural production, the result has often been 'the substitution of subsistence living for subsistence production' (Weitz, 1971). The existence of these farmers is even more precarious than that of those producing food-crops, because in addition to the low yields of traditional farming, they face the risks of price fluctuation due to changing market conditions.

Land and Labour

The two basic inputs of subsistence agriculture are land and labour. Whilst these two subjects will be treated in detail in Chapters II and V, respectively, a brief outline of the situation is indicated here. Basically, there are two groups of developing countries: (a) those in which it is still possible to extend the areas under cultivation, provided these areas are opened up by improved transport and marketing, are developed by major works of irrigation, drainage, or forest clearing, and the potential produce can be marketed; and (b) countries in which population increase already exercises such pressure on the available lands, that crop production can only be increased by improving agricultural technology.

As long as land is plentiful, it is the work-potential of the family that is the factor limiting the amount of food which can be produced.

As population increases, a point is eventually reached where all the available land resources that can be exploited with prevailing production methods are being cultivated. Under these conditions, because of population pressure, the size of many family farms is generally smaller than what a family could effectively cultivate—leading to chronic underemployment, even though there may be problems of coping with the work-load at periods of peak requirements. The situation is further aggravated by the fragmentation of the farms into small, scattered strips and the gradual depletion of soil fertility due to exhaustive methods of production.

In ten countries, with an average *per caput* income of US $145, the poorest 40% of the population received *per caput* income of only $50. In another ten countries, with an average *per caput* income of $275, the poorest 40% of the population had a *per caput* income of only $80. In the Indian subcontinent, with a population of more than 600 million persons, about 200 million subsist on incomes that averaged less than $40 a year (McNamara, 1973). These figures apply to the early 1970s but the situation does not appear to have improved much since.

Productivity of labour decreases, as more and more people must engage in farm work—irrespective of how low their productivity is and however intermittent the labour requirements are. Even if an equilibrium is eventually reached between the amount of land and the number of people deriving their living from the land, the critical characteristic of subsistence agriculture is low productivity. The result is 'grinding poverty, massive unemployment, drift to the cities, and a pervading atmosphere of unrest and irritation conducive to peasant risings, religious millennialism, and the empty-eyed apathy of those whose social circumstances make a mockery of hope' (Nash, 1973).

With further increases in population, a critical point may be reached when the agrarian structures themselves begin to crumble under the impact of continuing pressure of population on the land. Under shifting cultivation, this leads to reduction of the fallow and decline of soil fertility. In pastoral systems it leads to overgrazing and destruction of the productive potential of the pastures. In family smallholding systems with private ownership, it leads to intense subdivision and fragmentation of land; in landlord–tenant systems, it causes an increasing rent spiral.

Low Yields

Excess labour in relation to productive labour opportunities in agriculture not only signifies a low level of production per unit of labour, but is also a direct cause of stagnation in agricultural productivity. The lower the productivity of labour in agriculture, the less is it capable of producing an excess over the subsistence requirements of the workers involved in production, and, therefore, of creating capital required for the purchase of inputs that are essential for increasing yields.

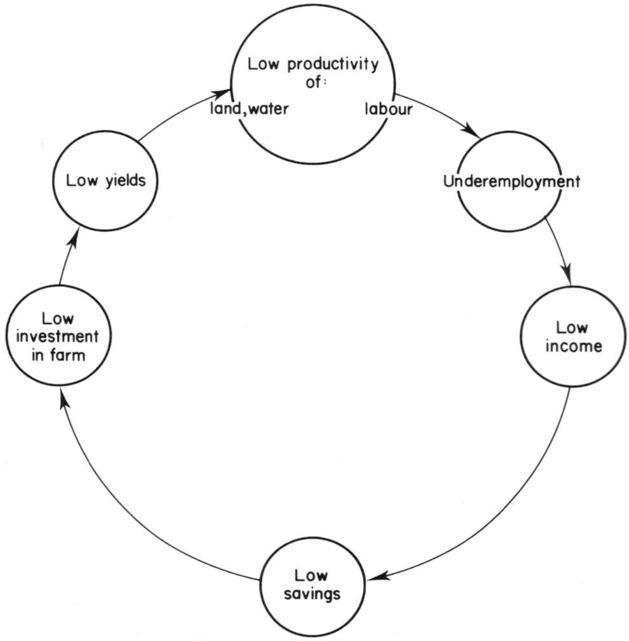

Fig. 1 Vicious circle perpetuating the low productivity of subsistence agriculture.

In brief: Subsistence agriculture is characterized by extremely limited capital resources, constancy in the use of traditional methods of production and in the commodities produced, and low productivity of land and labour. These characteristics tend to perpetuate the existing situation whereby agriculture produces barely enough for survival, and cannot therefore make a substantial contribution to economic growth. As a result, countries in which the majority of the population is engaged in subsistence agriculture, and which have no other important natural resources, are inevitably poor and their economies remain stagnant. Hence, population pressure and the resultant underemployment create a vicious circle: not only is the number of mouths to be fed from a given land-area increased, but the possibilities of increasing food production from this area are inhibited. Unrestricted population growth therefore produces conditions which make the problems of agricultural development and food supply almost impossibly difficult to solve (cf. Fig. 1).

ROLE OF AGRICULTURE IN DEVELOPING ECONOMIES

Agriculture has been largely neglected by policy-makers, economists, and planners, in many developing countries on the assumption that all, or anyway most, of their available resources should be devoted to the development of industry. Experience has shown, however, that development is not likely to occur if agricultural productivity is not increased as a prelude to industrial growth.

Agriculture must therefore make substantial contributions which will enable national economic growth to take place on a wide front. Specifically it should:

(a) Increase food-production considerably, so as to improve the existing nutritional levels, in quantity and quality, for a rapidly increasing population.

(b) Provide productive work for a rapidly increasing rural population.

(c) Produce export crops as a source of foreign currency.

(d) Support industrial development.

Improving Nutritional Levels

Rapid population growth and rising incomes create increased demands for food and clothing. The only alternative to increasing local production is imports for consumption, which then compete for scarce foreign exchange with the requirements for investment goods that are needed for the development of industry and agriculture. The most pressing problem of developing countries is, therefore, the need for food for their rapidly increasing populations.

FAO's Fourth World Food Survey (the data are cited in FAO, 1979) estimated that the available supplies of dietary energy *per caput* in the developing countries fell slightly between 1969–71 and 1972–74. It is also estimated that the number of undernourished people in these countries rose from about 400 millions to 450 millions, or a quarter of their total population (FAO, 1979).

On the whole, food production in the developing countries has shown a steady increase in the 1970s. Taking 1969–70 as a base, food production increased at approximately the same rate in developing and in developed countries; from 1975 on, the rate of increase was actually greater in the less-developed countries (LDCs) than in the developed countries; being 21% in the LDCs for the period 1969–70 to 1977, compared with 15% in the developed countries (Fig. 2). However, this relative advantage was completely cancelled by the considerably greater population increase in the LDCs. Food production *per caput*, after remaining stagnant from 1969–70 until 1974, increased in the LDCs by slightly more than 2% in 1975, and remained at this level until 1977. By contrast, the developed countries have shown a steady annual increase in *per caput* food production during the same period, so that the disparities between the two groups of countries, already great in 1969, have increased markedly since then (Fig. 3).

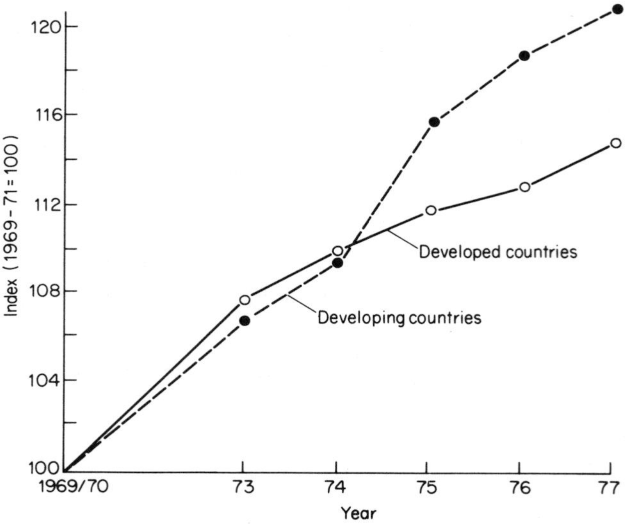

Fig. 2 Trends in food production in developed and developing countries (based on data from FAO, 1979).

To eliminate widespread malnutrition, the developing countries must double food production by the end of the century (NAS, 1975), which implies an annual increase of 3 to 4%. This estimate is based on about 2.5% annual population increase.

The actual situation is, however, not at all encouraging. Out of 96 Third World countries for which data are available, in no fewer than 45 countries the increase in food-production failed to keep abreast of population growth. As a result, *per caput* production of nearly 40% of the total population of the developing countries has actually decreased in recent years (FAO, 1976).

Kenya is one of the more prosperous African countries, and yet nutri-

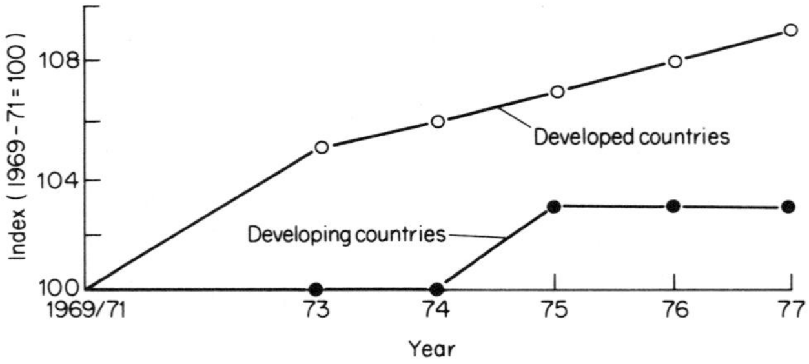

Fig. 3 Trends in *per caput* food production in developed and developing countries (based on data from FAO, 1979).

tional deficiencies in calories, proteins, and vitamins A, B_2, and C, in the diets of rural people in Kenya, are nationwide. In the preharvest period, 25–30% of rural families consume less than 60% of the estimated calorie requirement. Probably more than 25% of the children suffer from some form of malnutrition (World Bank, 1975).

In most of the developing countries, subsistence agriculture generally occupies a large proportion of the available land. Even in such a country as Nigeria, where agriculture is relatively market-oriented, subsistence crops still represent 75% by value of the agricultural output (Buchanan, 1971).

Rice, wheat, maize, sugar-cane, sorghum, millet, cassava, and cattle, are the main food sources in the developing countries, providing 75% of the calories and 65% of the protein consumed. For the poorest sectors they may account for 85 to 90% of calories and protein (NAS, 1975).

These subsistence crops are mostly characterized by a high level of carbohydrates, providing 70–80% of the calories consumed, and by low protein levels. The particular staple crops vary among the different environments: sorghum and millets in the lower-rainfall areas of Africa; rice in Asia, and to a lesser extent in certain areas of Africa and Latin America; maize in Mexico, the Central American countries Bolivia, and a number of areas in Africa; bananas and plantains throughout the wet areas of the tropics; and manioc (cassava) rather generally in Africa to the south of the dry northern areas, as well as in Indonesia, Brazil, and a number of the Central American countries. Sweet potatoes and yams are also important sources of carbohydrates in a number of countries.

When one considers that cereals have ten times the protein content of cassava, and five times that of yams, the poverty of a diet based mainly on roots and tubers is evident (FAO, 1966). A rough estimate of nutrient requirements *per caput* is a daily intake of 2700 to 2800 calories and about 40 grams of animal protein, provided by a varied diet including vegetables and fruits (Klatzmann, 1975).

The estimated average total protein intake for all underdeveloped regions is 58 grams per person per day, as compared with an average in developed countries of about 90 grams of which about half is of animal origin (FAO, 1966).

National statistics generally indicate the *average* supply of calories and proteins to the population. However, food distribution between the various sectors of the population must also be considered. The chief centres of undernourishment and poverty in developing countries are the rural areas. This is where most of the population still live, and where rapidly increasing population exerts its greatest pressure, making the living-standards still lower than previously. For example, it has been estimated that if Brazil had food supplies sufficient to provide 2800 calories *per caput* per day, only some 40% of the population would be consuming at about this level, while 50% would be receiving only 1600 calories; on the other hand, 10% would be consuming 10,000 calories (CIDA, 1966)!

Deficiencies in intake of various vitamins and minerals are also prevalent in many developing countries. As a result of the accumulated deficiencies, often in association with infectious and parasitic diseases, expectation of life in Africa is almost 50% lower than in advanced countries, and infantile mortality is five to ten times higher (Dumont, 1966).

Undernutrition and malnutrition are among the important reasons for the low productivity of labour in tropical countries. An avoidance of effort is the body's natural defence mechanism when suffering from under- or malnutrition. The traditional African diet may be more or less adequate for tribal life requiring little sustained effort, but is insufficient for a worker for whom a regular, sometimes considerable, output of energy is required (Inter-African Labour Institute, 1960).

An increase in the production of fruit and vegetables would provide vitamins and mineral salts for the diet of rural people and could help to improve the health, vitality, and hence production potential, of the farmers.

Wherever sufficient food is not produced to feed the rapidly expanding urban populations, it will have to be imported, with a resultant drain on foreign exchange and increased inflationary pressures. Notwithstanding the disproportionate size of the agricultural sector in the national economy, many of the developing countries are heavy importers of foodstuffs.

Food imports to Third World countries have shown a consistent trend to increase (Fig. 4). For these countries as a whole, the indices for food imports during the decade 1967–77 have increased by 71%; the corresponding figures for the different regions are: Far East, 18%; Latin America, 74%; Africa, 88%; Near East, 147% (FAO, 1979)!

In low-income areas, the tendency is to import the cheaper foodstuffs

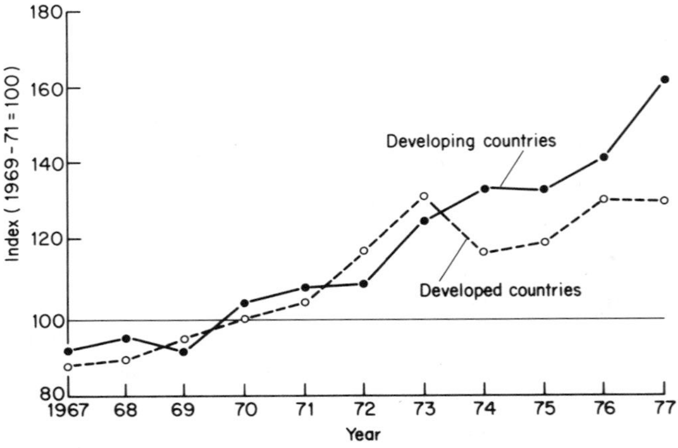

Fig. 4 Trends in value of food imports in developed and developing countries (based on data from FAO, 1979).

which, naturally, have lower food-values. The poorer the existing diet is, the stronger will be this tendency, so that even where pressing calorie-hunger is satisfied, protein hunger is increased (Palmer, 1972). If living standards of the majority are to be improved, it is therefore essential that agriculture be improved. Already in many cases, the proportion of total imports that is represented by foodstuffs is higher than in some of the industrialized countries of North-West Europe (Buchanan, 1971)!

In India, for example, in 1966 it was necessary to import 11 million tonnes of food just to keep the people alive. Spain, in 1965, was obliged to spend 570 million dollars on food imports.

In 1970–72, the food imports of the Third World countries averaged 14% of their total imports. A large part of these food imports consists of cereals, which are their main staple food. Many of these countries that were net real exporters before the Second World War, have become net importers in the last 20 years. In the 1960s and early 1970s, food aid provided about 45% of the total cereal import of the developing countries; supplies have, however, decreased sharply in recent years, because of the depletion of reserves and the steep increase in prices and transportation costs (FAO, 1976) (Fig. 5).

In Africa, the volume of food imports increased by 36% between 1970 and 1976, whilst the volume of agricultural exports declined by 9% (FAO, 1978).

Japan provides an interesting example of a country which improved its agricultural production under conditions of severe overpopulation, a *per caput* income approximating to the low levels of most underdeveloped countries, and small and fragmented farm units. Within a period of about

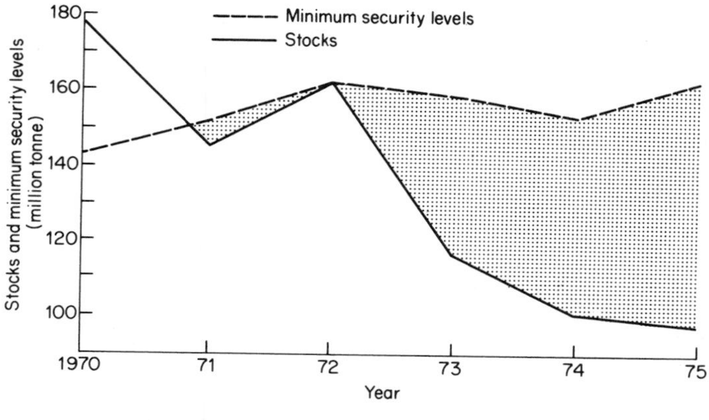

Fig. 5 World cereal carry-over stocks of wheat, coarse grains and milled rice, and minimum security levels (equivalent to 18% of world consumption) from 1970 to 1975. China and the USSR are not included (FAO, 1976). By courtesy of the Food and Agriculture Organization of the UN.

30 years, Japan was able to increase food production at a rate that was high enough to outstrip the growth-rate of the population, achieving an increase in *per caput* food supply by over 20%, and in output per farm worker by 106% (Johnston, 1951).

Provision of Work for a Rapidly Increasing Rural Population

Myrdal (1965) stresses the following points, which are no less important for being so obvious:

(a) For many decades, even a much more rapid process of industrialization than that achieved by most underdeveloped countries will not provide sufficient employment for the underutilized labour force in these countries. The main reason for this is that fully modern, large-scale industries provide relatively small additional labour demand and may even have a negative effect by competing with already established, more primitive labour-intensive industries.
(b) During a period of incipient industrialization, the labour force in all LDCs will increase by between 2 and 4% per year.

Therefore, there is every reason to expect an increase in the absolute numbers of the agricultural population in most of the developing countries for a long time to come, and agriculture will have to continue for many years to employ the majority of their population. This subject will be treated in detail in Chapter V.

Production of Commodities for Export

In many developing countries, production for export accounts for the bulk of the commercial agriculture, and is the major source of foreign exchange earnings, paying for imported inputs which may be vital to the overall economy. Crops for export usually constitute 'small islands of market-oriented production set in the midst of a sea of stagnating peasant economies' (Buchanan, 1971). Generally, with the exception of countries having important mineral resources (such as Venezuela with its large oil deposits), tropical agricultural commodities such as coffee, cocoa, rubber, bananas, tea, sugar, and palm products, account for 75%, and often more than 80%, of the total value of exports (McPherson, 1968).

Many of the developing countries rely on a single crop to provide the bulk of their agricultural exports. In Chad, cotton accounts for 80% of the country's exports; in Ghana, cacao accounts for 60% of exports; in Gambia and Senegal, groundnuts account for 90% of exports; and in Malaya, rubber accounts for 60% of exports. In Tanzania, Uganda, and Kenya, two or three agricultural products have provided over half the total export earnings. In the mid-1960s, coffee and cotton contributed over 80% of

Uganda's total export earnings, while sisal alone provided over half of Kenya's (Seidman, 1972). In most of West Africa, and in the Andean Republics of Latin America, three commodities account for 90% of their exports. This concentration on a very limited number of primary products for export makes the economy of the countries concerned highly vulnerable because they are dependent on widely fluctuating incomes resulting from great variations in the prices on the world market of the commodities produced.

The long-term trend has been for a gradual decline in the share of developing countries in world agricultural export earnings. This has gradually fallen from 46% in the mid-1950s, to 37% in the early 1960s and to 29% in 1975, with a slight recovery in 1977 to 33% (FAO, 1979).

Agriculture in these countries is therefore based on primitive production of food for their own consumption on one hand, and a dangerous reliance on a single crop for export on the other.

Supporting Industrial Development

Agriculture can contribute to industrial development in three ways (Malassis, 1975):

(a) by providing raw materials for certain industries—this implies agricultural growth and diversification;
(b) by providing workers and capital—this implies increased efficiency in agricultural production; and
(c) by providing a market for industrial products—this implies growth in farmers' purchasing power.

Provision of raw materials for industry: Local production of raw materials for its own industry gives a country a great advantage over those industrial countries which have to import their raw materials. A case in point is the textile industry. Nearly 60% of the total manufacturing output of Latin America is from industries that use chiefly agricultural raw materials: food, beverages, tobacco, textiles, footwear and apparel, leather, rubber, paper, and wood (Jones, 1971).

Transfer of Agricultural Labour to Other Sectors

In the long run, the policy objectives of the developing countries must be to achieve substantial industrial development with a concomitant reduction in the labour force engaged in agriculture. However, as long as employment problems remain acute, agriculture serves as a kind of 'residual storage-tank in which the bulk of the labour force can find some sort of subsistence until sufficient economic development has occurred to cause a structural transformation' (Shaw, 1971). Nevertheless, 'A marked rise in productivity per worker in agriculture must take place eventually as a precondition of the industrial revolution in any part of the world.' (Kuznetz, 1959).

A situation in which an excessively large proportion of the population is engaged in agriculture is incompatible with an improvement in the standards of living of the rural population, and with overall development of the national economy. Such a situation cannot be happily sustained in the long run. Gradually, as alternative opportunities are made available, more and more labour will need to be transferred from agriculture to industry and services if the economy is to develop. In order for such a transfer to be possible without a fall in agricultural output, substantial changes in production methods will be essential. In traditional agriculture, the demand for labour is highly seasonal, with peak labour periods—such as at the times of land preparation and sowing, of weeding, and of harvesting. Whilst 'underemployment' may characterize the agricultural sector as a whole, a shortage of labour may actually exist at certain periods of the year.

Therefore, in order to release labour permanently to industry and services, without a drop in agricultural production, it is necessary to raise output per man/day, mainly during the critical periods of labour requirements, and/or reduce seasonal variations in labour requirements. The former may be achieved gradually through the use of better implements, the introduction of animal power, and finally by mechanization. The second can be advanced by a well-planned diversification of production.

Provision of Investment Capital for Development

In the early stages of development, agriculture is generally the main, if not the only, source of capital for investment in agriculture itself, in industry, and in infrastructure.

Investment in Agriculture

There is a very wide range in investment in agriculture per unit of land and labour, both between developed and developing countries, and within each of these two groups, reflecting the intensity of agricultural production. In the developed countries, this varies from \$10 per ha in Australia to \$1000 in Japan, and from \$80 per agricultural worker in Portugal to over \$3000 in the USA.

In the developing countries, investment per ha has varied from less than \$1 in Ethiopia to \$150 in the Republic of [South] Korea, and per agricultural worker from \$1 in Kenya to over \$100 in Costa Rica (FAO, 1979) in the nineteen seventies.

The magnitude of the investments that are required to develop agriculture is exemplified by the estimated amount of capital required by Third World countries in Africa to achieve food production targets (Table I).

Table I Capital investments (in millions of US $) from 1975 to 1990 required to achieve food production targets (FAO, 1979, with permission).

Development of non-irrigated arable lands		1466
Irrigation		
development		8400
improvement		2255
Mechanization		8082
Livestock development		6531
	Total	26734
	Annual average	1783

Capital Transfer to Other Sectors

The transfer of resources from agriculture to other sectors may occur in either of the following two different ways (Kuznetz, 1964).

(a) *Compulsory:* Through taxation for the benefit of other sectors, so that the burden on agriculture is far greater than the services rendered by government to agriculture. This involuntary contribution by agriculture may be quite large in the early phases of economic growth.

The classical example of the role of agriculture in providing capital for industrial development is of course that of Japan, which imposed heavy taxes on their agriculture during that country's period of rapid industrialization.

In the mid-1870s, the land tax accounted for 86% of the tax revenue of Japan; in the mid-1890s, agriculture contributed 50% of the country's GNP but paid more than 80% of the taxes (Malassis, 1975).

Intervention by Government can take a number of forms apart from taxation. The most frequently adopted is establishing terms of trade which operate against agriculture but to the benefit of the urban and industrial sectors—such as keeping prices of agricultural commodities low in relation to locally manufactured goods, and manipulating tariffs and exchange rates where exports are mainly agricultural commodities (Griffin, 1972).

Another form of compulsory transfer of capital from agriculture to other sectors of the economy, is through the payment of exorbitant rents to the landowners, usurious interest to money-lenders, undervaluation of crops by unscrupulous traders, and unjustifiably high prices that have to be paid for inputs and services.

The transfer of resources from agriculture is not always used productively, and frequently only serves to supplement consumption by the more favoured sectors. Griffin (1972) mentions that in Pakistan, for example, out of a net transfer of 15% of the value of the agricultural output, probably as much as 80% was channelled into urban consumption.

These profits are frequently used to build opulent housing, to import luxury items, to travel abroad, and to salt away capital in foreign countries.

Even if agricultural surplus resources are channelled to industrial production, they are not necessarily productive. Industries are frequently inefficient, and many of these 'enterprises' can remain viable only by continuing to be subsidized by agriculture.

As long as the agricultural sector is very large in relation to industry, resources 'squeezed' out of agriculture can contribute to industrial growth (Griffin, 1972). However, in the long run, an excessive 'squeezing' of agriculture will not only impede agricultural development, but will restrict industrial growth as well.

When this point is reached, further growth by both sectors becomes possible only if agricultural production is increased substantially. Agriculture cannot serve as a significant source of capital investment in industry unless substantial amounts of public and private funds are invested in the infrastructure and services that are required for agricultural development—such as research, rural education, irrigation works, transport, etc.

(b) *Through savings:* When the rise in labour productivity in agriculture exceeds the increase in consumption levels of the rural population, a substantial fraction of the savings engendered by the increment of agricultural production can be used to finance capital formation in other sectors of the economy. This is particularly true when the increase in farm output and productivity requires only small capital outlays and modest increases in other inputs (Johnston, 1962). An increase in farm output and productivity also facilitates, indirectly, capital accumulation in the other sectors of the economy—by reducing the cost of food.

Providing a Market for Local Industry

Low-income or unproductive agriculture is a major constraint in developing a profitable home market for local industries, and industrial capacity may be greatly in excess of effective demand. The cash market of some developing countries is not larger than that of a moderately-sized European town. This makes it impossible to utilize production capacity fully and results in higher-than-normal costs of production. Rates of utilization of industrial capacity as low as 40–50% are common in the less-developed countries. This, in turn, makes it more difficult for industry to compete on the world market.

However, as the income from agriculture increases as a result of modernization, the rural population can provide a larger home market—a prerequisite for enabling domestic industries to achieve a volume of production and sales that can make them economically viable.

For example, the introduction of the high-yielding varieties (HYVs) of wheat and rice was the direct reason for a considerable increase in irrigated areas in a number of countries. As a result, a vast new demand for irrigation pumps and tubewells (deep, narrow wells in which the water is lifted by a turbine pump) was created. In India, the number of pumps produced by local industry increased from 67,000 in 1956 to 200,000 in 1967 and to about 400,000 pumps in 1970, creating employment for close to a quarter-million workers in the production of steel, the manufacture of the pumps, and their distribution (Brown, 1970).

In the Pakistan Punjab the rapid growth of the agricultural sector due to the 'Green Revolution' resulted in the development of a small-scale engineering industry, to supply key durable goods—mainly diesel engines, pumps, and strainers, but also various farm implements.

Early-maturing varieties of rice, which are harvested during the monsoon season, require artificial drying and therefore need grain-drying equipment. The increased use of chemicals for plant protection—essential for ensuring the profitability of the HYVs—has led to growth in demand for application equipment; harvesting and threshing equipment became necessary not only to overcome critical labour shortages but also to permit earlier ploughing for the next crop (Child & Kaneda, 1975).

The resultant development of industry 'occurred spontaneously, with no subsidies, no tax concessions, no special credit arrangements, no technical assistance or even recognition by official agencies' (Child & Kaneda, 1975). It provided an investment outlet for rural savings and made possible the development of entrepreneurial and managerial talent, the training of skilled and semi-skilled labour, and the productive absorption of surplus rural labour.

In brief, the failure to develop agricultural production potentials will impede economic progress in general, and industrialization in particular.

Conditions for 'Take-off'

The main merit of subsistence agriculture is that it provides food, shelter, and clothing, albeit at a very low level, to all members of the farm family, thereby avoiding their becoming a charge to society as a whole. Its main drawback is that it is self-perpetuating, and because subsistence agriculture produces little or no excess over the requirements of the producers, its contribution to the national economy is extremely small. Whatever excess may be produced, is generally swallowed up by taxes, rents, debts, and social obligations such as dowries, fiestas, etc.

The following examples illustrate the low level of income of subsistence farmers. In seven countries of Latin America, modal* annual campesino income was found to be equivalent to approximately \$300—except in a few regions with exceptional employment opportunities, or with a particularly favourable tenure system (Rosner, 1974). Family cash incomes are of course much lower. In Brazil's north-east, in much of Guatemala, and in the Andean highlands, typical family cash incomes are generally below the

*The mode is the value of a variable occurring most often in a series of data.

Table II Requirements of technical inputs (in millions of dollars) by agriculture of less-developed countries (FAO, 1969, 1976, with permission).

	Actual (1962)	Actual (1976)	Projected (1985)
Fertilizers	664	1837	7838
Chemicals for crop protection	180	777	2077
Tractors	575	1391	2675

equivalent of $100 annually (Barraclough, 1973). Subsistence agriculture is characteristically poor 'because the factors on which the economy is dependent are not capable of producing more under the circumstances'. Simply increasing the input of traditional factors of production with the existing state of the art, gives too low a rate of return to induce further investment (Schultz, 1964). Hence, a breakthrough is possible only by providing 'improvements in the quality of the inputs'—namely, new *agricultural inputs* with a relatively high pay-off. Virtually all of these 'new' inputs of potential promise must come from *outside* of traditional agriculture—whether they be improved varieties, fertilizers, equipment, pesticides, etc.—and their success depends on their being used efficiently.

The order of magnitude involved in the required increases in modern inputs is suggested by the estimates of FAO (1969, 1976) (Table II).

There are many opinions as to the preconditions which are essential if a 'take-off into self-sustaining growth' is to be achieved by developing countries, but three essential facts are generally acknowledged:

(a) Agriculture has a crucial role to play, and 'take-off' in agriculture is the first essential step.
(b) This role cannot be achieved without first transforming traditional agriculture.
(c) Traditional agriculture cannot itself supply the capital required to make its own transformation possible.

In order to make a breakthrough possible, some form of intervention, outside of subsistence agriculture, is essential. Factors such as the discovery and development of oil or mineral resources, national and/or foreign investments in agriculture or industry, and possibilities of gainful employment for redundant rural labour within or outside the country, are some of the possibilities that may contribute to moving subsistence agriculture towards modernization.

The Development of Improved Technology

Even where preconditions for 'take-off' in agriculture exist, the modernization process itself is a complicated one, involving profound changes in the factors of production and their relative importance.

Mosher (1966) has classified the facilities and services involved in the modernization of agriculture, in two main groups:

(a) the *essentials* which, as the name implies, *must* be present to enable a farmer to adopt an innovation; and
(b) the *accelerators*, which may be important to get an innovation adopted but are not indispensable.

Mosher (1966) lists five essentials: a market for farm products, constantly changing technology, local availability of supplies and equipment, incentives, and transportation; and five accelerators: education for development, credit, group action by farmers, improving and expanding agricultural land, and national planning. The absence of accelerator institutions is one of the characteristics of subsistence agriculture.

The number of accelerators required, and the intensity and complexity of their application, depend on the following three factors (Kulp, 1970).

The novelty and complexity of the innovation: The lesser the departure is from traditional practices, the smaller will be the importance of the accelerators. New crops, such as cassava, beans, maize, and sweet potatoes—introduced with the slave trade from the Americas—were adopted by African farmers and spread rapidly without the benefit of any accelerators.

Cost: Innovations that require substantial amounts of capital for their adoption, increase the importance of accelerators such as credit, planning, etc.

Profitability: The greater the proven returns from an innovation, the more inclined farmers will be to adopt the innovation, even without accelerators (such as credit, for example).

It is possible to group all the elements proposed by Mosher (1966)—essentials and accelerators—into four related functions which have to be simultaneously developed if efforts to transform traditional agriculture are to be successful:

(a) *Generation of new technology:* This has to be appropriate to the specific conditions of a region and its resources, and implies an effective organization for agricultural research.

(b) *Transfer of the new technology to the farmer:* This involves education and training—to make the farmer receptive of new ideas and capable of applying new technologies—and an effective extension service to provide the link between research and the farmer.

(c) *Provision of essential conditions:* Incentives are needed to motivate the farmer to change his methods of production, notwithstanding the risks involved. These include appropriate pricing, credit, land reform, and other measures. A complex *infrastructure* is essential to service agriculture and provide the necessary supplies and facilities for both production and marketing.

New social forms and structural changes in rural society are also needed—to enable the farmer to cope with the new complexities with which he will be increasingly faced, as he moves from traditional to modern agriculture.

(d) *Formulation of an appropriate strategy for promoting technological change:* The process of technological change is not well understood, even in advanced countries, and the problems of creating such a process in the thin economic atmosphere of developing countries has not yet been systemically investigated' (Edwards, 1973). Notwithstanding these limitations, technological change must be planned, decisions on strategy and priorities must be taken, and the economic and social consequences of these decisions must be anticipated. The chain of events which lead to a higher income for farmers is shown in Fig. 6, and the subjects involved will be treated in detail in Chapters VI to XI.

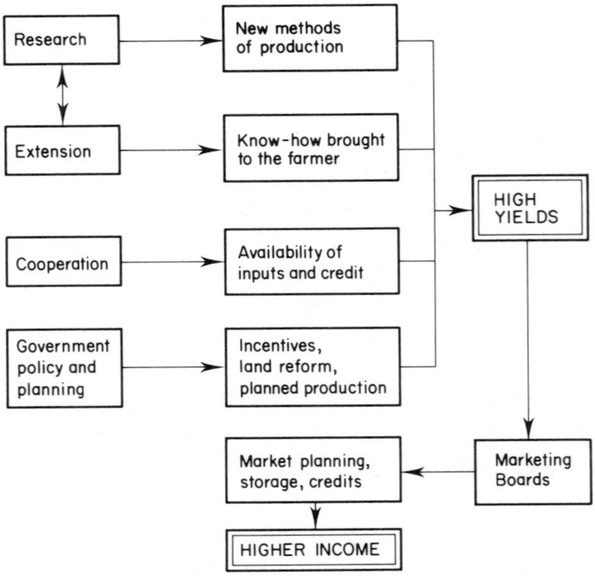

Fig. 6 Simultaneous activities required in order to achieve increased incomes for farmers.

Summary

Agriculture is the main economic sector of the developing countries, and most of the farmers in these countries are still producing at subsistence level. The critical characteristic of subsistence agriculture is low production and resultant human poverty. Because of the population explosion, labour is generally far in excess of productive labour opportunities in agriculture. These factors create a vicious circle, perpetuating a situation whereby agriculture produces barely enough for survival, and cannot therefore make a substantial contribution to economic growth. As a result, countries in which the majority of the population is engaged in subsistence agriculture, and which have no other important natural resources, are inevitably poor and their economies remain stagnant.

Until recently, agriculture has been largely neglected by policy-makers, economists, and planners, in most developing countries. Experience has, however, shown that overall development is not likely to occur unless agricultural productivity is increased as a prelude to industrial growth.

The potential contributions of agriculture to the economy are an increase in food supply, provision of productive work to a rapidly-increasing rural population, creation of capital for investment, and support of industrial development.

The transition from traditional, subsistence agriculture to modern, commercial agriculture cannot occur without some form of intervention from outside, which supplies the incentives and the means to make 'take-off' possible.

However, even where the preconditions for take-off are created, the modernization process is generally complicated and difficult. If efforts to transform subsistence agriculture are to be successful, four related functions have to be simultaneously developed:

(a) new technology has to be generated, implying an effective research organization;
(b) the new technology has to be rapidly transferred to the farmers, requiring an efficient system of education;
(c) the essential incentives and conditions have to be provided, in order to motivate the majority of the farmers to change their methods of production and to enable them to do so successfully; and
(d) an appropriate strategy for promoting the entire process must be devised and implemented.

References

Barraclough, S. L. (1973). *Agrarian Structure in Latin America.* Lexington Books, Lexington: xxvi + 351 pp., illustr.

Brown, L. R. (1970). *Seeds of Change*, Praeger, New York, NY: xv + 205 pp., illustr.

Buchanan, K. (1971). Profiles of the Third World. Pp. 17–44 in *Developing the Underdeveloped Countries* (Ed. A. M. Mountjoy). Macmillan, London & Basingstoke, UK: 270 pp., illustr.
Child, F. C. & Kaneda, H. (1975). Links to the Green Revolution: A study of small-scale, agriculturally-related industry in the Pakistan Punjab. *Economic Development and Economic Change,* **23**, pp. 249–75.
CIDA (1966). *Land Tenure Conditions and Socio-Economic Development of the Agricultural Sector, Brazil*. Organization of American States, Washington, DC [not available for checking].
Dumont, R. (1966). *African Agricultural Development*. Food and Agriculture Organization of the UN, Rome, Italy: vi + 243 pp. (mimeogr.).
Edwards, E. O. (1973). *Employment in Developing Countries*. Ford Foundation, New York, NY: 104 pp. (mimeogr.).
Eicher, C. K. (1969). Production is not sacred. *Ceres—The FAO Review*, **2**(3), pp. 36–9.
FAO (1966). *Agricultural Development in Nigeria 1965–1980.* Food and Agriculture Organization of the UN, Rome, Italy: 512 pp.
FAO (1969). *Tentative Indicative World Plan*. Food and Agriculture Organization of the UN, Rome, Italy: 672 pp.
FAO (1976). *FAO Trade Yearbook*. Food and Agriculture Organization of the UN: Vol. 31, Rome, Italy: vii + 356 pp.
FAO (1978). The state of natural resources and the human environment for food and agriculture. Pp. 3-1–3-65 in *The State of Food and Agriculture 1977*, FAO Agricultural Series, No. 8, Rome, Italy [not available for checking].
FAO (1979). *The State of Food and Agriculture 1978*. FAO, Agricultural Series, No. 9, Rome, Italy [not available for checking].
Griffin, K. (1972). *The Green Revolution: an Economic Analysis*. UN Research Institute for Social Development, Geneva, Switzerland: xii + 153 pp., illustr.
Inter-African Labour Institute (1960). *The Human Factors of Productivity in Africa* [not available for checking].
Johnston, B. F. (1951). Agricultural productivity and economic development in Japan. *J. Political Economy,* **59**, pp. 505–8.
Johnston, B. F. (1962). Agricultural development and economic transformation. *Food Res. Inst. Studies,* **3**, pp. 223–76.
Jones, G. (1971). *The Role of Science and Technology in Developing Countries*, Oxford University Press, London, UK: xviii + 174 pp., illustr.
Klatzmann, J. (1975). *Nourrir Dix Milliards d'Hommes?* Presses Universitaires de France, Paris, France: 268 pp., illustr.
Kulp, E. M. (1970). *Rural Development Planning*. Praeger, New York, NY: 664 pp., illustr.
Kuznetz, S. (1959). *Six Lectures on Economic Growth*. Free Press, Glencoe, Illinois: 122 pp.
Kuznetz, S. (1964). Economic growth on contribution of agriculture: Notes on management. Pp. 102–24 in *Agriculture in Economic Development* (Ed. C. K. Eicher & L. Witt), McGraw-Hill, New York, NY: vi + 415 pp., illustr.
McNamara, R. S. (1973). *One Hundred Countries—Two Billion People*. Praeger, New York, NY: 140 pp.
McPherson, W. W. (1968). Status of tropical agriculture, pp. 1–22 in *Economic Development of Tropical Agriculture* (Ed. W. W. McPherson). University of Florida Press, Gainesville, Florida: xi + 328 pp., illustr.
Malassis, L. (1975). *Agriculture and the Development Process.* UNESCO Press, Paris, France: x + 1123 pp. (2 vols).
Mosher, A. T. (1966). The extension process. Pp. 299–314 in *Getting Agriculture Moving* (Ed. R. E. Borton). Praeger, New York, NY: x + viii + 1123 pp (2 vols).

Myrdal, G. (1965). *The 1965 McDougall Memorial Lecture*. Food and Agriculture Organization of the UN, Rome, Italy: 213 pp.

NAS (1975). *Agricultural Production Efficiency*. National Academy of Sciences, Washington, DC [not available for checking].

Nash, M. (1973). Work, incentives, and rural society and culture in developing nations. Pp. 137–58 in *Employment Process in Developing Countries*. Seminar sponsored by the Ford Foundation, Bogota, Colombia: 556 pp. (mimeogr.).

Palmer, Ingrid (1972). *Food and the New Agricultural Technology*. United Nations Research Institute for Social Development, Geneva, Switzerland, Report No. 72.9, 85 pp.

Rosner, M. H. (1974). *The Problem of Employment Creation and the Role of the Agricultural Sector in Latin America*. University of Wisconsin, Land Tenure Center, Madison, Wisconsin: 101 pp.

Schultz, T. W. (1964). *Transforming Traditional Agriculture*. Yale University Press, New Haven, Connecticut: xiv + 212 pp.

Seidman, A. (1972). *Comparative Development Strategies in East Africa*. East African Publishing House, Nairobi, Kenya [not available for checking].

Shaw, R. d'A. (1971). *Jobs and Agricultural Development*. Overseas Development Council, Monograph No. 3, 74 pp.

Weitz, R. (1971). *From Peasant to Farmer: A Revolutionary Strategy for Development*. Columbia University Press, New York, NY: xiv + 292 pp.

Wharton, C. R. Jr (1969). The green revolution: Cornucopia or Pandora's box? *Foreign Affairs,* **47**, pp. 464–76.

World Bank (1975). *The Assault on World Poverty*. Johns Hopkins University Press, Baltimore and London: xi + 425 pp.

CHAPTER II

Natural Resources—Land and Water

The Need for Knowledge of the Basic Natural Resources

Knowledge of the basic natural resources of the developing countries and their significance for agricultural development is generally deficient. In Latin America, for example, until 1960, only about 6% of the total surface had been studied, and there is still very little detailed soil-mapping (Inter-American Committee, 1963). The situation in tropical Africa was quite similar.

Many development schemes have either been only partially successful, or have even failed disastrously, because of lack of basic information on the natural resources available in the region under consideration. Thus Wilde & McLoughlin (1967) mention two examples of the consequences of making large investments in irrigation and agricultural mechanization schemes, when the planners proceeded without adequate knowledge or consideration of the relevant factors.

In Mali (formerly French Sudan), an irrigation scheme was started in the late 1920s. Only subsequently was it found that the rainfall had been seriously underestimated. Nor had a prior soil study been undertaken, and moreover the topographical survey proved to be seriously inadequate. Only after investing large sums and efforts in an irrigation scheme for producing cotton, was it found that economic yields of cotton could be produced without irrigation. Also the fact that the irrigation network was laid out on the basis of inadequate topographic and soil information, resulted in serious problems of waterlogging and difficulties of effective drainage of large areas.

In an irrigation scheme in Kenya, the employment of basic irrigation, and the use of inappropriate water duties on a type of soil which proved difficult to drain, in combination with the method of irrigation adopted (basin irrigation), produced heavy infestion of Nut-grass (*Cyperus* spp.) which made over 20% of the area useless for further cropping.

It is thus abundantly clear that detailed topographical, geological, hydrological, soil, and ecological, surveys are an essential prerequisite of meaningful agricultural planning, and that their importance for the proper utilization of land and water resources for agricultural development cannot be overstressed.

Soil surveys can be carried out at various levels: (a) *broad reconnaissance*, at scales of from 1:100,000 to 1,000,000; (b) *reconnaisance* at 1:30,000 to 100,000, which are useful in selecting areas that have a potential for agricultural development; and (c) *detailed surveys*, from 1:1000 to 20,000, which are essential for rational land-use planning, settlement, and irrigation, schemes and for planning individual farms. For these purposes, details of the nature, distribution, and relationship between, the different soils is required in considerable detail (UN, 1963).

In recent years it has been shown that remote sensing from aerospace platforms can be used, with a minimum of ground-sampling, as a new tool for the mapping of major soil boundaries. As conventional soil-classes are based on both surfaces and subsurface soil characteristics, the technique has potential for augmenting, but not for replacing, traditional soil-mapping methods. Remote sensing may be of greatest benefit in developing countries in which soil surveys are not yet well established, as the most timely and economic means of obtaining land-use information.

General land-use surveys by remote sensing have been carried out in a number of developing countries during the period 1965–75: Project RADAM has provided 1:250,000-scale base-maps of geology, geomorphology, hydrology, soil, vegetation, and land-use potential, of the Amazon Basin; the Overseas Development Administration of the United Kingdom has provided maps of soils, vegetation, and land-use potentials, in Ghana, Kenya, Malawi, Nigeria, Fiji, and the Grand Cayman (Bauer, 1975).

Soils have been extensively surveyed in Africa by national, bilateral, and international, institutions, so that the general location of the major kinds of soils are now known. However, soil surveys and soil classification are not ends in themselves, but are intended to provide information that is essential for agricultural development and adequate soil management. The interpretation of soil surveys must make possible evaluations of soil suitability for various crops, of soil limitations, and of possible means for improvement.

This involves correlating the soil classification with the results available from the many agronomic experiments that have been carried out or are planned—an enormous but essential undertaking (Moutappa, 1973).

In a joint project of FAO, UNESCO, and the International Society of Soil Science, a *Soils Map of the World* is being prepared. FAO has started a 'soil data bank' with the object of providing information on land capabilities in different areas, based on the interpretation of soil survey data.

Whilst, for convenience, the two basic natural resources—soil and water—will be treated separately, it should be emphasized that, in any development programme, these two factors are completely interdependent, and cannot be considered as separate entities.

Land Resources

Importance of Land Resources in Development

When Malthus in 1798 proposed his thesis that population, if unchecked, increases geometrically, while food for subsistence increases only arithmetically, there were still vast areas in the world which could be developed relatively easily. This is no longer the case, whilst the actual cultivated area of the world is far from providing the calculated minimum *adequate* diet for the present world population.

Agricultural development is possible basically in two different ways which are, however, not mutually exclusive: bringing more land under cultivation, and increasing productivity per unit of land.

The first alternative is possible without changing traditional farming methods, whereas the second is entirely dependent on applying improved farming techniques. In this lies the essential difference between the significance of land for the traditional farmer and for the modern farmer.

The modern farmer can, to a large extent, increase production on a given area by using appropriate inputs, such as fertilizers, irrigation, drainage, etc., so that, in his case, land can be partly replaced by know-how and capital. For the traditional farmer, land is the most important means of production and his only guarantee of survival.

As long as the population pressure on the land is not excessive, even traditional agriculture can produce a sufficiency of food to maintain an acceptable level of supply. This explains why traditional agriculture has been able to provide subsistence to a farming population for a surprisingly long period of its history (Dandekar, 1969). But as growth of population overtakes the possibility of expanding land-use, the ability of traditional agriculture to produce even at a subsistence level decreases, and a state of continuous deterioration sets in.

This has actually resulted in loss of productivity of huge land-areas, in irreversible processes. As an example, the traditional form of land-use in the tropical humid regions—the bush-fallow system—has been surprisingly stable as long as a critical population density is not exceeded. When once this occurs, however, population pressure results in a shortening of the fallow period and a lengthening of the cultivation period. The result is a rapid decline in soil fertility, inability of the natural plant cover to re-establish itself, and finally destruction of the soil cover by erosion. In this way vast areas in Central Africa have been lost.

Another example is afforded by overgrazing—frequently a result of population pressure—which has devastated so many of the natural pastures in the world. This subject will be treated in detail in Chapter IV.

Availability of Additional Land Areas

The most recent assessments by a number of research groups indicate that the total extent of potentially cultivable land in the world is

3200–3400 million hectares, or about one-quarter of the total land surface. This is approximately three times the actual harvested area. In addition, there are 3600 million hectares available for grazing (Buringh, 1978).

The largest continuous area of unused and potentially arable land in the world in composed of the vast savannas and forests of humid tropical America, of which 391 million hectares are potentially arable (Nicholaides, 1979). Overgeneralizations about soils in the tropics have led to many misconceptions about the agricultural potential of these regions.

A general assumption is that most tropical soils are high in sesquioxides of iron, which harden irreversibly on exposure and drying, thereby making them unfit for cultivation when once the protective cover of vegetation is removed. According to Sanchez & Buol (1975), these so-called 'laterites (more recently renamed *plinthites*) are found at or near the surface in not more than 7% (by area) of tropical soils.

It is true that the majority of tropical soils (about 70% by area) are highly weathered, acid soils with a low cation exchange capacity and low capacity to hold bases. These *oxisols* and *ultisols* are inherently infertile, but are generally deep and well-drained and have good physical characteristics. Experience has shown that these two soil-types can be made very productive if managed properly (NAS–NRC, 1977). This subject will be treated in Chapter III.

The other important soils in the tropics are:

Alluvial soils: These are probably the most popular soils in the humid tropics for agricultural production; but they are liable to occasional flooding, and access-road construction is difficult and costly. Because of their level topography, they are ideal for irrigation, provided precautions are taken against salinity and a rise in the water-table. Large alluvial plains are mainly under rice-paddy cultivation—generally by monsoon flooding, or more rarely by controlled irrigation.

Volcanic ash soils (Andepts): These are slightly acidic, have a high nutrient content and favourable physical characteristics, and are excellent for field crops (Scharpenseel, 1977).

The second-largest areas of unused and potentially arable land, amounting in all to about 750 million hectares, are found in the dry regions. Their cultivation is dependent on the possibility of developing irrigation. Available quantitative data on surface and ground-water resources in the world are insufficient to enable an accurate assessment of total irrigation potential (FAO, 1978). Because of the uneven distribution of water resources, it is estimated that not more than one-third of these areas can be considered for agricultural production at acceptable costs with existing technology (Revelle, 1976).

A breakthrough in the costs of desalination, for example, could make possible a considerable expansion of the potential arable areas (Crosson & Frederick, 1977).

All available estimates, though inadequate, suggest that the world as a whole, and the LDCs in particular, have not yet used even half of their potential land resources, e.g. soils that have some minimum soil qualities and water supply (Crosson & Frederick, 1977).

This underutilization is generally due to limitations imposed by primitive hoe technologies or environmental constraints such as insufficient rainfall, infestation by diseases or pests, low fertility, etc.

Only about 11% of the world's soils offer no serious limitations to agriculture; Europe, Central America, and North America, have the highest proportions of these soils, whilst North and Central Asia, South America, and Australasia, have the lowest proportions.

The dominant feature of soils in South America is the low fertility status of about half the soils of the continent, centred on the Amazon basin and the central uplands; steep lands account for 11%.

In Africa, 44% of the soils are in arid or semi-arid areas, and soils with low fertility status account for 18% of the area (FAO, 1978).

In the world as a whole, however, as a result of population pressure, over 80% of the increased agricultural production since World War II, resulted from putting additional land under the plough without changing traditional methods, notwithstanding their frequently marginal character.

This trend is still evident in recent years, as shown in Table III.

In Latin America, the steep increase in the cultivated area is a function of the large areas of land that can still be developed, which are estimated as being several times as large as those actually cultivated. A second factor is the population explosion.

In addition, much of the land already used could be farmed more intensively. In seven countries representing two-thirds of the continent's agricul-

Table III Increase in areas under arable and permanent crops during the period 1961–65 (annual average) to 1977 (based on data from FAO, 1978).

	Areas in arable and permanent crops (million ha)		Increase (%)	Increase in population during 1970–78 (%)
	1961–65	1977		
Africa	188	208	10.6	24
Asia	436	458	5.0	18
South America	82	108	31.7	24
All developing countries	722	791	9.6	19
All developed countries	657	671	2.1	6.8
World	1379	1462	6.0	16

Table IV Arable land per head of population (calculated from data from FAO, 1978).

	Per caput agricultural population (ha)	*Per caput* total population (ha)
Africa	0.72	0.48
Asia	0.32	0.19
South America	1.40	0.46
All developing countries	0.42	0.26
All developed countries	4.39	0.59

ture, the big farms controlled 258 million ha. Of these, 20 million ha, though they were cultivable, were reportedly fallow, most of them under the existing primitive system of land rotation. In addition, there are extensive areas under pasture which could be intensively cultivated (Feder, 1961).

By contrast, in Asia, the increase in cropped land during the period 1961–65 to 1978 was only 5%—a reflection of the already heavy pressure on the land in this continent (see Table IV). Africa occupies an intermediate position, still having large reserves of land that can be developed, and with a population increase equal to that in Latin America.

In the developed countries, the increase in areas under cultivation is minimal—only 2.1% as compared with 9.6% for all developing countries during the period considered.

Population pressure and available land for agricultural production are reflected in the data presented in Table IV.

The developed countries have slightly more than 10 times the cultivated area per head of agricultural population than the developing countries.

In Latin America, the area currently available per head of agricultural population is on the average more than four times as great as in Asia, and two times as great as in Africa. As a general rule, the smaller the area per head of agricultural population, the larger will be the proportion allocated to subsistence crops.

In most developing countries, there are more or less large areas of population pressure, generally in the most favoured regions. The most severe and chronic cases of population pressure are in India, Bangladesh, and Java in Indonesia. In these countries there are virtually no areas left in which subsistence agriculture could expand.

In Bangladesh, more than half the holdings are of under one hectare and an equally critical situation exists in Java. The size of holdings is below the point where mini-farms are still viable. Under these circumstances, only increased off-farm employment can solve the problems of the mini-farms.

Part of the increase in arable areas in various parts of the world is due to the expansion of irrigated areas in what was formerly desert (Table V).

Table V Increase in areas under irrigation (millions of ha) during the period 1961–65 (annual average to 1977 (FAO, 1978, with permission).

	1961–65	1977	Increase (%)
Africa	5.8	7.8	13.4
Asia	100.0	128.8	28.8
South America	4.9	6.5	32.6
All developing countries	110.0	144.9	31.7
All developed countries	39.0	52.9	35.6
World	149.0	198.0	32.9

During the period 1961–65 to 1977, world areas under irrigation increased by about one-third (from 10.8% to 13.5% of the world's arable lands), the relative advance in the developing countries lagging only slightly behind that of the developed countries. The only exception is Africa, in which the irrigated areas, already small in absolute figures, increased by only 13.4%.

Problems

In considering the potential of expanding areas under cultivation, the environmental, social, and economic, costs of bringing new land and water resources into production must be taken into account. The following factors are considered as crucial to the success of the settlement of new lands: (1) selection of the right site, (2) selection of the right settlers, (3) physical preparation of the site before settlement, (4) adequate capital (whether supplied by the settler himself or the state), (5) organization of central services and project administration, (6) an adequate area per settler, and (7) secure and reasonable conditions of tenure (Lewis, 1966).

In general, most land that is relatively fertile and easy to cultivate is already being farmed, so that increases in cultivable land are usually due to the ploughing of marginal and difficult to cultivate land.

Whilst the development of the remaining potential land and water resources is technically feasible, it will inevitably entail greater costs than required formerly: in the case of land investments in clearing, levelling, terracing, or draining; and in the cost of water for transporting to greater distances, whilst surmounting natural obstacles, through pumping to greater depths than formerly and/or impounding in increasingly expensive reservoirs.

Some of the most fertile soils in the tropics, suitable for the development of intensive agriculture, are frequently swampy and waterlogged. They require drainage before they can be used, as well as a continuous heavy investment in water control for protection against flooding.

Diseases transmitted by various species of Tsetse flies (*Glossina* spp.) to humans and cattle are one of the major constraints to the settlement in general, and to the expansion of livestock in particular, in large parts of Africa south of the Sahara, despite some fertile and well-watered areas.

To control these flies is expensive, requiring removal of bush and tree vegetation that provides shelter, application of insecticides to their resting sites, and the creation of completely cleared belts that are too wide for the insects to fly over, so as to prevent re-infestation of the cleared sites. The fear of certain diseases, such as sleeping sickness (trypanosomiasis) or river blindness (onchocerciasis) may be so great that people abandon fertile areas of land in which these diseases are endemic.

Because capital costs of the development of new lands is high, feasibility studies should always precede such a programme. These studies can reveal whether the same investment will not yield greater returns if applied to increasing production in areas that are already being cultivated. If the feasibility study is favourable to colonization of new areas, a pilot project should be instituted to test the possibilities, solve problems, and train a nucleus of farmers.

Social and cultural problems can be no less worrisome than the economic cost of settling new lands.

One of the big advantages of irrigation schemes in desert areas, and of the large-scale clearing of forest areas, is that they usually involve land that is not settled, or is only sparsely settled, so that the planner can start with a 'clean slate'. Where an area is already settled, existing rights in land and property have to be cancelled in the process of creating a new system of land-use and ownership rights. If large numbers of people are involved, this may easily lead to a politically explosive situation. However, even development of uninhabited or sparsely inhabited areas is not without its problems. In Latin America, for example, the main possibility of developing new lands is by moving from the overpopulated highlands to the underpopulated coastal lowlands is strongly resisted by the people concerned.

One example of large-scale development of tropical forest land is that undertaken by Brazil, which intends to supplement its re-organization and investment programmes in the depressed North-East by the emigration of about 700,000 families to the trans-Amazonian colonization scheme (UN, 1973).

However, the massive, planned displacement of inhabitants to new lands, where it is necessary to create the infrastructures and public services, is not only an extremely costly undertaking, but may involve difficult human problems—such as result from the unfamiliar surroundings, the need to adjust to a different climate, the obligation to adopt new methods of production, and above all the insecurity resulting from the severance of family and wider social ties.

Two examples of such a situation are given by Gaitskell (1968): in West Africa, a big irrigation scheme was initiated near Sansanding on the Niger River, with the intention of imitating the Gezira project in the Sudan (see p. 61). The technical conditions were satisfactory, and the region was largely uninhabited. However, the assumption that farmers could be recruited from other areas turned out to be wrong. People just did not want to go there. The other example was even more negative. The Nigeria Agricultural Project was initiated by the British, also largely on the lines of the Gezira scheme in the Sudan, but without irrigation. This project was also started in a comparatively uninhabited area. The neighbouring chiefs were requested to encourage immigration to the new project. As a result, all kinds of malcontents and misfits were dispatched and they proved very unsuitable as farmers. The project ended in failure.

WATER RESOURCES

Increasing World Water Requirements

The estimated increase in world production, currently at an average rate of about 2% per annum, signifies that world water demand may double about every 35 years (UN, 1964). The increase in water requirements must be far more considerable in the dry regions than in the humid regions, for two main reasons: (a) water is the limiting resource in the dry regions and therefore determines the extent to which other resources can be developed, and (b) an increase in water requirements is concomitant with a rise in the standard of living. It is in the drier regions of the world that the standard of living tends to be lowest, and therefore the rise should be steepest.

Unfortunately, water resources are usually most abundant in regions in which they are least required, and scarcest in those regions in which they form the key to agricultural development and, also, to economic and industrial growth. Furthermore, water supply is generally most abundant during the periods of lowest requirement, and relatively scarce during periods of peak requirements.

Population pressures and development needs have generally created a situation whereby the development of a new water resource is most frequently possible only at the expense of previously available water supplies: this upland stream removal of water reduces the amounts available downstream, and may even reduce the quality of the remaining water to unsafe levels; the development of new wells affects the output of older wells that depend on the same underground water reservoir, and conveying water from one region to another may limit future development of the donor region.

The existing shortage of water supply in the dry regions does not necessarily imply a lack of water resources. In many of the dry regions there are potential water resources which could be developed with the necessary know-how and funds. Much additional water can also be made available by improving water conservation, by increasing the efficiency of use of existing water resources, and by a more rational allocation of water among competing demands.

Alternate Uses of Water

In developing countries in the arid regions, the bulk of the available water is used for irrigation. Unlike the situation in humid areas, where land has its own intrinsic value, in arid regions land is almost valueless without water; its value is determined by the amount of water that is available to the land.

The amount of drinking water needed each year by human beings and domestic animals is of the order of 10 tonnes per tonne of living tissue. Industrial water requirements for washing, cooling, and the circulation of materials, range from one to two tonne of product in the manufacture of bricks, to 250 tonnes per tonne of paper, and 600 tonnes per tonne of nitrate fertilizer. Even the largest of these quantities is small compared with the amounts of water needed in agriculture. To grow a tonne of sugar or maize under irrigation, about 1000 tonnes of water are 'consumed'. Wheat, rice, and cotton fibre, use about 1500, 4000, and 10,000 tonnes of water per tonne of crop, respectively (Revelle, 1966).

An economic evaluation of alternate uses of water shows that agriculture is far less productive in its use of water than are many other users. The production value of water for industry is frequently 100 times greater than for agriculture (Olivier, 1967). In a survey by the United Nations of developing countries in dry regions, it was found that, in addition to industry—mining, ports, tourist centres, and other economic activities, could afford to pay far higher costs for water than could agriculture (UN, 1964). In the United States, the water that will support one worker in arid land agriculture, will support about 60 workers in industry (Koenig, 1956).

However, factors other than the economic return per unit of water have to be considered, such as the need to provide for a minimum local production of food or the displacement of the agricultural population, while in certain cases security needs may be overriding. All these considerations, which are certainly legitimate, have to be weighed against the economic evaluation of alternate uses of water.

In the long run, economic considerations will be overriding in the determination of priorities for the use of water, which is a basic resource for which demand increases far more rapidly than supply.

As capital is one of the scarcest resources in a developing economy, errors in the economic evaluation of irrigation projects can have far-reaching consequences in retarding economic development.

Principal Water Resources

Irrigation water may be drawn from the natural flow of streams, from reservoirs which make it possible to regulate the flow of streams, from underground sources by means of wells, and increasingly from certain unconventional sources such as the use of effluents from sewage treatment and, in prospect, water desalination.

Rivers and Streams

The streams and rivers in arid regions are of three types:

(a) ephermal streams that originate within the region;
(b) streams that originate in a humid watershed, in close proximity to the arid lands, but usually from higher altitudes within the region; and
(c) perennial rivers that originate in distant humid climates and flow through the arid region.

In arid regions, springs with a copious and continuous flow of water are relatively rare. Stream-flows that originate within the region are usually intermittent in nature: torrential flows may occur for a few days or even hours at a time, to be followed by long periods during which the stream-bed is completely dry. Certain streams flow fairly continuously for a few months during the rainy reason, and then gradually dry up.

In most desert and semi-desert regions, flash-floods occur during which enormous amounts of water sometimes stream through the previously dry channels for short periods of time, varying from a few hours to a few days. These waters can be diverted, partially or entirely—depending on their volume and velocity—to inundate fields that have been specially prepared for this purpose. Their use will be discussed in Chapter III.

Many rivers originate in relatively close proximity to arid lands—usually at higher altitudes, where they are fed by melting snows and the water serves to irrigate the adjacent arid plains.

Other rivers originate in humid climates that are very distant from arid regions. If, like the Congo River, they flow at constant tropical latitudes, they remain within the humid region, and the enormous amounts of water which they collect are usually discharged into the ocean. This may account for a loss of two-thirds of the world's freshwater supply. If, however, they flow meridianally (from the Equator, pole-wards), as in the case of the Nile, their waters eventually reach an arid region and can be used for irrigation.

The seasonal flow of rivers to the arid regions fluctuates considerably. For example, the maximum discharge of the Nile in summer is about twenty times greater than in winter (Grigg, 1970). The variation is not only seasonal, but also from year to year. For example, the mean flow of the River Zambesi in southern Africa is *ca*. 3000 m^3 per second; in March 1949 the flow was only 840 m^3 per second, and in March 1952 it was 5300 m^3 per second (Kassas, 1972)! Attempts are accordingly made to regulate the flow of rivers so as to even out these fluctuations.

The development of many arid and semi-arid regions is frequently dependent on the transfer, by canals or other means, of water from streams flowing in more or less distant humid regions. Three examples are: transferring the waters of the Indus and its tributaries to the arid Sind and the semi-arid Punjab; transporting the waters of the Colorado River to the

Imperial Valley in California; and transporting the water from the sources of the Jordan in Upper Galilee in a system of open channels and closed large-diameter pipes to the Negev Desert in southern Israel.

In many cases, water of one river system is transferred through canals and tunnels to a different river basin, to supplement the water supply of the latter. For example, by tapping the headwaters of rivers flowing to the Amazon, it is possible to divert water through tunnels beneath the main watershed to the coastal plain. This has already been done on a small scale in Peru (Cole, 1965).

Ground-water

The ground-water 'reservoirs' are the source of water in wells and springs and the flow of streams during rainless periods. The volume of ground-water stored at depths of less than 800 m is generally estimated to be 3000 times as large as that contained in all rivers. The use of ground-water for irrigation has increased steeply in recent years. In India alone, the number of tubewells increased from 20,000 to 500,000 during the decade 1961–71 (Palmer-Jones & Carruthers, 1978).

Tubewells have a number of advantages over other sources of water supply: they are relatively easy to develop and their number can be increased rapidly in response to requirements; when they are situated on the land to be irrigated, the distribution system is short and inexpensive; by their use the farmers have improved control of the water-supply; health hazards are less than from surface water, provided the necessary precautions to avoid contamination are taken; and the danger of a rise in water-table with associated drainage problems is avoided.

The main disadvantages of tubewells are the fairly high capital and recurrent costs, and the possibility of overexploitation (Palmer-Jones & Carruthers, 1978).

In general, ground-water reservoirs are in dynamic balance with precipitation, evaporation, and drainage to the sea. However, because tubewells belonging to different owners tap the same aquifer, overexploitation is a frequent occurrence, resulting in a lowering of the water-table and, in more extreme cases, in irreversible damage to the aquifer as a result of the intrusion of salt water or land subsidence.

Though ground-water is the major source of water in dry regions, it has, as yet, received much less attention than have surface water resources.

Successful management of all the water resources of a watershed requires an important development of surface water and ground-water. Ground-water can supplement the supply of surface waters during the dry seasons, and surplus surface waters can be used to replenish the aquifer.

Excess water can be stored underground, to be used during the irrigation season. Underground water-storage has a number of advantages: the aquifer is replenished, evaporation losses are eliminated, and capital costs and maintenance charges are relatively low. One problem that may be encountered in certain situations, however, is that the water used for recharge may leach salts from the strate through which it passes and thereby become unfit for future agricultural use.

Non-conventional Water Resources

Re-use of water: The increasing demand for water in the dry region makes re-use, and sometimes repeated re-use, of return flows imperative. Before re-use is possible, treatment for quality improvement may be essential. The techniques of improvement vary according to the nature of the contamination and the purpose for which the water is to be used.

By constructing an irrigation return-water system at the low end of an irrigated field, it is possible to re-use water of a quality similar to that of the original water, and at relatively low cost. For this reason, a reservoir or sump is constructed which collects all surplus irrigation water, which is then re-pumped to the head of the field for re-use (Hagan *et al.*, 1967).

Use of effluents from sewage treatment processes: Untreated sewage has been traditionally used for irrigation in the Far East and south-eastern Asia, without any restrictions. In developed countries, the use of raw sewage is generally not allowed for irrigation of any kind, because of the risk of direct infection of the workers, transmission of disease by flies and other vectors, and contamination of the foods produced. Even partially treated effluent can present real health-risks, not only with fruit and vegetables that are eaten uncooked, but also with roots and tubers that are cooked. In the latter case it is the housewife, the kitchen, and the appliances, which may be contaminated.

However, the amounts of water used for waste disposal in developed countries are enormous. It has been reported that more than 80% of the total freshwater used in the USA is for waste disposal (UNEP/FAO, 1977)—hence the justification for making these effluents safe for re-use.

Biological treatment processes remove a substantial portion of the organic matter; many pathogenic organisms may be destroyed, though the water may still remain dangerous from the point of view of human health. As a rule, even a well-run biological sewage-treatment plant will not remove more than 90% of the pathogenic organisms that are present in the raw sewage (Shuval, 1967). Partially-treated effluents may be used with little risk for crops such as cotton, seed crops, tree nurseries, etc.

Under certain circumstances, there may be economic justification to use advanced technological processes that have been developed; these are based on physical, biological, and chemical, principles that make treated effluents safe for irrigation, without restriction, for all crops. Such a process typically consists of a number of stages.

In the first stage, biological treatment in oxidation ponds removes settleable solids such as sand, scum, and some suspended solids, from the raw wastewater. The turbid liquid obtained is treated with lime, which flocculates most of the suspended solids that consist mainly of organic matter. The lime-treated effluent is then maintained in open ponds, where most of the remaining solids settle and most of the nitrogen is released into the atmosphere in the form of ammonia (Idelovitch, 1977).

The use of the treated effluent for ground-water re-charge, rather than, immediate re-use, provides a considerable safety factor, because of reduction in the number of microorganisms and dilution.

Land Treatment of Wastewater

This method of effluent treatment consists of applying wastewater to the land, after it has been screened, settled, or comminuted (Bouwer & Chaney, 1974). Land treatment of wastewater disposes of the effluent without contaminating surface waters, and re-cycles the nutrients for plant growth. The methods used are simple, reliable, and have low energy requirements, as contrasted with the complex technology of advanced treatment plants. The soil is an effective filter. After movement through from one to several metres of soil, the water is generally free from microorganisms. Land treatment systems are of three types: overland flow systems, low-rate application systems, and high-rate application systems, and are as follows:

Overland flow systems: These are also called vegetation filtration systems, or spray–run-off systems, and are used where the soil is too impermeable or the content of suspended solids is too high to permit rapid infiltration, so causing most of the wastewater to run off.

Low-rate application systems: These consist in using the wastewater for irrigation at rates in accordance with crop requirements. All the wastewater applied infiltrates into the soil. In arid regions with high evapo–transpiration rates, drainage may be required to prevent salt accumulation.

High-rate application systems: The rate of application is much greater than crop requirements, and ranges from 0.5 to several metres per week. Infiltration periods are rotated with drying periods, to allow for a recovery of infiltration rates. The schedule of application and drying must be determined by local experimentation.

These systems require permeable soils and serve mainly as means to re-charge ground-water. The renovated water has about the same salt-content as the original wastewater.

For both low-rate and high-rate application, the wastewater can be applied by sprinklers or by gravity irrigation, using furrows, borders, or basins. In order to avoid contamination of ground-water by pathogenic microorganisms, a sufficient distance should be allowed between the land treatment facility and the point where ground-water is tapped for human consumption.

Certain kinds and sources of pollution are becoming increasingly important: (these include refractory organic materials, radioactive wastes, and inorganic salts. Many problems are therefore involved in the use of effluents.

Increasing Precipitation

The presence of clouds alone does not ensure precipitation. The droplets of moisture which constitute the cloud are frequently too small to reach the weight required for them to fall earth-wards are precipitation. For this reason only a small fraction, usually less than 10%, of the moisture in the atmosphere reaches the ground as precipitation (Schleusener & Grant, 1967).

A number of methods have been developed for artificially stimulating precipitation. They are all based on seeding the cloud with nuclei, on which the moisture will condense. Seeding suitable clouds with dry ice (solid carbon dioxide) reduces their temperature and triggers the process of formation of ice crystals.

Another approach, artificial nucleation, is based on the fact that dust or other particles may serve as nuclei of condensation, enabling rain droplets to form at temperatures that are somewhat above those normally required for ice crystal formation. In particular, crystals of silver iodide are known to be effective as an ice-nucleating agent at a temperature of -5 °C (Neuman, 1967).

Seeding of clouds can therefore be effective in causing rainfall only when conditions are not too different from those under which the precipitation would occur naturally, at which time seeding may 'tip the balance'—hence the difficulty of deciding unequivocally whether cloud seeding has been instrumental in causing rainfall or not.

In Israel, it is estimated that an average increase in rainfall of 18–19% over five seasons was achieved by cloud seeding. The gain in precipitation due to seeding generally occurred as the result of large increases in a few days of seeding, whilst on most days of seeding only minor or no increases were recorded. Similar results have been obtained in Australia.

Rain cannot fall from a dry atmosphere, and therefore even the most successful rain-making can, at best, only contribute to an increase in rainfall, but cannot consistently provide rain to arid regions.

Desalination

Practically untapped major sources of water for arid lands are sea-water and brackish lake- and ground-water. Methods of desalination which would

Table VI Number of functioning desalination plants and output.

	Number	Output (million gallon/day)[a]
Arab Peninsular and Iran	153	146
Africa	104	57
Asia (excluding Near East)	58	67.6

[a]1 gallon = 4.546 litre.

provide freshwater from these sources at an economical cost, would solve the problem of water for many areas in which no other solution is in sight.

A number of methods have been developed to produce freshwater from saline water. These conversion processes are based on distillation, passage through membranes, freezing, humidification, and chemical processes.

The methods that are most advanced technologically for converting water in the quantities required for irrigation, are based on: (a) distillation for conversion of water, such as sea-water, with very high salt content (35,000 p.p.m. total soluble salts (TSS) and more); or (b) passage through membranes for converting brackish water (3000 to 4000 p.p.m. TSS) into water that can be used for irrigation (500 p.p.m. TSS) (UN, 1964).

The use of desalination plants is spreading rapidly in those countries that have practically exhausted the potential of their conventional water supplies.

In 1975, the number of functioning desalination plants in selected regions was as shown in Table VI (Office of Water Research & Technology, 1975). Approximately 85% of the output in Table VI was from plants based on distillation, and the dominant type among them is the multi-stage flash system.

However, up to the present, the water obtained from desalination is almost exclusively used for industrial purposes, for promoting tourism, and for urban water-supply. Only insignificant quantities are being used for agricultural production—mainly under controlled conditions, such as in greenhouses.

The rapid expansion of water desalination stimulates research towards finding methods of reducing costs, so that eventually water may be produced that can be used economically in agricultural production.

There still remain the problems of energy cost. Under reasonable assumptions, desalination of enough sea-water to provide food for one person would require as much energy as is currently used on an average for all other purposes (Borgstrom, 1976). This does not include energy costs of distributing the water and of disposing of the salt water.

Summary and Conclusion

Many development schemes have failed, partially or entirely, because of lack of basic information on the natural resources that are available to the area to be developed. This situation is being gradually improved, by national and international efforts.

Agricultural development is possible in two different ways, which are not mutually exclusive: bringing more land under cultivation, and/or increasing productivity per unit of land.

The largest areas of unused and potentially arable land in the world, are the vast savannas and forests of the humid tropics, and the dry lands in areas where precipitation is lower than the minimum required for crop production.

All available estimates suggest that the developing countries, on the whole, have not used even half their potential land resources. This under-utilization is generally due to limitations imposed by primitive technologies or environmental constraints—such as insufficient rainfall, infestation by pests and diseases, low fertility, etc.

As a result of population pressure, over 80% of the increased agricultural production since World War II is due to putting additional land to the plough, without changing traditional methods. This has frequently led to the loss of productivity of huge land areas—which is usually irreversible—as a result of erosion, destruction of natural vegetative cover, over-grazing, salinization, etc., a subject that will be treated in detail in Chapter IV.

Part of the increase in arable land is due to the expansion of irrigation in formerly desert areas. It is projected that world water demand will double about every 35 years. The increased demand will be far greater in dry regions, where water resources are most limited, than in humid ones. Frequently, because of population pressures, the development of a new water resource is only possible at the expense of previously available water supplies.

There are, however, potential water resources in the dry regions which could be developed if the necessary know-how and capital resources were to be made available. Other possibilities are: improving water conservation, increasing the efficiency of use of existing water resources, a more rational allocation of water among competing demands, and developing unconventional water supplies (re-use of water, use of effluents of sewage treatment processes, increasing precipitation, desalination, etc.).

Capital costs of developing new lands are generally high, but social and cultural problems can also prove to be serious impediments.

Before 1950, opening new land and a gradual increase in the intensity of land-use enabled food production to keep pace with the relatively slow increase in food demand by a population whose growth-rate in most developing countries was under 1.3% per year (Stevens, 1977). But now it is estimated that, to sustain world population in the year 2000, an increase of 60% is required in agricultural production (FAO, 1978). It is expected that annual increases in cultivated land area will be at most 1% over the next 25 years. Therefore, the increase in food supply at the rate required can only come from increased productivity of the land, at a rate of 2.5% per year, over the next 20 to 25 years.

The poorest countries, because of their rapid population increase, lack of capital skills, and inadequate organization, will require even greater rates of increase (NAS–NRC, 1977).

In 1974, the World Food Conference estimated that $5 thousand millions per year for the next 20 to 25 years would be required in foreign exchange by the LDCs for investment in agricultural production in order to secure an adequate food supply (Hopper, 1976). This poses a challenge to developed as well as developing countries.

References

Bauer, M. E. (1975). The role of remote sensing in determining the distribution and yield of crops. *Adv. Agron.,* **27**, pp. 271–304.

Borgstrom, G. (1976). Food and energy in confrontation. *Proc. Amer. Phytopathological Society,* **3**, 28–34.

Bouwer, H. & Chaney, R. L. (1974). Land treatment of wastewater. *Adv. Agron.,* **26**, 133–76.

Buringh, P. (1978). The natural environment and food production. Pp. 99–129 in *Alternatives for Growth: The Engineering and Economies of Natural Resources Development*. Ballinger, Cambridge, Massachusetts: xii + 256 pp., illustr.

Cole, J. P. (1965). *Latin America, an Economic and Social Geography*. Butterworths, London, UK: xviii + 468 pp., illustr.

Crosson, P. R. & Frederick, K. D. (1977). *The World Food Situation*. Resources for the Future, Washington, DC: v + 230 pp., illustr.

Dandekar, V. M. (1969). Questions of economic analysis and the consequences of population growth. Pp. 336–75 in *Subsistence Agriculture in Economic Development* (Ed. C. R. Wharton, Jr) Aldine, Chicago, Illinois: xiii + 481 pp.

FAO (1978). *The State of Food and Agriculture 1977*. FAO Agricultural Series, No. 8, Rome, Italy.

Feder, E. (1961). Land reform in Latin America. *Social Order*, **11**(1), pp. 29–36.

Gaitskell, A. (1968). Problems of policy in planning agricultural development in Africa south of the Sahara. Pp. 214–38 in *Economic Development of Tropical Agriculture* (Ed. W. W. McPherson) Univ. of Florida Press, Gainesville, Florida: xvi + 328 pp., illustr.

Grigg, D. (1970). *The Harsh Lands: A Study on Agricultural Development*. Macmillan, London, UK: 321 pp.

Hagan, R. V., Houston, C. E. & Burgy, R. H. (1967). More crop per drop: Approaches to increasing production from limited water resources. Pp. 610–21 in *Water for Peace*, US Government Printing Office, Washington, DC: Vol. 3, vi + 968 pp., illustr.

Hopper, W. D. (1976). The development of agriculture in developing countries. *Sci. Am.*, **275**(3), pp. 196–205.

Idelovitch, E. (1977). Waste-water reclamation by advanced treatment. *Kidma*, **3**(2), pp. 30–5.

Inter-American Committee (1963). *Inventory of Information Basic to the Planning of Agricultural Development in Latin America*. Pan American Union, Washington, DC: 202 pp.

Kassas, M. (1972). Ecological consequences of water development projects. Pp. 215–35 in *The Environmental Future* (Ed. N. Polunin). Macmillan, London & Basingstoke, UK: ix + 660 pp., illustr.

Koenig, L. (1956). The economics of water sources. Pp. 320–30 in *The Future of Arid Lands* (Ed. C. F. White). American Association for the Advancement of Science, Washington, DC: ix + 453 pp., illustr.

Lewis, W. A. (1966). *Development Planning: the Essentials of Economic Policy*. Allen & Unwin, London, UK: 278 pp.

Moutappa, F. (1973). Soil mapping in relation to the use of fertilizers in the humid tropics. Pp. 71–82 in *Potassium in Tropical Crops and Soils*. International Potash Institute, Berne, Switzerland: 603 pp., illustr.

NAS–NRC (1977). *World Food and Nutrition Study: The Potential Contributions of Research*. National Academy of Sciences, Washington, DC: xxvi + 192 pp.

Neuman, J. (1967). Cloud seeding and cloud physics—a review of activities in Israel. Pp. 375–88 in *Water for Peace*. US Government Printing Office, Washington, DC: Vol. 2, vi + 970 pp., illustr.

Nicholaides, J., III (1979). Crop production systems in acid soils in humid tropical America. Pp. 243–77 in *Soil, Water and Crop Production* (Ed. D. W. Thorne & M. D. Thorne). Ari Publishing, Westport, Connecticut: ix + 253 pp., illustr.

Office of Water Research and Technology (1975). *Desalting Plants Inventory*, Report No. 5, US Department of the Interior, Washington, DC.

Olivier, H. (1967). Irrigation as a factor in promoting regional development. Pp. 266–81 in *Water for Peace*. US Government Printing Office, Washington, DC: Vol. 2, vi + 970 pp., illustr.

Palmer-Jones, R. & Carruthers, I. (1978). Agricultural water use. Pp. 237–64 in *Technology Transfer and Change in the Arab World* (Ed. A. B. Zahlan). Pergamon, Oxford, UK: xvii + 506 pp., illustr.

Revelle, R. (1966). Water. Pp. 1007–19 in *Getting Agriculture Moving* (Ed. R. E. Borton). The Agricultural Development Council, New York, NY: 2 vols, x + viii + 1123 pp.

Revelle, R. (1976). The resources available for agriculture. *Sci. Am.,* **235**(3), 165–80.

Sanchez, P. A. & Buol, S. (1975). Soils of the tropics and the world food crisis. *Science,* **188** pp. 598–603.

Scharpenseel, H. W. (1977). Soil—related determinants of cropping patterns. Pp. 53–60 in *Symposium on Cropping Systems Research and Development for the Asian Rice Farmer.* IRRI, Los Banos, Philippines: 454 pp., illustr.

Schleusener, R. A. & Grant, L. O. (1967). Weather variation and modification. Pp. 40–7 in *Irrigation of Agricultural Lands* (Ed. R. M. Hagan, R. H. Haise & T. W. Edminster). Amer. Soc. Agron., Madison, Wisconsin: xvi + 118 pp., illustr.

Shuval, H. I. (1967). Public health implications of waste-water utilization. Pp. 690–701 in *Water for Peace*. US Government Printing Office, Washington, DC: Vol. 3, vi + 968 pp., illustr.

Stevens, R. D. (1977). *Tradition and Dynamics in Small Farm Agriculture*. Iowa State University Press, Ames, Iowa: xiii + 266 pp., illustr.

UN (1963). *Science and Technology for Development*, Vol. 3 (Agriculture). United Nations, New York, NY: 309 pp.

UN (1964). *Water Desalinization in Developing Countries*. United Nations, New York, NY: viii + 325 pp., illustr.

UN (1973). *Implementation of the International Development Strategy* Vol. I. New York, NY [not available for checking].

UN (1974). *The World Food Problem: Proposals for National and International Action*. UN World Food Conference, Rome, Italy [not available for checking].

UNEP/FAO (1977). *Seminar on Utilization and Management of Agricultural and Indo-agricultural Wastes*. FAO, Rome, Italy [not available for checking].

Wilde, J. C. de, & McLoughlin, P. F. M. (1967). *Experiences with Agricultural Development in Tropical Africa*. Johns Hopkins Press, Baltimore, Maryland: Vol. 2, xii + 466 pp.

CHAPTER III

Land-use in Subsistence Agriculture and Potentials for Development

Throughout the historical period, certain traditional patterns of land-use have evolved that characterize subsistence agriculture in various regions of the world. These patterns are extraordinarily similar over very wide areas, despite the great diversification of habitats due to various combinations of topography, soil type, and climate. Land-use has also shown a surprising amount of stability, having remained practically unchanged in subsistence agriculture throughout the historical period and even to the present day.

We will consider here the land-use problems of those climatic regions in which most of the developing countries are found—the dry to semi-arid subtropical regions and the subhumid and humid tropics.

LAND-USE IN DRY REGIONS

Pastoralism

Undoubtedly the best use that can be made of desert areas, as such, is extensive grazing by livestock (Fig. 7). *Nomadism* and *transhumant pastoralism* are very ancient natural adaptations to the sparse vegetation, seasonal growth of the vegetation, and drying up of watering places. that are characteristic of arid lands, and are therefore the traditional form of land-use of enormous desert tracts in all the dry regions of the world. Even wild animals migrate spontaneously under these conditions, and, like them, the nomad pastoralist follows the rain, according to seasonal patterns, in seeking new pastures.

These pastures can be used in two ways: (a) *transhumance*, in which herds that are resident in one area, are moved for a few months to an adjacent area during the latter's productive period; (b) *long-range nomadism*, which involves the seasonal migration of the whole population with their herds over long distances along seasonal routes, alternating between different types of grazing areas, or 'moving with the rain' over vast expanses of the desert. This implies continuous tent-dwelling and a completely nomadic way of life for the whole tribe.

Fig. 7 *'Grazing' in the desert*. By courtesy of N. Tadmor.

Primitive pastoral methods can, at best, provide a very low standard of living for the pastoralists, who have traditionally always sought supplementary sources of income. They had a virtual monopoly of transportation over the traditional desert routes—such as the trans-Sahara slave trade from tropical Africa to the Mediterranean, and the ancient 'silk road' between China and the Middle East. Levies exacted from pilgrims and merchants travelling on the desert routes, tribute from cultivators working the nomad-owned land in the oases on a share-cropping basis, and plunder from raids augmented the income of the nomad tribes.

With the aura of romanticism that usually surrounds the nomad way of life in the desert, it is tempting to overestimate the degree of adaptation of the nomad pastoralist to his environment. A balance was indeed achieved, but at a very low standard of living for the vast majority of the tribal members. Though the nomad does not generally engage in the same back-breaking labour as the sedentary farmer, his life is arduous and knows many discomforts and dangers. The balance between livestock and the forage resources of the desert is also very precarious, notwithstanding the mobility of the pastoralist and his flock. A longer-than-usual period of drought may wipe out a large proportion of his herds, especially the young stock. After a period of relative abundance, the numbers of his stock may grow out of all proportion to the normally available forage resources, and overgrazing ensues before a balance is again achieved by an increased mortality rate. The nomads do not normally slaughter their animals for food—the herds being considered as non-expendable capital—thereby preventing any planned adjustment of livestock numbers to available forage. The nomads also suffer from many diseases.

The precarious balance between the nomad and his environment, achieved and maintained throughout centuries, has been disrupted in recent times. His essential role in the transport of merchandise and travel along the desert routes has become obsolete. Law and order have practically abolished his sources of income, derived from raiding and tribute. For these reasons, the problem of settling the nomads has become acute in many developing regions, and forms an integral part of development schemes in many countries.

Potentials for Intensification of Dry-land-Use

In the regions in which rainfall is not usually adequate for the production of conventional crops, the following measures can be adopted as a means of intensifying agricultural production:

(a) Supplying additional water, either through water-harvesting and spreading, tapping surface or underground water resources in the region, or by transfer of water from outside the region. These options will be discussed under 'irrigation agriculture' (pp. 51–69).

(b) Commercial exploitation of plants adapted to desert conditions.

(c) A more rational and scientific approach to the traditional use of desert areas for livestock production.

DOMESTICATION OF PLANTS FOR DESERT CONDITIONS

Desert Plants of Potential Commercial Significance for Arid Growing-conditions

The potentially most interesting arid-zone plants are those which contain constituents whose manufacture by the plant is not adversely affected by aridity or may even benefit from dry growing-conditions, and whose commercial value per unit weight is very high. Plants containing alkaloids, essential oils, and mucilaginous materials, belong to this category.

Plant species producing constituents of economic value, but whose yield is proportional to the vegetative growth of the plant (such as fibres or food reserves), and which have a relatively low intrinsic commercial value, are, *a priori*, less promising.

The number of desert plants containing *alkaloids* is relatively few, but their alkaloid content can be very high (Paris & Dilleman, 1960), with a number of the *Solanaceae* commonly the richest source. Thus, the most important alkaloids of desert plants are the hyoscine and hyoscyamine groups, steroidal alkaloids, ephedrine, and a variety of other alkaloids.

The second important group of constituents contained in many desert plants consists of *essential oils*, secretion of which appears to be a xerophytic character conferring a degree of protection against drought (Fluck, 1955).

The third group is composed of mucilaginous plants producing medicinal *gums*. The mucilage plays a part in water retention and is therefore an important xerophytic character; it has also been established that gum production is favoured by low atmospheric and soil humidity (Paris & Dilleman, 1960).

Prospects of Commercial Exploitation of Desert Plants and Animals

Harvesting from Natural Stands

The natural stands of most potentially valuable desert plants are too diffuse to permit harvesting to be economical, except possibly for people with a very low standard of living. Even vendors of plants of medicinal value that are relatively abundant in the desert may encounter difficulties when it comes to marketing, due to competition from cheaper sources. In other cases, the plants are not sufficiently abundant to provide an adequate and continuous supply. In the few cases of a seemingly abundant supply, such as that of Candelilla (*Euphorbia antisyphilitica*) plants (for wax) in Mexico (Pultz, 1956), or of *Hyoperimus* in Egypt, which is collected for its alkaloid constituent (Drar, 1954), excessive harvesting has rapidly led to a depletion of the natural populations.

Domestication

In the cases of a number of desert plants of potential economic value, a very large effort in plant breeding, management practices, and methods of processing, has been expanded as a preliminary step to domestication. Examples of this are Guayule (*Parthenium argentatum*) for rubber production, *Canaigre* (*Rumex hymenosepalus*) for tanning, and agave (*Agave spp.*) for fibre. Commercial production in the truly arid conditions that are characteristic of the native habitat of the plants concerned has usually failed to be economical. It is probably unrealistic to expect commercial cultivation in areas with less than 200 mm of annual rainfal.

Since the early nineteen seventies, the possibilities of domestication of Jojoba (*Simmondsia chinensis*), a native plant of the Sonora desert, is being actively investigated.

The plant can serve as a potential source of liquid wax, a possible substitute for sperm whale oil, and the oil-cake can be fed to livestock (Walters *et al*., 1979).

If the desert plants are grown under more favourable conditions of precipitation, and still more so if they are irrigated, the content of the valuable constituents is markedly reduced in most cases. This is not surprising in view of the fact that the production of these constituents by the plant is often a protective measure against aridity.

Up to the present, success in a few limited cases, such as the cultivation of Guayule, has been dependent on war-time shortages or other special economic considerations—such as the desire to be independent of outside sources, which has led to the extensive cultivation of the rubber-producing Kok-saghys (*Tarascacum kok-saghys*) in the USSR. When once the situation returns to normal, however, cultivation has usually ceased.

The potentialities of the world's flora seem sufficiently well known to enable the conclusion to be drawn that the likelihood of finding crops of economic significance that can be grown commercially on a large scale in the arid regions, without irrigation, is only slight. Whatever specific advantage plants that are native to arid lands may have, these are attenuated when the plants are grown under irrigation. With an assured water supply, there is a great choice of mesophytic crops, with high-yielding potentialities and commercial value; the xerophytic plants cannot compete under these conditions.

Improving Livestock Production by Ranching

Modern livestock production in regions of insufficient rainfall for crop production is known as ranching. The salient features of ranching are:

(a) The fencing of the area, or at least marking of boundaries making possible planned grazing procedures.
(b) The establishment of adequate watering points, which by proper spacing make it possible to reduce localized overgrazing. Self-feeding salt installations at the right locations also induce the livestock to graze the range more uniformly.
(c) Efficient management and utilization of the range. Ploughing and seeding with improved species of grass is rarely economic on a large scale. Therefore, grazing systems must be adopted that favour the most useful species and achieve maximum productivity throughout the year.

The application of fertilizers is generally not economically justified; resting parts of the range at regular intervals by proper grazing procedures is an essential alternative.

The single most important factor in ensuring efficient use of the range is to balance the number of grazing animals with the amount of annual growth that can be removed without impairing the productive capacity of the range. This can be achieved by: (a) maintaining stock numbers at a level lower than can be fed on the *average* forage production of the range; (b) creating forage reserves; and (c) reducing stock numbers rapidly during prolonged periods of drought.

An assessment of the carrying capacity of rangeland is generally obtained by arbitrary methods and is frequently misleading. Very little is yet known on proper procedures. Generally, on terrain with low erosion hazards, not more than 50–60% of the annual growth of the more palatable perennial plant species may be removed (Heady, 1965). Seasonal variations are balanced by leaving part of the production ungrazed during flush growth, to serve as a carry-over for the period of scarcity.

Prospects of Ranching

Economic prospects for cattle ranching are promising. According to the projections made by FAO (1970), the demand for beef, veal, mutton, and lamb, is expected to grow in the developed countries substantially faster than domestic supplies. The combined exportable supplies of exporting countries are likely to be below the effective import requirements at constant prices. Therefore, it should be to the advantage of developing countries to encourage an expansion in production, especially of beef and veal but also of lamb.

However, ranching is usually beset by a number of difficulties: Remoteness from market and poor communications cause either high freightage costs or require the transport of the cattle 'on the hoof', with a concomitant loss of weight and quality. Because of the low purchasing power of the population of the developing countries, most of the meat has to be exported. Therefore, quality and sanitary requirements are high and the control has to be stringent. To meet these requirements, considerable capital expenditure and a high professional level are essential. Because of the fluctuations in the production of the range, the animals gain weight for part of the year and then lose much of the gain. Drought years may either oblige ranchers to sell off at loss or result in considerable mortality.

For these reasons, modern ranching requires a high standard of management, very large areas per farming unit, and a considerable capital investment.

As a result, in most cases, ranching is in the hands of large companies or large landowners who are able to supply capital and technical knowledge when required, and are also able and equipped to export the beef stock economically and efficiently from the producing areas to the consuming countries.

Cooperative ranching has been suggested as a possible solution, with several cattle-owners combining their herds and other resources in order to make possible the provision of the essential capital development. Until recently, however, there are few, if any, examples of successful cooperative ranching in developing countries.

Rain-fed Agriculture in Semi-arid Regions

Characteristics

The semi-arid regions include the areas with a Mediterranean-type climate having winter rainfall and also the subtropical savannas with predominantly summer rainfall. In these regions, rain-fed crops can be grown with a reasonable degree of security—the risk of complete failure of the harvest is in general not greater than one in five, but the low and erratic rainfall limits the number of crops that can be grown. Cereal cultivation is the main source of livelihood—wheat and barley in the winter-rainfall areas, or maize, sorghum, and millets, in the summer-rainfall areas.

The amount of rainfall that has been set, somewhat arbitrarily, as the limit of regular dryland farming, is around 250–300 mm annually in areas with predominantly winter rainfall, and approximately 500 mm annually in areas with summer rainfall (Wallen, 1966).

It is on the fringes of the arid regions that arable cropping is most precarious, and undesirable methods of land-use have led to catastrophes. These regions are characterized by the irregular pattern of annual precipitation, which varies within very wide limits.

In years of sufficient precipitation, good yields can be obtained because of the build-up of fertility during the low-rainfall years. This serves as a temptation to plough up land for cropping—especially when years of high precipitation coincide with favourable marketing conditions for the grain.

Inevitably, years of under-average rainfall follow, and crops fail more or less completely. In addition to the economic distress involved, the misuse of the land leads to severe erosion—mainly by wind.

As one moves from the drier fringe towards regions with higher annual rainfall, the prospects for satisfactory crop production improve. However, even under these circumstances, occasional years of drought and crop failures are almost inevitable.

Experience of rain-fed arable cropping in various semi-arid regions of the world has shown that the ploughing-up of rangelands in high-risk regions having precipitation which is too scanty to ensure reasonable prospects of cropping, should be avoided. The 'dust bowl' created in the 1930s in the United States is the classic example of the dangers inherent in this type of land-use.

Potentials for Improvement

The main problem of the semi-arid regions is to adjust the farming system to the extremely variable rainfall—so that the maximum benefit can be derived from years with good precipitation, and drought years need not cause starvation or economic breakdown.

Many countries, in the early stages of development, may increase yields by improving certain simple cultural practices that do not require expensive, purchased inputs. More careful tillage, better stands of plants, timely sowing and weeding, can all contribute to improved yields.

By themselves, however, improved husbandry practices have limited effects on yield, and cannot bring about a marked expansion of production. Even an improved crop rotation without fertilizers will be largely inneffective. And yet, in the early stages of development, improvements in husbandry are extremely important, as they build up traditional agriculture to a point where a qualitative change becomes possible. They increase the farmer's ability to buy more costly inputs, such as fertilizers and insecticides, and enhance the effects of the latter. In order to achieve a breakthrough in yields, a 'package' of improved practices becomes essential.

Crop rotation, appropriate tillage methods, the use of improved and adapted varieties, in conjunction with fertilizer application and crop protection, have proved able to effect dramatic increases in yields, even in the areas with problematic moisture régimes.

Marginal Rainfall Areas

In the transition areas between aridity and semi-aridity, with just sufficient rainfall to make cropping possible, provided suitable measures are taken, alternating cereal and fallow, so as to conserve moisture from one year to the next, and thereby producing a crop on the combined precipitation of two seasons, is the procedure most generally adopted as a first step towards modernization. Although a high proportion of the rainfall stored in the soil during the fallow period is generally lost, the additional 80–100 mm or so may make the difference between success and complete failure of the succeeding crop in regions with marginal precipitation.

In Australia, in the wheat-growing regions with an annual precipitation of 250–375 mm, experience shows that the fallow–wheat sequence usually assures the farmer of a crop, whereas continuous wheat would generally be a complete failure. More important still, the additional moisture is generally a prerequisite for obtaining economic returns from the package of improved practices mentioned above.

An alternative solution, which has proved to be very effective, is based on an integrated crop–animal husbandry farming system. The key to this system consists of a ley rotation adapted to semi-arid conditions, based on alternating periods of two or more years of annual, self-seeding legumes—grazed by cattle or sheep—with one or more years of grain crops. This system has a built-in autoregulatory device ensuring a constant, albeit fluctuating, income. In the good-rainfall seasons, the legumes and the stubble of the cereal crop produce year-round grazing (provided the number of head of stock is properly adjusted to the carrying capacity), whilst good grain yields are harvested. In the poor-rainfall seasons, the relatively low production of the legumes is supplemented by grazing the cereal fields which have no prospect of producing a satisfactory grain-crop. What would have been a total loss in an exclusively grain-producing system, will give a fair yield of forage even in conditions under which no grain is formed.

This integrated system of animal husbandry and cereal cropping, widely used in the dry regions of Australia, also has the advantage that inputs are relatively low: the annual legumes are self-seeding, so that no tillage or only a minimum of tillage is required. Only phosphatic fertilizers are needed, and if they are applied in excess of the current crop requirements, the residues are not wasted. The legumes provide all the nitrogen that is required by the cereal crop; but, in a very good year, additional nitrogen may be top-dressed. The cereal can be seeded into the legume stand by means of a chisel-drill without additional tillage; excessive legume volunteer plants in the cereal can be controlled by herbicides. Grain volunteer plants in the legumes provide balanced grazing. Perennial weeds such as Johnson Grass (*Sorghum halepense*) and Couchgrass (*Cynodon dactylon*), are kept in check both by the grazing and by the competition of the cereal crop.

Adequate Rainfall Areas

In the areas in which annual rainfall is generally adequate to produce single annual crops, the main constraint encountered in subsistence agriculture is low soil fertility, due to exhaustive practices of crop production that have been continued throughout centuries with little or no restitution. As a result, yield levels are low, though surprisingly constant.

The traditional crop rotations practised are nearly all exhaustive of nutrients and make no contribution to soil fertility. The basic problem is therefore to devise a type of crop rotation that will raise the level of soil fertility, thus making it possible for each crop to benefit fully from the favourable moisture régime prevailing during its growing period, as well as from any application of modern inputs.

Leguminous crops that are not allowed to mature seed, but are used for pasture, soiling, hay, silage, or for turning under as green manure, have been shown to be highly effective for improving soil fertility. They also enhance the response of the cereals to other factors affecting production.

At first, green manures were considered the ideal solution for rapidly raising the level of soil fertility in these areas. On the whole, the results obtained in semi-arid conditions have been disappointing, and green manures have generally shown little, if any, benefit over yields derived from leguminous forage crops. Where green manures have been beneficial, the increase in yield rarely justified the loss of a cropping season or the risks and expenditure involved.

Under conditions in which the inclusion of forage legumes in the crop rotation increased the yields of the subsequent crop, residual soil moisture after cutting the legumes was generally high, and not much less than that prevailing after fallow.

The area devoted to a soil-improving crop may vary from one-third to one-sixth of the total area, with a tendency to devote as large an area as possible to the cereal: hence the plethora of three-, four-, five-, and six-years' crop rotations, with various sequences of crops, in which the cereals occupy an important position. The best overall results in a region of Mediterranean-type climate have been obtained with the following three-course rotation:

first year: winter cereals (wheat, barley, oats);
second year: summer crops (sorghum, maize, sunflowers, sesame, chick peas);
third year: hay crops (vetch, annual clover, etc.).

Simple practices for conserving soil moisture may increase yields in the semi-arid regions, and may even prevent crop failures. In northern Nigeria, yields of rain-fed cotton, grown in traditional agriculture, are around 165 to 220 kg/ha of cotton seed. Experimental work indicated that, by using improved varieties and fertilization, yields would be increased to a maximum of 660 kg/ha. More recent research has shown that mulching, in addition to improved varieties and fertilization, increases the percolation rate of water into the soil and thereby further increases yields considerably. These relatively high levels were practically independent of variations in rainfall; thus in the course of five years, with distinctly different rainfall patterns, yields per annum of 2400 kg/ha or more were obtained. Tie-ridging (when the furrows between ridges are blocked at regular intervals) can also increase cotton yields significantly by conserving and storing water from the early rains; as a result, the cotton plants are better able to survive the dry period which usually occurs after the early rains (Wilde & McLoughlin, 1967).

Irrigation Agriculture

Early Subsistence Irrigation

Sedentary agriculture was established early in large, contiguous areas along great rivers, such as the Nile, the Tigris, the Euphrates, and the Indus; in isolated pockets of well-watered oases of the deserts; more precariously on the floodplains of the stream-beds in the deserts; and on the semi-arid fringes of the desert in which rain-fed crops were produced.

This pattern of land-use has led to a virtual separation of arable cropping and animal husbandry. Conflicts between the nomad pastoralists and the sedentary cultivators have been a recurrent theme in the Old World throughout the historical period, as characterized by the biblical strife between Cain and Abel. This theme repeated itself in the New World, when the arid regions of the American West were first settled by the white man.

The lot of the subsistence irrigation farmer was always precarious and hard. Even when the water supply was fairly adequate and assured, traditional irrigated crop production was (and still is) associated with back-breaking labour in a difficult climate and under unsanitary conditions. In addition to typical agricultural operations—such as soil preparation, sowing, weeding, and harvesting—the farmer is burdened with additional heavy chores, of which the actual watering of the crops is the most onerous. The maintenance of the irrigation system—canals, ditches and levees—also involves unending drudgery. The irrigation farmer is often weakened by malaria, bilharzia, or other diseases that are endemic to the irrigated areas.

Control of the moisture supply to the crops may appear at first sight to make farming independent of the vagaries of weather, but actually it introduces a host of problems: irrigation increases the hazards of pests and favours the proliferation of diseases; the fragile soils that are typical of the arid regions break down easily under irrigation; and rising water-tables and salt accumulation imperil the very basis of farming. Primitive civilizations were not able, in the long run to cope with these problems. Even in modern times, irrigation schemes have failed disastrously when they did not adopt the necessary precautionary measures. An adequate solution to these problems is essential if the permanency of irrigated agriculture is to be assured; these problems will be discussed in detail in Chapter IV.

Principal Types of Irrigation Agriculture

Water-harvesting

A very crude form of irrigation, based on 'water-harvesting' has been practised since the earliest times in the deserts of the Old and New Worlds. By this means the insufficient precipitation is supplemented by run-off water from higher-lying areas.

Rainfall in dry regions usually occurs in short, violent downpours, which the hard-baked and relatively impermeable soil is not capable of absorbing. The water runs off the stoney, bare hillsides and even a relatively light rainfall is capable of producing a significant 'water crop', carrying a certain amount of silt. A rainfall of 10 mm on a watershed may cause a flood of 30,000 m^3/h, over a period of from four to five hours. There are two possible alternatives: storing the run-off water behind dams (Fig. 8), or storing it underground in the valley bed and in level plots on the banks of the ephemeral streams. The latter alternative is the more effective and enduring of the two. A rainfall of a few millimetres on the catchment area may permit the storage, in a restricted area, of the equivalent of several hundred millimetres of rainfall (Fig. 9).

Where seasonal floods are very variable, with years of no floods interspersed with years of floods of destructive force, the agricultural economy is extremely unstable, and subsistence has to depend to a large extent on food-gathering and fishing. Where floods are more dependable, agriculture is the main source of subsistence.

Under certain conditions, water-harvesting is the only method for producing crops in desert areas with no other water resources.

Fig. 8 Run-off from high-intensity rainfall collected in a small reservoir (tank) and used for assuring a second crop for at least part of the land surrounding the tank. Indian scientist measuring losses due to evaporation and seepage. By courtesy of the International Crops Research Institute for the Semi-Arid Tropics (Hyderabad).

Micro-watersheds

Modern technology is attempting to develop more effective methods of reducing the ratio between donor and recipient areas. Instead of the traditional methods of conveying the water from barren hillsides to adjacent relatively level land, the new approach aims at creating a micro-relief within a more or less level field. Systems have been developed in which run-off water from parts of a field are concentrated in strips in which crops are planted. The crop is sown in narrow strips between wide intervals that are ridged as artificial miniature watersheds. These are later compacted to increase run-off of water to the crop rows. The relative widths of the watershedding strips and of the crop-producing strips depend on the amount of annual precipitation that can be expected. The usual ratios are from 2:1 up to 6:1 (Fig. 10). This system is considered to be more efficient than fallowing, in which water is conserved from one season to another (Kemper, 1964).

Another method, called 'catchment basins' has been developed for use on relatively even-sloping plains (with slopes up to 5%). The area is

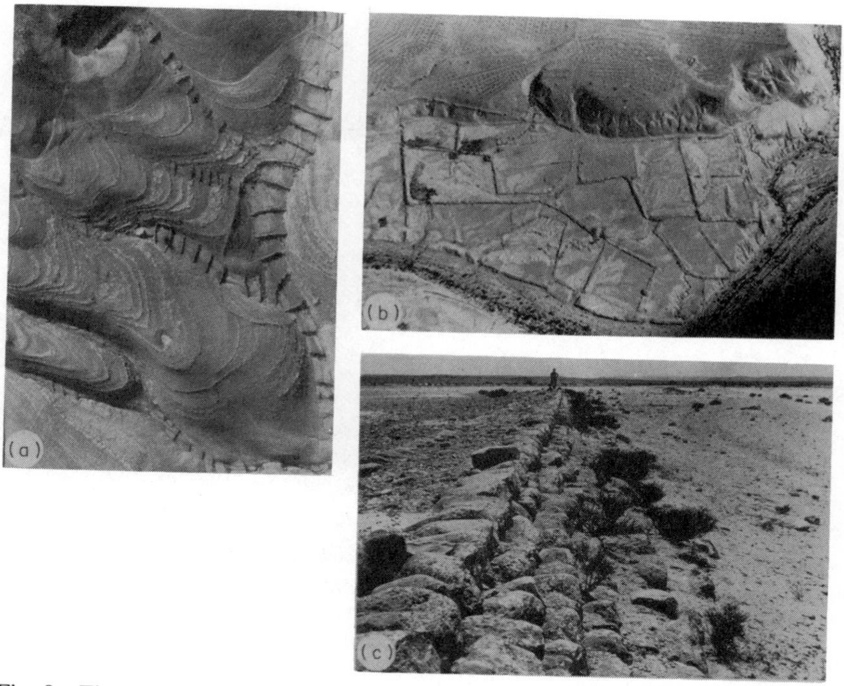

Fig. 9 Floodwater harvesting and spreading in Nabatean farming: (a) Crop production in tributary valleys; the well-preserved low terraces for slowing down the run of water are clearly seen in this aerial photo; (b) aerial view of Nabatean catchment area (in the background) and water dispersal system; (c) Nabatean diversion dam in a wide wadi. Note the areas in the background that produce the 'water-crop'. By courtesy of N. Tadmor.

divided by small earthen strips into gently-sloping rectangular plots. The entire run-off collects at the lowermost corner of each plot, in a planting basin. The method is most suitable for orchard trees (Shanan & Tadmor, 1976) (Fig. 11).

The advantages of these methods are their simplicity of construction and relative low cost. The water does not need to be transported for long distances, and its cost is low.

Various methods for increasing the run-off from the water-supplying strips are being investigated in various parts of the world. The two main approaches used are: (1) ground covers of plastic films, rubber, or metal sheeting materials; and (2) waterproofing and stabilizing soil surfaces by spraying with various materials (Myers, 1967).

The main disadvantages of the various films are the high cost and susceptibility to damage by winds. Research is continuing on developing new, less expensive and more durable materials.

Fig. 10 Micro-watershed, with a catchment–cultivated area ratio of 6:1, in a region with an annual rainfall of 100 mm (Shanan and Tadmor, 1976). Courtesy of the Centre for International Agricultural Cooperation, Ministry of Agriculture, Rehovot.

Fig. 11 Three-year old tree in catchment basin, in area with 100 mm rainfall (Shanan and Tadmor, 1976). Courtesy of the Centre for International Cooperation, Ministry of Agriculture, Rehovot.

Floodwater Farming

Floodplains of the great rivers usually lie several metres above the river bed. The floodplain is divided into basins, whose area varies from some hundred square metres to several square kilometres. When the river floods, the plain feeder-canals lead water to the basins, which are then submerged to a depth of a metre or more. After $1\frac{1}{2}$–2 months, excess water is drained off—either into an adjacent basin downstream or into the bed of the subsiding river.

The annual flood brings with it large quantities of suspended materials, and therefore usually leaves a thin layer of fertilizing material on the soil surface of the basin. As soon as possible, a crop is sown without any preliminary ploughing. This is basically a subsistence economy—the basin system being technically a very primitive form of agriculture, similar in many respects to traditional rain-fed crop production in semi-arid areas. In both cases, only a single crop can be grown annually. On the other hand, the system shows a degree of permanency that is most rare for irrigated agriculture: the fertility of the soil and its productive capacity have been maintained under this system for thousands of years. Several factors contribute to the biological soundness of the system: the thorough annual flushing of the soil, combined with excellent natural drainage, prevents salinization—the foremost problem of irrigated agriculture. The silt deposited annually provides plant nutrients in quantities that are adequate in view of the extensive form of cropping practised, thereby preventing soil exhaustion. Where the annual fallow period, which follows the harvesting of the cereal, takes place during the hottest months of the year, the soil is completely dried out and, as a result, weeds are destroyed and the population of pathogens in the soil considerably reduced. Under this régime, the soil which had been packed by sowing, weeding, and harvesting operations, usually develops deep cracks whilst drying out, so that air is able to penetrate to a considerable depth and the soil structure is improved on rewetting.

The system is, however, not entirely without its dangers; successful cropping depending on the time of flooding and on its extent. If the flooding is too late, the growing-season of the winter crops is shortened, with resultant reduced yields. If the flood is too low, the cultivable area is correspondingly reduced; if it is too high, the basins are transformed into marshes and epidemics occur. The historic examples of floodwater irrigation are those practised in the valleys of the Nile, Tigris, and Euphrates, and the Indus and its tributaries.

An example of basin irrigation in the New World is the Laguna area, a wide and fertile basin of approximately $1\frac{1}{2}$ million ha, along the rivers of Nasas and Aquareval in Mexico, in which agriculture depends wholly on irrigation. In the past, the water arrived in huge floods, and irrigation consisted of flooding 100-ha squares, in consecutive order along the ditches. It was only after the completion of the Cardenas Dam, that the floods were controlled and perennial irrigation was made possible (Hopkins, 1969).

Perennial Irrigation

The basin system makes only subsistence farming possible. Many cash crops, such as sugar-cane, indigo, cotton, vegetables, and rice, require a regular water-supply throughout the growing period when the water level in the river may be at its lowest and its flow is greatly reduced. These crops can be grown only under conditions of controlled irrigation throughout the year. Perennial irrigation also makes it possible to grow more than a single crop annually.

The beginnings of perennial irrigation can be traced to the ponds and small lakes left behind after the floodwaters had retreated. Summer crops could be grown in the neighbouring fields and watered, months after the flood, by using buckets filled by hand. Gradually, this method was replaced by primitive mechanical devices for lifting water.

A subsoil water reservoir, whose level ranges from 4–7 m below the surface, according to the time of the year, is generally fed underground from the river. This aquifer can be tapped by wells, whose water complements the river-floods and makes year-long irrigation possible. However, the water is not silt-laden, and the continuous cropping made possible by this system, creates serious problems of impaired soil fertility. Animal manure or soil from long-extinct settlements, that is rich in phosphorus and potash, have to be applied regularly.

Potentials and Problems of Irrigation Farming

Importance

Irrigation is considered, and rightly so, to be the first essential step in the development of *dry regions*, and a powerful tool for progress even where rain-fed cropping is possible. What is, however, often lost sight of, is that irrigation is *not* identical with modernization. In many cases irrigated agriculture is still the epitome of backward methods of production.

The main importance of irrigation in the *semi-arid areas* is to make possible a more intensive succession of crops, a greater variety of crops, and higher yield levels, than would otherwise apply. Replacing a rain-fed crop, such as wheat, with an irrigated crop, such as cotton, makes it possible to treble and even quadruple the net income from a unit area.

Only about 10% of the world's cultivated area is irrigated. A number of countries are, however, heavily dependent on irrigation; for example Egypt, where the whole of the cultivated area is irrigated, Peru (75%), Japan (60%) Iraq (45%), Mexico (41%), and Pakistan (38%) (FAO, 1969).

The potential importance of irrigation in a country with an erratic or insufficient rainfall, can best be exemplified by the following data. In Mexico, only a small fraction of the total land is irrigated. Thus, only 0.5% of all farm units (which are irrigated) produce 32% of the total value of agricultural production, whereas at the other end of the scale, 50% of all farms produce (with rain-fed cropping) only 4% of the total value (Stavenhagen, 1969).

Even where annual rainfall is relatively abundant, irrigation may be important. Large areas of the subhumid tropics have more or less long dry seasons. In Bangladesh, for example, average annual rainfall varies from about 1250 mm in the western part, to 6250 mm in the north-eastern part of the country, but most of this rain falls during the 4–5 months of the monsoon season. Here, the main objective of irrigation is to allow year-round cropping. For perennial crops, such as bananas, irrigation during the dry season is profitable because it increases yields considerably—even in high-rainfall areas such as are found in the Caribbean region (McPherson & Johnston, 1967).

The development of irrigation also has an impact on the development of non-agricultural sectors of the economy, such as commerce, services, and transportation; it contributes to the balance of payments, and provides a market for the industrial production of agricultural equipment and supplies.

Lack of Success of Irrigation Projects

Investment in irrigation projects in Asia, Latin America, and Africa, increased rapidly after World War II. 'Water supply decisions have largely been made in the political arena rather than in the market place, and this to exploit public romanticism—for making the desert bloom' (Clark, 1970).

Most irrigation projects are concerned with major irrigation structures: diversion dams, storage reservoirs, canals, etc.. These provide water, but without a coordinated development plan for the conservation of the watershed above the project, the provision of drainage, and the necessary infrastructures, as well as the inputs required for the effective use of the water, these projects cannot be fully productive.

Unfortunately, some of these aspects are frequently neglected. The *authorities* are concerned mainly with the prestige derived from spectacular large-scale structures; and the *irrigation engineers*, who are concerned exclusively with diversion, storage, and conveyance of water, usually have little understanding of the effective application and use of the water on the individual farm (U.S. President's Science Advisory Committee, 1967).

The majority of the countries introducing irrigation, especially in the developing countries, still continue to make fundamental errors in planning and implementing irrigation projects (Kovda, 1977):

(a) They disregard the implications of the natural soil situation in each country, the level and quality of the irrigation water, the salinity of

the soil up to the ground-water table, and the conditions of natural drainage.
(b) They neglect to install deep drainage in order to reduce costs of development.
(c) They effect excessive application of water.
(d) They allow substantial losses of water in the national irrigation canals and fields.

As a result, irrigation efficiencies in developing countries are extremely low: it is estimated that only 40% of the diverted water actually reaches the field, and this is used with an irrigation efficiency of 30–40% at the best—so that an effective utilization of water of only 10–20% is by no means uncommon (Garbrecht, 1979), and the productivity of water bears little relationship to its cost (Crosson, 1975).

Irrigation agriculture *can* be the most productive form of farming, but it would be a great fallacy to suppose that irrigation, as such, is equivalent to intensive agriculture and is capable of increasing productivity and raising the standard of living when applied under primitive conditions.

Because of its high capital and operating costs, irrigation agriculture must be intensive. Successful irrigation involves far more than applying water to the land; it becomes an effective tool for increased production, maintained on a high level, only when applied in conjunction with appropriate changes in crops and varieties and employment of the most intensive and up-to-date cultural practices, ensuring high yields per unit of water.

A typical example of inadequate results from irrigation is that of the Plain of the Punjab and Sind regions of West Pakistan, which is watered by the Indus and its five tributaries, carrying more than double the flow of the Nile. In this region, more than 70% of the population of 30 million people live from farming, and produce food and fibre for a total of 50 million people. Only half of the flow of the rivers is used to irrigate about 9 million ha—the largest single irrigated area in the world. The underground aquifer contains an amount of freshwater that is estimated to be 10 times the annual flow of the rivers (Revelle, 1966).

In spite of this enormous potential for developing a prosperous agriculture, the majority of the population lives in poverty, and food has to be imported into the region. This is the result of poor drainage, that has caused waterlogging and salt accumulation, and of problems of land tenure and poor farming practices. About one-fifth of the area is already damaged by salinity and waterlogging. The canals lose so much water by seepage, that the actual amount delivered into the fields is sufficient to irrigate only half of the land during each season. The shortage of water causes a tendency to under-irrigate in summer, acerbating salinity problems. The situation is further aggravated by primitive cultivation methods, the use of unselected seeds of low-yielding varieties, and inadequate fertilization.

Similar problems exist in the valley of the Tigris–Euphrates, albeit on a smaller scale. It is estimated that with adequate leaching and drainage, the adoption of multiple cropping, and better farming practices, including the use of fertilizers, total agricultural production could be increased at least fivefold in these two regions (Revelle, 1966).

Irrigation facilities used for traditional farming are seldom adequate for producing high-yielding crops and for double-cropping. First priority should be accorded to renovating existing irrigation systems and increasing efficiency in the use of water in these systems. Developing new irrigation schemes should come next.

The need for adapting farming practices to the potentialities of irrigation farming can best be illustrated by the disappointing results obtained in a large number of irrigation projects in developing countries, notwithstanding substantial investments.

Many irrigation schemes in East Africa—Mubuku in Uganda, Mbarali in Tanzania, and Perkerra, Galole, Ishiara, Wei Wei, and Tavlota, in Kenya—have either failed outright or have operated over a long period of time without recovering their costs (Chambers & Moris, 1973).

In a West Nigerian settlement scheme, unsuitable crops were recommended by 'experts' and prescribed by the Ministry of Agriculture and Natural Resources. In tree-crops plantations, crops that had failed to grow were re-planted year after year just because they were prescribed crops! Many farmers were discouraged by these inevitable setbacks (Oni & Olayemi, 1973).

The greatest agricultural investments in the Middle East have been for the exploration and development of water resources. And yet, the overall contribution of irrigation to agricultural output has been small. This is particularly because poor irrigation techniques and inadequate drainage have caused salinity problems.

Lack of provision of technical and financial assistance for field levelling and field drainage was responsible for disappointing results of irrigation schemes in Iraq (Stippler & Darwish, 1966).

Six institutional bottlenecks in the Near East prevented the full impact of irrigation schemes on agricultural production (FAO, 1962). These were:

(1) Lack of incentive to investment as a result of the agrarian structure and systems of land tenure.
(2) Shortage of institutional credit. More than 85% of all farm credit is provided by moneylenders who charge exorbitant interest. Farmers have to sell their products immediately after harvest at prices that are dictated by their creditors.
(3) Inadequate supplies of fertilizers, insecticides, etc. Supplies are either not available locally or at the right time, or not in sufficient quantities or at reasonable prices.
(4) Poor marketing facilities, lack of storage, and no effective guaranteed prices for basic commodities.

(5) Inadequate extension service: the present ratio is 1 extension worker to 8000 farmers, in contrast to a minimum desirable level of 1 to 1000. Standards for training of extension workers are also unsatisfactory.
(6) Absence or inadequacy of farmers' organizations.

There is a general agreement, among both Indian and foreign experts, that disappointing results from irrigation projects in India were due to the lack of relation between water inputs and other inputs, and the lack of effective coordination between the departments of irrigation, of agriculture, and of agrarian reform. A contributing factor was the inappropriate size of land holdings. But the main reason was that areas under irrigation were increased without concern for intensification of production methods.

Some of the drawbacks and failures of irrigation schemes are common to all types of settlement schemes. The main defects of farm settlement schemes in Nigeria, for example, are held to be (FAO, 1966):

(a) They are uneconomic because they have been conceived on far too lavish a scale in terms of the amount of land cost per settler. It is therefore probably that the sums invested can never be recovered by Government, and, more important still, the high investments per settler render it impossible to make any significant impact on the magnitude of the problem.
(b) Many mistakes have been made in selecting the settlers.
(c) Hasty large-scale implementation, without previous testing on a pilot scale, is frequently a cause of failure.

Kulp (1970) stresses the importance of pilot testing before project implementation as the only way to find out how a project will fit into the farm system, the rural institutional systems, and the rural development system, because of the many social imponderables involved. He proposes that the pilot tests should be as realistic as possible and yet should be carried out under high-level supervision, rather than that of an ordinary field cadre. This is important because of the innumerable unforeseen problems that are bound to crop up and require high-level judgement.

After completion of the pilot test, the project should be undertaken in several different locations that are representative of the general conditions under which the project is to be applied. At this stage, scales should be economic, implementation realistic, and all participating institutions should be involved. The field cadres should be armed with standard detailed operating procedures. On the basis of the experience gained during this stage, the project can be expanded as rapidly as the necessary resources can be increased.

Chambers (1969) suggests that the failures of settlement schemes in Africa in general, and of irrigation schemes in particular, have generally not been due 'to lack of any one of the factors of production—land, labour, or capital—but to the inability to combine them into an economic unit'. Hence, 'weaknesses in managerial ability may be an important underlying reason for the poor performance of many schemes'.

The only two really successful irrigation schemes in Africa—Gezira in the Sudan and Mwea in Kenya—are characterized by their effective organization, adherence to a prescribed set of rules, and semi-military implementation.

Settlement schemes based on irrigation and mechanization require disciplined and carefully-scheduled operation which depends on formalized organization and able management. Besides poor management, many irrigation schemes in Africa have been characterized by 'over-ambitious plans, headlong implementation, uneconomic mechanization, and ignorance and neglect of social factors in settlement' (Chambers, 1969). These shortcomings are magnified when political decisions overrule technical considerations in dictating the pace of development of a project, as has frequently happened, aggravating organizational stress and conflict.

Successful Irrigation Projects

The classic example of a successful large-scale development scheme based on irrigation is the Gezira Project in the Sudan. This scheme was established on a large clay plain, lying between the White and Blue Niles. The flat land is extremely well suited to surface irrigation. Environmental conditions are very favourable for growing long-staple cotton, a cash-crop of high value. The scheme has some 70,000 settlers and covers 720,000 hectares (Millikan & Hapgood, 1967).

Large-scale development was preceded by pilot-scale tests based on research, carried out by an efficient organization that was specially set up for the purpose. The social controls imposed on the settlers were designed to protect them against antisocial actions but not to suppress individual initiative. The controls consisted of family holding tenancy rather than ownership; prescribed rotation, and a prohibition on mortgage and fragmentation. Organized services were provided, such as a research station, seed farms, supply of fertilizers and pesticides, and availability of machinery for cultivation and pest control, as well as technical supervision of the above.

In due course, much of the management was gradually transferred from a hired bureaucracy to farmers' organizations. Limitations of tenancy to 12 ha per family protected the farmers against domination by landlords, prohibition of fragmentation ensured an adequate level of income, and profit-sharing gave each tenant a stake in the success of the scheme. The cropping system was laid out in such a way as to enable the use of modern machinery, so that the farmer could benefit from the economies of scale of the supporting services (Millikan & Hapgood, 1967).

Another example of successful development of irrigation agriculture is the Mwea Irrigation Settlement [Scheme] in Kenya, described in detail by Chambers & Moris (1973).

Location

The project is situated at the point of transition between the well-watered, high-potential upland zone and the surrounding belt of drier, 'marginal' farming lands, in an expanse of open plains, some 500 square kilometres in area, lying between Mt Kenya and the Tana River. In the course of five decades, land-use had shifted from wildlife to stock grazing and then to extensive farming and finally, at present to intensive farming.

Water for the project is derived from two rivers which flow out of the highest-rainfall zone on Mt Kenya, and the supply is therefore fairly stable. Excellent natural drainage is provided by the small Murubara River, which runs through the centre of the project. The mean annual rainfall of 975 mm is sufficient to enable rain-fed crop production.

The soils can be divided into two main groups: black, montmorillonitic clay—an impermeable, poorly-drained soil that cracks on drying, and becomes sticky and putty-like when wet—and reddish-brown lateritic clay–loams, which are highly porous and generally steeply sloped. The project is located near major road and rail connections.

The Beginnings

Initially, the project suffered from many of the drawbacks encountered in most irrigation schemes. Flow records from the rivers supplying the water were available for only three years. No detailed soil surveys had been carried out. Water requirements were unknown, and methods of cultivation and irrigation had still to be evolved. Agricultural and engineering expertise were lacking. The major project was started in 1954 without even an accurate topographical survey. Responsibility for the scheme was divided between three government departments (Administration, Agriculture, and Public Works), leading to a power struggle and lack of cooperation. Because of an emergency situation, work had to be found for thousands of people, so the Administration was impatient to get the scheme started on a large scale, and exerted continuous pressure on the staff.

It is therefore not surprising that the first results were disappointing. The maize and beans grown on the black soils rotted, and the rice was badly damaged by Quelea birds (*Quelea erythropus*). In the course of two years, over £360,000 of expenditure was incurred, but only 340 ha had been fully prepared; many potential settlers had left, and still no assured cropping systems had been developed for the red and black soils, respectively.

Four years after its initiation, the scheme appeared to be 'an extravagant

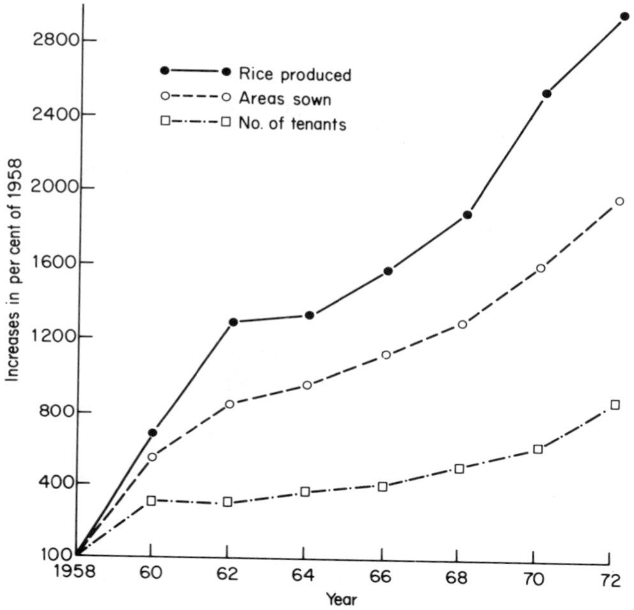

Fig. 12 Growth and performance of the Mwea irrigated rice settlement in Kenya (based on data from Chambers & Moris, 1973). Note: (a) that the rate of increase of the areas irrigated is greater than that of number of tenants—indicating an increase in the area irrigated by each tenant; and (b) that the rate of increase of rice produced is greater than that in area, indicating a steady increase in yields.

and unsuccessful experiment', and the Treasury demanded cuts in the estimates for expansion.

However, during the subsequent five years (1958–62), a number of decisions had been taken and procedures adopted which enabled the scheme to develop into a secure and viable organization and economic system, and by 1972 the scheme was producing some 70–80% of Kenya's rice consumption and providing relatively high incomes to tenants (Fig. 12).

The Mwea System and Management

Since its inception, a system of management has been evolved, based on precise rules and regulations, which has contributed in no small measure to the success of the scheme.

Production: Irrigation is reserved exclusively for the black soils; water is drawn from free-flowing rivers that do not require water-storage; and delivery is effected by gravity-fed, surface irrigation into standardized field-units incorporating $\frac{1}{4}$-ha flood-basins in the normal field size.

The irrigation system excludes rotation, fallow, intercropping, livestock, and subsistence crops, and all the scheme's activities concentrate on a single crop—paddy rice. The scheduling of most production functions is done centrally, with a clear specification of formal tasks to specialized sub-units. The scheme controls the supply of all inputs: planting materals, plant protection chemicals, fertilizers, etc.

All field operations needed to synchronize the planting of paddy to fit a staggered schedule have been mechanized by the scheme, which also organizes the collection, drying, and marketing, of the crop on all tenants' behalf. The tenants carry out nursery preparation, sowing, transplanting, care of the growing crop, harvesting, threshing and bagging, and cleaning feeders and drains.

The red soils are allocated to the tenants for growing rain-fed subsistence crops. This is additional to four-acre irrigated holdings on black soil.

Staffing

Staff at all levels are hired on contract, and are not part of the usual civil service structure or subject to standard government procedures. They each have precisely specified tasks. Promotion is based on evaluation of personal performance by quantified results, providing a major incentive for maintaining efficiency

The ratio of field staff to tenants is high, and there are parallel organization hierarchies responsible for water control and husbandry advice.

Research: The research station was at first part of an organization that was separate from the management of the scheme, and until 1960 there was a marked lack of communication and liaison between them. There was considerable ignorance about what crops to grow, and about aspects of crop management. Initially many crops were investigated.

During the process of consolidating and organizing the settlement, the research station came directly under the control of the scheme, and investigations were concentrated on a narrower range of problems with clear immediate economic value to the scheme, such as improvements in the existing production process (Chambers, 1969).

National Irrigation Board: Specialized functions related to irrigation policy, hydrology, engineering, and crop marketing, are performed for the scheme by the National Irrigation board, which was established in 1966 and made responsible for the development, control, and improvement, of all national irrigation schemes.

Selection of Settlers: New tenants enter the scheme once a year, and are recruited from landless individuals proposed by local committees within the district. No tests of aptitude or agricultural competence are applied. The scheme has the right to veto any candidate on the grounds of physical inability. The most successful tenants have been found to be between 35 to 40 years of age, married, and with two or more children.

Allocations: Each tenant received the right to four one-acre (0.405 ha) irrigation units, and is not eligible to expand this holding unless it has been well-managed. He also receives a house in one of the compact villages created by the scheme, and a short-term bank loan that is arranged through the scheme.

The choice of four-acre-sized holdings is based on an estimate of the area a family is capable of operating, using its own labour (except at peak periods), and that is capable of producing sufficient income to assure an acceptable standard of living. About £100 per annum was considered adequate for this purpose. In actual practice, a settler's family devotes approximately eight months of the year to working on the irrigated area, and the remaining four months on its red soil plots and other affairs.

Net income for 1972 was £162 per tenant. This enabled savings and investments in commercial enterprises, such as acquisition of shares in the Mwea Rice Mill Company.

Control: A high degree of control is exercised by the management over the tenants. Rules have been legalized, providing specific instructions for a wide range of activities, and include the provision that a tenant should devote his full personal time and attention to the cultivation and improvement of his holding.

A carefully graded series of penalties is applied to tenants who fail in their obligations—ranging from fines and cut-off of irrigation water, to eviction as the ultimate sanction (Chambers, 1969).

Incentives are also used: better tenants become head cultivators and receive a monthly honorarium; they may develop 'extra fields', in addition to their four acres (1.62 ha); they may be assigned as 'seed-growers', receiving a bonus for their rice seed, etc. Finally, the settler can transmit the holding to a designated heir providing he complies with the rules.

The overall system of control is semi-military, and the success of Mwea has been largely attributed to the close supervision which has protected the tenants from failure. Actually, over the past 10 years, the rate of evictions of tenants has been only 0.68% per annum—about half of them for poor husbandry, indicating the overall effectiveness of the measures adopted.

Accounts: Separate accounts are maintained for each tenant, who is paid annually, according to his paddy deliveries to the scheme, after deduction of overhead costs corresponding to the acreage of the tenant's holding. The tenants are allowed to retain a fixed amount of paddy for their own use.

In brief, the organization and operation of Mwea are characterized by 'a differentiation of specialized functional tasks; the precise definition of rights and obligations and technical methods attached to each functional role; a hierarchaic structure of control, authority, and communication; the location of knowledge at the top, where the final reconciliation of tasks is made; and a tendency for operations and working behaviour to be governed by the instructions and decisions issued by superiors' (Chambers, 1969).

Reasons for Success of the Mwea Scheme

The technical success of the Mwea Scheme is only partly the result of its well-organized production system; it has also been favoured by particular features of Mwea's natural, social, and economic, environment.

The most important of the advantages, particular to Mwea, are the uniformity of the terrain and high fertility of the soil; the assured supply of water at the required time; the lack of serious crop diseases; an assured market for the crop; and a local population willing to accept the discipline imposed by the management (Moris & Chambers, 1973).

Drawbacks of the Mwea Scheme

Economic: The reliance on a single crop—rice—has been made possible by a quasi-monopoly granted by the Government. Cheaper rice with greater consumer acceptance could have been available to the Kenya consumer if it could have been freely imported.

Production based on a single crop is extremely vulnerable to unpredictable changes in market conditions, incidence of pests and diseases, climatic hazards, etc. The scheme's exclusive reliance on monocropping is accompanied by an ignorance of any alternative crop to grow on the black soil (Moris & Chambers, 1973).

Socio-economic: The almost complete control over settler activities, the far-reaching sanctions, the absence of legal rights to the holding, or of the assurance of the inheritance of the holding—all these lead to settler insecurity. Apart from its social implications, this feeling of insecurity 'can be economically harmful, discouraging full participation in the scheme and commitment to long-term investment activities, and making it less likely that a settler will work with the energy which is often associated with a sense of ownership of the land' (Chambers, 1969).

A major drawback of the Mwea system is that it is unbalanced. The exclusive concern with a high level of technical achievement has led to neglect of the problems of human welfare. As a result, housing is unsatisfactory, overcrowded, smoke-filled, and with a lack of privacy, nutritional standards are low, pure water for household purposes is not provided, and endemic parasitic diseases, such as malaria and bilharzia, are on the increase (Moris & Chambers, 1973).

The annually renewable production license does not give the settlers a feeling of security and permanence; in the absence of a secure legal claim to his holding, the Mwea settler tends to direct his savings to outside investments. This, in turn, leads to considerable absenteeism among tenants.

The lot of women has been worsened on the settlement. The replacement of subsistence crops by a cash-crop has caused a deterioration in the situation of the women through the loss of control over subsistence income, as the men have the sole right to the family's earned income from the cash-crop. It has also led to a breakdown in the traditional division of work between men and women within the household, leaving the women with heavier loads of work than they would have had traditionally. As a result, life on Mwea is not as attractive for women as in other communities.

Now that the physical expansion of the scheme appears to be reaching its limits, management is seriously discussing some of the basic welfare services which have been requested for many years by the settlers (Moris & Chambers, 1973).

The semi-military régime under which the scheme is managed, and to which most of its success is attributed, is probably its most disquieting feature. Much depends upon whether the present situation is considered a transitional phase, during which the settlers are educated towards becoming independent farmers, or whether this discipline is a built-in feature of the scheme, without which it would eventually break down, as has occurred in so many other irrigated settlement schemes in Africa. Its continuance would inevitably perpetuate the apathy and dependence of settlers lacking initiative, and drive the others to search outlets for their initiative and enterprise outside the settlement.

Small-scale Irrigation Projects

In addition to large-scale irrigation projects, water resources can be developed at the level of the individual farm or village. Low-lift pumps and tubewells can, in the aggregate, make considerable contributions to the development of a region,

Priority for small-scale projects is justified for the following reasons: Their cost per hectare is usually low (one-fifth on the average in India as compared with the large-scale projects); a greater proportion of the work can be carried out by the future beneficiaries; and the full allocation of the water is more rapidly and easily achieved. This rule appears to apply even more to flood-control measures, as in the Niger and Senegal deltas, than to irrigation schemes proper (Dumont, 1966).

In Pakistan, in the course of the Second Plan, 25,000 private tubewell installations were made by farmers (Bucha, 1968). These not only provided much added water, but made a substantial contribution to overall productivity by reducing the level of the ground-water in the region and leaching salts out of the root zone.

In Iran, a review of previous agricultural development plans led to a change in emphasis to encourage small irrigation schemes rather than the large-scale river development projects that had been favoured previously. This shift to small schemes made it possible for the Iranians to administer these irrigation schemes themselves and so lessened their dependence on foreign experts (Gittinger, 1965).

The potential for smallholder irrigation in Africa south of the Sahara, has not been sufficiently exploited. Whilst floodplain cultivation of river banks and shorelines of lakes is quite common, little effort has as yet been made to dig wells even where the water-table is sufficiently close to the surface. For example, in the Casamance area of Senegal, lowland rice is cultivated only during the rainy season; the water-table is less than two metres from the surface, and it would be possible to grow another crop during the dry season, when farmers have little work, if this potential for irrigation was effectively used (Eicher *et al.*, 1970).

With tubewell irrigation, losses from evaporation in the reservoirs, and of seepage from the canals, is avoided. Because of insufficient control of the farmer over the water supplied by the canal system, he will rarely grow more than one crop a year, whilst tubewell irrigation has been found to encourage the practice of double-cropping. Because of the possibilities of double-cropping, and of the sale of excedent water to neighbours, tubewell owners in India have been able to recover their capital investment in from two to three years (Dasgupta, 1977).

However, the development of tubewell irrigation has not been without its drawbacks. In India, under existing circumstances, it has led to a concentration of water resources in the hands of the richer farmers. It is, however, reported that bamboo tubewells, introduced in Bihar State (India) in 1968, have brought the tapping of ground-water within reach of small farmers. Even those with one-eighth of a hectare have installed bamboo tubewells and have been able to irrigate their entire holdings (FAO, 1978).

Because tubewells are generally privately owned and operated, their spacing and depth, and the quality of water pumped, must be controlled by a public agency if the water-table is to be maintained at an economic level, and salt damage avoided. The quality of the pumped water must also be periodically monitored.

In communist China, during the earlier years of the régime, many large water conservation projects 'were carried out with great enthusiasm and fanfare. It soon became evident that many of the projects did not perform as satisfactorily as expected'. Subsequently, emphasis was given to small and medium-sized projects, such as digging of wells and ponds, construc-

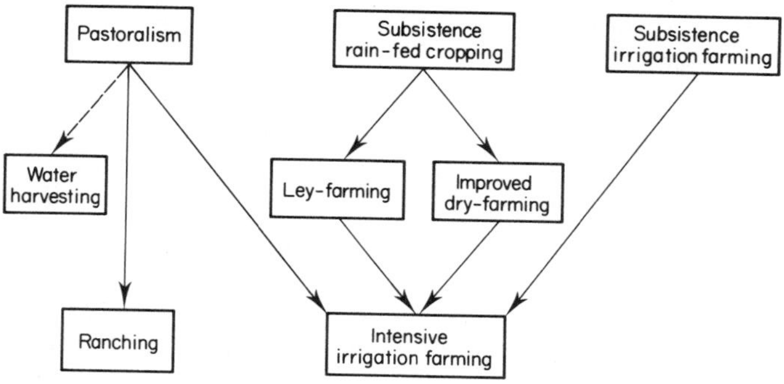

Fig. 13 General pathways of change from traditional land-use in the dry subtropical and tropical regions.

tion of dikes, ditches, reservoirs, and embankments, and harnessing of small rivers. The intention was to form, gradually, coordinated networks (Kuo, 1972).

A characteristic of semi-arid regions is that much of the precipitation occurs in short, intensive rains, conducive to run-off. Much of this water can be collected and subsequently used for irrigation (Fig. 8).

Changing Patterns of Land-use

The general pathways of change that are followed in the dry subtropical and tropical regions is shown graphically in Fig. 13.

Land-use in the Tropics

There are four major ecological zones in the tropics (US President's Science Advisory Committee, 1967):

(1), (2) The desert and semi-desert areas, accounting for 16% and 11% respectively. Land-use in these areas has been discussed earlier in this chapter.

(3) The savanna and associated grasslands, which account for 49% of the land-area.

(4) The evergreen forests, covering 24% of the land-area.

Approximately one-quarter of the tropics has an altitude of more than 900 m—the tropical highlands, with their special ecological problems.

Greenland (1974) distinguishes the following four phases of increasing intensity of land-use.

Phase I: This is characterized by an essentially nomadic existence, accompanied by very limited and sporadic cultivation. It constitutes the simplest form of shifting cultivation, as both homes and fields are shifted at relatively frequent intervals.

Phase II: This occurs when the homes of the cultivators become relatively fixed, but the cultivated area continues to be shifted at frequent intervals. It is the form of land-use known as shifting cultivation in Africa and Asia, or as slash-and-burn in Latin America.

Phase III: This appears when one or more fields, within a system of shifting cultivation, are continuously cultivated, and is a transition stage between phases II and IV.

Phase IV: This is the stage at which all the land is continuously used. It may involve tree crops, commercial crops, mixed cropping, pastures, or ley-farming and crop rotation.

In the alluvial plains of the river deltas, where inundations occur or perennial irrigation is possible, phase IV may occur from the beginning, and may be maintained indefinitely—provided soil salinization and/or excessively high water-tables do not occur (Greenland, 1974).

Shifting Cultivation

Shifting cultivation has been defined as a continuing agricultural system in which the cropping period is shorter than the fallow period (Sanchez, 1973). The two main types are bush-fallow in forested areas and grass-fallow in the savanna areas.

Grass-fallow is a less satisfactory system than bush-fallow, because, in it, fertility levels are lower and leaching is more severe.

Shifting cultivation (also called in Latin America *'roza y quema'*—slash-and-burn) is a farming system whereby the production of annual crops and a number of semi-perennial crops is alternated with periods of vegetative fallow. In the humid tropics, it generally consists of clearing patches of primeval or second-growth forest, by cutting down the smaller trees and vines and burning as much of the woody growth as possible (Fig. 14) Crops are then planted, usually with a planting stick, among the charred stumps. The soil is not cultivated before planting, but with the onset of the rainy season, the crops are planted in a layer of porous soil that is rich in organic matter. The cleared land is usually free from weeds and the soil has a reserve of plant nutrients in addition to those contained in the ash of the burned plant material—hence yields are at first quite fair.

Because of the removal of the protection of forest cover, and the consequent exposure of the soil to sun and rain, a rapid loss of the accumulated organic matter occurs. The considerable leaching of plant nutrients and their additional removal by the harvested crop, also cause a rapid decline

Fig. 14 Cleared forest, shortly after burning (Trinidad). Photograph by courtesy of U. Levy.

in soil fertility. Yields therefore decline steeply as fertility is exhausted and the clearing is invaded by grasses and other weeds.

Duration of Cropping Period

Abandonment occurs when the farmer estimates that he will have to expend more labour on weeding than on clearing a new site. An additional consideration is that clearing can be effected in the dry season, when there is little else to do, whilst weeding must be carried out during the growing season. Natural growth is allowed to take over, which rapidly suppresses the growth of weeds that could no longer be controlled during cultivation. At Instituto Interamericano de Ciencias Agricolas (IICA), on an area in which *Panicum maximum* had become dominant within five months after clearing, this grass was completely suppressed by bush-growth 18 months after abandonment (Moody, 1977).

Shifting cultivation is widely practised all over the tropical regions of the world. (Fig. 15). It is estimated that over 200 million people live by this system, on close to half of the total area in the tropics that is suited to the production of food-crops (Nye & Greenland, 1960). It has often been condemned as being wasteful of soil resources that are suitable for agriculture. The Dutch in Java called it 'robber economy'. But this need not be the case if the length and intensity of the cropping period, and the length and type of the vegetative fallow, are properly adapted to the local limitations of soil and climate.

Fig. 15 Mixed subsistence cropping in recently cleared forest area, (Paraguay). Photograph by courtesy of U. Levy.

Taking into account the prevailing levels of crop production, shifting cultivation is an entirely rational system that can be maintained indefinitely—as long as the biological equilibrium is not upset through overshortening of the fallow period. Labour requirements are relatively low: the heavy work involved in clearing is mainly done by fire, while soil fertility is regenerated by nature, and weeds are suppressed by the forest vegetation during the fallow period.

Shifting cultivation is based mainly on root and tuber crops (manioc [cassava], sweet potatoes, yams, and taros), cereal crops (mainly maize, sorghum, and millets), pulses (such as beans, peas, and cowpeas), and cucurbits (such as melons, pumpkins, calabasters, etc.).

If the farmers are more or less sedentary, these crops can be supplemented by tree crops and vegetables grown in home gardens. A few domestic animals are kept—mainly goats, sheep, and chickens. Additional sources of food are derived from hunting, food-gathering, and fishing.

So long as abundant land is available, shifting cultivation yields better returns in terms of kilograms of grain per man-hour of labour input, than does settled agriculture when using equally primitive methods of production (Clark & Haswell, 1971). Shifting cultivators in Tanzania were able to produce 100 kg of rice with 16 man-days of labour, whilst semi-permanent farmers in the same area required 21 days (Collinson, 1972).

The long distances between plots of shifting cultivation, however, create difficulties in the collecting and transport of crops and in using manures for maintaining soil fertility. Because of the continuous changes in the siting of plots, and the occasional changes of hut sites, there is no incentive to make roads, or to invest in permanent improvements or buildings. As fertility can be restored without effort during the fallow period, the tendency is to 'mine' the land and then move on to clear other sites, when declining yields and soil erosion make this necessary.

Planned systems, such as the 'couloir' system, have been developed and tested in Zaire (the former Congo), to replace the haphazard traditional methods. The 'couloir' system involves small, individual-owned farms, each of which consists of parallel strips of forest of about 100 m width. These strips run east and west in order to reduce the effect of shade on yields of light-sensitive crops. The use of mixed and succession crops, including semi-perennials and annual crops, helps to keep the soil shaded most of the time. The strips are cropped systematically, so that the strips or 'corridors' represent all stages of cropping and fallowing. The duration of the cropping period is strictly determined so as to allow a cultivated field to revert to forest before soil fertility is completely exhausted (Coene, 1956).

The 'couloir' system, however well conceived, has not survived in Zaire after independence, because it required compulsion and supervision to ensure strict adherence to rules by whole communities.

Shifting cultivation is far less effective in maintaining soil fertility in the savanna zones than in the tropical forest zones. The grass and brush vegetation of the former is only capable of storing relatively minor amounts of nutrients during the fallow period, so that the contribution of litter and ash deposited in the soil is very small. Re-growth is retarded by annual burning, which is generally resorted to in order to provide a flush growth of grass for the grazing cattle. Though leaching, due to rainfall, is less serious than in the humid tropics, the rule is 'a vicious circle of poor fallows leading to poor crops' (Nye & Greenland, 1960).

A widespread and destructive form of shifting cultivation known as the 'Chitimen system' is largely practised in the Zambian and south Katanganese wooded savannas. Because of the low soil fertility, the standard practice of clearing the brush and burning *in situ* does not ensure acceptable yield levels. Therefore, the farmer cuts down and burns all the bushes over an area of from ten to fifteen times as great as the area to be sown (Jurion & Henry, 1967). Under this system, the areas required to maintain a family at subsistence level are very large, and the extensive resulting denudation of the soil accelerates erosion processes.

Length of the Fallow Period

The influence of the 'fallow' period on the regeneration of the natural vegetation, and hence on the restoration of soil fertility, depends mainly on the length and intensity of the cropping period, the duration of the fallow, and the soil type.

In the humid tropics, with well-distributed rainfall, the climax vegetation is forest or woodland; anything like complete regeneration after clearing and cropping may take 40 years or more. When tall forest growth again occurs, the land will generally have recovered most of its original fertility. If the fallow period is 10–20 years, the result will be a secondary forest. Longer and more intensive cultivation periods and shorter fallow periods will eventually prevent the regeneration of forest, so the character of the vegetation changes beyond recognition; instead of a dense, primeval rainforest, grasslands often develop, with coarse grasses and scattered shrubs. This process is hastened by stump eradication, weeding, and burning, to produce typical man-made savanna.

Under these conditions, the woodlands and thickets generally become grasslands. In the Philippines, for example, about 18% of the total land-area is covered with grassy vegetation as a result of shifting cultivation and fires. When once the vegetation of an area has been changed from forest to grass, and the latter is subjected to overgrazing, the final result will be a semi-desert. The land will no more be of agricultural use to the shifting cultivator. The only remaining benefit that can accrue from the land will be from hunting or poor grazing (Ferwerda, 1970).

In practice, the duration of the fallow periods required for restoring soil fertility is extreme variable. It is mainly determined by the environment and the character of the soil, its durability under cultivation, and the rapidity with which fertility is restored under the conditions of the fallow. In wet tropical regions, soils of poor durability and very low regenerative capacity predominate. For this reason, systems with long or very long fallow periods were prevalent. There are, however, also relatively limited areas of soils of high fertility and rapid regenerative power. On such soils, short fallow periods and virtually permanent cultivation systems are possible. For example, on good alluvial soils and on deep volcanic soils, an almost indefinite sequence of annual crops can be grown. In areas of somewhat above-average fertility, the restorative interval between cropping sequences need be no longer than the period of cultivation. In contrast, on poor lateritic soils, recovery of fertility may take 20 or more years after only two or three years of cultivation. Between these extremes lies a complete range of gradations (FAO, 1962).

Mixed Cropping

Mixed cropping is a characteristic of most shifting cultivation. It is a method which attempts to make the most of the potentialities of the environment. By planting together a number of crops with varying planting and harvesting times and growth habits, plant nutrients in different soil layers are better exploited and light-energy is more effectively intercepted. Plants of the same species compete more intensively with each other than do some plants of different species—mainly because of the latters' differences in root systems and periods of peak water requirements, so that a limited water supply can be used more efficiently in a mixed-cropping system than in pure stands. The risks due to diseases, pests and climatic factors, can be reduced and also better distributed in mixed stands, while weeds are more effectively smothered (Baldy, 1963).

In monocultures light supply is in excess of requirements during the early growth stages, and is frequently a limiting factor at periods of maximum growth. During the early stages the open spaces not only waste light energy, but also encourage the growth of weeds.

Mixed cropping generally gives a larger total yield than the pure stand of each crop does in the aggregate; it provides an insurance against complete failure, and it reduces soil erosion—particularly if one of the two associated crops has a trailing habit.

Crop mixtures of early and late varieties are sometimes sown by farmers in developing countries—the former to provide insurance against early cessation of the rains, the latter to make possible higher yields when rainfall is abundant (Hall *et al*., 1979).

In research on mixed cropping in the Zaria province (northern Nigeria) under conditions of traditional farming, the following interesting results were obtained: mixed cropping required 62% more labour over the year than pure stands; and the cash value of the crops obtained exceeded that of pure stands by 62%. As a result, income per hour of work was similar in both systems, but variability in income was much less in the mixed cropping system (Norman, 1970).

An interesting example of double insurance provided by sorghum and maize mixtures is described by Collinson (1972). As recently as the 1940s, sorghum, mainly because of its drought-resistance, was the preferred grain in many parts of Sukamaland. The variety grown was particularly susceptible to bird damage. As the bird problem increased, the sorghum was interplanted with maize, which is much less drought-resistant than sorghum, but is not attractive to birds. Tastes gradually changed, and as maize became the preferred grain, it was interplanted with sorghum as a drought-insurance.

Interplanting with cassava provides an ideal famine reserve crop. It is highly productive and can be stored *in situ* in the ground for up to three years, so that it needs to be harvested only when there is a food crisis (Collinson, 1972).

The Multi-storey System of Mixed Cropping

The crops can be divided on a height basis into 10 groups (Lagemann, 1977):

Crop	Height (m)
Oil palms, coconuts	20–25
Breadfruit, raffia palms, avocadoes	12–20
Mangoes, colanut	8–15
Orange, papaya	5–10
Bananas, plantain	3–8
Yams	3–6
Maize	1.5–2.5
Cassava, peppers	1–2
Groundnuts, cucurbits, vegetables	0.1–0.3

The closer to the ground the denser the leaf cover. All the trees are pruned so that only a few branches remain on the top, and more light can reach the lower storeys. The branches are used for firewood, food for goats and material for compost.

Rappaport (1971) gives an example of mixed cropping in which the crops grow in a more or less vertical order. The uppermost layer consists of the fronds of bananas, below these are the more or less erect leaves of sugar cane; then follow taro leaves, which project above a mat of sweet potato leaves, which covers the soil surface at ground level. Sweet potato tubers are produced just below the soil surface, cassava roots lie deeper and the yams exploit the lower layers.

A notable example of mixed cropping, practised by a primitive people, is the maize–beans–squash complex, which originated in Mexico during the period 1500–900 BC. It was so successful that it spread out from Mexico and became the basis of practically all prehistoric Indian agriculture in all parts of America.

The complex of three crops exploited the soil and light-energy most effectively: the beans climbed on the maize stalks, exposing their leaves to the light without excessive shading of the maize leaves; the squash grew prostrate on the ground and choked out weed growth. Mixed cropping also produced a highly-balanced diet: the maize supplied most of the carbohydrates and certain amino-acids in which beans are deficient; the beans supplied the bulk of the protein, and also phosphorus, iron, and the vitamins riboflavin and nicotinic acid. The squashes added calories and an increment of fat (Stakman *et al*., 1967).

In mixed cropping plant densities are very high, partly because the plants are not vigorous as a result of the low fertility of the soil, and partly in order to protect the soil against the beating rain. Crop density is also related to population pressure and soil fertility: the more densely populated the village and the lower the soil fertility, the greater the number of crops grown together in one field (Lagemann, 1977).

Phased Planting

Mixed cropping is generally based on phased planting throughout the year, giving a continuous sequence of growth and harvesting. This allows for a favourable distribution of work throughout the year and for a regular supply of fresh food to the household.

For example, the Kikuyus in Kenya practice the following sequence of planting and harvesting: at the beginning of the rainy season, maize and pigeon peas are sown broadcast; after the seeds have germinated, an early variety and a late variety of beans are sown in the stand of maize and pigeon peas; after these have emerged, the area is weeded and seedlings of sweet potato are planted into the mixed stand. The early-maturing beans are the first to be harvested, followed by maize and the late-maturing bean variety. Sweet potatoes and pigeon peas ripen next, and are harvested as required (Ruthenberg, 1971).

Improving Productivity of Shifting Cultivation

Greenland (1975) suggests that the intelligent application of modern low-cost, low-energy techniques can be expected to more than double yields of shifting cultivation, without affecting the viability of the system.

In order to provide soil protection during production, and to facilitate regeneration, the use of narrow and not too large plots, parallel to the contour, is suggested. Other measures proposed by the author include: Pollarding or trimming of the larger trees in preference to clear felling; prevention of accidental fires; minimum disturbance of the soil; mulching of weeds and crop residues.

One obvious possibility for increasing productivity in shifting cultivation is the use of moderate amounts of fertilizers. Most of the trials of the FAO Fertilizer Programme in West Africa have been laid out on farmers' fields under shifting cultivation. The results from 715 trials on maize in Ghana have shown that a modest application of fertilizers (22.5 kg/ha of NP, occasionally NPK) achieved the same yield increases as 7–10 years fallow, and maintained these yield levels for three years, whilst those on fallow declined steeply. Starting from an average base yield of maize of 700 kg/ha in the year before the fallow, both fertilizers and fallow gave the same yield increase—400 kg/ha in the first year; in the second year the increment from fertilizers remained unchanged whilst that from fallow was halved, and in the third and final year of the experiment, the increment from the fertilizers was still 400 kg/ha, whilst in the fallow plots it had dropped to nil (Hauck, 1971).

The levels achieved and maintained by fertilization under conditions of shifting cultivation were modest, but this was mainly due to the lack of other improved practices, apart from fertilizer application.

From the author's own experience, weed competition is probably the major constraint to improving yields, and in particular in preventing a greater response to fertilizers. Rarely is the shifting farmer able to keep up with the luxuriant growth of weeds under tropical conditions. An inexpensive, general purpose pre-planting herbicide, that could keep the plots weed free for a few weeks after planting, could be a major contribution to improving productivity of shifting cultivation.

Another challenge for agricultural research is to devise ways of increasing the yielding ability of mixed cropping. This would involve breeding varieties that are adapted to this form of production and selection of appropriate mixes for maximum yield (Grandstaff, 1978). These aspects are treated in more detail in Chapter VIII.

Summary of Mixed Cropping

Mixed cropping has the following advantages:

(a) The multi-storey cropping systems assures the best utilization of sunshine;
(b) weed growth is suppressed because of the close leaf canopy;
(c) damage from heavy rainfall is reduced to a minimum;
(d) phased cropping assures a continuous supply, throughout the year, of food, fruits and fibre; labour requirements are well distributed throughout the year, and labour productivity is high.

The main drawback of mixed cropping, as practised in traditional farming systems, is the difficulty of applying crop-specific inputs, such as fertilizers, disease-, insect- and weed-control chemicals.

Improved methods of mixed cropping have been developed (in particular at several of the international research centres) that have given excellent results. These aspects will be treated in Chapter VI.

Transition from Shifting Cultivation to Permanent Farming Systems

To agronomists from temperate countries the long vegetative fallows, serving no other purpose than restoring soil fertility, may seem a terrible waste of land and labour. Countless efforts have been made in the course of the last century, by agricultural experiment stations, private plantations and state farms, to replace the vegetative fallow by useful crop plants.

Many agricultural settlements in the tropics lead a miserable existence or have been abandoned, because governments and foreign technical aid organizations have overestimated the land-use capability. It is wise to remember that the highly skilled farmers of south-east Asia, who have been familiar with the permanent cultivation of irrigated rice on the same land for approximately 4000 years, continue to practice shifting cultivation on non-irrigable forest land.

Shifting cultivation is therefore still the most widespread farming system in the humid tropics. The system is suitable for a society needing only subsistence, as long as land remains plentiful and the number of people small.

The system of shifting cultivation is successful as long as: (a) the cultivation period is not too long, resulting in a level of fertility too low to enable a relatively rapid regeneration of the natural vegetation; and (b) the fallow (or resting) period is sufficiently long to redress the impairment of soil productivity that occurred during cultivation, and the build-up of weeds. All this, of course, is provided practices are followed that do not provoke significant soil erosion (Greenland, 1974).

When these conditions can no longer be maintained because of increasing population pressure, the system breaks down and a permanent type of farming system becomes an unavoidable step forward.

Semi-permanent Cultivation

In the transition from shifting cultivation to continuous cropping subsistence farmers have adopted a spatial organization of cropping (Lagemann, 1977).

The land around the dwellings—the compounds (200 to 1000 square metres)—has a high density of trees and a large range of food crops; fertility is maintained by manure from sheep, goats, poultry and night soil. On the fields on the perimeter of the compounds, oil palms and other useful trees dominate. Food crops are grown in an intensive bush-fallow system, with fallow periods of one to five years, according to population pressure.

The distant fields are used to produce most of the starchy food in an extensive bush system, under a low density of trees.

From the land under fallow the villages obtain fruit from oil palms, breadfruit, wood and mulching material for the compounds. In the near and distant fields, most crops are planted as soon as possible after the early rains, so that the soil is generally well covered at the time of heavy rainfall.

Agri-sylviculture

Agri-sylviculture is a system that aims to avoid the destructive effects of shifting cultivation, after the population has reached a critical level, by an integrated resource management plan in which agricultural crops are grown in combination with timber production (Roche, 1974). The system has a number of vernacular names, of which the best known is taungya. The following procedure is generally adopted in Nigeria (Enabor, 1974): farmers are allocated plots ranging from 0.4 to 1.25 ha; the farmer clears the bush, burns the debris and prepares the site for planting. He then plants his crops as soon as the rains begin in accordance with a list and at a spacing determined by the forest department. The forest department soon after interplants forest trees at its expense. In the first year, the farmer tends both his crops and the young trees. From the second year, the forest department tends the trees, but the farmer may be permitted to continue cultivation for a second and sometimes third year. Thereafter, he is given a new allocation of land, provided he has proved himself during the period of his tenancy. The forest department, in deciding what crops may be planted, mainly considers the competitive effect on the forest trees. The crops are predominantly annual and biennial food crops, such as yam, cassava, groundnuts, maize, beans, pineapples, tomato, pepper and millet. Tree crops and food crops with spreading crowns, such as plantains, are excluded.

Agri-sylviculture is conceived as a transitional system, to be initiated when shifting cultivation has to be abandoned because of population pressure and 'until tropical economies become self-sustaining, until they are able to afford the fertilizers, machinery and implements that are necessary for tropical agriculture, until storage facilities are improved, and until new genetic strains suitable to tropical conditions are evolved' (King, 1968).

Permanent Cultivation

In the Subhumid Tropics

The changes that occur in the transition from shifting to permanent cultivation in Senegal have been described by Charreau (1974).

Fields become permanent, stumps are completely removed and cultivation by oxen is introduced. At first, ploughing is only resorted to every four to five years, to incorporate grass-fallow, green manures or crop residues. Otherwise, crop residues are burnt and only superficial cultivation is practised. In the year prior to fallowing, millet or sorghum is sown, cut for forage and the ratoon crop, after first being grazed, is ploughed under. The recommended cropping sequence in Senegal, during this stage, is: fallow or green manure (1–2 years); groundnuts; pearlmillet or sorghum; groundnuts or cowpeas.

With further population pressure, cropping becomes continuous, without resting periods of fallow or green manures. The incorporation of crop residues, including straw of shortseason cereals, is frequently resorted to and has been shown to have the same beneficial effects as the green manures described above.

In the Humid Tropics

The key to continuous arable cropping in the humid tropics is devising suitable farming systems that:

(a) provide adequate cover either permanently or during the rainy season or seasons;
(b) maintain soil fertility; and
(c) make efficient weed control possible.

Perennial Crops

The first move away from shifting cultivation towards permanent land-use has usually been the planting of perennial tropical plants as forest undergrowth. For this purpose, the existing forest growth is not cleared but is only thinned out. Much of the coffee and cocoa farming, as well as coconut and oil palm production in traditional farming is still produced in this way. The next step is the systematic establishment of tree plantations. This has generally been successful because the ecological effects of the plantations are somewhat similar to those of the original vegetation that they replace. They protect soils fairly well and they also have the necessary root depth to enable them to re-cycle plant nutrients.

The main ecological disadvantage of perennial crops is that they form pure stands in which specific diseases and insect invasions can quickly assume epidemic proportions.

Mixed Plantations

Permanent mixed plantations have been recommended by a number of authors cited by Watters (1971). Such mixtures are somewhat analogous to forest associations, and are likely to be the most ecologically stable and longlasting. At Turrialbu (Costa Rica), for example, with a mean annual rainfall of 2639 mm, higher net returns were obtained from mixed planting of cocoa and rubber than from rubber alone. The two crops form a two-storey structure that is complementary, and they therefore do not compete for light or space. Another frequent combination is bananas under coconut palms. Plantations may also provide grazing either in the early stages only (oil palms) or even under the mature stands (coconut palms).

Initially, plantations of many perennial crops for cash were mainly carried out by large, commercial estates. Increasingly, almost all types of perennial crops are also being grown on smallholdings. In the latter case, the cash-crop is usually integrated with food-crop production—yams, manioc, rice, maize, beans, etc. for household use—and mixed cropping is the rule. However, because marginal land can often be used for perennial crops, a separate area may be allocated to them. For example, bananas, tea and rubber are suitable for steep slopes, tea and rubber for stony terrains, date-palms for saline lands, etc.

Establishment

The small farmer can begin with a few trees, and expand his plantation progressively with otherwise under utilized family labour and using small amounts of cash only. The perennial crops therefore constitute a form of saving and security for old age.

By contrast, the establishment of commercial plantations requires a high initial investment, high costs of maintenance of the plantation for a number of years until bearing, and of investment in an infrastructure including processing plants, etc..

Large estates also usually prefer pure stands, on which they can apply management practices that are specifically suitable to the crop. This does not, of course, preclude the growing of cover- and green-manure crops between the trees.

The successful establishment of commercial plantations is clearly dependent on sound planning, based on a thorough study of the region, and effective implementation. Essential preliminaries to planning are needed: soil survey, classification and mapping; forest inventories, land-capability surveys, hydrological surveys, climate data, demographic data, etc.. Before implementation, crop rotation experiments, variety and fertilizer trials, testing methods of forest clearance, and the initiation of pilot schemes are essential.

Advantages and Disadvantages

Perennial crops in smallholdings give rise to stationary living and to investments in infrastructure, such as tracks, irrigation systems, improved housing, etc.

Whilst perennial crops are generally the first to which modern inputs are applied by the subsistence farmer, the standard of husbandry on most of the smallholdings is still very low, and many operations are either neglected or delayed. However, they can compete with commercial estates because of their low costs of production, use of redundant family labour and the smaller risks involved.

Perennial crop production on estates is generally carried out at a high level of technology: they have—over the smallholders—the advantages of a more efficient organization of delivery of the crop to the processing plant, more efficient processing of product and better access to markets (Ruthenberg, 1971).

However, cash outlay is considerably greater on the plantations in establishment, infrastructure and considerable recurrent expenditure; their income is more affected by yield fluctuations due to diseases, pests or wind damage; or because prices drop on the world market. They are competitive as long as labour is cheap and plentiful; most operations on the plantation cannot be mechanized, except in sugar and sisal production.

The advantages of combining nucleus estates with a surrounding ring of smallholders are discussed in Chapter X.

Tree Crops

These include oil and coconut palms, cocoa and rubber. Manual work required for maintenance, weeding and harvesting is relatively low in relation to other perennial crops.

Oil Palm

Smallholding production has been based either on semi-wild palm groves, which range from secondary forest containing some oil palm, to open palm bush with various degrees of plant density (Zeven, 1967).

Oil palms are the most recent of the tree crops to be established as commercial plantations, following the development of heavy yielding, early bearing, and highly productive varieties and technical improvements in the processing of the fruit. In many parts of Africa, the natural groves are re-habilitated by replacing the old palms by superior varieties.

The Malaysian Government regards oil palms as the only proven crop that can be expanded on a large scale, and is making substantial investments in research. Originally, most of the oil palm was grown on estates. However, in more recent years the Malaysian Federal Land Development Authority (FELDA) has organized successful land settlement schemes of smallholders, such as the Jenka and the Johore Land Settlement Projects, in which an average size of holding of four hectares is considered viable for oil palm. By 1973, the smallholder area accounted for about 50% of the total area, and is still increasing.

Modern techniques have been adopted by smallholders surrounding nucleus estates in the Ivory Coast (Zeven, 1967).

The oil palm gives the highest yields of oil at the lowest costs and world prices have been relatively stable since the Korean War (Dumont, 1966). However, the establishment of oil palm plantations requires a high capital investment. Returns are relatively high and risks from disease and adverse conditions are low. Protection of the soil is satisfactory. Labour input is highly intensive, but employment is regular throughout the year.

Coconuts

The coconut palm is the most widely planted perennial oil-producing tree in the world, and can be planted in the unfertile sandy coastal belts, especially if potassium fertilizer is used (Dumont, 1966). It is, however, confined to the humid tropics, mainly the lowland coastal regions (Williams, 1975).

Most coconuts for commerce are produced by smallholdings, though there are also occasional large plantations. Costs of establishment of commercial plantations are high; the interval between planting and the first harvest is fairly long, and returns are lower than with other tree crops. However, recurrent costs are moderate; though relatively few workers are required, labour costs are the main item of expenditure. The labour requirement is fairly evenly spread over the entire year (Ruthenberg, 1971).

Where rainfall is sufficient, coconuts can be intercropped or intergrazed, provided fertilization is adequate. In the Philippines, intercropping with papaya and pineapple more than doubles the income from the land. Bananas are also used extensively (Gomez, 1974).

Cultivation of annual and bi-annual crops, such as maize, peas, beans and manioc, in young coconut plantations is quite common in India and Ceylon, and may be continued for several years. In this way, the young plantations can be established with minimal cost and do not require special care; the care given to the food crops also protects the young palms from weeds, and prevents regeneration of the bush until the trees are tall enough to suppress weed growth. New, early-maturing, wind-resistant dwarf hybrids are available, making possible yields of over three tonnes of copra per hectare, after four years (Fremond, 1968).

In the West Indies, in particular in Jamaica, a virus disease called lethal-yellowing has practically wiped out the original coconut plantations, which are being gradually replaced by a resistant Malayan dwarf variety.

Cocoa

Cocoa, in its natural habitat in the Latin American rainforests, is an understorey tree. Whilst temporary shade is necessary for the young seedlings and during establishment of the plantation, it is now generally accepted that under favourable growing conditions, optimal yields are obtained under full sunlight and adequate fertilization. This has led to the use of crops that provide temporary shade, during the period of establishment. Bananas are used for this purpose in Ghana. Tall coconuts, which provide about 50% shade, have also given favourable results.

There are two major groups of cocoa varieties, the 'Criollo' type which originated in Central America, and is of superior quality and the 'Foresteros' type, of Amazonian origin, which is of lower quality, but is widely grown because of greater hardiness, vigour and yield. From the Criollo type, have been developed the Amelonado cocoas of West Africa and the Bahia region of Brazil, which are particularly suitable for the manufacture of milk chocolate (Williams, 1975).

Rubber

Most rubber estates are found in Malaysia, Sumatra, Sri-Lanka and Liberia. Costs of establishment are relatively high; income is lower than is the case with oil palm, tea or coffee, but somewhat higher than that of coconut plantations. The main item of cost is labour, with no period of pronounced peak requirement, as tapping of the latex proceeds throughout the year. The soil is well protected. Considerable progress has been made in breeding and cultural techniques which have improved yields and offset the decline in market price of latex, especially as the cost of tapping remains unchanged irrespective of yield levels.

Because the rubber tree is very undemanding, it is also grown widely by smallholders, who usually interplant with fruit trees, coconuts and arable crops (Ruthenberg, 1971).

In South-East Asia, rubber is relatively free of serious diseases, but in South and Central America, the establishment of rubber plantations is almost impossible because of the prevalence of a devastating leaf disease, caused by the fungus *Dothidella ulei* (Williams, 1975).

Shrub Crops

Tea and coffee are typical shrub crops. Tea leaves and coffee beans are processed immediately after harvest, and as they are not bulky, can be transported at low cost. The two crops are labour-intensive, requiring a high input of manual labour, mainly for weeding, pruning and harvesting.

Coffee

There are three main types of coffee: '*arabica*' is the most widely grown and commands the best prices. It is suitable for high altitudes (700 to 2500 m); '*robusta*', indigenous to West Africa is hardier and easier to grow than '*arabica*' and suited to altitudes from sea-level to 700 m; and '*liberica*' of poor quality but very hardy (Pollock, 1971).

Coffee can be grown on a wide variety of soils, provided they are sufficiently deep and well-drained. The mineral requirements of coffee are high, and the crop removes considerable amounts of nutrients. After the initial soil fertility is exhausted, yields decline rapidly unless heavy dressings of fertilizers are applied (Malavolta *et al*., 1962).

Coffee production is well suited to smallholdings, which produce most of the world's crop. The costs of plantation on small farms are low, gross returns are generally high and the crop is highly labour-intensive. The introduction of coffee into subsistence farming does not cause any marked seasonal peaks in labour requirements. The coffee may be grown under the cover of the natural forest (West Africa, Ethiopia), in mixed stands with bananas and other food crops (East Africa, Latin America), and in pure stands (Brazil). Commercial plantations have gained in importance in recent years; the capital investment required is lower than for most other plantation crops. Manual labour accounts for about half the cost of production. Using modern technology, yields can be very high and of high value (Ruthenberg, 1971).

Tea

Tea requires high rainfall and slightly acidic soils. It is mainly a plantation crop. The phases of production are highly sophisticated and a high-quality product is dependent on rigid timing of harvesting and processing within a few hours of plucking. The processing itself is done in factories, excepting in Japan and Taiwan, where smallholders process their own tea using traditional hand methods.

Part of the dense forest in the hills to the south of lake Victoria, an area with 1500 to 2000 mm annual rainfall, has been cleared and planted to tea, which has become one of Kenya's most important export crops and an important source of employment. Capital investment for elaborate soil conservation work was very high (Pereira, 1973). Over the first eleven years, the mean annual evapo–transpiration (1300 mm) was similar for the tall rainforest and for the continuous canopy of tea and shade trees that replaced it. The water yield was not significantly affected by the change in land use. Annual run-off was very low, and after the plantation had developed sufficiently, amounted to 1% of the annual rainfall, the same as under the original forest cover (Blackie, 1972).

The Kenyan experience has shown that tea plantations, provided they are planned and developed with full soil conservation at a professional engineering level, can give the same adequate protection on fairly steep slopes in a high rainfall area, as was previously provided by forests that were preserved against fire, felling and grazing (Pereira, 1973).

Bananas and Plantains

Bananas are one of the most important and ubiquitous food crops in the tropics, and are used in a number of ways—as fresh fruit, for cooking, processing and for brewing beer, Bananas are also an important export crop. The crop is easy to establish and cash investment is relatively low.

Bananas are often planted on newly cleared forest land. They are grown on a wide variety of soils but heavy clays with poor drainage have to be avoided. Because of the superficial root system, the banana plant is unable to draw on water in the lower profiles, and irrigation is often practised in order to obtain maximum yields, even in humid tropical areas (Williams, 1975).

Bananas afford excellent protection to the soil during most of the year and produce large amounts of organic matter which is left on the field. However, the removal of the fruit from banana plantations causes a very heavy drain on soil nutrients which must be replaced by fertilization (Ferwerda, 1970).

Harvesting begins one year after planting and yields can be very high. In terms of carbohydrates, yields are almost as high as those obtained from tropical tuber crops and their cash value is greater. The supply of fruit is regular throughout the year, an important consideration when the crop is grown for one's own consumption. Labour requirements are high, but are fairly well distributed throughout the year, without marked peaks.

Bananas are equally well adapted to production on large and medium estates—mainly for export—and on smallholdings. In the latter case the bananas serve both as a cash-crop and for subsistence. The stalks, leaves and reject fruits are frequently used to feed pigs and cattle, whose manure, in turn, contributes to maintaining fertility of the plantation (Ruthenberg, 1971).

The main drawbacks of banana production are the susceptibility of the plant to a number of diseases, and the high cost of control. It is also easily damaged by strong winds.

Perennial Field-crops

In many areas, the trend towards the establishment of perennial commercial field-crops, such as sugar cane, bananas, plantains, pineapple and sisal, has been very successful. These crops may be grown on the same land over many years, provided soil fertility is maintained.

Sugar-cane

Sugar-cane production is possible in a number of ways: on commercial plantations combining cane production and a sugar factory; on large and medium-sized plantations producing cane for sale to factories; by small and medium farmers producing individually and processing the cane in a cooperative-owned factory; and by smallholders, who produce cane as a cash-crop to supplement their subsistence farming. These latter usually surround large sugar-cane plantations which absorb part of the family labour of the smallholders.

In well managed sugar-cane plantations, the soil is protected by the growing crop for most of the time and is bare only during relatively short periods once every four or five years. Though sugar-cane takes up large quantities of nutrients from the soil, the exported product contains no mineral nutrients. Large quantities of organic matter are produced and left in the field; however, burning of the leaves to facilitate mechanized harvesting reduces these amounts considerably.

The most favourable moisture conditions for sugar-cane production are an ample water supply during most of the growing period, with a short dry season prior to harvesting. Water stress at this period favours accumulation of sucrose in the stalks (Williams, 1975).

Though sugar-cane is grown in monoculture, good yields can be maintained for long periods under good management. It has been the object of considerable research and plant breeding efforts. It requires a high labour input, but as operations can be fully mechanized, machinery is progressively displacing manual labour in the commercial plantations (Ruthenberg, 1971).

Pineapples

The cultivated forms of *Ananas comosa* are self-sterile and propagated vegetatively, producing seedless fruit. They can be grown on a wide range of soil types, provided they are well drained. They are not suited to calcareous soils in which they suffer from lime-induced chlorosis.

Whilst fair yields are possible on virgin soils without fertilizers, heavy fertilization is essential for heavy cropping. The return of the residues to the soil plays an important role in maintaining soil fertility.

Irrigation is seldom practised in pineapple production (Williams, 1975).

Pastures

Both shifting cultivation and plantations may be followed by grassland. In southern Brazil, for example, old coffee plantations may be allowed to run to grass, after the old trees are chopped down.

In many parts of Africa and Latin America, the gradual degradation from natural forest to grass is often encouraged in order to provide grazing for cattle. As a result, there is a gradual transition from shifting cultivation to animal husbandry.

Huge areas of forest land in Panama have been converted from shifting cultivation into partially improved grassland, which appear to be ecologically stable and capable of supporting cattle (Watters, 1971).

A common practice in Nicaragua is for landowners to lease out woodland to landless campesinos who clear the forest, and plant and harvest an annual crop. Thereafter, they move on, whilst the cleared land is taken over by pasture which is then used by the owners for livestock grazing. This land rarely reverts to crop production (Taylor, 1969) (Fig. 16).

For single-family settlers, Sanchez (1977) recommends the following procedure: slash and burn annually an area of 1–4 ha that can be managed by the family; grow a cash-crop or a mixed crop on part of the land and place the remainder under a grass–legume mixture; gradually increase the area under pasture until an economic beef cattle unit can be achieved.

Pastures provide excellent protection for the land against erosion and loss of organic matter. They can provide a viable alternative to shifting cultivation especially in subhumid regions, if they are properly managed and utilized.

Replacing arable cropping by ley-farming requires considerable investments in land clearing pasture establishment, fencing, implements, stock animals, buildings, etc. Most of these areas are used for extensive ranching by larger landowners, with little effort at grassland improvement or high-level animal husbandry.

Creating viable units for family farms involves the use of improved, productive pastures and well managed animals for beef or dairy production. Such intensive enterprises will be mainly justified in areas with access to markets that can guarantee a stable demand and adequate prices.

Fig. 16 Swiss-brown × Creollo cattle grazing on weed growth in recently cleared forest area, (Nicaragua). Photograph by courtesy of U. Levy.

Improved Pastures

Improved pastures can be based on grasses, legumes or grass–legume mixtures.

Grass Pastures

The most intensive form of pasture production in the tropics is based on well fertilized and adequately managed grass swards (Fig. 17). High-producing tropical grasses are available from a large number of genera including: *Brachiaria*, *Cenchrus*, *Chloris*, *Cynodon*, *Digitaria*, *Eragrostis*, *Eriochloa*, *Hyparrhenia*, *Melinis*, *Panicum*, *Paspalum*, *Pennisetum*, *Setaria*, and *Tripsacum* (Geus, 1977).

Amongst the most productive are elephant or Napier grass (*Pennisetum purpureum*), Guinea grass (*Panicum maximum*), Rhodes grass (*Chloris gayana*) and *Paspalum* spp.

Improved grass pastures are propagated vegetatively and are expensive to establish. They also require considerable outlays of cash for fertilizers. They are however highly productive and the stands, if properly managed, last indefinitely. The pastures can be used for grazing or for soiling. The grazing system removes about half the amount of forage, as does cutting; but as grazing is selective and the leafy parts are mainly consumed, the quality of the forage is far higher.

Fig. 17 Grazing on raid-fed sown pasture in forest-cleared area, (Dominican Republic). Photograph by courtesy of U. Levy.

Under the cutting system, amounts of fertilizer nutrients equivalent to those removed by the forage have to be applied in order to keep the soil in productive condition. In grassing systems, not only is less forage removed, but 80% of the NPK* consumed is excreted by the grazing animals (Geus, 1977).

Legume Pastures

A number of productive legumes are available for pastures from within the genera *Centrosema*, *Desmodium*, *Leucaena*, *Macroptilium*, *Phaseolus*, *Pueraria*, *Stylosanthes*, and *Vigna* (Geus, 1977).

Legumes do not need complex grazing management systems; if the overall stocking rate is kept sufficiently low, they can be grazed continuously with no adverse effect (Jones, 1972). They require regular and fairly heavy phosphorous fertilization, which is important not only for the forage but also for the animals.

Supplemental calcium and phosphate are almost always required for range cattle in tropical America in order to avoid the occurrence of broken bones, because of the deficiencies of these elements in the soil (Sanchez and Buol, 1975).

*NPK stands for nitrogen, phosphorus, and potassium fertilizers.

Grass—Legume Mixtures

At first sight, the most favourable solution would appear to be the use of grass–legume mixtures. In Pucallpa, Peru, an improved grass–legume pasture, when fertilized with 19 kg of phosphorus per hectare and mineral supplements were supplied to the stock, gave 350 kg of live-weight gain per hectare per year as compared to 79 kg from unfertilized pasture and no mineral supplement (Nicholaides, 1979). However, grass–legume pastures have certain inherent drawbacks that make their use problematic. Grass and legume have different grazing management and fertilizer requirements. The grasses must be grazed or cut fairly frequently otherwise they become unpalatable and of low nutritive value; the legumes must either be grazed very lightly or given long periods for recovery between grazings. If cutting or grazing frequency is high the legumes will die out, if low the luxuriant growth of the grasses will choke them out. If nitrogen fertilizer is applied, the legumes suffer; if none is applied overall production is low.

In brief, the easiest and most effective solution is to have separate grass and legume pastures. The latter are most profitably sown as a soil cover in coconut, oil palm and similar plantations, especially in the early years of establishment.

Annual Crops

When bush-fallow is replaced by rotational cropping systems, stability of production can only be achieved by replacing the nutrients removed by the crops, maintaining a favourable soil structure and preventing soil erosion. The principles involved to achieve these interrelated aims are well known, their successful application under the varied ecological conditions of the tropics however have not yet been sufficiently studied.

The early experiments in replacing shifting cultivation with continuous cropping in east Nigeria concentrated on maintaining soil fertility by green manures, composts and animal manures. A six-year experimental rotation, including these green manure crops, was a complete failure. Whilst yields of yams and maize had not declined on the bush-fallow plots (two years cropping–four years bush), yields on the continuously farmed plots had fallen to nil (Lagemann, 1977).

Experiments with heavy dressings of compost and of farmyard manure showed that continuous cropping was possible, albeit at the same low level of yields as bush-fallowing. The amounts of organic manure required to maintain even these levels could only by provided for relatively limited areas.

In Uganda, the Department of Agriculture recommends a 3:3 rotation: three years in arable crops followed by three years of rest, preferably grass for grazing. Research has shown that this rotation can maintain a medium to low level of soil fertility, which tends to diminish over a number of cycles of the rotation (Jones, 1976).

Limitations of Annual Crops

In temperate climates, the development of the annuals is generally in phase with the seasonal variation of radiation. Their leaf area index is generally low in spring, and achieves a peak in summer, coinciding with maximum daily radiation.

Because of the lack of seasonal variation of radiation in the tropics, apart from its frequently low intensity, the most effective use of the available radiation is obtained from plants with a relatively constant leaf area index, i.e. perennial crops, plantation trees and mixed cropping. There are very few annuals that can achieve the productivity of sugar-cane, for example (Chang, 1968).

Cereals

The major tropical grain cereals are rice, maize, sorghum and millets; their nutritive value is similar to that of the temperate grain cereals (wheat, barley, rye), though the protein content of rice tends to be somewhat lower. Many attempts are being made to introduce wheat production into various regions in the tropics, as in India, Central Africa, Brazil, Paraguay, etc., though the ecological problems involved are considerable. Apart from rice, which we will treat separately, maize is the most important cereal grain crop of the tropics.

Maize has its origin in a semi-arid area, but it is not a reliable or productive crop under conditions of limited rainfall. Because of consumer preferences it is however grown on large areas in Africa and Latin America with marginal rainfall instead of the better adapted sorghums and millets. Consequently, considerable efforts are being invested in breeding drought-resistant varieties of maize, though achievable yield levels under dry conditions are low. In the better rainfall areas, the introduction of hybrid maize and synthetic varieties—generally in conjunction with fertilizers and improved management—has raised yields considerably.

Sorghum and millets are the basic food plants of the dry tropics. With the advent of hybrid sorghum, this crop can now compete in productivity with maize, even under favourable conditions.

The millets are an important source of food in the dry regions of the Old World—in India, Afghanistan, Iran, Turkey and the African Sahelo–Sudan region, where they are grown on soils too poor or in climates too dry to support any other crop. In the New World, they are relatively minor crops, and then are mainly sown as forage crops.

Cotton

Since the Second World War, radical changes have taken place in the production, processing and commerce of cotton. Many countries that formerly produced insignificant amounts of cotton, such as the Latin American countries, have become important suppliers of fibre to the world market. Others, such as India and Pakistan, which formerly produced for home consumption only, have also become major exporters. The principal change that has taken place in cotton production and processing since the former colonies achieved independance, is that the traditional distinction between countries which produce and those which process the raw material, has all but disappeared. Countries such as India, Pakistan and Egypt, have now developed their own textile industries and have become competitors on the world textile market.

For many LDCs, developing a textile industry, using home-grown cotton as a raw material, was the easiest and most logical step in the industrialization of these countries.

Grain Legumes

Grain legumes are a major component of lowland tropical cropping systems. Most of the tropical grain legumes have evolved under high stress conditions or are not genetically capable of responding to favourable growing conditions (Rachie & Roberts, 1974).

More than a dozen species are grown, of which pigeon peas (*Cajanus cajan*), cowpeas (*Vigna unguiculata*), various types of mung beans (*Phaseolus mungo*), soybeans and peanuts are most suited to the warm-weather lowlands.

A large number of secondary pulses are also grown with localized use. Often a farmer will grow a poorly adapted, low yielding species or cultivar because of preference for taste or a ready market.

There are two major groups of grain legumes: the pulses and the leguminous oil seeds.

Among the pulses, pigeon pea has considerable potential for use over a wide range of tropical conditions, from subhumid to semi-arid. It is especially valuable as a perennial crop of three to four years in mixed cropping and bush-fallow systems.

The greatest potential for relatively high yields and income is provided by the second group—the leguminous oil seeds.

Mung beans: These are frequently grown as catch crops following a main crop, such as rice, utilizing the end-of-season rainfall or residual moisture. They are also grown intercropped with maize, cotton or sugar-cane.

Soybeans: With more than double the yield potential, 60% more protein and 20 times more oil than indigenous legumes, soybeans have not been accepted in the African tropics, notwithstanding their exceptional potential and the repeated attempts at introduction since the 1920s. The major deterrent in many tropical regions is a lack of understanding of cultivation and use of this crop.

Peanuts: These are more important than all other tropical legumes combined. They made an enormous impact as a combined cash and subsistence crop in peasant farming, in particular in Senegal.

There are comparatively few well adapted grain legumes for the very humid tropics (Rachie & Roberts, 1974).

Roots and Tubers

Roots and tubers can survive under conditions in which cereals are seriously damaged and therefore provide additional insurance against adverse climatic conditions. In 1974, when harvests of cereals were exceedingly low, many poor people survived in India on cassava. Though rice and wheat have been improved by breeding, are grown on the best land and get most of the irrigation water, fertilizers and other inputs, the root and tuber crops generally give more uniform and higher yields than the grain crops (FAO, 1977).

A marginal addition of 10–20% of cassava flour to cereal flour, could make a considerable contribution to making many developing countries self-sufficient in food-grains, without causing problems of consumer acceptance.

Homestead Gardens

Limited areas of more or less permanent cultivation of a mixture of annual, semi-perennial and perennial crops have developed in and around villages in the wet tropics. Permanent land-use is made possible by means of large quantities of manures, composts and household refuse, as well as confined grazing. The fertility of the homestead soil is maintained and even improved at the cost of the fertility of other land. This is no serious matter as long as these gardens are small.

Intensive Vegetable Production

Where population density is very high, labour cheap, and there is access to markets, intensive vegetable gardening may become very important.

One example of highly intensive vegetable gardening is that practised in Hong Kong, where about one-half of the vegetable growers produce seven to nine crops per year, while 80% raise over four crops (Luh, 1970).

In Java, about 20% of the holdings are so tiny and so intensively cultivated that they are more in the nature of gardens than farms. Composts from household refuse and harvest residues, night soil, green manure crops, sediment from rivers and canals, ashes, oil-cake, fish and more recently, chemical fertilizers, are applied to maintain soil fertility.

This form of land-use is extremely labour-intensive. The highest recorded labour input is in irrigated vegetable holdings of the Chinese-type farming, reaching 10,000 to 15,000 man-hours per hectare with a production of around eight crops per year. The vegetable production is often combined with the rearing of pigs and poultry (Ruthenberg, 1971).

Wet-rice Systems

In the land-hungry, overpopulated areas within the rainy tropics and subtropics of East and South East Asia, the staple crop is rice.* Initially rice growing is developed in the valley bottoms, but with increasing population pressures, it is frequently grown on steep mountain slopes that have been converted by hand labour, over the course of centuries, into narrow terraces.

Most rice, however, is grown in the alluvial plains, on flat, banded fields, under a system of uncontrolled annual flooding that limits production to a single crop a year. Under these conditions there is no loss of soil fertility by erosion and little or none by leaching. The decomposition of organic matter proceeds at a slower rate than on dry land under the same climatic conditions. Crop residues and weeds are ploughed in, and not, as a rule, destroyed by burning. The irrigation water may add nutrients and soil to the field. Any decline in soil fertility is almost entirely due to crop removal. As long as most of the crops are grown for subsistence, these losses may be compensated for almost entirely if the crop residues, household refuse and human and animal excreta are returned to the land as manure or compost. This may explain how rice in South-East Asia could be grown for centuries in succession on the same land without fertilizing. Yields have usually been low (1000–2000 kg paddy per ha), but until recently there was little evidence that they were limited to a large extent by nutritional deficiencies. The majority of the tall, leafy, late-maturing tropical *indica* cultivars rarely support much more than 20 kg N per ha. Lodging is a major cause of low yields in these cultivars.

In paddy rice production there are two types of land-use: *monoculture*, in which 90% or more of the cropped area that is suitable for rice production (i.e. excluding upland and tree crop area) is planted to rice both in the wet and dry seasons' and *mixed farming*, where at least one crop, such as sugar-cane, jute, tobacco, maize or wheat, in addition to rice is of major economic importance. These other crops can be sown in rotation, such as rice followed by wheat or maize, or on different portions of the farm area

*Ecologically similar areas are to be found in the humid tropics and subtropics of Latin America.

during the same season, such as rice in combination with sugar-cane, jute, or tobacco. The mixed farming type is likely to have more favourable conditions, even for rice production, such as better irrigation and drainage facilities (Barker & Anden, 1975). However, in many situations, large areas of paddy fields remain idle after the rice crop because farmers are unable to obtain the inputs needed for a second crop (Villareal & Lai, 1977).

Objectively, multiple cropping can be practised where temperature conditions are favourable to plant growth throughout the year, or at least over a long growing season. The actual intensity of cropping is then mainly dependent on a reliable supply of water, either from rainfall or through irrigation. In Asia, multiple cropping is generally, though not exclusively, built around the staple crop rice.

Associate crops with rice are generally dry-land crops which depend on residual rainfall, and may occasionally benefit from chance winter rainfall. They are sown either sequentially or as intercrops with rice. Where monthly rainfall is at least 200 mm for four to five months, the rice crop can generally be followed by another crop or an intercrop. Where six to eight months of high rainfall can be expected, a double rice crop (using early-maturing varieties) can be introduced, and a drought-tolerant crop can follow the rice to utilize residual moisture and the rain that falls towards the end of the rainy season (Harwood & Price, 1976).

Under very favourable conditions of soil, water, temperature and good management, more than two crops, in addition to rice, can be grown annually. It is reported that in the Dalat area in Vietnam, the vegetable growers raise between six and nine crops a year (Luh, 1970). (Evidently, these include very short-season crops, such as lettuce).

However, because of its unique ability to grow under submerged conditions, rice is frequently grown on land that is unsuitable for other crops and monoculture is then the usual practice.

When high-yielding, early-maturing varieties of rice are grown, the field may be too wet after harvest to plant a dry-land crop, and residual moisture may be insufficient for another crop of rice. Under these circumstances ratooning the rice, which takes 45 to 60 days from harvest to harvest, may be a desirable solution (Moody, 1977). Where only a single crop of rice is grown annually, late-maturing varieties, which are potentially higher yielding, are preferred.

The main drawback of this system of land-use is the underemployment of available labour, of which only about half is fully used, with peak periods of labour requirements at planting and harvesting.

If water is deficient during half of the year, the land lies fallow in the off-season and may supply some grazing. A frequent practice under these conditions is to alternate two or three years of rice cropping with a similar period of grazing the land.

Integration of Wet-rice and Fish Production

Fish generally abound in the water distribution systems in the paddy rice regions. Fish may be raised in a pond located on the homestead, in the growing rice (provided there is a sufficient and continuous depth of water), or the land may be used alternately for rice cropping and for fish production.

Paddies can be managed so as to accommodate fish production concurrently with the growing of a rice crop. Stocking paddies with fish is widely practised in Japan, Indonesia, the Philippines, Vietnam, India, Tanzania, Madagascar and Italy. The most commonly used species are *Cyprinus carpio* (carp), *Trichogaster pectoralis* and various species of the genus *Tilapia*. Management practices vary from country to country, depending on climate and local conditions (Daget, 1977).

Rice and fish production are mutually beneficial: the fish feed on phytoplankton, water weeds and insects which would otherwise compete with the crop; their droppings provide a source of nutrients for the rice crop. Properly selected species help to control snail and mosquito larvae. Trials in Madagascar indicate that fish yields of 20–30 kg/ha can be obtained in 120 days when no additional food is supplied, 80–200 kg/ha when manure, but no additional food is supplied, and 200–400 kg/ha when both manure and daily food rations are supplied (Moreau, 1972).

The system also has some shortcomings: most rice varieties require a water depth of 5–10 cm; in such shallow waters fish are an easy prey for birds. Fish and rice have different development cycles and draining the fields may harm the fish (Kassas, 1977). Pesticides applied to the rice may adversely affect fish yields and marketability (White, 1978). Most of these problems can be overcome by proper management, in particular the provision of trenches and artificial depressions in which the fish can find refuge when the water level in the field becomes too low.

LAND-USE IN HIGH-ALTITUDE REGIONS OF THE TROPICS

Characteristics

Within the tropics exist areas of high-altitude plateaux and valleys, with special characteristics, problems and land-use methods. Such areas are found in East Africa, Asia and South America. In Latin America, it is the vast extent of highlands which determines the environmental conditions and land-use of a great part of the region. This includes the mountain system of the Andes and its northern prolongation into Central America and Mexico, the Brazilian plateau, and isolated highlands in southern Venezuela.

The Andean region is typical for the problems encountered and land-use prevalent in high-altitude regions in the three countries in which most of the region lies. In Ecuador, Peru and Bolivia, more farmers live in the highlands than in the lowlands, notwithstanding the far less favourable ecological conditions of the former. Because most of the development in recent decades has been in the coastal lowlands, the farmer' income in the highlands is also far lower. In Peru, for example, with an average *per caput* income of $179, the *per caput* income is three times greater in the lowlands than in the highlands (Hapgood & Millikan, 1965). The difficult terrain of the region and its fragmentation into relatively small zones makes communication between the zones, as well as with their potential markets, difficult and costly. The road system is generally poor and limited, and in many areas the only means of communication are pack trails. This is probably one of the major constraints preventing the transformation of subsistence agriculture into modern, commercial agriculture.

The extraordinarily varied topography of the region and the differences in exposure to moisture-bearing currents causes climate to show sudden changes and wide contrasts with resultant variations in vegetation and agricultural production. For example, in contrast to the eastern sides of the Andes, the western slopes and outridges are barren wastes, since the humid winds which blow from the East deposit their moisture on the eastern watershed.

Ecological Zones

Broadly speaking, the region is divided into three main ecological zones according to altitude; with considerable variation in temperature over a very short distance.

(a) *Lower Altitudes (up to 2000 m)*

Temperatures are generally relatively high, frosts unlikely and the growing season is generally long. In valley bottoms, the soils are deep, flat and generally fertile. Land-use is, on the whole, similar to that of the lowlands

(b) *Higher Elevations (between 2000 and 4000 m)*

In the so-called '*tierra fria*' the danger of frosts increases, average temperatures are lower and the growing season becomes shorter with increasing altitude. Rainfall is very variable and its incidence irregular. A large proportion of the crops is grown on very steep slopes. For example, in the Andean region of Venezuela, it is estimated that 70% of the crops are grown on slopes steeper than 35° (Duckham & Masefield, 1970). On the slopes, the soils are generally shallow and frequently with little inherent fertility. Soil erosion is a major problem. The valleys are frequently deep-hollowed with very steep sides, cut by glaciers or by rivers flowing over the plateaux. Many small agricultural communities are dispersed throughout the region, with a greater concentration of population on the larger valley floors. The majority of the farm units are small and fragmented, with as many as 50 'fields' per hectare. Most of the land is cultivated by hoes, while oxen may be used on level or almost level land. The main crops are maize (up to about 2500 m); wheat (up to 2000 m); barley and potatoes (up to about 3200 m). A number of special, and little-known crop plants are also grown, among which are to grain species of *Chenopodium* grown for grain: 'quinoa' (*Chenopodium quinoa*) and 'kanahua' (*Ch. pollidicuuli*); and two tuber crops, 'ulluco' (*Ullucus tuberosas*) and 'apoi' (*Arracha esculente*) (Duckham & Masefield, 1970). Much of the better land belongs to large haciendas, where land is generally underutilized. Communal lands are the joint property of the villages and can be parcelled out for individual cultivation or for grazing by the animals of the whole village.

Farm methods, especially those on the smaller farms, have changed very little during the last few decades, and progressive overcrowding of the Andean region is making it more and more difficult for small-scale farmers to apply certain types of technological improvements (Hopkins, 1969).

A few of the possibilities for improving agriculture—besides land reform and adequate infrastructure—include improved crop rotations, terrace construction for soil conservation, planting of trees where cultivation is not possible and irrigation.

Because of the erratic rainfall, irrigation can be a major factor in increasing the productivity of the valley bottoms. These regions, on the whole, are generally well endowed with permanent streams fed by snow accumulation on the mountain peaks. There are also enclosed physiographic basins where underground water can be tapped by tubewells.

Very High Altitudes

The plateau-like tracts located above 3000–3500 m, can be divided into the drier altiplano and the humid paramos.

Extensive areas of range are found in both regions and grazing mainly by sheep but also by llamas and alpacas, is the major form of land-use.

Much can be done to improve agricultural production on the high-altitude regions as outlined above; it is, however, doubtful whether an appreciable increase in the standard of living in these regions (excepting the bottom valleys) can be achieved on the basis of a purely agricultural economy. The steep slopes, the fragmentation of holdings due to relief, the thin soil and continued losses from erosion, the harsh climate, remoteness from markets, all militate against the adoption of solutions that can be adequate in other areas. In the high-altitude regions—more than elsewhere—significant progress depends on integrated development of a variety of economic resources: agriculture, industry, homecrafts, tourism and others.

SUMMARY

The Dry Regions

Desert areas have traditionally been used for extensive grazing by livestock. Without irrigation, this continues to be the best possible use of these areas, if improved ranching methods are adopted. The outlook for domestication of plants adapted to desert conditions is bleak.

The Semi-arid Regions

These include mediterranean-type climates with winter rainfall, and subtropical savannas with predominantly summer rainfall. Cereal cultivation is the main source of livelihood: wheat and barley in the winter rainfall areas; maize, sorghum and millets in the summer rainfall areas.

Crop rotation, integrated animal husbandry and arable cropping, appropriate tillage methods, the use of improved varieties, in conjunction with fertilizer application and crop protection have proven able to effect dramatic increases in yields in these areas.

It is in the marginal areas that the transition from grazing to arable cropping, generally under increased population pressure, is most dangerous and may have catastrophic results, mainly through wind erosion.

Irrigation Agriculture

Different methods of *water harvesting* may provide a viable solution for dry regions in which no other source of water is available. *Conventional irrigation* can be a powerful tool for agricultural development, not only in the dry regions, but even where rain-fed cropping is possible. The development of irrigation can also have an impact on the development of non-agricultural sectors of the economy.

However, whilst irrigation agriculture can be the most productive form of farming, disappointing results have been obtained in a large number of irrigation projects in developing countries, mainly because of fundamental errors in planning and implementation.

One of the few successful irrigation projects in Africa is the Mwea irrigation settlement in Kenya, the reasons for its success and its shortcomings are therefore described in detail.

The advantages of small-scale irrigation projects, in which water resources are developed at the level of the individual farm or village, are being increasingly appreciated. Their cost per hectare is usually low, a greater proportion of the work can be carried out by the beneficiaries and allocation of water is easier.

The Tropics

The main characteristics of *traditional shifting cultivation* are (Charreau, 1974):

(1) Progressive clearing of the land, with little, or no removal of stumps.
(2) The shifting cultivation is confined to the fields, the village sites are permanent, as well as cultivation on the perimeter of the villages.
(3) Cycles of cropping alternate with fallows; the relative lengths of cropping and fallows depending on population pressure.

(4) Livestock may be present in variable numbers, but there is no true mixed farming. Without animal power, there is no real tillage and crop residues are not incorporated into the soil.
(5) Poor control of weeds, inadequate control of diseases and pests.
(6) Mixed cropping is the rule.

As long as shifting cultivation is general in a region, land-use is very homogeneous. However, once the system begins to break down under population pressure, different pathways of change, leading to different land-use forms appear.

The transition to permanent farming systems presents many problems, that have as yet generally not been completely solved. The main difficulties encountered are to provide effective protection to the soil, the maintenance of soil fertility and the control of exuberant weed growth.

The first move away from shifting cultivation towards permanent land-use has usually been the planting of perennial crops, in particular tree crops, because the ecological effects of plantations are most similar to those of the original forest cover. Perennial field crops and pastures fulfil a similar role. The transition to rotational cropping still poses most unsolved problems, excepting in the case of the wet-rice systems, which are a case unto themselves.

The different pathways of change from shifting to permanent cultivation have been very clearly presented graphically by Ruthenberg (1971) (Fig. 18).

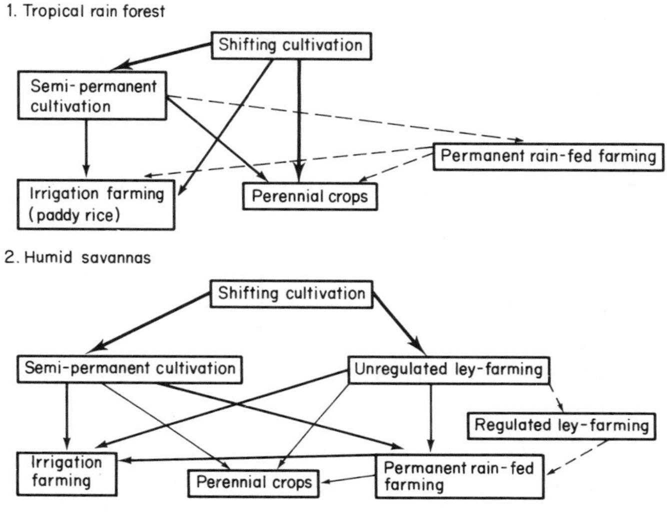

Fig. 18 General pathways of change followed by indigenous smallholders from shifting cultivation to permanent farming systems in the humid tropics (Ruthenberg, 1971). By courtesy of the author. The relative importance and feasibility of each evolutionary path is indicated by the thickness of the arrows.

The main objectives for wet-rice cropping systems are:

(a) *To increase the productivity* of production. This has been made possible by the new short, sturdy, early-maturing hybrid cultivars developed by the International Rice Research Institute in the Philippines, these respond well to nitrogen fertilizing and with an adequate package of improved practices may yield up to 7000 kg and more of paddy per hectare. Multiple cropping is another development made feasible by new varieties and practices (cf. p. 96).
(b) *Diversification of human diet.* Nutrition based mainly on rice is deficient in proteins and vitamins. Diversification of the human diet is therefore of considerable importance. Hence, the need for a shift from rice monoculture to diversified crop production, including rainfed and irrigated crops in a variety of crop combinations.

An important contribution to the improved diet could be the expansion of the practice of raising fish in the rice paddies, already prevalent in certain areas.

References

Baldy, C. (1963). Cultures associées et productivité de l'eau, *Ann. Agron.*, **14**, 489–534.

Barker, R & Anden, T. (1975). Factors influencing the use of modern rice technology in the study areas. Pp. 17–40 in *Changes in Rice Farming in Selected Areas of Asia*, International Rice Research Institute, Los Baños. 377 pp., illustr.

Blackie, J. R. (1972). Hydrological effects of a change in land use from rain forest to tea plantation in Kenya. Paper presented at *Symp. Rep. and Exp. Basins, Wellington, N.Z.*

Bucha, M. K. B. (1968). Agricultural development in Pakistan. Pp. 41–50 in *Strategy for the Conquest of Hunger*. Rockefeller Foundation, New York: 131 pp., USA.

Chambers, R. (1969). *Settlement Schemes in Tropical Africa: a Study of Organization and Development*. Routledge and Kegan Paul, London: 294 pp.

Chambers, R. & Moris, J. (Ed.) (1973). *Mwea, An Irrigated Rice Settlement in Kenya*. Weltforum Verlag, Munich, FRG: 539 pp., illustr.

Chang, J. H. (1968). Agricultural potential of the tropics. *The Geographical Review*, **58**, 333–51.

Charreau, C. (1974). *Soils of Tropical Dry and Dry-wet Climatic Areas and their Use and Management*. Cornell University, Ithaca, NY: 434 pp., illustr. (mimeogr.).

Clark, C. (1970). *The Economics of Irrigation*. Pergamon Press, Oxford, UK: 155 pp.

Clark, C. & Haswell, Margaret (1971). *The Economics of Subsistence Agriculture*. Macmillan & Co., London: xiii + 267 pp., illustr.

Coene, R. de, (1956). Agricultural settlement schemes in the Belgian Congo, *Trop. Agric.*, **33**, 1–12.

Collinson, H. (1972). *Farm Management in Peasant Agriculture*. Praeger Publishers, New York: xxvi + 444 pp., illustr.

Crosson, P. R. (1975). Institutional obstacles to expansion of world food production. Pp. 17–22 in *Food: Politics, Economics, Nutrition and Research* (Ed. P. H. Abelson). Am. Assoc. Adv. Sci., + 202 pp., illustr.

Daget, J. (1977). La production de poissons de consommation dans les écosystemes irrigués. Pp. 295–30 in *Arid Land Irrigation in Developing Countries* (Ed. E. B. Worthington). Pergamon Press Oxford: xi + 463 pp., illustr.

Dasgupta, B. (1977). *Agrarian Change and the New Technology in India*. UNRISD, United Nations, Geneva: xxvii + 408 pp.

Drar, M. (1954). Plants for raw material in the deserts of Egypt. Pp. 70–76 in *Proc. Symp. on Scientific Problems of Land Use in Arid Regions, Heliopolis:* 219 pp., illustr.

Duckham, A. N. & Masefield, G. B. (1970). *Farming Systems of the World*. Chatto and Windus, London: xviii + 542 pp., illustr.

Dumont, R. (1966). *African Agricultural Development*. Food and Agriculture Organization of the UN, Rome, Italy: vi + 243 pp.

Eicher, C. K., Zalla, T., Kocher, J. and Winch, F. (1970). *Employment Generation in African Agriculture*. Research Report No. 9, Institute of International Agriculture, College of Agriculture and Natural Resources, Michigan State University: xl + 593 pp., illustr.

Enabor, E. E. (1974). Socio-economic aspects of taungya in relation to traditional shifting cultivation in tropical developing countries. Pp. 191–202 in *Shifting Cultivation and Soil Conservation in Africa*. FAO Soils Bull. No. 24.

FAO (1962). *FAO Africa Survey: Report on the Possibilities of African Rural Development in Relation to Economic and Social Growth*. Food and Agriculture Organization of the UN, Rome, Italy.

FAO (1966). *Agricultural Development in Nigeria 1965—1980*. Food and Agriculture Organization of the UN, Rome, Italy.

FAO (1969). *Smaller Farmlands Can Yield More*. Food and Agriculture Organization of the UN, Rome, Italy: vii + 73 pp., illustr.

FAO (1970). *Provisional Indicative World Plan for Agricultural Development*. Food and Agriculture Organization of the UN, Rome, Italy.

FAO (1977). *The State of Food and Agriculture, 1976*. FAO Agricultural Series, No. 4, Rome, Italy: vi + 157 pp.

FAO (1978). *The State of Food and Agriculture, 1977*. FAO Agricultural Series, No. 8, Rome, Italy.

Ferwerda, J. D. (1970). Soil fertility in the tropics as affected by land use. Pp. 317–29 in *Proc. 9th Congr. Int. Potash Inst., Antibes:* 489 pp. International Potash Institute, Berne.

Fluck, H. (1955). The influence of climate on the active principles in medicinal plants, *J. Pharm.* (*London*), **7**, 361–83.

Fremond, Y. (1968). *Coconut—Palms Selection*. Report 8/10 to the Conference on Agricultural Research Priorities for Economic Development in Africa, Abidjan: mimeogr.

Garbrecht, G. (1979). Increasing of irrigation efficiencies under the conditions of developing countries. P. 3 in *Abstracts: Dialogue on Development. Towards the 21st Century*. Assoc. Engineers and Architects in Israel, Tel-Aviv: 151 pp.

Geus, J. de (1977). *Production potentialities of Pastures in the Tropics and Subtropics*. Centre d'Etude de l'Azote, Zurich: 54 pp., illustr.

Gittinger, J. P. (1965). *Planning for Agricultural Development: The Iranian Experience*. National Planning Association, Washington DC: xi + 123 pp.

Gomez, A. A. (1974). Intensification of cropping systems in Asia. Pp. 93–99 in *Interaction of Agriculture with Food Science* (Ed. R. MacIntyre). International Development Research Centre, Ottawa: pp. 166.

Grandstaff, T. (1978). The development of swidden agriculture (shifting cultivation) *Development & Change*, **9**, 547–579.

Greenland, D. J. (1974). Evolution and development of different types of shifting cultivation. Pp. 5–13 in *Shifting Cultivation and Soil Conservation in Africa*, FAO, Soils Bull. No. 24. Rome.

Greenland, D. J. (1975). Bringing the Green Revolution to the shifting cultivator. *Science*, **190**, (4217), 841–844.

Hall, A. E., Foster, K. W., & Waines, J. G. (1979). Crop adaptation to semi-arid environments. Pp. 148–179 in *Agriculture in Semi-Arid Environments*, A. E. Hall, G. H. Cannell, & H. W. Lawton, eds. Springer Verlag Berlin, Heidelberg, New York. xvi + 340 pp. illustr.

Hapgood, D. & Millikan, M. F. (1965). *Policies for Promoting Agricultural Development*, Massachusetts Institute of Technology, Cambridge, Mass.: 312 pp.

Harwood, R. R. & Price, F. C. (1976). Multiple cropping in tropical Asia Pp. 11–40 in *Multiple Cropping* (ed. P. A. Sanchez and G. B. Triplet). Am. Soc. Agron. Spec. Publ. No. 27: viii + 378 pp., illustr.

Hauck, F. W. (1971). Soil fertility and shifting cultivation. Pp. 131–138 in *Improving Soil Fertility in Africa*, Food and Agriculture Organization of the UN, Rome: v + 145 pp.

Heady, H. F. (1965). Rangeland development in East Africa. P. 73–82 in *Ecology and Economic Development in Tropical Africa* (Ed. D. Brokenshaw). Univ. California Inst. Int. Studies, Berkeley: 268 pp.

Hopkins, J. (1969). *The Latin American Farmer*. U.S. Department of Agriculture, Washington, DC: 137 pp.

Jones. E. (1976). Soil productivity. P. 43–76 in *Agricultural Research for Development* (Ed. M. H. Arnold). Cambridge University Press, London: x + 333 pp., illustr.

Jones, R. J. (1972). *The Place of Legumes in Tropical Pastures*. ASPAC Techn. Bull. Nr. 9.

Jurion, F. & Henry, J. (1967). *De l'Agriculture Itinérante à l'Agriculture Intensifiée*. Institut National pour l'Etude Agronomique du Congo: 497 pp., illustr.

Kassas, M. (1977). Discussion and conclusions. Pp. 335–340 in *Arid Land Irrigation in Developing Countries. Environmental Problems and Effects* (Ed. E B. Worthington). Pergamon Press, Oxford: xi + 463 pp., illustr.

Kemper, W. D. (1964). Micro-water-sheds, *Agric. Res*., **13** (2), 10–11.

King, K. F. S. (1968). *Agri-silviculture* (*The Taungya System*). Department of Forestry, University of Ibadan, Bull. No. 1.

Kovda, V. A. (1977). Arid land irrigation and soil fertility; problems of salinity, alkalinity, compaction. Pp. 211–36 in *Arid Land Irrigation in Developing Countries. Environmental Problems and Effects* (Ed. E. B. Worthington). Pergamon Press, Oxford: xi + 463 pp., illustr.

Kulp, E. M. (1970). *Rural Development Planning*. Praeger Publishers, New York: 664 pp., illustr.

Kuo, L. T. C. (1972). *The Technical Transformation of Agriculture in Communist China*. Praeger Publishers, New York: xx + 266 pp., illustr.

Lagemann, J. (1977). *Traditional African Farming Systems in Eastern Nigeria*. Weltforum Verlag, Munich, FRG: 269 pp., illustr.

Luh, C. L. (1970). Report on vegetable production survey in South-east Asian countries. Pp. 12–13 in *Seminar on Food Problems in Asia and the Pacific, Honolulu, Hawaii*.

Malavolta, E., Haag, H. P., Mello, F. A. F. and Brazil, Sobro M.O.C (1962). *On the Mineral Nutrition of Some Tropical Crops*. International Potash Institute, Berne: 155 pp., illustr.

McPherson, W. W. & Johnston, B. F. (1967). Distinctive features of agricultural development in the tropics. Pp. 180–233 in *Agricultural Development and Economic Growth* (Ed. H. M. Southworth & B. F. Johnston). Cornell University Press, Ithaca, NY: xv + 608 pp., illustr.

Millikan, M. F. & Hapgood, D. (1967). *No Easy Harvest*. Little, Brown & Co., Boston, Mass.: xiv + 178 pp.
Moody, K. (1977). Weed control in multiple cropping. Pp. 281–93 in *Symposium on Cropping Systems Research and Development for the Asian Rice Farmer*. International Rice Research Institute, Los Baños, Philippines: 454 pp., illustr.
Moreau, J. (1972). Perspectives offertes par la riziculture à Madagascar. *Terre Malgache*, **14**, 227–42.
Moris, J. & Chambers, R. (1973). Mwea in perspective. Pp. 439–83 in *Mwea, an Irrigated Rice Settlement in Kenya* (Ed. R. Chambers & J. Moris). Weltforum Verlag, Munich, FRG: 539 pp., illustr.
Myers, L. E. (1967). New water supplies from precipitation harvesting. Pp. 631–40 in *Water for Peace*. US Government Printing Office, Washington, DC: Vol. 2, vi + 970 pp., illustr.
Nicolaides, J. III (1979). Crop production systems on acid soils in humid tropical America. Pp. 243–277 in *Soil, Water and Crop Production* (Ed. D. W. Thorne and M. D. Thorne). Ari Publishing Co. Inc., Westport, Connecticut: ix + 253 pp., illustr.
Norman, D. W. (1970). Cultures mixtes. Paper presented at *Seminar on Traditional African Agricultural Systems and Their Improvement, Ibadan, Nigeria*. University of Ibadan, Ibadan.
Nye, P. H. & Greenland, D. J. (1960). *The Soil Under Shifting Cultivation*. Commonwealth Agricultural Bureau, Harpenden, UK: vi + 156 pp., illustr.
Oni, S. A. & Olayemi, J. K. (1973). Determinants of settlers success under the Western Nigerian Farm Settlement Scheme. *The Nigerian Agr. J.*, (10), 226–39.
Paris, R. & Dilleman, G. (1960). Medicinal plants of the arid zones with particular reference to the pharmacological aspects, *Arid Zone Res.*, **13**, 55–91.
Pereira, H. C (1973). *Land Use and Water Resources*, Cambridge University Press, London: xiv + 246 pp., illustr.
Pollock, N. C. (1971). *Studies in Emerging Africa*. Butterworths, London: 342 pp., illustr.
Pultz, L. M. (1956). Problems in the development and utilization of arid lands plants. Pp. 414–18 in *The Future of Arid Lands* (Ed. G. F. White). Amer. Assoc. Adv. Sci., Washington, DC: ix + 435 pp., illustr.
Rachie, K. O. & Roberts, L. M. (1974). Grain legumes of the lowland tropics. *Adv. Agron.*, **26**, 2–132.
Rappaport, R. A. (1971). The flow of energy in an agricultural society. *Sci. Am.*, **225**, 116–32.
Revelle, R. (1966). Water. Pp. 1007–19 in *Getting Agriculture Moving* (Ed. R. E. Borton). The Agricultural Development Council, New York, NY: 2 vols, x + viii + 1123 pp.
Roche, L. (1974) The practice of agri-silviculture in the tropics with special reference to Nigeria. Pp. 179–90 in *Shifting Cultivation and Soil Conservation in Africa.* FAO, Soils Bull. No. 24.
Ruthenberg, H. (1971). *Farming Systems in the Tropics*. Clarendon Press, Oxford: xiv + 313 pp. illustr.
Sanchez, P. A. (1973). *A Review of Soils Research in Tropical Latin America*. Tech. Bull. North Carolina, Agr. Exp. Sta. 219.
Sanchez. P. A. (1977). Advances in the management of oxisols and ultisols in tropical South America. Paper presented at *Int. Sem. on Soil Environment and Fertility Management in Intensive Agriculture*, 10–13, Oct. Tokyo.
Sanchez, P. A. & Buol, S. (1975). Soils of the tropics and the world food crisis. *Science*, **188**, 548–603.
Shanan, L. & Tadmor, N. H. (1976). *Micro-Catchment Systems for Arid Zone Development*, Hebrew University, Jerusalem and Centre of International Agricultural Cooperation, Ministry of Agriculture, Rehovot: 128 pp., illustr.

Stakman, E. C., Bradfield, R. & Mangelsdorf, P. C. (1967). *Campaigns Against Hunger*, Harvard University Press, Cambridge, Mass.: xvi + 328 pp., illustr.

Stavenhagen, R. (1969). A land reform should answer the questions it raises, *Ceres*, **2** (6), 43–7.

Stippler, H. H. & Darwish, M. J. (1966). *Land Tenure and Land Utilization in Shakha 8*, Ministry of Agrarian Reform, Baghdad, Govt. of Iraw & Ford Foundation.

Taylor, J. R. Jr. (1969). *Agricultural Settlement and Development in Eastern Nicaragua*. Land Tenure Research Paper, No. 33, Land Tenure Centre, Madison, Wisconsin.

US President's Science Advisory Committee (1967). *The World Food Problem*. Rep. Panel World Food Supply, Washington, DC: Vol. 3, xxii + 332 pp., illustr.

Villareal, R. & Lai, S. H. (1977). Developing vegetable crop varieties for intensive cropping systems. Pp. 373–390 in *Symposium on Cropping Systems Research and Development for the Asian Rice Farmer*. International Rice Research Institute, Los Baños, Philippines: 454 pp., illustr.

Wallen, C. C. (1966). Arid Zone meteorology. Pp. 31–50 in *Arid Lands A Geographical Appraisal* (Ed. E. S. Hills). UNESCO, Paris: xviii + 461 pp., illustr.

Walters, P. R., MacFarlane, N. & Spensley, P. C. (1979). *Jojoba, an Assessment of Prospects*. Tropical Products Institute, London: vi + 32 pp.

Watters, R. F. (1971). *Shifting Cultivation in Latin America*. Food and Agriculture Organization of the UN, Rome, Italy: 305 pp., illustr.

White, G. F. (Ed.) (1978). *Environmental Effects of Arid Land Irrigation in Developing Countries*. MAB Tech. Notes 8, UNESCO, Paris.

Wilde, J. C., de & McLoughlin, P. F. M. (1967). *Experiences with Agricultural Development in Tropical Africa*. Johns Hopkins Press, Baltimore, Maryland: Vol. 2, xii + 466 pp.

Williams, C. W. (1975). *The Agronomy of the Major Tropical Crops*. Oxford University Press, London: xix + 228 pp., illustr.

Zeven, A. C. (1967). *The Semi-Wild Oil-Palm and its Industry in Africa*. Agricultural Research Report G89, Wageningen, Holland.

CHAPTER IV

The Deterioration of Natural Resources

Agricultural production is by its nature an extractive process, it is however only in relatively recent times that the realization has crystallized that appropriate measures have to be taken if agricultural resources are not to be irrevocably destroyed by negligent use of land and water by man.

Agricultural development increases the potential for resource deterioration and even destruction. The construction of reservoirs, the development of irrigation projects, mechanization, the use of chemicals for increased production or plant protection, react with environment in a number of ways that may have disruptive effects.

In most countries of the world, but especially in the tropics and subtropics, enormous areas of land have been damaged by clearing of forests, erosion, salinization and sodication, and by desertification.

Buringh (1978) estimates the areas lost to cultivation annually as approximately 500,000 hectares (10 hectares every minute!) of which 50% are ascribed to erosion, 30% to salinization and 20% to desertification and degradation. Other estimates of these losses are much greater still.

In addition, the growth of the infrastructure of transport and communication, increasing urbanization, the development of industry (all of which are concomitant with rural development), have consequences for the ecological system (United Nations, 1971).

The solution of the problems of deterioration of natural resources in the LDCs not only involves technical problems, but also has social, economic and political aspects. Factors such as population control and distribution, the development of off-farm economic opportunities, incentives to adopt control measures and appropriate laws and regulations are involved.

These problems will therefore, in all probability, not be solved in the near future. Because the negative impacts are not felt immediately and only become serious after a number of years, they are frequently disregarded—especially as the costs involved in preventing damage may be high. Hence there is a tendency to ignore these problems until an irreversible situation has been created, or cure has become prohibitively costly.

SOIL EROSION

Economic development depends, in the last analysis, on the proper use and conservation of the soil. Population pressure, destructive tillage methods, faulty irrigation practice and overgrazing can cause the deterioration of the soil and this is frequently an irreversible process.

Erosion *per se*, is a *natural process* contributing to soil formation by wearing down mountains and building up the soil in more level lands. Proceeding at a slow pace, it may, therefore, be very beneficial. However, it becomes a catastrophic process when it is excessively accelerated by man. The main factors of accelerated erosion are:

(a) The ploughing up of farmland, in marginal semi-arid regions, which is generally conducive to wind erosion. The classical example is, of course, the creation of the dust-bowl in the 1930s in the dry, central parts of the United States.

(b) The destruction of vegetative cover on slopes, in particular in the humid regions, leading mainly to water erosion. This is generally the direct result of population pressure causing a shortening of the fallow period in shifting cultivation, and overgrazing in pastoral agriculture.

(c) Incorrect tillage methods (e.g. ploughing down the slope), usually as the result of excessive parcellation between heirs in successive generations (Fig. 19).

Fig. 19 Faulty tillage methods leading to severe erosion. Courtesy of Israel Soil Conservation Service.

Table VII Amounts of soil loss in different ares in West Africa (Charreau, 1972. Reproduced by permission of Agronomie Tropicale).

Country	Slope (%)	Mean yearly rainfall[a] (mm)	Soil losses (tonnes/ha) under: Forest	Cropping	Base soil
Upper Volta	0.5	850	0.1	0.6–8.0	10–20
Senegal	1–2	1300	0.2	7.3	21.3
Ivory Coast	4.0	1200	0.1–0.2	0.1–26.0	18–30
Ivory Coast	7.0	2100	0.03	0.1–90.0	108–170

[a]Means over 15 years of study.

An appreciation of the amounts of soil losses involved under different slopes, amounts of rainfall and vegetation cover are shown in Table VII.

Of particular interest note the perfect protection offered by forest cover, even on a slope of 7% and extremely high rainfall; by contrast losses on bare soil increase by 100–200 times when the forest cover is removed even on relatively flat land (0.5%) with a low rainfall régime (850 mm).

Removal of the forest cover, on steep slopes in a high rainfall area, leads to catastrophic results. It is important to note that cropping can provide fair protection or lead to heavy erosion, depending on the type of crop.

Soil erosion is frequently the culmination of a process that begins with the loss of soil fertility and impaired soil structure. Soil erosion in turn further reduces soil fertility by selectively removing the smallest and lightest particles, thereby reducing the proportion of soil colloids and increasing that of the large, inert particles of sand. Thus, a vicious circle is created: the fertile topsoil is gradually removed, leaving the far less fertile subsoil exposed; because the subsoil is relatively compact, water infiltration into it is slow and limited; tillage becomes more difficult and less efficient; biological activities in the soil are slowed down, as a result of impaired soil structure and deficiencies of available nutrients. Even long after the addition of fertilizers and manures, the subsoil remains less fertile than the topsoil that was removed by erosion.

The subsoil in turn is even more susceptible to erosion than the topsoil because of its poor structure and low organic matter content. Run-off increases, gullying results, and the field may soon be in a condition which makes cultivation uneconomical.

The damage is not confined to eroded fields. Both wind and water transport the eroded soil selectively; the first to be deposited are the coarse sandy particles, which may cover to a great depth fields that were previously fertile, whilst the light fertile particles remain in suspension for great distances, and usually end up in deserts or in the oceans.

Besides impairing the fertility of cultivated soils, soil erosion causes the discharge of considerable amounts of silt into river systems, the rapid filling up of reservoirs with sediments, and also increases the danger of floods.

Much land formerly under cultivation has been abandoned owing to soil erosion, which is widespread in many developing countries. South America has been called the 'vanishing continent'. It is estimated that a quarter of the total land under cultivation, past and present, has lost its topsoil through erosion (Benham & Holley, 1960). A survey of soil erosion in Chile, on 1,200,000 ha, for example, showed that only 12.6% was not affected, while 40% had been badly damaged (Pawley, 1963).

All the destructive processes of soil erosion tend to increase when marginal areas of low rainfall are cultivated for field crops. Though annual rainfall is low, very high intensities occur, and combined with the poor plant cover, result in increased erosion losses.

The ploughing of great areas of grassland for wheat and maize cultivation in Argentina, has cause wind erosion on a large scale and has brought about a situation very similar to that of the dust-bowl in the United States (Benham & Holley, 1960).

In the humid tropics the erosion of even a thin surface layer may have very serious consequences if it uncovers a laterite, which forms a concrete-like surface when subjected to periodic drying and may make large areas unfit for cultivation.

Soil erosion is extremely severe in many hilly or mountainous parts of the Andean regions; the silt carried by the Orinoco River, for example, discolours the waters of the Atlantic up to a distance of 150 km from the shore.

Where mechanization is used for large-scale re-settlement schemes in tropical forest areas, the resulting complete removal of trees and shrubs compounded by the battering effect of high-intensity rainfall on the bare soil can increase the potential damage due to water and wind erosion manifold.

Soil erosion should be controlled. Methods such as ploughing on the contours, the construction of terraces, strip cropping, cover-crops, etc., have been developed which are very effective under most conditions.

Paradoxically, a conflict of interests may result from soil conservation measures: the more effective the fight against erosion in the upper reaches of the watershed, the greater is the reduction in run-off and the lower the amounts of water that reach reservoirs downstream. Thus, control of grazing to preserve a grass cover was effective in preventing excessive trampling and soil exposure and in reducing soil erosion but also resulted in a reduced waterflow from the area.

Protected forest areas cost money for policing and fire protection. In spite of much research in East Africa, commercial non-destructive harvesting of the natural forests has given too low yields to pay for the protection.

The investment in effort required for effective soil erosion prevention measures may be quite considerable and the effects are mainly long term. It has therefore frequently been difficult or even impossible to obtain the willing cooperation of subsistence farmers in the execution of soil-conservation processes.

The major problem is that the areas requiring adequate measures are usually remote from the political centres and are frequently difficult to assess. For a detailed treatment of the subject, the reader is referred to a recent publication by FAO (1977).

Overgrazing

In humid regions, both tropical and temperate, pastures not only conserve soil fertility, but may even improve it. Nutrients and humus accumulate under the sward, conditions are favourable for biological activity, a good soil structure is preserved and the vegetative cover prevents erosion.

In the drier zones, the ecological balance is extremely fragile and is easily upset by overgrazing, which in most of the drier parts of the subtropical and tropical regions, has seriously damanged the carrying capacity of rangeland.

The main sources of forage supply to domestic animals in the world, and in particular in the developing countries, are the natural pastures or rangeland, which account for nearly one-quarter of the world's land-area (FAO, 1978).

Areas under forest or woodland are also frequently used for browsing or grazing as well as areas in the arid regions used for nomadic grazing. Most of the grazing lands are unsuitable for arable cropping, either because rainfall is insufficient or they are waterlogged or too steep, shallow or stoney. The productivity of pastures varies in the range $\frac{1}{5}$–$\frac{1}{3}$ ha per animal unit on well-managed and fertile pastures in temperate zones to 50–60 ha per animal unit in arid areas.

Between 1955 and 1976 cattle numbers in the developing countries increased by 34% and sheep and goats by 32%.

Overgrazing first results in the replacement of nutritious and palatable species by others of lower feeding value and palatability, and finally by the devastation of the vegetation.

The direct effect of overgrazing on the ground cover is compounded by certain other practices, such as the cutting for fuel of surviving trees and bushes, uncontrolled burning of dead grass, and the use of animal manure as fuel.

The reclamation of grazing land by re-seeding with suitable species including grasses, legumes and shrubs, and proper grazing management is possible but costly and difficult to implement in areas of nomadism. The production of forage reserves, in particular under irrigation on relatively limited areas, or integrated grazing–arable systems can also provide viable solutions.

Improving the grazing lands in neighbouring regions with better precipitation can also relieve grazing pressure in the more marginal areas.

A solution adopted in Senegal as a means of reducing the risks of overgrazing is to restrict stock-raising by pastoralists in the dry areas to the production of young animals. At the time of weaning these are sent to the cultivated areas, where they are fed on grown forage and crop residues, and used as a source of traction, manure, milk and meat. Upon reaching maturity or after being retired from work, the animals are fattened and sold.

DESERTIFICATION

Houérou (1976) defines desertification as 'a combination of factors leading to a more or less irreversible reduction of the plant cover resulting in the conversion into desert landscapes of tracts that did not formely have desert characteristics'.

Desertification occurs on the fringes of all the hot deserts in the world, on all continents, generally where rainfall is between 100 to 200 mm, occasionally in even somewhat more humid situations.

The areas already affected, or likely to be affected, are shown in Table VIII.

The risk of desertification is a function of both the inherent vulnerability of the region and of human or animal pressures on its resources.

As much as 95% of the total land-area in the arid and semi-arid zones is exposed to the risk of desertification, with most of the arid zone classified as high risk. However, desertification is not confined to dry areas, and the hazards can be great even in the subhumid zones.

Desert encroachment is expanding at an alarming rate in the Sahelian belt of Africa, from the Sudan to Niger.

Kassas (1972) estimates that in the Sudan alone, the desert has encroached, within the last 15–20 years, over land-used areas in a belt of no less than 150 km in width. A similar situation exists in India and elsewhere in Central Asia. In Southern Africa, subtropical rainforest has been devasted and replaced by desert and semi-desert vegetation.

Table VIII Areas of existing deserts (in thousands of km^2) and of those liable to desertification (FAO, 1978, with permission).

Continent	Existing extreme desert	Degree of desertification hazards[a]			Total	Share of land area (%)
		Very high	High	Moderate		
Africa	6178	1725	4911	3741	16555	55
North and Central America	33	163	1313	2854	4363	19
South America	200	414	1261	1602	3478	20
Asia	1581	790	7253	5608	15232	34
Australia	—	308	1722	3712	5742	75
Europe	—	49	—	190	238	2
Total	7992	3449	16460	17707	45608	(35)

[a]The rapidity with which desertification is likely to take place if existing conditions do not change.

The expansion of desertification on the fringes of semi-arid areas is the direct result of excessive population pressure and the concomitant rise in the numbers of livestock. As primitive animal husbandry is the dominant land-use of the drier areas, overgrazing is the primary factor in desertification.

The process of desertification has been re-constructed by Bryson (1972) in his studies on the Rajputana desert, an area once occupied by a high culture with an agricultural base—the Indian civilization. The ramains of this civilization as well as palaeobotanical studies indicate that the region was not always as desertic as it is now, but probably a kind of savanna. Bryson describes the history of the area and the probable course of events that led to desertification as follows.

During the time of the Indian civilization, the Sambhar salt lake held freshwater and the vegetation of the area was adapted to much moister conditions than at present. Then the lake became salty and about 1000 BC dried up completely. By the fourth century AD after a long period of dryness, an extensive culture established itself in the area, followed by widespread nomadism. By the seventh century many dust storms were reported and by 1000 AD the desert had spread considerably, a process that was accelerated in the recent past.

Nocturnal temperatures, with clear skies and over vegetation-covered areas, are lower than they are over compact, bare desert soil covered by a pall of dust. With the high dew-point of the Rajputana desert, heavy dew would form on the vegetation, helping the plant cover to grow and hold down the soil. As a result there would be less dust in the air, less subsidence and more frequent showers. Hence, the vegetative cover maintained a favourable climate, which in turn stimulated plant growth. As a result of overgrazing, the vegetative cover was destroyed, and the earth which was no longer protected was exposed to wind erosion. Because of the effect of a high dust content of the air on radiative cooling, the air over the desert subsides more than it would do otherwise, and as a result there are less showers. In brief, the desertification process described is basically a man-made phenomenon.

In northern Uganda, at the top of the headwater catchments of the Nile, the savanna country receives an annual rainfall of 500–750 mm, adequate to maintain an open woodland of well-developed trees with a rich flora of grasses. This productive grassland has been reduced to thornscrub with desert grasses, as a result of persistent overgrazing. Run-off from the bare and compacted soil is so great that roads and bridges are persistently damaged by torrent flows, whilst water in the boreholes becomes increasingly scarce. Efforts to mitigate the situation by creating additional watering points through building earth dams and drilling boreholes resulted in a rapid increase in the cattle population around the new water sources and an extension of the area of destruction (Pereira, 1973).

Overpopulation in the dry regions leads to overstocking in areas that are inherently fragile ecosystems. Combined with a lack of range management, the result is the destruction of plant cover, large-scale erosion, and finally—in many cases—desertification. The process may be accelerated by the incidence of a series of more than usually dry years, such as recur periodically in the dry regions.

Desertification can be halted only by adjusting stock numbers to the carrying capacity of the land. This is basically a political and social problem, requiring a policy of land management and its effective implementation. The framing of the policy is dependent on a knowledge of the potential productivity of the land. Certain technical measures can be taken to improve carrying capacity, such as the planting of drought-tolerant shrubs and trees, the creation of forage reserves from seasonal surplus vegetation or from small irrigated areas, control of bush fires, improved grazing management, etc. However, in the medium-and long-term perspective, these measures will probably be ineffective if the population explosion is not brought under control (Houérou, 1976).

Burning and Its Effects

Burning of bush and grass in savannas and steppes either occurs spontaneously by lightning or periodically by man, for agricultural purposes or to facilitate hunting. Fire has always been the most simple and easy way to get rid of excess vegetation. The cultivator uses fire to clear the land for crop production and to provide plant nutrients. For the pastoralist fire is a means of removing impalatable vegetation, preventing brush encroachment over the pasture and reducing parasites and the vectors of disease. Hunters have used fires to facilitate their movement around the country, and sometimes to drive their quarry (Worthington, 1972).

Burning can be damaging: it destroys the surface cover of vegetation and of organic matter, baring the soil surface to the effects of rain, wind, trampling by stock and game and thereby reducing its resistance to erosion and breakdown in structure. Some useful trees such as oil palms are damaged and a gradual build-up of fire-tolerant, low-productive species occurs (Lagemann, 1977).

Burning has been a major factor in the change from forest to savanna. To maintain the productivity of the savanna for grazing, periodic burning is practically essential.

Burning in shifting cultivation has not been harmful as long as the fallow period has been sufficiently long to allow the original forest vegetation to recover. With a shortening of the fallow beyond a critical limit, fire has been a contributing factor to the changeover from forest to brush, and from brush to grasses.

In brief, when properly managed, the advantages derived from burning may outweigh its drawbacks. Otherwise it may be a major contributor to the degradation of vegetation and to erosion.

The sudden removal by fire of the protective cover of vegetation in general and of forests in particular, may have considerable effects on the watersheds by increasing erosion and water run-off.

After a wild-fire destroyed the chaparral scrub in the Arizona foothills, run-off increased from 4% to 40% of a seasonal rainfall of 600 mm, and soil losses increased 1000–3000% (Glendening *et al.*, 1961).

Whilst overland run-off and stream flow increase markedly, infiltration is reduced and becomes insufficient to re-charge the aquifer. As a result, the flow from springs is reduced or ceases entirely, and the water-level of the aquifer is lowered. Shallow wells may dry up completely (Pereira, 1973).

Population Pressure and Shifting Cultivation

In parts of the tropical regions in Africa and Latin America, high population densities have caused impoverishment of the land. In these regions, shifting cultivation is the most widespread system of land utilization. It has proven to be a relatively safe method for conserving soil, vegetation and fauna, whilst providing subsistence to the population, as long as the population density remained below a critical figure.

According to Wilde & McLoughlin (1967), an average population of 20 per km^2 may be compatible with maintaining yield levels on typical lateritic soils by using the traditional bush-fallow in the forest zone. On fertile soils and with an ample and well-distributed rainfall, the critical density may be around 40–50 per km^2. On the other hand, where rainfall is scant and soils are poor, the critical density will be around 10–15 per km^2.

After mature forest growth has been cleared by burning, the soil is usually loose and weed-free, so that it is easily cultivated by hoeing. When, as a result of increasing population density, intervals between successive cultivations become so short that 'bush-fallow' results, the soil, after clearing, is usually much harder and more difficult to cultivate with hoes; weed infestation is also more serious. Finally, when the fallow is very short, the land becomes infested with perennial weeds, which are extremely difficult to control by hoeing.

With the reduction of the fallow period, fertility also declines, gradually at first, but the process soon becomes a vicious spiral. As yields fall, more land must be cultivated, the fallow period is further reduced and the rate of degeneration increases, so that the soil is no longer capable of producing even the low yields considered as a bare minimum. The system of shifting cultivation therefore becomes self-destroying once the 'critical population density' has been exceeded.

When this stage is reached, farmers are forced to change from shifting to sedentary farming, first on a portion of the land nearest to their homes and then in gradually expanding circles.

The transition to sedentary farming can only be self-sustaining if measures are adopted to replace the nutrient-restoring and soil protection effects provided by the plant cover in traditional shifting cultivation.

On steeper slopes, however, the process of soil deterioration caused by the breakdown of shifting cultivation as a result of excessive population pressure is practically irreversible.

Large tracts of former tropical forest in Central Africa and along the Amazon in Latin America have been completely denuded of vegetation and the soil cover destroyed by erosion. These are the inevitable results of laissez-faire; whilst pressure builds up inexorably, lands are abandoned by the population that can no longer be sustained. If similar areas are not to be irretrievably lost in the future, regulatory measures, such as licensing of land-use, must be taken before the situation is allowed to become critical.

Another irreversible process following the removal of the forest cover is *laterization*, a specific problem of the tropical areas. The so-called laterites (or plinthites), when exposed to the alternate pounding by heavy rainfall and drying under a tropical sun, become hard and compacted and are lost to cultivation. However, the extent of the areas prone to damage by laterization has been greatly exaggerated (Sanchez & Buol, 1975).

Forest Resources and Deforestation

FAO (1978) distinguishes between two types of forest: *closed forest*, where tree crowns cover more than 20% of the ground, and which have a more or less typical forest environment, and *open forest*, which represents all other areas carrying some type of woody vegetation, but which do not have a true forest environment.

Besides their direct economic contribution—supply of wood, game, other types of food—forests have a number of additional crucial roles: they protect the soil against degradation and erosion, restore soil fertility in shifting cultivation, ensure a continuous flow of clean water, reduce the danger of flooding, and provide protection against dessicating winds or excessive temperatures.

Most forest areas of the developing countries are situated within the intertropical region. Moving from the Equator pole-wards, there are four parallel belts of tropical forest:

(a) the moist evergreen forest belt;
(b) the moist deciduous forest belt;
(c) the transition belt between permanently humid and dry; and
(d) the dry forest belt, made up mostly of savannas and open forest.

The dry tropical forest belts end to the north in the Mexican, the Sahara and the Asian deserts, and to the south in the Chilean, South African and Australian deserts (FAO, 1978).

There has been a large increase in recent years in destruction of forest cover resulting in soil erosion, flood damage, and silting.

In the Himalayas, population pressure has caused the clearing of forests even on very steep slopes, up to 2000 m. Huge landslides occur during the rainy season, sometimes over two kilometres wide. The problems in the Andean and African highlands are very similar.

The rate of destruction of the humid tropical forest has been estimated to be as high as 1.5–2% per year, which would signify the complete disappearance of all humid tropical forest within 50 years (FAO, 1978).

Reliable estimates, which may be considered fairly representative, are available for specific cases of forest clearing for agricultural purposes.

In the Azuero peninsula in Panama, out of a total mountainous forest area of 215,000 ha, 42% were cleared within 18 years. In northern Thailand, 58% of an area of nearly 40,000 km^2 of forests were cleared in 56 years (followed by increased floods in the watershed area). In the Ivory Coast, only one-third of the 15 million hectares which existed at the beginning of the centry is left. In Mindanao (the Philippines), one million hectares of forest were cleared between 1960 and 1971 (FAO, 1978).

In the humid tropical belt and the transition belt, the main factor involved in the destruction of the forest is clearing for agricultural purposes, mainly by shifting cultivation.

Almost two-thirds of the land under shifting cultivation is in upland forests, where serious erosion follows the baring of the land. The process can be irreversible if the bedrock is exposed.

The adverse effects of deforestation in the highlands on the lowlands, such as siltation of reservoirs, devastating floods, reduced underground water resources, etc., must also be considered.

In the dry forest belt, most of the destruction is due to excessive and indiscriminate felling combined with destructive grazing and fires. Each of these processes has been discussed in detail separately.

The cutting of forests does not always have to cause irreparable damage. As long as population pressure is not excessive, the complex primary forest may be succeeded by a more simple, but equally useful secondary forest, as a result of shifting cultivation.

If, under increasing population pressure, the forests are replaced by suitable farming systems for productive land-use as discussed in the previous chapter, environmental damage may be kept within acceptable limits. Unfortunately, the destruction of the forest is all too often unplanned and has devastating effects, frequently causing irreversible damage.

Critical areas, on steep slopes or of especially erodable soil, should not be cleared. Forest reserves should also be maintained for the survival of aboriginal communities or the conservation of some ecosystems, including wildlife. Other areas should be conserved for properly managed, commercial exploitation of the natural forest, or for replacing with commercial trees.

However, the problems of preventing the destructive effect of forest removal, are mainly social and political.

Depletion of Wildlife and Genetic Resources

The potential economic importance of wildlife for the developing countries will be discussed in Chapter VIII, but it is rarely recognized as a significant renewable natural resource (FAO, 1978).

In Africa, which is probably the richest of the continents in fauna, most of the wildlife is found in the savannas and the rainforests. In many countries, in particular in West Africa, game has disappeared, mainly because of uncontrolled hunting. The situation is better in parts of East Africa, but poaching is causing increasing depredations. In the Near East, most wildlife has almost completely disappeared, also because of indiscriminate hunting.

The conservation of wildlife resources involves legislation with strict supervision, controlled harvesting and culling to prevent overgrazing, and captive breeding to restore wild populations where these have become extinct.

An ambitious project has been undertaken in the Southern Negev of Israel, to restore the formerly indigenous fauna mentioned in the Bible, by re-introduction and captive breeding of locally extinct species and promoting the comeback of those species of which limited numbers have still remained.

Modern crop production is based on high yielding varieties of uniform genotype which frequently completely displace the traditional 'landrassen' characterized by their wide genetic diversity. Long-term breeding programmes may require genes present in the traditional varieties for presently unforeseen needs.

A somewhat similar situation exists with respect to wild relatives of crop plants, which can provide a source of genes for resistance to diseases, pests and adverse conditions, as well as characteristics of possible economic importance. These sources of genetic variability are less endangered than the traditional cultivated varieties, but their fate depends on that of the ecosystems of which they form part. Where such ecosystems are destroyed by overgrazing, fires or destruction of forests, valuable genetic material may be irretrievably lost.

Awareness is growing of the need to conserve genetic resources, and the Food and Agriculture Organization of the United Nations has been particularly active in stimulating interest in this aspect of conservation of natural resources and in organizing effective international and national action. In 1974, following a recommendation of the 1972 United Nations Conference on the Human Environment, the International Board for Plant Genetic Resources (IBPGR) was established. Action is being taken to establish world networks dealing with the collection, conservation and use of crop germ plasm. Institutions have been designated to be responsible for maintaining the world's major collection of seeds of the principal food crops. Priorities have been established for the collection of crops, and for work in those areas showing significant genetic diversity. National programmes will be needed to participate in these international efforts and to carry through any implications for particular countrys (FAO, 1978).

Large-scale Surface-water Storage

River basin development projects are of major importance for economic and social development and are frequently an essential part of development programmes as sources of water for irrigation and of electric power for industrialization. However, many environmental problems may arise as a result of the implementation of these projects, such as the spread of water-borne diseases, the drying up of downstream fisheries, the inundation of valuable agricultural and forest land and the displacement of population. Some of these adverse effects emerge gradually, others are more rapidly evident. Some, after they have occurred, are practically irreversible. Therefore, planning must take into account the possible negative effects of these schemes, and attempt to avoid or mitigate them. Evidence is increasing that many of the spectacular works undertaken in developing regions to store river water on a large scale were ill-conceived. They are also always expensive and extremely wasteful of water.

Losses of Water by Evaporation

Evaporation is very high in the arid regions which are to be served and in which the reservoirs are frequently located, and can amount to as much as 125 cm annually (Stamp, 1961). As much as three-quarters of the water stored may be lost by evaporation (Dixey, 1966), and the remaining water becomes increasingly saline. As a result of the extremely high rate of evaporation, the total yield of a watershed in an arid region is actually decreased when large reservoirs are built. There is even a point at which the building of additional reservoirs causes a net deficit in usuable water, when aggregate evaporation losses exceed the amounts of usable water gained (Burgy *et al.*, 1967).

No techniques have yet been developed for reducing evaporation losses from very large reservoirs. For the enlarged Aswan reservoir, the loss by evaporation has been estimated as equivalent to 10% of the annual flow of the Nile (Addison, 1961).

Impairment of Water Quality

The salinity of the water flowing into the reservoir may vary considerably in the arid regions, depending on rainfall distribution, the salinity of the watershed and the changing rates of evaporation.

All water evaporating from reservoirs and lakes must leave a residue of salts which increases the salt content of the stored waters. When the rate of evaporation exceeds the rate of water inflow, the situation may deteriorate at a fairly fast rate, the loss of water being compounded by its increasing salinity.

The salt content of the water in the reservoir will be at its lowest after the main flood flow during the rainy season and will reach its highest concentration towards the end of the dry season. The higher the surface: volume ratio, the greater the rate of evaporation and, as a result, the higher the salt concentrations.

Useful Life-span of Reservoirs

'Reservoirs are mortal' (Addison, 1961) and once filled up with sediment the site is irretrievably lost for further use as a reservoir. The life-span of reservoirs has frequently proven to be far shorter than planned, especially when the necessary measures to protect the catchment area are not taken or are inadequate.

The cutting of trees and destruction of natural vegetation on mountain slopes increases the rate of erosion and results in the siltation of rivers, reservoirs and irrigation canals.

The incidence of floods in India, Pakistan, Thailand, the Philippines, Indonesia, and Malaya has increased considerably in recent years because of the rise of river beds due to silting with soil washed down from the denuded slopes of the watersheds. In Colombia, billions of dollars will be lost in hydroelectric benefits over the next several decades because of accelerated sedimentation of reservoirs due to spreading deforestation.

The life expectancy of the $600 million Mangla reservoir in Pakistan, originally estimated as 100 years, has been reduced by half because of an increase of the rate of sedimentation resulting from clearing the steep slopes of the watershed for farming and from overgrazing (Brown, 1977).

The Ksob dam in Algeria became first partially then completely unusable within the space of 10 years, well before the additional harvests had been able to pay off the very high construction costs (Dumont, 1966).

Experience in South Africa has shown that reservoirs have filled up with sediment within a dozen years. In the Levant and in Sinai, relatively small reservoirs have filled up in a still shorter period. It is true that certain reservoirs have a longer life expectancy: in the USA, the great reservoirs are expected to remain effective for one or two centuries. However, even in the USA there are examples of reservoirs that have silted up completely within a few years. New Lake Austin on the Colorado River in Texas lost 95.6% of its capacity in 13 years, and the Grand Reservoir on Toulumeme River in California lost 83% in 36 years (Kassas, 1972).

Invasion by Water-weeds

Invasion of water reservoirs by water-weeds may also serverely impair their effectiveness.

The water hyacinth (*Eichhornia crassipes*), originally introduced into Africa for its decorative value, has become a serious pest of waterways and reservoirs because of its remarkable powers of spreading. Thousands of miles of the Congo River are infested; it is also by now well established on the Nile. Enormous expenditure has as yet failed to give satisfactory control (Ivens, 1967).

Another weed that invades reservoirs in Africa is *Salvinia auriculata*, a free-floating, branched fern, introduced from South America. Under favourable conditions the plants can cover large areas with a thick layer of vegetation. First observed in East Africa in 1957, it now covers hundreds of square miles with a mat up to 25 cm thick; in particular in Lake Kariba, the Kitali dam in Kenya, and in Lake Victoria (Ivens, 1967).

The major effects of weed-invasion of reservoirs have been summarized by Little (1969): they form large mats which may block hydroelectric installations, feeder streams and irrigation outlets; they make fishing more difficult and by inducing deoxygenation increase fish mortality; they reduce the effective capacity of the reservoirs; they provide excellent breeding-grounds for many disease-transmitting vectors. The association of aquatic weeds with the spread of certain vectors of disease in man and animals is one of their major negative effects. *Bulinas* and *Biomphalaria* snails, which are intermediate hosts for schistosomiasis, flourish under the protection of aquatic weeds. The development of vegetation like *Pistia* and *Polygonum* favours the genus *Mansonia*; weed-infested waters provide a suitable breeding place for culex (White, 1977). Polunin (1972) adds that there is a possible loss of photosynthetic activity of phytoplankton and benthic Algae at lower levels, due to interference with the light supply and a concomitant loss of biological productivity in the body of water.

Environmental Effects

After high-level dam construction, infiltration of subsoil water streams may extend for ten or even hundreds of kilometres both up- and downstream of man-made lakes, causing a rise in the level of ground-water. If the rise is stabilized beyond a critical level of 1.5–2 m, alkanization and salinization may result. Extensive erosion and landsliding may also occur along the banks of reservoirs, causing the loss of agricultural land (Kovda, 1977).

Rivers normally carry considerable amounts of sediments, of which part is deposited on the irrigated land, contributing to its fertility, and part is deposited in the deltas and counteracts wave erosion from the sea. After construction of the Aswan dam, the reduced sediment rate no longer fulfils these roles effectively and the coastline is suffering heavy damage.

Fishing in the formerly highly-productive estuaries and coastal waters is seriously affected, and—in the case of the Nile—has dropped from 18,000 tonnes in 1965 to 500 tonnes in 1968 (Dorst, 1972).

Health Hazards

A tragic and costly side-effect of the construction of large dams in many developing countries is the steep increase in the incidence of schistosomiasis, a debilitating intestinal and urinary disease caused by the larvae of a blood fluke. The disease is estimated to afflict 250 million people, and surpasses malaria as the world's most prevalent infectious disease (Brown, 1977).

All these adverse side-effects must be foreseen and their economic implications evaluated at the time of planning and designing projects.

Adverse Effects of Irrigation

Irrigating arid soils is *not* the equivalent of creating a humid environment; irrigation has specific, far-reaching effects on soil fertility and crops which have no counterpart in the conditions encountered in rain-fed farming.

Also, subhumid and arid ecosystems are inherently unstable—small changes in their management can cause large changes in the ecosystem (White, 1977).

Irrigation makes possible considerable increases in crop production, but it affects soil fertility in a number of ways which may seriously impair its productivity on a sustained basis.

Irrigation water, unlike rainwater, contains quite considerable amounts of salts in solution. Certain of these salts, such as potassium sulphate or nitrates, can be of direct benefit to plants; others, such as calcium sulphate, can contribute to the improvement of soil structure. On the other hand, salts such as sodium chloride, or compounds containing boron, may have detrimetnal effects on the soil or the plant.

Even high-quality water contains appreciable amounts of undesirable salts. For example, water with a chlorine content of 200 ppm, which is considered a low safe level, can add anually, under normal irrigation practice, up to 3 tonnes per hectare of NaCl (this does not even take into account other salts that are present in the water and inevitably get added to the soil).

It will be easily realized that constant additions of this magnitude can cause considerable changes in the soil's characteristics and fertility. Kovda (1977) estimates that not less than 20–25 million hectares of land have deteriorated into saline, infertile soils as a result of faulty irrigation.

The negative effects of irrigation on soil fertility are not confined to the addition of harmful salts, especially when irrigation methods are primitive or faulty. The development of land for irrigation farming frequently makes large-scale levelling operations necessary, during which the subsoil may be exposed over fairly large areas. The subsoil is usually of very low productivity, because of poor physical condition and marked deficiencies in plant nutrients. For example, it has long been known that the semi-arid loess subsoils in Nebraska produced very poor crops of cereals, and that only legumes—provided they are supplied with phosphorus and potassium—could grow well under these conditions.

Whilst it is possible to build up the fertility of subsoils gradually by growing a succession of legumes, this is usually a lengthy process: 20 years after a subsoil was exposed in Ohio, its colour was still noticeably lighter than that of the topsoil that had been treated in the same way. Also, under conditions of drought, the crops growing on the subsoil were affected much more quickly than those growing on topsoil (Bachtell *et al.*, 1956). The cumulative loss of income after 18 years of cropping due to the poor yields of the crops grown on the exposed subsoil, as compared to those grown on the soil in its original condition, was equivalent to the value of the land.

Changes in soil texture may result from the addition of silt carried by irrigation water and from the downward movement of clay in the soil profile. In one area in southern Victoria, Australia, it was found that about 14% of the clay content of the surface loam had been washed downwards by irrigation during a period of nine years (Penman, 1940). If the water is saline, sandy loams may become progressively heavier and less permeable, as occurred in the Salt River Valley of Arizona after 26 years of irrigation (Pratt, 1959).

One of the major dangers resulting from faulty irrigation management and inadequate drainage is a rise in the level of the ground-water-table, which may reach several metres a year. In many cases, the water-table has risen in the course of ten years from a depth of 25–30 m to within 1–2 m of the soil surface. When the water-table comes to within 2 m of the soil surface, the volume of soil available to the root system becomes restricted, salts rise to the surface by capillary and the problem becomes acute. In extreme cases, the land may become wholly unfit for further irrigation farming.

Waterlogging may also become serious, where river irrigation is practised, as a result of the percolation of water from a dense network of rivers and canals; for example, transit losses due to seepage from the canals amounted to 36% of the water entering the canals in Utter Pradesh and Punjab in India, and 43% in the case of the Karz-Kum canal in USSR (White, 1977), resulting not only in serious losses of water, but also rendering large areas unfit for cultivation because of waterlogging.

Whilst there are no reliable estimates of the damage caused by salinity

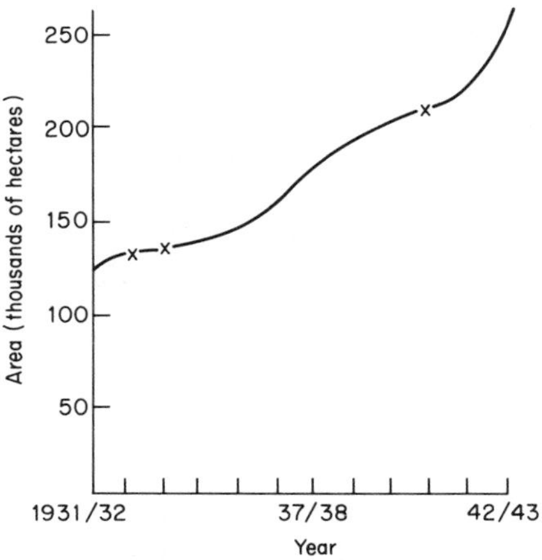

Fig. 20 Areas damaged by salt accumulation and waterlogging in prepartition Punjab (after Bharadwaj, 1961, by courtesy of Unesco).

and rising water tables in all developing countries, a few examples will illustrate the enormous potential for damage resulting from these causes.

The area of salt-affected and waterlogged soils amounts to 50% of the irrigated area in Iraq, 23% in Pakistan, 80% in Punjab (Pakistan), 50% of the Euphrates Valley in Syria, 30% in Egypt and over 15% in Iran (El Gabaly, 1977).

In the Punjab in a single year (1942–43), more than 280,000 ha of land were rendered useless as a result of salt accumulation and waterlogging. About one-third of the irrigated area in Peru and at least half of the irrigated land in the north-east of Brazil are affected by salinity and waterlogging. In the latter area, the yields in some of the affected areas are lower than before they were irrigated! (Eckholm, 1976) (Fig. 20).

Small wonder, then, that many of the civilizations that were based on irrigation ended in disaster, and that large-scale failures are not exceptional even in the twentieth century.

Excessive Exploitation of Ground-water Resources

There are two basic approaches to ground-water utilization: (a) the *concept of 'safe yields'* which is based on the rule that annual ground-water withdrawals should not cause undesirable effects, such as sea-water intrusion, land subsidence, increased pumping requirements, etc., and (b) the *concept of 'mining' ground-water*: in this case, withdrawal exceeds annual re-charge, sometimes considerably.

Planned overdraft of ground-water may be justified in the initial stages of development in order to sustain agricultural development, during the period of construction of dams, reservoirs, canals, etc.

In some cases aquifers are so great, and have such huge reserves, that a very long period of time is neccesary before any assessment can be made of what constitutes a safe rate of withdrawal. Such a situation occurs in deserts where waters have accumulated underground, during a more humid period of the recent geological past; these waters are no longer replenished and are therefore called 'fossil' water. Typical examples are the ground-water reservoirs of the Libyan and Arabian deserts. Withdrawing water from such reservoirs is indeed a 'mining' operation.

A justification for 'mining' water is when it is first used in order to develop and strengthen the economy, so that at a later stage it will be possible to bring in water from other regions or to develop additional water resources in the region. This really constitutes a 'planned overdraft'. However, excessive pumping will sooner or later cause a deterioration—qualitative, quantitative or both—of ground-water resources.

Land Subsidence

Excessive pumping may lead to a subsidence of the land surface, resulting in compaction and the reduction of pore space in the aquifer. The loss of underground pore space is a permanent reduction of the underground storage capacity and is therefore an irreplaceable loss.

Peat lands under irrigated cropping are particularly subject to subsidence as a result of the oxidation of the peat. The surface of the peat lands in the Sacramento–San Joaquin delta of California has sunk by approximately two metres between 1962 and 1966 (Penman, *et al.*, 1967).

Salt-water Invasion of Aquifers

When a coastal aquifer is open to the sea, sea-water forms a wedge underlying the fresh-water flowing seaward.

Excessive ground-water withdrawal will cause a lowering of the water-table. When the level of the water-table becomes lower than that of the saline water, the latter is drawn into the aquifer. This is a frequent occurence in coastal areas. It is an accepted rule-of-thumb that decreasing the water-level one metre by pumping will lead to a rise in the level of the zone of junction between salt- and fresh-water of *40* metres. Hence, overutilization of the aquifer in coastal regions, which leads to a lowering of the water-table, can have a disastrous effect by salinizing the aquifers. Drawing in salt-water is even more detrimental than simply depleting an aquifer. An emptied aquifer may eventually be re-charged, but in a polluted aquifer withdrawal of water has to cease even when plenty of water is still available. Reclamation is also more costly and difficult.

The dangers inherent in the uncontrolled exploitation of ground-water resources are exemplified by the experience in the Hermosillo coast of Sonora de Mexico, as described by Alcantara (1976). During the 1950s, the easiest way to obtain more profits from wheat and cotton was to drill new wells, and after this was prohibited in 1954, to extract ever-increasing amounts of water from existing wells to extend the holdings further into the desert. Farmers did not, as a rule, level their fields or use concrete canals. They simply emptied large streams of water into long unlined canals, which lost 20–25% of the water by percolation, and used the water inefficiently on land that had not been properly prepared for irrigation. By 1956, the volume of water extracted on the Hermosillo coast was double the amount which could be renewed by natural recuperation and progressive salinization of the land had begun. By 1963, an alarming drop in the water-table forced government to undertake steps to curtail pumping. In view of the near-total exhaustion of the water supply, the federal government has been petitioned to construct a multi-billion peso hydraulic system to bring water several hundred kilometres overland from Sinaloa.

Effects of Chemicals used in Agriculture

Pesticides

The great increase in the use of pesticides has become a major threat to the environment. The most serious and widespread problems are caused by insecticides; fungicides, nematocides, herbicides and rodenticides appear to induce less serious problems (Wurster, 1972).

Insecticides can be divided into very *stable compounds*, mainly the organochlorides (DDT, Aldrin, Dieldrin, Isodrin, Endrin, Chlordane, Telodrin, Heptachlor, Strobane, Toxaphene, Mirex, etc.), and *non-persistent* insecticides, mainly organophosphates (Parathion, Systox, TEPP, Malathion, Chlortion, Dibrom, Ronnel, Dipterex) and carbamate compounds, which rapidly break down into non-toxic products.

The effect of the stable compounds may extend for great distances beyond the sites of their use, and may persist for many years, accumulating in enormous amounts in the biosphere. It has been calculated that 1,500,000 tons of DDT are circulating in the biosphere (Dorst, 1972). Because of their low solubility in water and high solubility in lipids, the organochloride compounds are not 'lost' by dilution in water, soil or air but, on the contrary, become increasingly concentrated in the tissues of living organism with each step in the food chain. This biological concentration causes levels of organochlorides in organisms that can be millions of times higher than those found in the surrounding inorganic environment (Wurster, 1972). Because of their broad spectrum of biological activity their potential danger to the environment is considerable.

Certain major effects of contamination by organochlorine compounds, such as extensive mortality among birds, toxicity to fish, reduced photosynthetic activity in certain species of phytoplankton, have been conclusively demonstrated. In insects, their main effect has been to induce resistance in harmful insects and destruction of the numerous predators and parasites of the former. Paradoxically, whilst the immediate effects on harmful insects are dramatically successful, in the long run they destroy the biological balance between phytophagous insects and their natural enemies, so that ultimately they are actually benefiicial to the harmful insects (Wurster, 1972).

Other effects, such as inducing cancer and mutaganesis in humans, have not been conclusively proven, but because of the hazards involved, the use of DDT and other organochlorine compounds has been banned in many developed countries. They are mainly replaced by organophosphorous and carbamate compounds.

Whilst these have the advantage of being quickly degraded and do not accumulate in the food chain, some of them are extremely toxic to humans, and all have the same limitations as the stable insecticides in that they disrupt the biological balance between phytophagous insects and their natural enemies, to the ultimate benefit of the former.

The implications are obvious: whilst it is impossible to ban the use of insecticides, more attention must be given to avoiding, or at least mitigating, their destructive effect on the environment. The first step is to replace the stable insecticides by the non-persistent types, and to reduce their use to the bare minimum by combining chemical control with cultural and biological control methods in a system of integrated control (see Chapter VIII).

However, for the developing countries, these apparently obvious solutions present intractable problems. Firstly, the stable insecticides play a major role in the control of important disease vectors; secondly the non-persistent insecticides are much more expensive to buy, so that only a minority of farmers at best can afford their use. Thirdly, the high toxicity to man and animal of many of the most common organophosphates makes them extremely dangerous to use. This danger and the requirement to adopt integrated control methods, require a level of sophistication that the majority of farmers have not yet attained. Finally, much location-specific research in developed countries on integrated control needs to be carried out before effective methods can be proposed for adoption.

In summary: insecticides have been used with long-term negative effects on the environment and have frequently increased the difficulties involved in control by creating resistant strains and disrupting the delicate balance between the insect and its enemies. However, there is at present no practical alternative to the use of insecticides in the control of vectors for eradicating certain major diseases. The most effective insecticide for this purpose, because of its very persistence, is still DDT. Other solutions proposed for developed countries, such as the use of non-persistent insecticides and integrated insect control, are still difficult to adopt in developing countries, because of high cost, lack of knowledge of appropriate methods, and need for training of the majority of farmers.

With few exceptions most herbicides are relatively non-toxic to mammals, but little is known of their long-term effect on the environment.

Persistence in the Soil

An important characteristic of several commonly used chemicals for pest, disease, and weed control is their persistence in the soil. Soil fertility depends to a considerable extent on the maintenance of the equilibrium between the various components of the complex populations of microorganisms in the soil. The disturbance of this precarious equilibrium by the residual effect of chemicals applied to the soil may affect soil fertility adversely (Audus, 1964).

There are vast differences in the persistence of chemical pesticides in the soil: some are completely broken down within a few days after application, while others may persist in toxic concentrations for years. Persistence of chemical residues in the soil depends on many factors: the nature of the pesticide, soil type, soil moisture and temperature, rate of application, the formulation used, and the activity of microorganisms (Lichtenstein, 1966). The stable chlorinated hydrocarbons may persist in the soil for years; the organic phosphorus compounds are usually decomposed rapidly, but occasionally they persist from one season to the next. Compounds incorporating copper, lead, and arsenic, persist for long periods (Weisner, 1963). DDT, benzine hexachloride, and chlordene amongst others, are toxic to the microorganisms that are active in nitrification, and they can accumulate in the soil. However, no drastic effects are usually experienced with these pesticides if care is taken not to exceed the normal rates required for pest control, thereby avoiding a build-up of residues in the soil, such as may occur through excessive or imprudent use. Herbicides break down more easily than do insecticides. However, they can alter the ecology of aquatic ecosystems sufficiently to destroy populations of animals associated with plants (White, 1977).

2,4-D, the ubiquitous weed-killer, decomposes rapidly in the soil, and has been shown to have a favourable effect on soil microorganisms. Amitrole, Endothal, and Simazine, do not appear to have an appreciable effect on soil microorganisms at normal application rates. Certain chlorinated organic compounds can be metabolized only by specific genera of soil organisms, e.g. Simazine is decomposed only by fungi and actinomycetes. The rate of biodegradation increases in soils that are rich in organic matter (Audus, 1964).

From the foregoing it is clear that slow, persistent buildup of toxic materials, resulting from the widespread use of agricultural chemicals, can have far-reaching effects on soil fertility. These will depend upon the physical, chemical and biological environment within the soil (Webber & Elrick, 1966). The processes will be different under arid and semi-arid conditions—with and without irrigation—as compared with those of humid regions. Knowledge of these effects is still very limited, however.

Fertilizers

Over a century ago, J. Liebig propounded his theory that the only role performed by organic manures was to provide plant nutrients. A school of thought has arisen, inspired by Howard (1940), Balfour (1945) and many others, that ascribes most of the ills of modern agriculture to the use of chemicals in general, and to chemical fertilizers in particular.

The detrimental effect on soil fertility that has been ascribed to chemical fertilizers is supposedly due to the poisoning of the soil micro- and macro-organisms, thereby causing an increase in the incidence of diseases and pests of plants, and indirectly affecting the health of the farm animals and humans consuming the products grown with chemicals.

The addition of fertilizers to the soil has both beneficial and detrimental effects: on one hand it promotes plant growth by providing nutrients, but on the other it hinders plant growth by increasing the osmotic pressure of the soil solution and hence increasing the difficulties which the plant encounters in supplying its needs for water and nutrients. In arid regions, for dry-land crops growing under conditions of high soil-moisture tension, this increase in osmotic pressure may be decisive under certain circumstances and more than outweigh the advantage of the additional nutrient supply.

When the long-term effects of various types of fertilizers on the physical and chemical properties of an irrigated soil are investigated, far-reaching differences may be observed. Aldrich and Martin (1954), when comparing different nitrogen carriers, found marked physical and chemical changes in the soil after 16 years of differential treatment. In particular, the rate of water percolation was apt to be greatly affected. The use of sodium nitrate and ammonium sulphate brought about structural breakdown resulting in reduced macropore space. In the sodium nitrate plots, the poor physical conditions appeared to be due to an unfavourable calcium–sodium ratio. In the ammonium sulphate plots, the cause of the structural breakdown was apparently the dispersing action of the ammonium ion. The low pH, caused by the continuous use of ammonium sulphate and due to the accumulation of hydrogen in the exchangeable complex, inhibits the ability of soil organisms to nitrify the ammonium. No such unfavourable effects were found to result from the use of urea or calcium nitrate.

Ammonium sulphate has long been the main nitrogen carrier used in the humid tropical regions of the world. Among the reasons for its popularity are its excellent storage properties and the beneficial effect on many tropical soils of the residual sulphate. However, it is now generally accepted that the use of ammonium sulphate should be curtailed and replaced by other nitrogen carriers on inherently acid soils of low cation exchange capacity in the tropics, unless adequate liming is resorted to (Olson, 1972).

Where continuous use of large rates of application of certain types of fertilizer is apt to cause difficulties, the answer is not to reduce or abandon fertilizer application, but to change the type of fertilizer used or to apply appropriate soil amendments.

Salinity and alkalinity are major problems of crop production under irrigation, and the opinion is frequently voiced that heavy fertilization increases these problems. When it is realized, however, that the quantity of salts added to the soil in irrigation water, even of satisfactory quality, easily amounts to 5 tonne/ha/year, it is clear that the effect of fertilizers on salinity, under normal irrigation practice, is usually not of overriding importance, unless excessive quantities of unsuitable fertilizers are applied and provided periods of severe moisture stress are avoided.

Unbalanced fertilization may also cause difficulties: excess fertilization with one element may hinder the uptake of other essential nutrients.

It therefore follows that it is not the use of fertilizers *per se* that may have an undesirable effect on soil fertility, but their incorrect use. By choosing the appropriate carrier, avoiding excessive application or an unbalanced supply, and using proper application methods, it should be possible to avoid or mitigate the negative effects mentioned above.

Insofar as organic matter is essential to soil fertility, fertilizers can play a considerable part in increasing the amounts of organic residues to be incorporated in the soil. Green manures, grown with fertilizers, are sometimes the only economical source of organic matter that is available.

Finally, the addition of fertilizers is frequently indispensable, when ploughing under bulky organic manures or crop residues, in order to avoid the setback to the cultivated crop which would otherwise result from the fixation of plant nutrients by soil microorganisms.

Fertilizer Use and Environment

Pollution from fertilizers can be due to: (a) the production process; and (b) actual use in agriculture (Nelson, 1975).

The mining of phosphates, for example, produces waste slimes that are stored in ponds, which present hazards from dike breakage and pond overflow. During the manufacture of fertilizers various compounds are released as gases, fumes and dusts. The technology already exists for reducing these sources of pollution.

The heavy use of fertilizers in modern agriculture, which has increased fivefold in a quarter-century (Harre *et al.*, 1971), and the indisputable fact that only part of the applied plant nutrients can be accounted for in the crops harvested, has inevitably given rise to apprehension as to the possible detrimental effects of fertilizer residues on the environment, in particular on the quality of water resources, with potential hazards to human health.

Potential Contribution of Fertilizers to Pollution

The three main plant nutrients, applied at high rates in intensive agriculture, and therefore potentially dangerous pollutants, are: phosphorus, potassium and nitrogen.

The following figures give an indication of the amounts of plant nutrients that are lost by leaching in Senegal (Charreau, 1972): nitrogen (N), 5 to 30 kg/ha; phosphorus (P_2O_5), 0.1 to 0.5 kg/ha; potassium (K_2O), 10 to 20 kg/ha; calcium (CaO), 40 to 150 kg/ha; magnesium (MgO), 15 to 40 kg/ha; and sulphur (S), 3 to 40 kg/ha.

The amount of fertilizer leached is a function of the concentration of the mineral element in the drainage water (C) and the volume of drainage water (V). The addition of fertilizers increases C slightly, but because of improved vegetative cover an increased evapotranspiration, V is drastically reduced. This explains the apparent paradox that fertilization can actually reduce losses by leaching (Charreau, 1974)!

Phosphorus

In most agricultural soils, with the exception of sands and peats, the phosphorus applied in water-soluble fertilizers is rapidly converted in the soil into forms with very low solubility. As a result, phosphate ions have extremely low mobility and it is highly improbable that significant amounts will be leached to the depth of the ground-water. However, it is a fact that many water resources have been polluted with excessive amounts of phosphorus, but the main contribution is from the use of detergents with high phosphorus contents, and not the fertilizer used in intensive agriculture.

Potassium

Though most potassium fertilizers are soluble in water, in temperate climates, the potassium applied rapidly becomes immobilized in the soil as a result of adsorption by the soil particles. Though somewhat more mobile than phosphorus, it is extremely unlikely that potassium ions will be leached downwards to appreciable depths in most agricultural soils of the temperature climate. In most tropical soils, potassium is more mobile and is subject to leaching.

Nitrogen

Under normal soil conditons favourable to crop production, the nitrogen applied in fertilizers, whether in the form of ammonium, ammonia, urea, or other organic compounds, is rapidly nitrified into highly soluble nitrates, which are not fixed by the soil particles, and can move rapidly through the soil.

Therefore, *a priori*, nitrogenous fertilizers would appear to be the most likely potential source of contamination, especially as the amount of fertilizer nitrogen used in the world has increased at a phenomenal rate—far more than that of the other plant nutrients. The estimated world consumption in 1945 was 30 million tonnes of nitrogen (FAO, 1975) and had risen to 45 million tonnes by 1976 (Peter, 1978).

It is undeniable that considerable loss of nitrogen occurs beyond the root zone from fertilizer nitrogen applied at the high levels required by modern agriculture. For these reasons ecologists are increasingly contending that the high level of fertilizer use is contributing significantly, through run-off and leaching, to the pollution of surface- and underground-water (Commoner, 1970).

Deleterious Effects of Excess Nitrogen

Human Health

A direct relationship has been established between the levels of nitrates in water and human health; hence, legislation in most countries defines the maximum allowable nitrate content of drinking water. However, it is not the possible toxicity of NO_3 that is the only cause for concern; the nitrate content is simply a useful indicator of: (a) the amounts of toxic nitrites that may be introduced in the blood stream through the reduction of nitrates after ingestion; and (b) the degree of fecal contamination of the water leading to enteric infection. The latter is by far the more important consideration and has no relation whatsover to fertilzer application.

Eutrophication

Eutrophication is a process in which the levels of nutrient salts in water are increased. The levels of the nutrients and their proportions in relation to each other determine largely the type of vegetation found in rivers and lakes.

A steep increase in plant nutrients in the water may cause a proliferation of certain algae, upsetting the biological balance of inland waters such as lakes and reservoirs. The rapid multiplication of these algae may prevent or considerably reduce the penetration of sunlight to the subsurface water, preventing the photosynthetic process in algae which normally develop in these lower layers. As a result, the production of oxygen is reduced; and anaerobic conditions are created. The organic remains of the increased mass of organic matter produced by the proliferating algae of the surface waters, which sink towards the bottom, are decomposed by bacteria and further acerbate the anaerobic conditions. Undesirable changes occur in the colour, taste and odour of the water, making it unfit for drinking purposes.

The main nutrients involved in eutrophication are nitrogen and phosphorus. Both have to be present at a certain level for the process described above to occur (Mellanby, 1972). Whilst most of the phosphorus that contributes to eutrophication comes from detergents, the role of fertilizers in supplying the nitrogen involved is more controversial (see below).

Atmospheric Pollution

The possibility has been raised that part of the nitrogen applied to the soil will undergo denitrification producing nitric oxide and nitrogen dioxide which may rise to the upper atmosphere where it may attack the ozone layer.

Reducing the ozone layer, which shields the earth's surface against ultraviolet radiation, could increase the incidence of skin cancer (Crosson & Frederick, 1977).

Actual Contribution of Fertilizers to Pollution

It is extremely difficult to prove or disprove a direct relation between the large amounts of fertilizer nitrogen used in intensive agriculture and the increased pollution of many water resources, whether ground-water, streams or lakes.

It is usually not possible to determine the amount of nitrate that is leached from the soil; in addition, fertilizers applied are not the only possible source of the leached nitrate. The decomposition of organic matter in the soil, and the fixation of atmospheric nitrogen by bacteria or algae, are processes that are highly beneficial to agriculture and which are enhanced by good crop management practices. These can contribute amounts of residual nitrates far greater than those derived from the fertilizers applied.

It is also almost impossible to identify the source of nitrates found in water. Sediments due to soil erosion and sewage effluent, whether treated or raw, the wastes from modern feedlots on which large numbers of livestock are concentrated, food processing wastes, industrial effluents, are all sources of contamination that have increased concurrently with the augmented use of fertilizers in agriculture.

Further, the rate of water re-charge is extremely low. It may take decades before a possible effect of fertilization on nitrate pollution of aquifers is detected. On the other hand, a dangerous situation could be created and it would take decades to reverse the process (Viets & Hageman, 1971).

The few studies that have been made on changes of nitrate content of ground-water and of river-waters draining out of well-farmed land in the course of 10–20 years have given conflicting or inconclusive results.

Rises in nitrate levels in inland waters have been directly related to increases in fertilizer application in the surrounding arable land for the Missouri River and for rivers in Illinois (Commoner, 1970).

During the period 1934–63, during which fertilizer use on the irrigated valley land adjacent to the Upper Rio Grande of New Mexico increased manifold, a substantial decline in the nitrate content of the river water actually occurred (Bower & Wilcox, 1969).

On the other hand, ground-water investigations in California showed that in at least two areas, the principal sources of nitrogen pollution of ground-water were directly related to the increased use of nitrogen fertilizers in irrigation agriculture (Ward, 1970). Bingham *et al.* (1971), in a study in a citrus watershed in California, estimated that the loss of nitrogen in the drainage water amounted to 45% of the applied fertilizer nitrogen.

Conversely, it can be argued that an adequate application of fertilizers, by increasing root proliferation and the amount of organic residues remaining in the soil, will improve soil protection by the crop, and therefore reduce erosion, a major contributor to the pollution of streams, lakes and reservoirs.

Chemical Degradation

Chemical degradation of the soil occurs not only following injudicious use of fertilizers or as a result of sodification or alkalinization. It will also occur when the nutrients exported by crops and lost by leaching are not returned to the soil in adequate amounts. The humid tropics in particular are susceptible to leaching of soil nutrients. 23% of the world's soils suffer from mineral stress; the proportion is as high as 59% in South-East Asia and 47% in South America (FAO, 1978).

Conclusions

In brief, there does not yet appear to be sufficient reliable experimental data to state authoritatively whether fertilizers do or do not contribute to the pollution of water resources (FAO, 1972). Common prudence indicates that all conceivable precautions for minimizing pollution of the environment should be taken. Fortunately, there is no conflict between this requirement and farmers' own direct interests.

Crop management aimed at maximum efficiency in fertilizer use will minimize the amounts of fertilizer residues that might contaminate the environment.

Amongst the management procedures proposed are (Parr, 1972):

(a) applying fertilizers at rates below those required for maximum yields, namely, at the point of greater economic return (cf. pp. 390–391);
(b) proper timing of application to coincide as closely as possible with maximum crop demand;
(c) alternating deep-rooting crops with crops that have received high rates of nitrogen;
(d) Maintaining cover crops wherever possible;
(e) Use of nitrogen carriers with coated granules enabling a controlled release of nitrogen.

In any case, for the developing countries as a whole, the danger of pollution resulting from the use of fertilizers appears, at least for the foreseeable future, to be negligible, because the levels of fertilizer applied are still far below those required for optimum production. For those areas in which the 'green revolution' has resulted in the use of high levels of fertilization, in particular of nitrogen, it is safe to assume that little or no damage will accrue to the human environment if the amounts used are within the limits imposed by economic considerations, and sound managment practices are adopted.

Urban Intrusion

Urbanization is increasing rapidly in the developing countries; the population big cities in the developed world grew about two and a half times from 1920 to 1960, but more than eight times during the same period in the developing countries (Pollock, 1971). Moreover, many cities in the developing countries are increasing far more rapidly than industrial expansion and therefore lack the firm base of industrial growth that would contribute to economic growth sufficiently to compensate for the loss of productive capacity of the land due to urban encroachment.

Hence, the loss of farmland to expanding cities is on the increase in most parts of the world. There are three particular spatial aspects to this diminution: selective encroachment, fragmentation of agricultural land, and the effects of the advancing city in adjacent agricultural land—'urban shadow' (Gregor, 1970).

One reason why urbanization is frequently at the expense of the best agricultural soils is that many urban centres originate as service centres for farming communities.

The fragmentation of farming land due to the activities of real-estate 'developers' can cause a considerable disruption of agricultural activities and accelerate land transfer to urbanization. Manifestations of the 'urban shadow' are, for example, an increase in tax assessments designed to pay the growing urban population nearby, the negative effects on investments in agricultural development as a result of the anticipation of urbanization; increased pollution of the air; increasing competition for scarce water resources, etc.

Even in villages, loss of land for agricultural production can assume serious proportions. In overpopulated villages in Bangladesh, for example, 30–40% of the total area is devoted to homesteads, kitchen gardens, tanks, trees, embankments, footpaths, roads, etc. (Dumont, 1973).

The most commonly adopted solution is 'zoning', whereby the areas to be used for urbanization, industry and agriculture are clearly defined. Rarely do local authorities withstand the economic and political pressures brought to bear by real-estate developers.

Possibly the only radical solution to this problem of urban encroachment is for government to acquire title to agricultural land, and lease it to farmers with stipultions as to its future use. Where the overall national interest dictates the transfer of agricultural land to urbanization, this, too, should be a decision made at national level, and not by individuals submitting to economic pressure.

Public Control of Natural Resources

Frequently, the farmers who cause environmental damage are either not aware of the damage resulting from their mismanagment of natural resources, or do not take the long-term view of possible damage in the future, or do not care because the social cost is borne by others or by society as a whole. For these reasons one or more public institutions are required, to establish and enforce rules for the management of land and water (Crosson & Frederick, 1977). Devising such an institutional structure, in which all conflicting interests are represented, may be very difficult in developing countries.

Enforcement of rules and regulations by legal means would require a large investment in monitoring and enforcement, and it is doubtful if it would be particularly effective. Many of the solutions adopted in developed countries, such as converting the land-use of soils prone to erosion, subsidies to farmers to withhold such lands from crop production, or government expropriation, are usually not feasible in the LDCs.

Summary

Traditional agriculture is by nature an extractive process. Though yields are meagre, the nutrients removed by the crops are not returned to the soil.

Straw and other crop residues are fed to animals, and their dung is frequently burnt for fuel or dissipated when grazing on uncultivated land. The only returns to the land are the droppings of cattle and sheep when grazing on stubble or weeds in the fields after harvest.

As long as population pressure on the land is not excessive, a certain balance is maintained between nutrient removal in the crops and restitution by slow release from soil reserves or recycling through forest growth, and the addition of atmospheric nitrogen through rainfall and biological fixation. As a result, yields remain fairly stable, albeit at a very low level, provided rainfall is adequate. The land itself is only superficially disturbed by the hoe and the primitive tillage technology used, and is generally fairly well protected by a cover of crops, pasture plants, weeds, scrub or forest, according to the farming system practised. Erosion can thereby be kept indefinitely within acceptable limits.

However, as soon as a critical population density is exceeded, a breakdown occurs in the delicate balance between land-use and its permanence. In the semi-arid regions, population pressure causes an expansion of cultivation to more marginal areas, destroying the natural plant cover and exposing the soil to wind erosion in the frequent years of crop failure; in the drier areas, overgrazing by excessive numbers of livestock leads to a progressive degradation of the grasslands and finally to desertification. In the subhumid and humid tropics, the progressive shortening of the fallow period finally prevents the restoration of the protective cover of trees and shrubs, and catastrophic soil loss ensues.

Where irrigation is practised, the process of soil degradation is entirely different. The soil becomes unproductive, mainly as a result of the continual additon of large amount of salts and the rise of the water-table. Primitive technologies have generally been unable to cope with these problems. Population pressures cause an excessive exploitation of ground-water resources, with their consequent depletion; concomitantly, salt-water intrusion into aquifers and land subsidence cause irreversible damage to the aquifers.

In modern agriculture, soil losses through erosion can be kept within bounds by appropriate technology, and the main problem becomes increasingly one of pollution of the environment through the use of agrochemicals, which are essential for achieving a high level of production and for protecting the crops against the depredations of diseases, pests and weeds. Awareness of these problems is increasing and appropriate measures for preventing, or at least reducing, pollution are being devised.

For the developing countries, the breakdown of ecosystems and the resulting loss of land as a result of the uncontrolled population increase constitute a grave threat to their very survival. Whilst the measures that need to be taken are known, the solutions adopted in developed countries are generally not feasible in the LDCs. For the latter, the main contraints are not technological, but social and political, and hence they are particularly intractable. However, the gravity of the situation in many developing countries makes it imperative to adopt without delay the measures necessary to prevent a further deterioration of their natural resources before it becomes too late.

References

Addison, H. (1961). *Land, Water and Food. A Topical Commentary on The Past, Present and Future of Irrigation, Land Reclamation and The Food Supplies They Yield.* Chapman and Hall, London: xii + 284 pp., illus.

Alcantara, Cynthia H. de (1976). *Modernizing Mexican Agriculture; Socioeconomic Implications of Technological Change 1940–1970*, UN Research Institute for Social Development, Geneva: xvii + 350 pp.

Aldrich, D. G. & Martin, J. P. (1954): A chemical-microbiological study of the effects of exchangeable applications on soil aggregation. *Proc. Soil Sci. Soc. Am.*, **18**, 176–81.

Audus, L. J. (Ed.) (1964). *The Physiology and Biochemistry of Herbicides*. Academic Press, New York: xxii + 553 pp., illustr.

Bachtell, M. A., Willard, C. J. & Taylor, G. T. (1956). Building fertility in exposed subsoil. *Res. Bull. Ohio Agric. Exp. Statn.*, 782.

Balfour, E. B. (1945). *The Living Soil.* Faber and Faber, London: 258 pp. illustr.

Benham, F. & Holley, H. A. (1960). *The Economy of Latin America.* Oxford University Press, London: x + 169 pp., illustr.

Bharadwaj, O. P. (1961). The arid zone of India and Pakistan. *Arid Zone Res.*, **17**, 143–73.

Bingham, F. T., Davis, S. & Shade, E. (1971). Water relations, salt balance and nitrate leaching losses of a 970 acre citrus watershed. *Soil Sci.*, **112**, 410–18.

Bower, C. A. (1961). Prediction of the effects of irrigation waters on soils. *Arid Zone Res.*, **14**, 215–22.

Bower, C. A. & Wilcox, L. V. (1969). Nitrate content of the Upper Rio Grande as influenced by nitrogen fertilization of adjacent irrigated lands. Proc. Soil Sci. Soc. Am., **33**, 971–3.

Brown, L. R. (1977). Population and affluence: growing pressures on world food resources. Pp. 25–53 in *Tradition and Dynamics in Small-Farming Agriculture* (Ed. R. D. Stevens). The Iowa State University Press, Ames: xiii + 266 pp., illustr.

Bryson, R. A. (1972). Climate modification by air pollution. Pp. 134–54 in *The Environmental Future* (Ed. N. Polunin). Macmillan, London: ix + 660 pp., illustr.

Burgy, R. F., Fletcher, J. E. & Sharp, A. L. (1967). Watershed management. Pp. 1089–107 in *Irrigation of Agricultural Lands* (Ed. R. M. Hagan, R. H. Haise and T. W. Edminster). Am. Soc. Agron., Madison, Wisconsin: xxi + 1180 pp., illustr.

Buringh, P. (1978). The natural environment and food production pp. 99–129 in *The Alternatives for Growth: The Engineering and Economics of Natural Resources Development* (Ed. H. J. McMains & L. Wilcox). Ballinger Publishing Co., Cambridge, Mass.

Charreau, C. (1972). Problèmes posés par l'utilisation agricole des sols tropicaux par des cultures annuelles. *L'Agr. Trop.*, **27**, 905–29.

Charreau, C. (1974). *Soils of Tropical Dry-wet Climatic Areas and their Use and Management*. Cornell University Press, Ithaca: 434 pp., illustr. (mimeogr.).

Commoner, B. (1970). Threats to the integrity of the nitrogen cycle: nitrogen compounds in soil, water, atmosphere and precipitation. Pp. 70–95 in *Global Effects of Environmental Pollution* (Ed. S. F. Singer). Reidel, Dordrecht: 218 pp.

Crosson, P. R. & Frederick, K. D. (1977). *The World Food Situation.* Resources for the Future, Washington, DC: v + 230 pp., illustr.

Dixey, F. (1966). Water supply, use and management. Pp. 77–102 in *Arid Lands: A Geographic Appraisal* (Ed. E. S. Hills). UNESCO, Paris: xviii + 461 pp., illustr.

Dorst, J. (1972). What man is doing. Pp. 67–86 in *The Environmental Future* (Ed. N. Polunin). MacMillan, London: ix + 660 pp., illustr.

Dumont, R. (1966). *African Agricultural Development*. Food and Agriculture Organization of the UN, Rome: vi + 243 pp.

Dumont, R. (1973). Rural development in Bangladesh. Pp. 172–95 in *Rural Development and Employment* (Ed. C. Gotsch). Ford Foundation Seminar, Ibadan, Nigeria: 774 pp.

Eckholm, E. P. (1976). *Losing Ground: Environmental Stress and World Food Prospects*. W. W. Norton & Co., New York.

FAO (1972): Effects of intensive use of fertilizer on the human environment. *FAO Soils Bull.,* No. 16: viii + 360 pp., illustr.

FAO (1975). *Annual Fertilizer Review*. Food and Agriculture Organization of the UN, Rome.

FAO (1977). Soil conservation and management in developing countries. *FAO Soils Bull.,* No. 33: 212 pp., illustr.

FAO (1978). The state of natural resources and the human environment for food and agriculture. Pp. 3-1–3-65 in *The State of Food and Agriculture 1977*. FAO Agricultural Series No. 8, Rome.

Gabaly, El., H. M. (1977). Problems and effects of irrigation in the Near East region, Pp. 239–50 in *Arid Land Irrigation in Developing Countries* (Ed. E. B. Worthington). Pergamon Press, Oxford: xi + 463 pp., illustr.

Glendening, G. E., Pase, C. P., & Ingebo, P. (1961). *Proc. 5th A. Arizona Watershed Congr.* US Forest Service Tem., Arizona.

Gregor, H. F. (1970): *Geography of Agriculture: Themes in Research*. Prentice-Hall, Englewood Cliffs, NJ: ix + 181 pp., illustr.

Harre, E. A., Garman, W. H. & White, W. (1971). The world fertilizer market. Pp. 27–55 in *Fertilizer Technology and Use* (Ed. R. A. Olson). Soil Sci. Soc. Am., Madison, Wisconsin.

Houérou, H. N. le (1976). Can desertification be halted? pp. 1–15 in *Conservation in Arid and Semi-Arid Zones*. FAO, Conservation Guide Nr. 3. Food and Agriculture Organization of the UN. Rome, xi + 125 pp. illustr.

Howard, A. (1940). *An Agricultural Testament*. Oxford University Press, London: xv + 253 pp., illustr.

Ivens, G. W. (1967). *East African Weeds*, Oxford University Press, Nairobi: xiv + 244 pp., illustr.

Kassas, M. (1972). Ecological consequences of water development projects. Pp. 215–35 in *The Environmental Future* (Ed. N. Polunin). MacMillan, London: ix + 660 pp., illustr.

Kovda, V. A. (1977). Arid land irrigation and soil fertility; problems of salinity, alkalinity, compaction. Pp. 211–36 in *Arid Land Irrigation in Developing Countries. Environmental Problems and Effects* (Ed. E. B. Worthington). Pergamon Press, Oxford: xi + 463 pp., illustr.

Lagemann, J. (1977). *Traditional African Farming Systems in Eastern Nigeria.* Weltforum Verlag, Munich, FRG: 269 pp., illustr.

Lichtenstein, E. P. (1966). Persistence and degradation of pesticides in the environment. Pp. 221–9 in *Scientific Aspects of Pest Control.* National Academy of Sciences, Washington, DC: xi + 470 pp., illustr.

Little, E. C. S. (1969). The water hyacinth—beautiful but a menace. *East African Standard*, 16992,6.

Mellanby, K. (1972). Unwise use of chemicals other than pesticides, Pp. 335–44 in *The Environmental Future* (Ed. N. Polunin). MacMillan, London: xiv + 660 pp., illustr.

Nelson, L. B. (1975). Fertilizers for all-out food production Pp. 15–28 in *All-Out Food Production: Strategy and Resources Implications* (Ed. W. P. Martin). Am. Soc. Agron., Madison, Wisconsin: v + 67 pp., illustr.

Olson, R. A. (1972). Effects of fertilizer use on the human environment. Pp. 15–83 in *Effects of Intensive Fertilizer Use on the Human Environment.* Swedish Int. Dev. Auth. & FAO, Rome: viii + 360 pp., illust.

Parr, J. R. (1972). Chemical and biochemical considerations for maximizing the efficiency of fertilizer nitrogen. Pp. 53–86 in *Effects of Intensive Fertilizer Use on the Human Environment*, Swedish Int. Dev. Auth. & FAO, Rome: viii + 360 pp., illustr.

Pawley, W. H. (1963). *Possibilities of Increasing World Food Production.* Basic Study No. 10, Freedom from Hunger Campaign, Food and Agriculture Organization of the UN, Rome.

Penman, F. (1940). Soil changes under irrigated pasture. *J. Dept. Agric. Victoria,* **37**, 83–100.

Penman, F., Minashima, N. G., Kononova, M. M. & Malinkin, N. P. (1967). Some effects of irrigation and drainage on soils. Pp. 504–62 in *Irrigation and Drainage of Arid Lands.* FAO/UNESCO, Paris: 663 pp., illustr.

Pereira, H. C. (1973). *Land Use and Water Resource in Temperate and Tropical Climates.* Cambridge University Press, Cambridge: xiv + 246 pp., illustr.

Peter, A. von (1978). The economies of fertilizer use and fertilizer resources. Pp. 479–99 in *Potassium Research—Review and Trends.* International Potash Institute, Berne: 499 pp., illustr.

Pollock, N. C. (1971). *Studies in Emerging Africa.* Butterworths, London: 342 pp., illustr.

Polunin, N. (Ed.) (1977). *The Environmental Future*, Macmillan, London: ix + 660 pp., illustr.

Pratt, P. F. (1959). Chemical changes in an irrigated soil during 28 years of different fertilization, *Hilgardia,* **28**, 381–420.

Sanchez, P. A. & Buol, S. (1975). Soils of the tropics and the world food crisis. *Science,* **188**, 598–603.

Stamp, L. D. (1961). Introduction. *Arid Zone Res.,* **17**, 17–24.

U.N. (1971). *Development and Environment.* Norsted and Soener, Stockholm.

Viets, F. G. Jr and Hageman, R. H. (1971). *Factors Affecting the Accumulation of Nitrate in Soil, Water and Plants: Agricultural Handbooks* No. 413. US Department of Agriculture.

Ward, P. C. (1970). Existing levels of nitrates in waters—the California situation. Pp. 14–26 in *Nitrate and Water Supply: Source and Control. Proc. 12th Sanitary Engineering Conf., University of Illinois Urbana.*

Webber, L. R. & Elrick, D. E. (1966). Research needs for controlling soil pollution. *Agric. Sci. Rev.*, **4** (4), 10–20.

Weisner, J. B. (1963). Use of pesticides. *Residue Rev.,* **6**, 1–22.

White, G. F. (1977). The main effects and problems of irrigation. Pp. 1–72 in *Arid Land Irrigation in Developing Countries* (Ed. E. B. Worthington). Pergamon Press, Oxford: xi + 463 pp., illustr.

Wilde, de, J. C. and McLoughlin, P. E. M. (1967). *Experiences with Agricultural Development in Tropical Africa.* Johns Hopkins Press, Baltimore: Vol. 1, xi + 254 pp.

Worthington, E. B. (1972). Sustained biological productivity. Pp. 398–410 in *The Environmental Future* (Ed. N. Polunin). Macmillan, London: xiv + 660 pp., illustr.

Wurster, C. F. (1972). Effects of pesticides. Pp. 293–310 in *The Environmental Future* (Ed. N. Polunin). Macmillan, London: ix + 660 pp., illustr.

CHAPTER V

Human Resources

THE PROBLEM

Population Increase in the Developing World

The demographic explosion that is occurring in the developing countries is generally discussed in terms of the difficulties of producing food in adequate quantities for the continuously increasing world population. Another, no less important aspect is that it has also created a degree of unemployment that is becoming increasingly unmanageable, with resultant social, political and economic problems. Unless the population explosion is brought under control, the possibility of creating sufficient work opportunities, capable of absorbing all the rapidly expanding manpower of the developing countries, appears to be illusory.

The graph in Fig. 21 gives an indication of the increase in the economically active population, actual and estimated, for the period 1950 to 2000. For the developed countries, the increase is 66% (from 386 million in 1950 to an estimated 642 million in 2000); for the developing countries it is a frightening almost threefold increase (from 679 million in 1950 to an estimated 1946 million in 2000).

It is pertinent to stress that industrial production is becoming more and more automated and capital-intensive. Therefore, greater capital investments are required to create employment in industry than was the case in the past.

Developing countries today are squeezed at both ends in comparison with those that developed earlier: at one end the labour force is increasing faster and at the other the capacity to industry to absorb labour is much less (FAO, 1973).

The levels of un- and underemployment in the LDCs are shown in Table IX. In the developing countries, more than half the potential work force is underemployed or unemployed; this contrasts with the less than 4% unemployed in the developed countries.

The rural underemployment situation is numerically worse, because it results from large families sharing the little work provided by small farmers, or from landless labourers who find employment only during peak seasons (Lele & Mellor, 1972).

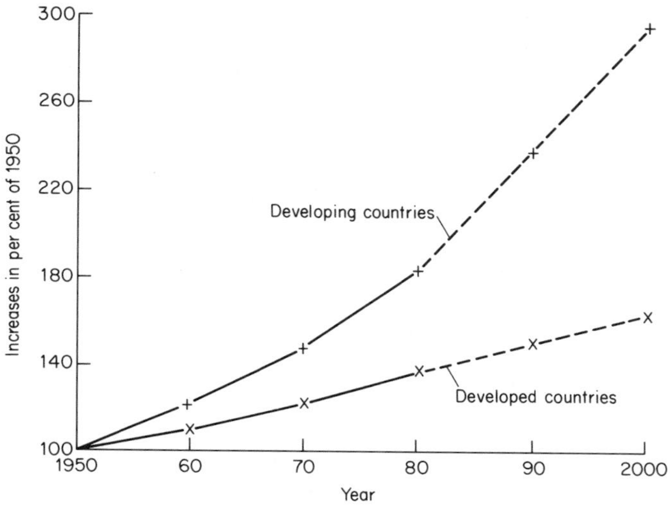

Fig. 21 Economically active population increase (1950 = 100) (based on data from FAO, 1973).

ILO estimates that in order to eliminate unemployment and underemployment in the developing countries (excluding China), 784 million productive jobs will have to be created between 1975 and the year 2000. This is six times greater than the number of additional productive jobs required in the developed countries during the same period (Hopkins, 1980).

Even if family planning programmes should prove to be successful, the problem for this generation is already acute; the labour force for the next 15 years has already been born and immediate solutions to expand employment opportunities must be found, quite apart from the need for population control as an essential long-term measure.

Table IX Levels of employment in developed and developing countries (excluding China) (Hopkins, 1980. Reproduced by permission of the International Labour Office).

	Developed countries	Developing countries	Developed countries	Developing countries
	Millions		% of total work force	
Employed	425.6	316.7	96.2	41
Underemployed and unemployed[a]	16.6	458.8	3.8	5.9
Total work force	442.2	771.5	100.0	100

[a]'Underemployed' in a developing country is defined by Hopkins (1980) as 'not producing enough to equal one's marginal productivity'.

The following example will illustrate the magnitude of the problem with which many of the developing nations are faced: in a single week in 1970, roughly 100,000 more new workers needed jobs in India, than the number of older workers who retired to make room for them (Tobias, 1971). Therefore, in ten years, India needs to provide new employment opportunities for new workers who number as many as the total combined labour force of Britain, France and West Germany (Shaw, 1971).

Underemployment in Agriculture

In the 1950s it was assumed by most economists that disguised unemployment in the agriculture of developing countries was widespread. The corollary of this assumption was that migration of labour from agriculture to industry could be achieved with no loss to agricultural output. More recently, a number of economists have tried to show that there is little reliable evidence to support the existence of more than token disguised unemployment or underemployment in underdeveloped countries (Eicher *et al.*, 1970).

Underemployment has been defined as a situation in which it is possible to reduce the labour force without a loss of output, although no change in production techniques or capital investment has occurred (Kamarck, 1965).

It is a fact that when the opportunities for outside work increase suddenly and labour is withdrawn from agriculture, production may fall, unless changes in production methods occur. Schultz (1964) cites two examples: the building of a road in Peru down the east slope of the Andes, which withdrew labour from farms adjacent to the road, and an upsurge in construction in the city of Bel Horizonte, Brazil, which drew workers from the nearby countryside. In both cases, agricultural production fell as a consequence of withdrawal of labour.

The fall in agricultural production in these and similar cases does not necessarily disprove the existence of un- or underemployment in agriculture. If labour is withdrawn at a time of seasonal peak demand for labour during relatively short periods, this is bound to have a negative effect on agricultural output. The undisputable fact that a large proportion of the rural labour force is un- or underemployed during most months of the year still remains (cf. pp. 146–147).

This underemployment may be quite effectively hidden, because on subsistence family farms, the entire labour force is employed, whatever its marginal productivity, as long as no employment opportunities are available elsewhere.

Labour Productivity In Agriculture

Low agricultural productivity *per caput* is one of the characteristics of developing countries. This is illustrated by the data presented in Figs 22 and 23, showing the relation between the average density of the agricul-

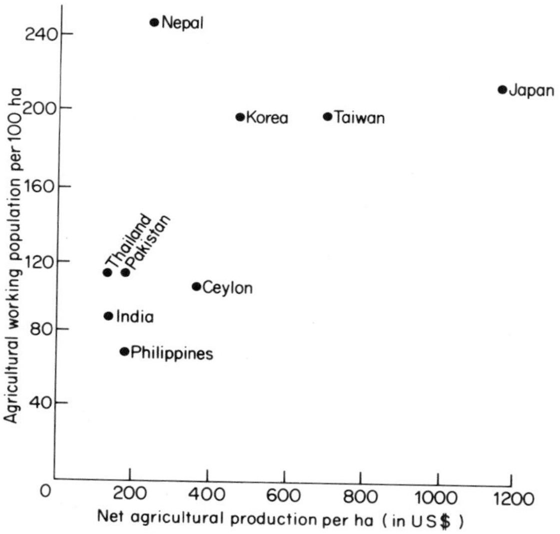

Fig. 22 Relation between average density of the agricultural working population and the output per hectare (based on data from FAO, 1973).

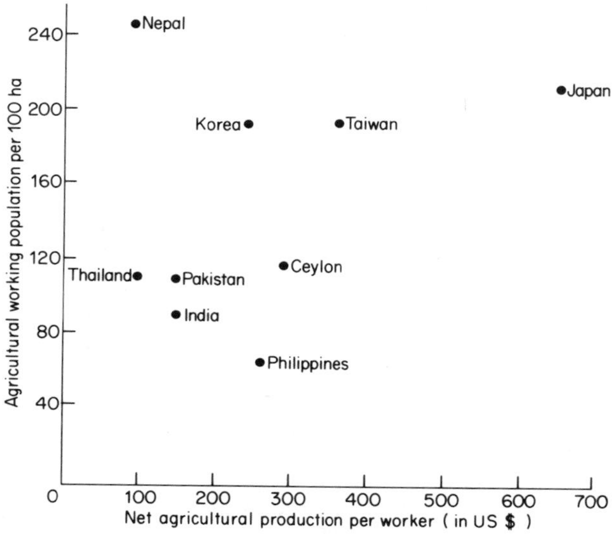

Fig. 23 Relation between average density of the agricultural working population and the output per worker (based on data from FAO, 1973).

tural working population on one hand, and the output per hectare, and per worker respectively on the other, for nine Asian countries. These figures give an idea of the wide span of levels of labour productivity between different countries, as influenced mainly by differences in agricultural productivity per unit area.

In respect to production per worker, there does not appear to be a clear-cut connection between labour intensity and output per worker.

Very high labour intensity was related to very low output in Nepal, whilst the opposite is true for Japan. Nepal and Thailand have approximately the same levels of productivity per worker, but the input in labour per hectare is more than twice as high in Nepal as in Thailand.

A comparison of the figures for Japan with those for the Philippines and India for example, show that though agriculture was about 2.5–3 times more labour-intensive in Japan, productivity of labour in the latter country was more than twice as high as in the Philippines and more than four times as high as in India.

In respect to output per hectare, Japan has $2\frac{1}{2}$ times higher productivity than Korea, and about twice that of the Philippines, at approximately the same level of labour intensity, reflecting without doubt the more intensive and effective use of inputs. All three countries have far higher productivities combined with intensive labour input compared with the other countries, with a single exception. Nepal is the only country to combine a high input of labour with low productivity per hectare—reflecting a low level of management and use of inputs.

In brief, a highly labour-intensive agriculture can be productive in terms of output per worker, provided output per hectare is maintained at a high level. This can only be achieved by efficient management and the use of modern inputs.

Relation Between Total GDP and the Share of Agriculture in National Income

As income (GDP) *per caput* increases, the share of agriculture in the total GDP decreases, as shown in Table X for a number of selected countries in different continents.

It has also been established that in the developing countries, *per caput* GDP in agriculture is always lower than per *caput* GDP for the nation as a whole (Table XI).

The ratio between per *caput* GDP for the population as a whole and that of the agricultural population is remarkably uniform in the developing regions of the world, the latter being about half of the former.

An indication of the amount of labour and drudgery involved in traditional farming is given by Giles (1963): labour requirements in India to produce, harvest and market a hectare of paddy are roughly 875 man-hours; 1125 for a hectare of maize; and 1000 for an equal area of irrigated

Table X GDP *per caput* and share of agriculture (FAO, 1978, by permission); annual average 1970–75.

	GDP *per caput* (US $)	Agriculture's share in GDP (%)
Ethiopia	100	44.2
Pakistan	160	32.2
Kenya	220	27.9
Thailand	350	18.7
Jamaica	560	26.1
Cyprus	1240	15.6
Iraq	1250	12.4
Israel	3790	4.9
Japan	4450	5.7
USA	7120	3.3

Table XI *Per caput* GDP and agricultural GDP per person engaged in agriculture (in US $) (FAO, 1970, by permission).

Region	(1) *Per caput* GDP	(2) Agricultural GDP per head of agricultural population	(2) as % of (1)
Africa south of the Sahara	88	43	49
Asia and Far East	87	46	53
Latin America	294	138	47
Near East and N.W. Africa	174	73	42

wheat. With full mechanization these crops can be produced with less than 25 man-hours per hectare (not including irrigation).

If we consider the amount of work involved in producing a given quantity of a commodity, the disparity is still greater. Dumont (1957), for example, reports that the Turumbi of Zaire require 970 man-hours to produce 1000 kg of rice and 666 man-hours for an equal quantity of maize; the corresponding figures for the United States are 7.5 and 4.4 man-hours.

The gross farm output (net of seed and feed) per male worker of India is only 2.3% of that of the United States and 20.7% of Japan's for 1957–62 (Hayami, 1969).

Factors Causing Low Labour Productivity

Low labour productivity in subsistence agriculture is due to the following factors:

(a) cyclical nature of agricultural production;
(b) population pressure;

(c) low yields;
(d) environmental and institutional factors;
(e) socio-cultural factors.

Economic development and modernization of the rural areas require an increase in the productivity of labour in agriculture. To achieve this aim, progress must be made on three fronts:

(a) Mitigate underemployment in agriculture by adopting labour-intensive farming systems.
(b) Raise production by adopting improved practices.
(c) Create alternative sources of employment in industry and services.

Seasonal Underemployment

Whilst the overall number of people engaged in traditional agricultural production may be excessive, thereby causing hidden or overt underemployment it is frequently impossible to reduce the number markedly, without causing serious disruptions in production or a decrease in the areas under production. This apparent paradox is due to the considerable seasonal fluctuations in demand for labour which is a characteristic of agricultural production in general. However, whilst in modern agriculture it is possible to overcome this limitation by mechanization and/or by hiring labour, there is no such solution for traditional agriculture. A 'floating' labour force of sufficient size to cope with seasonal requirements is usually available in rural areas; however, traditional farmers seldom have the cash required to pay for such labour.

Therefore, in subsistence farming, though the total number of hours worked over the year may be small, there are seasonal peak periods during which a shortage of labour may seriously impair the efficiency of production, especially if most farms in a region follow the same pattern of production. Hence, it is possible to have a situation of underemployment in agriculture during most months of the year, together with seasonal bottlenecks (Fig. 24). These seasonal inequalities in the demand for labour are one of the reasons why so many operations in traditional agriculture are either not completed in time, or are done in a slipshod manner, with consequent losses of yield and income. For example, early sowing may be an extremely important factor in obtaining satisfactory yields, because of the need to use seasonal rainfall efficiently, or to profit from favourable temperatures, or to avoid killing frosts, etc. Delays in sowing may cause serious losses in regions with short wet seasons. For example, in Tanzania, a month's delay in sowing maize may result in yield reductions of up to 50–80% (Arkhurst & Reedharan, 1965). But early sowing may be impossible because of labour bottlenecks at the critical period. Similar situations may arise in relation to timely weeding; unduly prolonged harvesting may

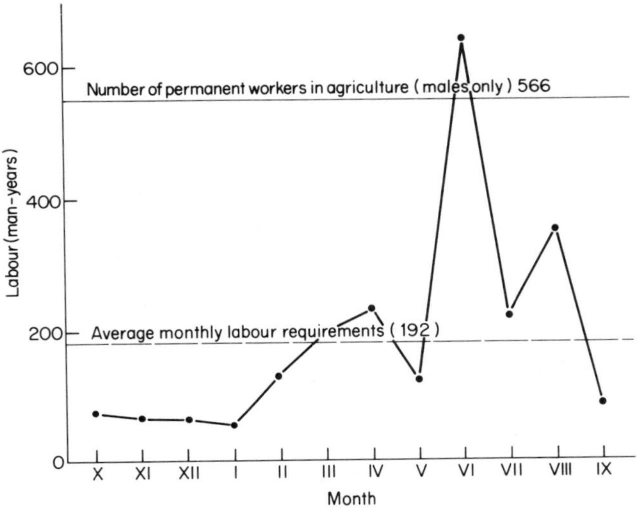

Fig. 24 Employment in agriculture and labour requirement in the Arab village of Tireh in 1950. (Subsistence farming based mainly on wheat, rain-fed vegetables, olives and vineyards.) Note that the labour force is fully employed only one month in the year. (From Arnon & Raviv, 1980).

also cause considerable increases in losses due to the depredations of birds or rodents, or to unfavourable weather conditions.

Environmental Factors

Labour output is frequently low in the tropical climates. There is no doubt that these climates, and in particular, the humid tropics, are not an ideal environment for physical labour, especially of the backbreaking kind associated with the use of primitive tools and equipment, common in subsistence agriculture.

However, there is every reason to assume that the main reason for low labour output is the generally low standards of nutrition and the many endemic diseases (to a large degree due to poverty, ignorance and lack of hygiene) prevalent in these regions; whilst the climatic constraints are not in themselves an insuperable obstacle to productive labour. Malnutrition certainly restricts the amount of work that can be performed, in particular during critical periods such as planting, weeding and harvesting, during which a far above-average effort is required of all members of the farm household. This leads to a bottleneck in production, which in turn perpetuates low farm income with the resulting malnutrition.

It has been shown that village women in India have approximately one-third fewer red corpuscles than women of that age, build, weight and size ought to have (Deutsch, 1971). Blood parasites and insect-borne diseases of humans and domestic animals may also effectively prevent cultivation of large tracts of land or their use as pastures, thereby increasing pressure on cultivable land—hence the importance of improving health services to the rural population. Some developing countries have been fairly successful in this respect by using low-cost approaches to mass medical care, such as paramedics, rural clinics, disease-vector eradication campaigns, etc. (Grant, 1973). This, in turn has led to an increase in population pressure, and an acerbation of employment problems.

Socio-cultural Factors

Attitudes Towards Agricultural Work

In many societies the low status accorded to agricultural work is such that even quite small landowners do not engage in the actual work of tillage. Labour is elaborately graded and the more onerous tasks such as digging, hoeing, and carrying earth are relegated to the lowest strata among the landless (Beteille, 1971).

There is no doubt that the attitude to work can be an important factor determining labour output. The example of China is an interesting object lesson in how work motivation can be raised by getting people really interested in their work and willing to exert great efforts in order to achieve well-defined goals. According to the Chinese, this has been achieved by altering work environments, changing incentives, and providing education, technical training, and health services (Gurley, 1973).

Division of Labour Between Sexes

Community traditions may result in a rigid division of labour between the sexes. Traditionally, certain kinds of farm work are considered 'man's work', for example, the clearing of new land, whilst operations such as sowing, weeding and harvesting, are 'woman's work'. As a result the man may be 'underemployed' at times when actual work requirements may be very high (cf. p. 154).

Agricultural Policies and Employment

Government policies—economic, social and agricultural—can directly or indirectly contribute to un- or underemployment in agriculture.

The lack of success of many settlement schemes in West Africa is ascribed by Spencer (1977) first and foremost to the fact that the motivation is mainly to derive political advantages from these projects, leading to frequent political interference in purely technical matters. It results in over-centralization, lack of cooperation between interested institutions, and numerous technical errors. Feasibility studies are not carried out and detailed programming is neglected. Considerations of economic profitability are subordinate to possible political advantages.

Subsidies

The tendency to replace animal power and human labour by machinery is frequently encouraged by subsidizing the purchase, import and use of big tractors. In many developing countries, the import of tractors is possible at the official exchange rate, which is frequently seriously overvalued. The real cost of a tractor in Indian Punjab has been shown to be half the cost to the farmer in the USA (Billings, 1972). In addition, special credit arrangements are made available to farmers for the purchase of machinery, at very low interest rates. Rebates on fuel and the subsidizing of training centres for tractor drivers are further incentives to premature mechanization. Calculations made in the Philippines, Pakistan, India and Ethiopia indicate that if prices were in accord with the real costs of mechanization to the economy, in certain cases it would be found to be unprofitable, apart from its adverse effect on employment (Gotsch, 1973).

A level of custom duties or taxation needs to be imposed on the import of labour-displacing machinery, that will balance the private profitability of investment in these inputs with their social cost to the economy. Further measures that can be adopted are: removing subsidies, avoiding allocations of scarce foreign exchange, and raising license fees.

Self-sufficiency Policies

In a number of countries, the desire to achieve self-sufficiency in basic food crop production has had detrimental effects on employment. Government support schemes for cereal production in certain countries, for example, have discouraged crop diversification and prevented the expansion of areas under vegetables, pulses, etc., which have a higher labour requirement than cereals (Gotsch, 1973).

Anti-export Agricultural Policies

Export crops are the most easily taxed, providing an important immediate source of income for the government. Such a policy, however, discourages large-scale production, and reduces employment opportunities in the rural areas.

Social Policies

In some countries, legislation has been enacted requiring the payment of minimum wage rates. Whilst socially justified, the direct consequence of this policy is that commercial plantations and estates find it profitable to replace labour by machinery and/or to adopt labour-saving practices, such as chemical weed control, instead of labour-intensive practices, such as the use of the machete for chopping weeds (Eicher *et al*., 1970).

Poorly Conceived Government Investment Schemes

Many of the large land settlement schemes, irrigation projects and state farms established in Uganda, Nigeria, Tanzania, Sierra Leone have been poorly conceived, planned and executed. They are generally capital-intensive, highly mechanized, poorly managed. As a result, they have created unemployment problems.

THE PROVISION OF EMPLOYMENT OPPORTUNITIES

An acceptable standard of living for farm families cannot be achieved without transforming the large mass of marginally productive labour in agriculture working for poverty level incomes, into a fully productive labour force.

A logical strategy aimed at solving the underdevelopment problem of the rural areas could encompass the following objectives:

(a) Create a labour-intensive agriculture with a high level of productivity.
(b) Develop industries, construction and services in rural and urban areas, so as to provide new employment possibilities capable of absorbing redundant labour from agriculture.

Agriculture as a Source of Productive Employment

Figure 25 shows the trends of the share of agriculture in the economically active population, actual and estimated, during the period 1950 to 2000.

Though the share of the agricultural work force in the developing countries will decline markedly as a percentage of the total (from 79 to 43%) it is estimated that the number of people whose livelihood is directly derived from agriculture is likely to increase (from 537 million to 836 million, an increase of 56%).

By contrast, in the developed countries, there will be a marked decrease, both in the share of agriculture in total employment (that will drop from 38.2 to 5.5%), and in the number of people employed in agriculture—from 147 to 35.3 million. Of the world's total labour force in agriculture, it is estimated that by the end of the century, 96% would be in the developing countries! The comparable figure for 1970 is 68% (FAO, 1973).

Because of the large proportion of the population in developing countries that depends on agriculture as a source of livelihood, the high rate of

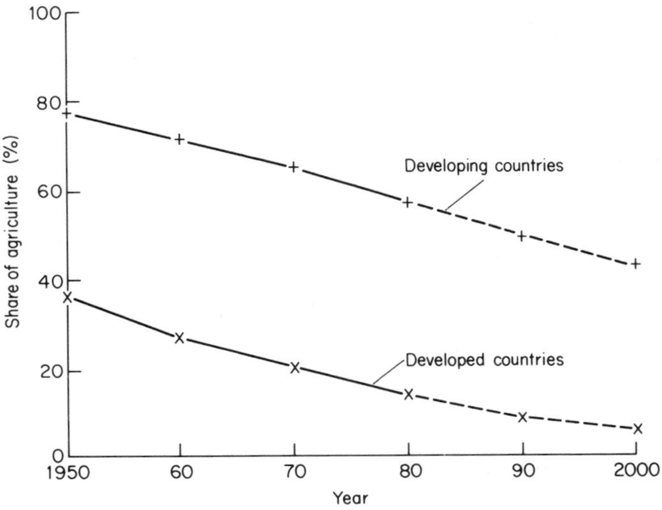

Fig. 25 Share of agriculture in economically active population (based on data from FAO, 1973).

population growth, and the extremely low increase of employment opportunities in industry, there is a general consensus that the most important single factor which can affect the ability of a developing country to absorb a growing labour force into productive employment *in the present stage of development* is in the agricultural sector itself. This situation is not likely to change until the end of the century.

Agricultural development must, therefore, be aimed at creating a labour-intensive technology. This is exactly the opposite aim of that encountered in developed countries, which have had to solve the problems of continously decreasing quantities of manpower available to agriculture. However, increasing employment in agriculture without at the same time increasing production levels, will result in perpetuating the low productivity of labour in agriculture, with its attendant evils. Hence, agriculture must be labour-intensive *and* produce high yields.

Modernization of Agriculture

At a certain stage of development, the introduction of modern technology can increase labour requirements in agriculture. The higher yields that are made possible by the adoption of improved varieties, together with fertilizer application and crop protection measures can result in a considerable increase in labour requirements for harvesting, threshing, cleaning and transporting the larger harvests.

The processing, marketing, storage and distribution of the larger crops produced, as well as the increased demand for agricultural inputs and consumer goods also generate additional employment possibilities. That such a process can occur is shown by the example of Japan, which was very successful in improving labour-intensive agriculture with very little land per agricultural worker. By increasing productivity of labour and land, a growth of farm output was achieved which greatly exceeded the cost of the additional inputs required. The gains in agricultural productivity made it possible to meet the food requirement of an expanding population, to increase foreign exchange earnings, and to promote the simultaneous growth of industry and agriculture (Johnston & Cownie, 1969). The pattern developed by the Japanese may have considerable relevance to other developing countries during a certain stage of their development.

In a comparison between labour utilization on selected farms engaged in similar types of farming in India and in Japan, it was found that 50% more labour per land unit is employed in Japan, labour utilization is twice as high and operating expenses are eight times as high. The value of fertilizers alone is sixteen times as high. The gross production is ten times as high on Japanese farms as on Indian farms. However, it should be noted that, even in Japan, with its high labour output in agriculture as compared with many other Asian countries, productivity per man in agriculture is far lower than in industry of modern services (Mellor, 1963).

Another instructive example is provided by Taiwan. During the period between 1911 and 1965, population pressure halved the average farm size from about 2 ha to only 1 ha. This was, however, accompanied by an intensification of farming due to irrigation, improved varieties and diversified cropping (including vegetables, fruit, livestock), which resulted in a quadrupling of total agricultural production. Notwithstanding the widespread use of small-scale machinery, the number of workers in agriculture rose by 50% and the number of days worked per worker increased one-third. As a result, the total amount of work invested in agriculture during this period more than doubled.

By contrast, in Mexico, during the period from 1950 to 1960, agricultural production increased considerably, but 80% of the increase came from only 3% of the farms. The number of landless farm workers increased by 43%, while their working days declined by almost one-half from a meagre 194 days to 100 days per year, as a result of labour-displacing modernization on large farms. Due to the higher yields, productivity per worker increased by 250% (Grant, 1973).

The problems involved in creating a labour-intensive technology, producing high yields, and in particular the effects of the Green Revolution on employment will be discussed in Chapter VIII.

Intensification and Diversification

The change from rain-fed to irrigated agriculture involves a large increase in labour requirements, and can do much to eliminate seasonal peaks of labour. In contrast with rain-fed farming much work is required during the growing season for irrigating and weeding, and during the periods between crops for the repair of irrigation channels, equipment, etc. Irrigation also enables multiple-cropping, and under favourable conditions more than one crop a year can be produced. Whilst in 1940, only 18% of the land-area in Taiwan was under multiple-cropping; by 1969, virtually all crop land was producing two crops per year (Brown, 1970).

In the Mysore State of India, farmers are producing three crops of maize every 14 months, and at Los Baños in the Philippines, research workers are regularly harvesting three crops of rice per year (Brown, 1970).

The intensification of agriculture, if properly planned, can also mitigate or eliminate seasonal unemployment. An outstanding example is provided by Taiwan: after efficient water control had been achieved, and the production of several annual harvests became possible, agricultural work was transformed from seasonal into sustained employment (Lefeber & Datta-Chaudhuri, 1970).

The choice of commodities produced can have as much effect on labour requirements as techniques of production. Certain crops, such as cereals, cotton, sugar-cane, can be grown either in a labour-intensive agriculture or a completely mechanized system; beef production generally requires little labour, whilst dairy farming, poultry and pigs are labour-intensive.

Replacing a paddy–wheat rotation by sugar-cane, increased labour requirements in Gujerate (India) by about 100 man-days per hectare per year. A single sugar-cane factory, with a crushing capacity of 1600 tons per day and requiring 3600 ha of sugar-cane, created 70,000 man-days of employment within the factory and 351,000 man-days on the farms (Desai & Schluter, 1973).

The shift from traditional crops into the production of strawberries and out-of-season vegetables for export, has been a major factor in increasing employment opportunities in Arab villages in Israel. Notwithstanding complete mechanization of all operations in which this is possible, labour requirements have increased eightfold in villages which have substituted strawberries for the traditional cereals, olives and vines (Fig. 26).

Diversification of agriculture is also able to contribute to increased employment opportunities and better distribution of labour requirements throughout the year. The integration of crop production and dairy farming, where conditions make this possible, is an example of diversification conducive to a better seasonal distribution of labour requirements.

In agricultural economies that do not justify large-scale mechanization, the seasonal peaks of labour can be mitigated by planning suitable rotations, which include crops with different growing seasons and different periods of peak labour requirements. A proper distribution of labour requirements may be one of the basic factors in deciding whether or not to adopt new crops, and in the planning of crop-rotation.

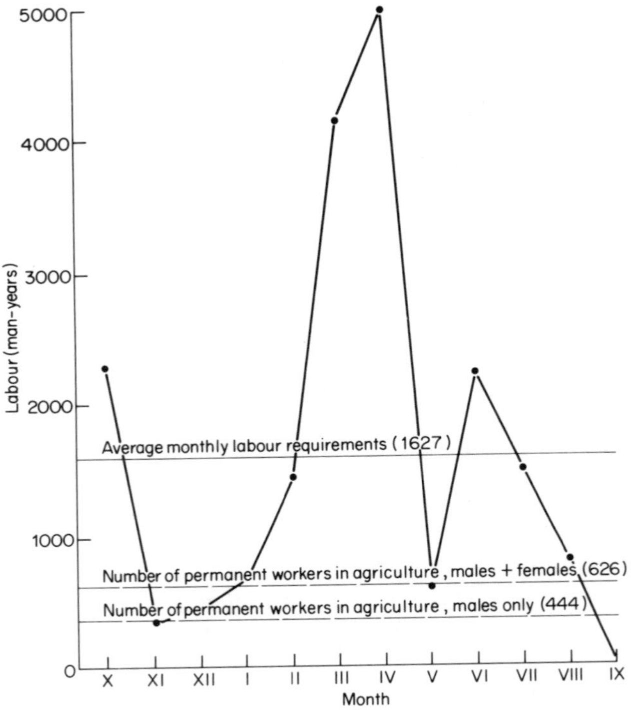

Fig. 26 Employment in agriculture and labour requirement in the same village as in Figure. 25. Subsistence crops have largely been replaced by irrigated out-of-season vegetables and strawberries under cover. Note that: (a) labour requirements have increased eightfold; the number of male workers employed permanently in agriculture is adjusted to the lowest monthly requirements; all other labour requirements are supplied by family and hired labour. Employment outside agriculture is available for all village labour.

Female Labour Participation

In most underdeveloped countries, girls and women are an essential part of the farm labour force. The nature of the work done by women is largely determined by physical and economic factors, as well as by the concepts and social values of different societies.

Sex differentiation in work tasks may well be the result of traditions that were justified at a given period, but that have become irrelevant due to changed conditions. For example, the allotment of specific tasks to women in Buganda was based on the need for the men to be available for Kabaka's army (Fullers, 1964). The carrying of loads by women allowed the men to use their weapons freely (Collinson, 1972).

Physical and Economic Factors

In general, the smaller the farm, the lower the income, and the more primitive the technology is, the greater is the proportion of women in the farm labour force. The female work participation rate naturally increases when males find part-time work off their own farms. Conversely, with a rise in farm income there is generally a decrease in the participation of women in the farm labour force.

The division of labour between sexes, based on cultural traditions is however not immutable and may change in response to economic stimuli (Kamuzora, 1980).

In Gambia, for example, when rice changed from being purely a food crop to a cash crop, men became involved in growing the crop, an activity which had formerly been a women's prerogative (Lele, 1975).

An analysis of the 1961 census data showed that the inter-regional and inter-district differences in Punjab in participation rates of women in descending order of importance: (a) the proportion of workers engaged in agriculture; (b) literacy among females; and (c) gross value product per worker.

The wide differences in employment patterns of women in agriculture due to economic factors are illustrated by the situation pertaining to two contrasting regions in the Punjab. In the poorer hilly regions, women take part in all field operations, including ploughing, irrigation is done almost entirely by women; in sowing, transplanting, weeding, reaping and winnowing, women actively help men; they also carry produce to the market on their heads. By contrast, in the more prosperous central districts, very little use is made of female labour: their work is generally confined to cotton picking, harvesting maize and groundnuts and stripping sugar-cane before crushing (Billings & Singh, 1970).

An extreme case is where women take complete charge of the family farm, whilst the able-bodied adults seek outside employment. In the case of the Torga, a tribe living on the western shores of Lake Nyasa, 60–70% of the adult males are absent from their homesteads at any given time, working abroad (Velsen, 1960), and as a result most of the farm work is done by women and males too young or too old to leave the village.

In Chile, women do not generally work in the fields, and only 5% of the average family's female labour supply (12 to 64 years) is utilized in farming operations, compared with about 30% of the male labour supply. Female labour is used primarily for care of animals and some work in the family garden. Young rural women in Chile are attracted to cities to a greater extent even than young males. In consequence, rural families have a disproportionate number of male to female adults (Donald, 1976).

Land-use

The relative employment of male and female labour is also influenced by the traditional forms of land-use.

Where *shifting cultivation* is practised, most farm tasks are usually done by women. The tree felling is done by men, whilst the removal and burning of brush, the sowing in the ashes, the weeding of the crop, its harvesting and storage are all done by women.

In more densely populated regions were draught animals are used, ploughing and sowing are generally done by men, whilst women perform the hand operations. In regions of intensive cultivation of irrigated land, both men and women must work hard to support a family on a small piece of land (Boserup, 1971).

Employment Outside Agriculture: Transfer of Redundant Agricultural Labour to Other Sectors

As long as employment problems are acute, agriculture serves as a kind of 'residual storage tank in which the bulk of the labour force can find some sort of subsistence until sufficient economic development has occurred to cause a structural transformation' (Shaw, 1971). Any attempt to reduce farm labour before alternative opportunities are available is socially unjustifiable and only increases the number of unemployed in the towns, landlessness and underemployment in the rural areas.

The improvement of the national economy depends ultimately on industrialization and expansion of services, which must then draw on the reserves of manpower available in agriculture. Conversely, a high proportion of population in agriculture is evidence of inefficient farming or of structural defects in the economy which prevent full realization of a country's economic potential.

The very large agricultural population of the developing countries constitutes an enormous reservoir of manpower for the development of other sectors of the economy. The transition from traditional to modern farming sooner or later involves problems of re-allocation of this labour force. After industry and services have taken the labour they require, what remains stays in agriculture. Therefore, the rate of reduction of the percentage of people engaged in agriculture is a fair index of economic progress. Part of the apparent decline in farm employment is actually no more than a form of specialization in many of the tasks formerly performed by farmers, such as milling, dairying, maintenance of equipment, etc.

Japan, Greece, Israel, Mexico and Taiwan are examples of countries in which increased productivity of agriculture created an economic surplus and freed labour for other activities, thereby enabling industrial growth to take place and structural changes to occur in the national economies.

Out-migration from Rural Areas

In most of the underdeveloped countries of the world, there is considerable migration from rural to urban areas, even where employment opportunities exist in agriculture and urban unemployment is rampant.

In Africa, the population of many capital cities is doubling in size every few years (Thomas, 1970), and their ability to provide adequate employment for the increased work force is certainly not increasing at this rate.

In Latin America, approximately five million families live in urban shanty towns and slums, and the population of the latter is increasing at an estimated annual rate of 15% (Thiesenheusen (1969). Much of this increase is due to migration from the rural areas and as a result, it is estimated that the proportion of the labour force engaged in agriculture has been reduced from 63% in the 1930s to 41% in the 1960s. However, the exodus from the rural areas has not eliminated redundant labour in the rural sector nor have all the workers who left agriculture been absorbed in the productive employment in the urban sector (Prebisch, 1971). In other words, part of the relatively unobtrusive redundance of workers has been transferred from the rural areas to the cities where it becomes highly visible.

Migration from rural to urban areas continues in spite of open unemployment in the latter, because even when the probability of obtaining employment is only 1:3, the average wage in the urban areas is two to three times the average agricultural income* (Harris & Todaro, 1970).

Certain governments in Africa (Tanzania, South Africa) have attempted to control migration from the rural areas, and others are considering the institution of such a policy (Harris & Todaro, 1970). This can only be effective if employment opportunities are created within the rural areas, as outlined below.

Effects of out-migration on the Agricultural Sector

Research on the impact of rural–urban migration has shown that the latter may have contradictory effects. In certain cases out-migration from rural areas, by decreasing underemployment in agriculture, has resulted in increased productivity and wage increases. It has also led to land consolidation and reduced cost of land (Nicholls, 1964). The flow of remittances from urban centres to the rural areas was also beneficial (Johnson, 1971).

In other cases, negative effects have been noted, such as a fall in agricultural output, especially in relatively land-abundant areas in Africa, Latin America and some South Asian countries (Schultz, 1964) (cf. p. 142). Out-migration may also result in labour shortages in plantations and commercial farms.

The effects of rural–urban migration will depend on many factors such as the stage of economic growth, available resources, population density, and technological innovation.

Where agricultural productivity is low, due to excessive population pres-

*This does not reflect a high level of urban wages but the low productivity of agricultural labour.

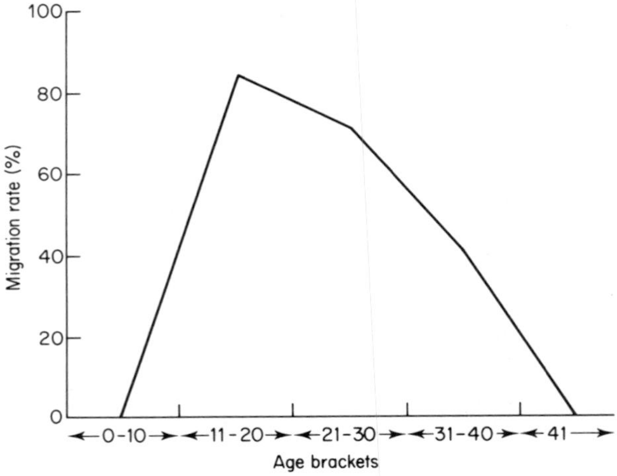

Fig. 27 Rate of rural–urban migration by age of the rural family members in a selected community in western Nigeria. Based on data from Essang & Mabawonku (1975). The figure clearly shows the age-selective character of rural migration, namely, of the age-group that usually provides the labour for the most heavy operations in agriculture.

sure and uneconomic size of holdings, out-migration from the rural areas may increase the productivity of labour. The opposite effect may occur where population density is low, and/or agriculture is particularly dependent on male manual labour (such as bush-clearing, land preparation, etc.) unless technological, labour-saving innovations are adopted.

Migration is also selective. It removes from agriculture the most productive age-group.

In western Nigeria, it was shown that only 2.3% of the family members in the 0–10 year age bracket and none in the 41 year and over age bracket had migrated to urban areas, whilst 85% of the 11–20 and 70% of the 21–30 year groups had migrated (Essang & Mabawonku, 1975) (Fig. 27).

It may also be safely assumed that it is the young people with most initiative and education who will tend to leave the villages, and the loss of this element may be a serious constraint to the modernization of agriculture.

Off-farm Employment whilst Conserving Links with the Village

A special case occurs when family members found employment outside their villages, either seasonal or permanent, whilst retaining their links with native villages. This type of employment may become the catalyst for changes in agriculture by providing the capital, incentives and motivation and changes in outlook required for change to become possible.

A case in point is the transformation of Arab traditional agriculture in Israel. At the time of the establishment of the State of Israel, over 85% of the Arab rural population derived their livelihood from agriculture. Because of the seasonality of farm work, in particular in the rain-fed crop production which was the mainstay of the majority of the villages, the labour force was fully occupied only during one or two months of the year (Figure 24).

The levels of production were extremely low due to centuries of exhaustive cropping and privitive farming practices. The two characteristics of subsistence agriculture—underemployment and low productivity—proved to be mutually self-perpetuating.

The first step in breaking this vicious circle was the elimination of un- and underemployment in the villages. This became possible when all redundant Arab labour was absorbed in the rapidly expanding industry, building and services of the country.

However, outside employment opportunities for redundant village labour *per se* is not necessarily conducive to change in agriculture. It generally leads to the abandonment of the villages by the youngest and most active members of the population, leaving the villages in a continued state of stagnation—with agriculture either unchanged, or worse off than previously.

What characterized the outside employment of the Arab villagers in Israel was the continued and intimate link they maintained with their villages, to which the majority of the workers returned daily.

However, even outside employment opportunities for villagers who maintain their links with their villages is not a new phenomenon, nor has it necessarily resulted in improvement in agriculture or village life. For example, the overwhelming majority of the Tonga tribe, who live in the western shores of Lake Nyasa, find work in Rhodesia and the Union of South Africa. At any given time, 60–75% of adult males are absent, working abroad. They leave their families behind them and return to their villages after varying periods of time. The amount of remittances sent in from abroad in a given year, was more than the cash income of the district from all other sources, e.g. wages, government building, export of produce, etc. Yet it amounted to no more than £1 per head of population (including women and children) or about £5 per labour migrant per annum (Velsen, 1960).

Describing the situation in East Africa, Gulliver (1955) writes: 'The profits of labour migration are so low that there is relatively little effect upon the economic conditions of the home area. A returning migrant may bring with him sufficient in cash and kind to last him and his family a year or so, and sometimes during this period he relaxes his efforts at home until his savings are finished. He has learned little or nothing which he can or wishes to practice at home.'

Hence, another essential factor in the process of change is that the income from outside work must be sufficiently remunerative to enable the accumulation of capital that can be invested in the family farms.

The institution of a common 'family purse' was the rule in Arab villages. This, of course, facilitated investment of savings in the family farm. There was no conflict of interests between those members of the family who worked outside the village and those that continued to work on the farm. Developing the family farm, which could provide a haven in time of depression and incipient unemployment was considered by all a sound investment. A situation had therefore developed in the Arab villages with assured employment for all, and savings at an unprecedented scale. The reduced work force available for farming made mechanization of the farms unavoidable; and the capital needed for this purpose was amply available.*

However, mechanized farming is an expensive process and traditional farming can certainly not pay the costs involved. Expenditure cannot become modern, whilst income remains primitive. The only possible solution open to the Arab farmers was to grow high-value crops and adopt modern agricultural techniques. Thus, through the force of circumstances, Arab farmers became receptive to modern methods of production proposed to them by the agricultural extension workers. Once the farmers realized that the only possibility open to them was modernization of agriculture, progress was rapid; with the first successes traditional distrust of innovation disappeared, and very soon a level of sophistication and yields was achieved by the Arab farmers that is the equal of that of modern farming in developed countries.

This somewhat oversimplified description of the process of transformation of Arab agriculture in Israel, (Fig. 28) has important implications applicable to situations elsewhere, wherever there is a possibility of interaction between an advanced sector that can absorb rural labour and a village sector in need of modernization that can supply the labour. Examples of such situations are:

(a) When two sectors, one advanced and one traditional, exist interspersed in the same country; e.g. an advanced urban sector and a traditional rural sector; modern industry and primitive agriculture; plantation agriculture and subsistence agriculture.
(b) When a country is divided geographically into regions that are highly modernized and others that are backward. The classic example is northern and southern Italy.
(c) When an advanced country absorbs migrant labour on a temporary basis from developing countries. Cases in point are in Europe—Common Market countries on one hand; Turkey, Greece, Yugoslavia, etc. on the other; in America—the USA and Mexico,

*A very similar process occurred in Japan during the period 1954–68 (cf. pp. 366–368).

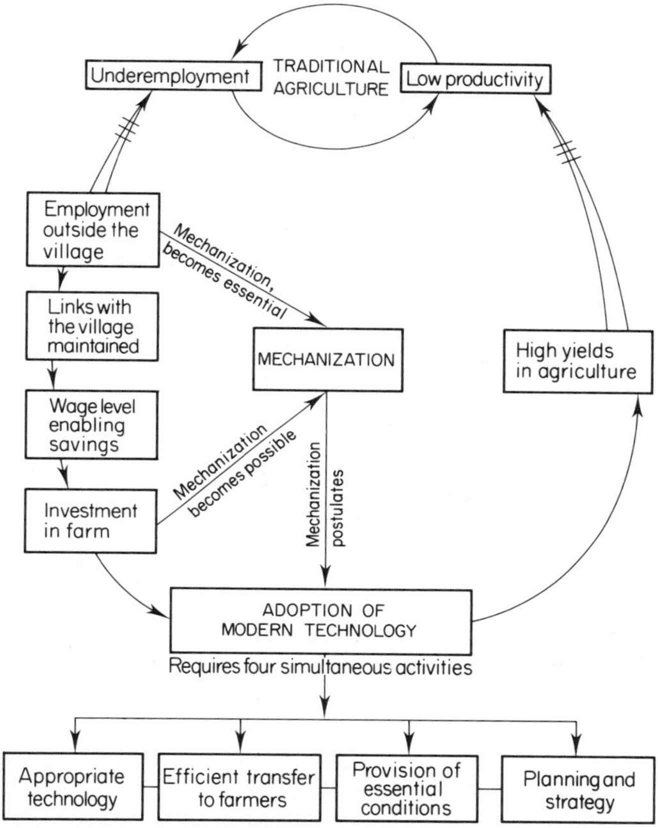

Fig. 28 Process of modernization of agriculture: chain of events triggered by employment opportunities outside agriculture (Arnon & Raviv, 1980).

respectively; in Africa—the Republic of South Africa and surrounding African states.

In all these cases, redundant labour can transfer on a temporary basis from the underdeveloped to the developed sector, without severing its home links.

Experience has shown that the simple existence of such a situation is no guarantee that the process of modernization described above will actually occur. On the contrary, in the majority of cases, temporary out-migration of labour from rural regions has made little or no impact on the agricultural development of the labour-donor regions. Such an impact is dependent on a complex of factors—technical, social, cultural and economic in addition to those mentioned above—and the absence of one or more of these factors may easily lead to a continuation of stagnation rather than progress (cf. p. 4).

The temporary exodus of predominantly rural labour to developed areas within their own country, or to other (developed) countries, could be a powerful tool for the modernization of agriculture in the home villages of the migrants, if their savings could be channelled constructively for this purpose.

Since World War II there has been a massive movement of manpower: (a) from developing to developed countries; and (b) within developing countries from rural to urban areas.

In *Europe alone* the number of migrant workers is estimated at about 8 million, representing about 10% of the labour force of the donor countries. 55–80% of this labour force is male and the majority of the married men leave their families at home, indicating that their links with home remain intact (Hume, 1973).

This migration undoubtedly has great benefits for the developing countries:

(1) To create employment for 8 million workers would require an investment by the donor country equal to the total flow of official aid (about \$60 billion) from the West and Japan over the full decade 1960–70.
(2) The flow of funds from migrants in Europe to their families is well over \$2.5 billion annually which is in excess of the total worldwide annual lending by the World Bank.
(3) Skills: most migrant labour is unskilled or semi-skilled; but between 65% (France) and 75% (Germany) of these originally unskilled workers are engaged in industry and construction.

In brief, employment opportunities, whilst undoubtedly benefitting part of the population, can leave agriculture in the home country unaffected or even worse off through selective desertion of the villages.

In many of the developing countries, such as Spain, Portugal, Greece, and Turkey in Europe, Mexico in Latin America, etc., the temporary exodus of predominantly rural labour within or outside the country could play the same catalysing role as that observed with Arab agriculture in Israel.

In most of the major labour-receiving countries there are already well-established official recruiting institutions for migrant labour, operating through formal bilateral agreements with the donor countries. These could serve as a framework for a concerted policy to achieve the modernization of agriculture in the developing countries which provide the migrant labour, on the lines mentioned above.

The appropriate unit on which such a planning policy should be based is *a region slated for development.* The programme should aim at providing certain essential elements needed for integrated comprehensive regional development. Some points which could be included in such a policy are, for example:

(1) selective recruitment of migrant labour;
(2) an employment programme (including planned training) which will impart the professional skills required for modernization of agriculture and for the industries planned for the labour-donor region;
(3) an educational programme, which will not only benefit the recipients for their future; but will also provide meaningful content to their lives during the period they are in a strange, alien and usually (for them) dreary environment;
(4) provision for the institutional channelling of the savings of the migrant workers to productive enterprises envisaged in the development plan.

Such a scheme could be one of the most productive forms of aid provided by developed to developing countries and by advanced to backward sectors in the same country.

A development that is remarkably similar to that which occurred in the Arab villages of Israel, is described by Dasgupta (1977) for the Punjab in India.

Agricultural progress in the Punjab started well before the advent of the Green Revolution. Two features of the Punjab's socio-economic life which was far in advance of that of most other states in India facilitated this process:

(a) A large proportion of the villagers have found work in industrialized countries—the United Kingdom, Canada, the United States—and have maintained their contacts with their home villages, as evidence by the volume of remittances sent over the years for the purchase of land and farm development. In addition, by tradition, a large proportion of the villagers are engaged in the army, and remit their pay to their families in the village.
(b) The technological experience that the Punjab workers gained in the army, transport, trade and industries, was an important factor in the purchase of farm machinery with a concomitant growth in servicing. The number of thrashers, for example, increased from none in 1947, to 20,000 in 1964 and 80,000 by 1970.

We have seen that such a process of mechanization, in order to remain viable, must be followed by the adoption of yield-increasing technology.

This process, primed by the money derived from outside employment, resulted in the Punjab being the first state to introduce the new high-yielding varieties *HYV's*. From being a food-deficit state, Punjab rapidly reached the stage where it is generally described as the 'granary of India'.

Credit facilities are more generous and much less obstructive to small farmers than in other states. Cooperatives have been established in all the villages.

The foregoing examples are an indication that the temporary exodus of predominantly rural labour to developed areas within their own country, or to other (developed) countries, could be a powerful tool for the modernization of agriculture in the home villages of the migrants, if their savings could be channelled constructively for this purpose.

Part-time Farming

In many countries fragmentation of a large proportion of the farms has reduced their size to levels at which they can no longer provide a living to the farm family even at subsistence levels. The only viable alternative to abandoning the farm is to find off-farm sources of income for some, or all members of the family, who remain available for farm work during seasonal peaks in labour requirements.

That this solution is an economically viable and desirable alternative to the liquidation of very small farms, is shown by the example of Japan, where two-thirds of all farm households are classified as part-time farms. Part-time farming in Japan has enabled families with tiny holdings to enjoy acceptable standards of living (FAO, 1966).

Part-time farming is on the increase in many developing countries. In Egypt, 8% of the total working time of members of farm households is spent on non-agricultural work; in the Republic of Korea, non-farm income accounts for 17–20% of farm household income. There is little doubt that the proportion of off-farm income would increase if more employment opportunities were available.

Interestingly, similar trends for supplementing farm income with off-farm employment are evident in the industrialized countries. In the United States, off-farm sources of income account for about half the net farm income of the agricultural labour force (Paarlberg, 1973).

Industry as a Source of Employment

Modern industry is generally capital-intensive and even in those countries that have been able to increase industrial production, the new work opportunities have not been able to keep pace with the annual increases in the labour force. It has been estimated that even if industrial production and mining in Africa should increase twenty-fivefold by 1985, the additional labour force could not be absorbed even with the related employment created in trade, construction and services (Walle, 1970).

Gaitskell (1968) gives an example to illustrate this situation. In Tanzania, the 1964–69 development plan postulates a 7% growth rate in industry, thereby providing an estimated 116,000 new jobs in the period.

However, the estimated number of new entrants seeking jobs is 1,150,000. In a single year for a typical age group of 250,000 youngsters, of whom 47% receive no schooling, only 23,000 jobs will be available in organized employment. The situation in Kenya is similar. There, the estimated annual number of those who complete the eight-year standard school programme is 150,000; 15,000 of these can find places secondary schools, and the rest must compete for 35,000 paid jobs arising per annum in the economy. The remaining 100,000 must find other self-employment. If these are representative figures for Africa, the position in overcrowded countries in Asia is far worse.

In Latin America, industry was able to absorb, between 1925 and 1960, only a little over 5 million of the 23 million added to the urban labour force during that period, and the tendency is for industry to become less rather than more labour-intensive (Thiesenheusen, 1969).

Rural Integration of Industry, Agriculture and Services

Industrialization does not necessarily imply removing labour from a rural environment. On the contrary, an appropriate rural–urban balance of employment opportunities within a country is of primary importance. One of the most promising approaches for overcoming the backwardness of the rural areas and reducing out-migration from these areas is the concept of comprehensive integrated regional planning (see Chapter XI). Edwards (1973) stresses the need to extend planning, infrastructure, appropriate technology, and complementary resources and services to the rural areas as a matter of the highest priority. Rural industrialization is one solution for channelling new productive resources to the countryside, and thereby absorbing part of the rural labour force without displacing it.

The economic disadvantages of higher transport costs of raw materials and finished products resulting from a decentralization of industry and its dispersal in rural areas is compensated for by lower real cost of labour and housing, and avoidance of the social disadvantages of urban overcrowding and slums (Schickele, 1971).

The kind of industries to be fostered in the rural areas is also of importance: medium- and small-scale industries (processing agricultural products, producing simple consumer goods such as textiles, furniture, etc.) and improved agricultural tools employ relatively large numbers of workers in relation to the capital invested.

This pattern has been adopted in Taiwan; in the early 1960s, 64% of its industrial employment was in the countryside. By contrast, Colombia, under roughly comparable circumstances, only had 25% of its industrial employment outside the principal cities (Grant, 1973). Successful development of the rural sector is to a considerable extent dependent on transforming traditional agriculture. The commercialization of farming with its attendant use of new technological inputs creates many new employment opportunities outside farming, such as the provision and distribution of seeds, fertilizers and equipment and credit, the transport, processing and marketing of agricultural produce, and the provision of goods and services to farmers with improved income levels. This is illustrated by statistics for the United States, where only 6% of the population is directly engaged in agriculture. In 1962, of 65 million people employed in the United States, 25 million produced for and serviced farmers, 9 million processed and distributed farm products, and 250,000 scientists directly served agriculture. The career demand generated by agriculture was for 15,000 graduates per annum: 2000 on the actual farms and ranches, and 13,000 in research, extension, conservation, and commercial operations. 40% of all jobs in the United States were in 500 distinct occupations relying on agriculture (FAO, 1962).

Much of the activities associated with agriculture can be located in small regional towns in which medium- and small-scale industries can also be established. These rural growth centres will, in turn, provide markets for agricultural produce and a source of employment, seasonal and otherwise, for part of the farm family's labour force, increasing overall labour productivity on the farm and providing additional income that can be used for investing in the farm (cf. pp. 160 and 483).

China, after experimenting with the Russian capital-intensive approach in the early and mid-1950s has developed a highly labour-intensive agriculture similar to that of Japan, based on rural integration of agriculture and industry, an extensive system of social services, such as education and health, and a cooperative structure of communes, with subordinate groups at village level called production brigades. As a result, increasing numbers of people in the rural areas have been involved in industry; the rural areas have been transformed into self-reliant agrarian–industrial–cultural local economies with acceptable living conditions. Severe imbalances between urban and rural areas, and between one region and another have thereby been avoided (Gurley, 1973).

However, decentralization of industry has its costs, mainly a dilution of public infrastructure. Nor is it easy to counteract the attractions of the metropolitan centres, and in Latin America, for example, attempts at such decentralization have generally been frustrated (Rosner, 1974).

Rural Works Programmes

Public works are frequently resorted to, in order to provide relief when critical situations arise as a result of natural disasters such as drought or flooding. Once the immediate calamity passes, relief works are usually abandoned, generally without having been completed, or after having been carried out in a makeshift way. As a result, they have no lasting effect on the economy of the region because they are not integrated in the development plans of the region, or properly designed or executed (Shourie, 1973).

Special public work programmes may be justified even in normal times, as a means of supplementing the meagre income of the poorest segments of the rural population, and of mitigating excessive fluctuations in demand for labour. Unlike industrial employment, rural works programmes can be planned to employ labour only during the four to five months of the slack season in agriculture.

There is no doubt that in many underdeveloped countries it is possible to increase the rate of rural capital formation by mobilizing un- or under-employed labour, especially during periods of seasonal unemployment or after natural disasters. The use of excess labour supply in many overpopulated countries can help to create a productive rural infrastructure such as roads, domestic water supply, electricity, irrigation canals, drainage systems, soil conservation works, flood embankments, afforestation, etc., which can in itself be a major contribution to agricultural development (Lefeber & Datta-Chaudhuri, 1970). The engineering corps could train technical personnel on a large scale to act as local engineering foremen, tractor drivers, maintenance workers, etc. The activities that can be carried out in the public works programmes have different short- and long-term effects on employment.

According to a study by Krishna (1974), the development of irrigation is the most productive in terms of continuing employment. The extension of irrigation increases labour demand per hectare in crop production from a weighted average of 64 man-days to 115 man-days, an increase of approximately 80%.

It must be stressed that at best these public works programmes have been able to alleviate economic distress, but have not provided more than a partial solution to unemployment. In East Pakistan, a public works programme that created 173,000 man-years annually, for five years, with a benefit–cost ratio of 4:1, represented an annual reduction in agricultural employment of only 3.4% (Thomas, 1968)!

In India, government attempted to create durable capital assets by harnessing surplus rural labour in the so-called Rural Works Programme, which had a target of 100 work-days per year for 2.5 million people.

Another programme, The Drought-Prone Area Programme was to provide work for agricultural labourers in areas vulnerable to drought. The target was to provide 21 million jobs, mainly in road construction, irrigation, land reclamation and conservation (Dasgupta, 1977).

However, past experience in India with public works schemes has not been very favourable: the quality of the works executed was bad, their administration was inefficient and associated with corruption, so that there was no substantial increment in productive capacity nor a significant transfer of income commensurate with the enormous outlay involved.

A common feature of large-scale government public works schemes is that 'because of the way they are run, the people participating do not feel responsible for the outcome in terms of creating of productive assets like roads or irrigation dams' (Dasgupta, 1977).

This goes far to explain why a small public works, carried out at community level by voluntary labour, on projects of direct benefit to the community, such as access roads, public buildings, etc., can provide an effective and constructive source of employment. Though voluntary, these projects require funding—cash for wages, tools, and supervision. The use of surplus food, made available by the US Public Law 480, and the World Food Programme, is one form in which funds were made available for this type of project (Rosner, 1974).

The productive use of otherwise redundant labour is 'one of the most interesting, original and successful features of Chinese rural development' (FAO, 1979). The government, considering latent manpower as unused production capital, has made it a major goal to avoid rural unemployment and seasonal underemployment by organizing labour-intensive work projects that improve the production or material conditions of the villages themselves.

Mainly during the slack season, the Chinese have mobilized for many years all available labour resources in the villages, not otherwise required, for infrastructural improvement under the heading 'Farmland Capital Construction'. During the seven years (1971–78) it has operated it has involved 100–160 million participants annually, with average individual inputs of 40 to 100 days. Annual averages of 5.3 million hectare of crop land have been levelled and freed of stones; 0.66 million hectare have been terraced; 1.6 million hectare have been brought under permanent irrigation; 5 million hectare of forests and shelter belts planted, and countless roads, canals and dams created. In 1977 alone, a total of 1.26 million projects were executed, ranging from huge land reclamation and improvement schemes, involving the labour force of several provinces (i.e. the Haiho River scheme) to the building of a road or small dam. At first these projects were mainly seasonal; many of them are now implemented by year-round work with a permanent labour force of 28 million. Specialists in farmland capital construction have gradually been formed. The use of machines is increasing, as formerly large manpower surpluses are being gradually absorbed by the communes and in other tasks.

Of particular interest is the rumuneration system which ensures that farmland construction labour is placed on the same income basis as other members of the commune. Workers on farmland construction participate equally in the distribution of the net proceeds of all collective activities, on the basis of work points allotted according to the quantity and quality of their work. Where the state investment budget pays for a given project, the total sum is transferred to the commune treasury and credited to collective revenue.

This approach (so-called 'walking on two legs') has enabled China to achieve high rates of economic growth and provide the rural sector with important infrastructural assets (FAO, 1979).

Youth Services

Most developing countries are faced with the problem of the growing number of young people leaving villages and migrating to towns, with little or no prospects of employment. Youth services can provide a short-term , partial but still important contribution to the solution of this problem.

A solution that has been adopted in many countries is to organize unemployed youth that has congregated in the cities into semi-military organizations, such as civil conservation corps, workers brigades, development service corps, etc. Besides engaging on the type of public works mentioned above, the development service corps may be used to open up new areas and to lay out new farms in which they can be settled.

The young people in the youth services are engaged in programmes aimed at: (a) training in agriculture, with eventual settlement on the land; (b) carrying out works relating to rural infrastructure, housing, public work projects, etc.; and (c) providing services to the community, such as education and health. The first two programmes aim at providing employment opportunities to undereducated youth, and the third is mainly based on semi-skilled, skilled and educated youth (Rossillion, 1967).

The youth programmes are generally organized as a branch of the army; the ranks, disciplinary methods and uniform are more or less the same; they also include a certain amount of military training. Initially, they generally aim at creating potential community leaders rather than large-scale enrollment of unemployed youth, and therefore the services usually rely on volunteers from among those conscripted for the normal military service. Some of these services include both men and women, whilst in most countries there are seperate branches for women.

Agricultural Training and Settlement

A number of countries in Africa and Latin America have established special units in their armed forces to provide vocational training in agriculture with the objective of leading to their re-settlement.

The 'Service Civique' of the Ivory Coast is fairly typical. The service has established a number of training farms, in which soldiers, after completing their first year of compulsory military service, receive an additional year of agricultural training within a military framework. It was hoped that military conscription and discipline would help overcome the negative attitudes towards manual work prevalent among many young men in African societies (Shabtai, 1975). Based on conscription, the first attempts were not a success; most of the conscripts, after completing their military service, did not return to their villages, but drifted to the towns.

In a new approach, the Service Civique was granted sufficient autonomy within the Ministry of Armed Forces to select its recruits from villages in the vicinity of the training camps; by taking 10–20 recruits from the same village, homogeneous groups were created, working in a familiar environment. The period of training in the farm camp, including minimal military training was reduced to six months; as a longer period led to estranging the young people to village life. At the camp, the recruits receive agricultural training in accordance with the type of commodities produced in their villages; they also receive some basic general education: reading, writing, simple arithmetic and 'citizenship'. On return to their respective villages, a plot of land is made available to each conscript. He remains a soldier on active duty for a further period of $2\frac{1}{2}$ years, during which time he is required to cultivate his plot according to the instructions and guidance of a resident agricultural instructor:

Regional Development Operations

In many Latin American countries the 'civic action' programmes consist of participation of the army in regional development projects, particularly in difficult or remote areas, such as opening up new areas to settlement by road and bridge building, forest clearance, irrigation structures, etc. Tens of thousands of youths take part in these programmes. They frequently receive vocational training before leaving the service, so as to prepare them for civilian life.

An example of army participation in development operations in Latin America, is the role of the military in the colossal task of driving roads through the mountains to connect the eastern seaboard to the Amazon basin, as a prerequisite to opening up the fertile jungle lands on the eastern slope of the Andes. This work is being done by six army engineer battalions and an engineering company. In the wake of the road-building, settlements are established in which discharged soldiers are encouraged to stay on as civilian colonists (Hanning, 1967).

In Kenya, the National Youth Service has undertaken the following tasks (Uhlig, 1967):

(a) construction of irrigation works and dams;
(b) planning of native and wildlife preserves;
(c) construction of roads and housing;
(d) afforestation and forestry fire-watching services;
(e) construction of drainage canals;
(f) sanitary installations in towns and rural areas;
(g) opening up land through tse-tse fly control.

Providing Services for the Community

A number of developing countries are using the youth service as a means of alleviating acute shortages of trained personnel in teaching, health, technical services, etc., in the rural areas. The services rendered to the community also have an important educative effect on the young graduates, technicians and doctors involved in the programmes, by exposing them to the social and economic realities of their country (Rossillion, 1967).

In Ethiopia, for example, students cannot graduate from university before having completed one year's service in a rural community, primarily as teachers in the rural areas. In Iran, besides the Education Corps, in which young graduates serve as village teachers during the period of their military service (cf. pp. 216–217), a 'Health Corps' has been established for young doctors and other members of health professions, and a 'Rural Development Corps' for young engineers and technicians.

In Tanzania, the National Youth Service has a special branch in which young graduates, such as teachers, engineers, doctors, etc., must serve for two years before being eligible for public employment (Rossillion, 1967).

Summary

The demographic explosion that is occuring in the developing countries, is not only creating problems of food supply, but is also creating a degree of unemployment that is becoming increasingly unmanageable. As a result, traditional agriculture is characterized by chronic underemployment, in which large families must share the little work provided by small farms, and landless labourers find employment only during peak seasons.

A concomitant characteristic of traditional agriculture is low productivity per *caput*, caused by the cyclical nature of agricultural production, population pressure, low yields; and environmental, institutional and sociocultural factors.

Productivity in agriculture can be improved by introducing labour-intensive farming systems, raising production by improved practices and creating alternative sources of employment in industry and services.

Each of these issues is discussed in detail.

References

Arkhurst, B. C. & Reedharan, A. (1965). Time of planting—a brief review of experimental work in Tanganyika, 1956–62. *East African Agr. Forest. J.*, **30**, 189–201.

Arnon, I. and Raviv, M. (1980). *From Fellah to Farmer*, Settlement Study Centre, Rehovot, and Agricultural Research Service, Bet Dagan, 228 pp, illustr.

Beteille, A. (1971). The social framework of change. Pp. 114–164 in *Regional Development—Experiences and Prospects in South and Southeast Asia* (Ed. L. Lefeber and M. Datta Chaudhuri). Mouton, Paris: 278 pp.

Billings, M. H. (1972). *Tractor Subsidiation Practices in India and Other Less Developed Countries*. USAID, New Delhi.

Billings, M. H. & Singh, A. (1970). Mechanization and the wheat revolution, effects on female labour in Punjab. *Econ. Political Weekly*, **5** (52).

Boserup, E. (1970). Women's role in agricultural development, *Development Dig.*, **9** (2), 97–122.
Brown, L. R. (1970). *Seeds of Change*. Praeger, New York: xv + 205 pp., illustr.
Collinson, M. P. (1972). *Farm Management in Peasant Agriculture*. Praeger, New York: xxvi + 444 pp., illustr.
Dasgupta, B. (1977). *Agrarian Change and the New Technology in India*. UNRISD, United Nations, Geneva: xxvii + 408 pp.
Desai, G. M. and Schluter, M. G. (1973). *Generating Employment in Rural Areas*. Paper presented at *Ford Foundation Sem. on Technology and Development in Developing Countries, New Delhi*: 22 pp. (mimeogr.)
Deutsch, K. W. (1971). Developmental change: some political aspects. Pp. 27–50 in *Behavioral Change in Agriculture* (Ed. P. Leagans and C. P. Loomis). Cornell University Press, Ithaca: xii + 506 pp.
Donald, G. (1976). *Credit for Small Farmers in Developing Countries*. Westview Press: 286 pp.
Dumont, R. (1957). *Types of Rural Economy*. Praeger, New York: xii + 553 pp., illustr.
Edwards, E. O. (1973). *Employment in Developing Countries*. Ford Foundation, New York: 104 pp. (mimeogr.).
Eicher, C. K., Zalla, T., Kocher, J. & Winch, F. (1970). *Employment Generation in African Agriculture: Res. Rep., No. 9*. Inst. Int. Agric., Michigan State University: 66 pp.
Essang, S. M. & Mabawonku, A. F. (1975). Impact of urban migration on rural development: theoretical considerations and empirical evidence from Southern Nigeria. *The Developing Economics*, **13**, 137–49.
FAO (1962). *Africa Survey—Report on the Possibilities of African Rural Development in Relation to Economic and Social Growth*. Food and Agriculture Organization of the UN, Rome: 168 pp.
FAO (1966). Agricultural development in modern Japan. *Agricultural Planning Studies No. 6*. Food and Agriculture Organization of the UN, Rome.
FAO (1970). *Provisional Indicative World Plan for Agricultural Development*. Food and Agriculture Organization of the UN, Rome.
FAO (1973). Agricultural employment in developing countries. Pp. 127–74 in *The State of Food and Agriculture*. Food and Agriculture Organization of the UN, Rome: xii + 222 pp., illustr.
FAO (1979): *The State of Food and Agriculture, 1978: FAO Agric. Ser No. 9*. Food and Agriculture Organization of the UN, Rome.
Fullers, L. A. (1964). *The Kings Men*. Oxford University Press, London.
Gaitskell, A. (1968). Importance of agriculture in economic development. Pp. 46–58 in *Economic Development of Tropical Agriculture* (Ed. W. W. McPherson). University of Florida Press, Gainesville, Fla: xvi + 328 pp., illustr.
Giles, G. W. (1963). Opportunities for advancing agricultural mechanization in India and Southeast Asia. *ASAE Paper No.* 63–154. American Society of Agricultural Engineers.
Gotsch, C. (1973). Economics, institutions and employment generation in rural areas. Pp. 218–83 in *Rural Development and Employment* (Ed. C. Gotsch). Ford Foundation Seminar, Ibadan, Nigeria: 774 pp. mimeogr.
Grant, J. P. (1973). *Growth from Below—A People-Oriented Development Strategy. Dev. Pap. No.* 16. Overseas Development Council: 32 pp.
Gulliver, P. H. (1955). *Labour Migration in a Rural Economy*, East African Institute of Social Research: x + 48 pp., illustr.
Gurley, J. G. (1973). Rural development in China, 1949–1972, and the lessons to be learnt from it. Pp. 306–56 in *Rural Development and Employment* (Ed. C. Gotsch). Ford Foundation Seminar, Ibadan, Nigeria: 774 pp.

Hanning, H. (1967). *The Peaceful Uses of the Military Forces.* Praeger P., New York; xxvi + 325 pp.

Harris, J. R. & Todaro, M. P. (1970). Migration, unemployment and development: a two sector analysis. *Amer. Econ. Rev.* **60**, 126–42.

Hayami, Y. (1969). Sources of the agricultural productivity gap among selected countries. *J. Agric. Econ.*, **51**, 564–75.

Hopkins, M. J. D. (1980). A global forecast of absolute poverty and employment. *International Labour Rev.*, **119**, 565–577.

Hume, I. M. (1973). Migrant workers in Europe. *Finance and Development*, **10** (1) 3–6.

Johnson, G. (1971). The structure of rural–urban migration models. *Eastern Afr. Econ. Rev.*, **1** (1).

Johnston, B. & Cownie, J. (1969). The seed–fertilizer revolution and labour force absorption. *Am. Econ. Rev.*, **59**, 569–81.

Kamarck, A. M. (1965). Notes on underemployment. Pp. 78–85 in *Economic Development in Africa* (Ed. W. W. McPherson). Basil Blackwell, Oxford: xvi + 328 pp., illustr.

Kamuzora, C. L. (1980). Constraints to labour time availability in African smallholder agriculture: the case of Bukoba District, Tanzania. *Development & Change*, **11**, 123–135.

Krishna, R. (1974). Unemployment in India., *Teaching Forum No.* 38. Agric. Dev. Council, New York: 14 pp.

Lefeber, L. and Datta-Chaudhuri, H. (Eds.) (1970). *Regional Development, Experiences and Prospects, Report No.* 70.21. United Nations Research Institute For Social Development, Geneva: 278 pp.

Lele, U. J. (1975). *The Design of Rural Development: Lessons from Africa*, Johns Hopkins University Press, Baltimore.

Lele, U. J. & Mellor, J. W. (1972). *Jobs, Poverty and the 'Green Revolution': ADC Reprint.* The Agricultural Development Council, New York: 7 pp.

McNamara, R. (1973). *One-Hundred Countries—Two Billion People.* Praeger, New York: 140 pp.

Mellor, J. W. (1963). The use and productivity of farm family labour in early stages of agricultural development. *J. Fm. Econ.*, **45**, 517–34.

Nicholls, W. H. (1964). The place of agriculture in economic development. Pp. 11–44 in *Agriculture in Economic Development*, (Ed. C. Eicher and L. Witt). McGraw-Hill, New York: vi + 415 pp., illustr.

Paarlberg, D. (1973). *Farm Policy Implications and Alternatives, National Agricultural Outlook Conference.* US Dept. of Agric. Washington, DC.

Prebisch, R. (1971). *Change and Development: Latin America's Great Task*, Praeger, New York: xxxi + 293 pp., illustr.

Rossillion, C. (1967). Youth services for economic and social development: a general review. *Int. Lab. Rev.*, **95** (4), 1–12.

Rosner, M. H. (1974): *The Problem of Employment Creation and the Role of the Agricultural Sector in Latin America.* University of Wisconsin, Land Tenure Center, Madison, Wisconsin: 101 pp.

Schickele, R. (1971). National policies for rural development in developing countries. Pp 57–72 in *Rural Development in a Changing World* (Ed. R. Weitz). MIT Press, Cambridge, Mass.: 587 pp.

Shabtai, S. H. (1975). Army and economy in tropical Africa. *Econ. Dev. & Cultural Change*, **23**, 687–701.

Shaw, R. d'A (1971). *Jobs and Agricultural Development, Monogr. No.* 3. Overseas Development Council, New York: 74 pp.

Schultz, T. W. (1964). *Transforming Traditional Agriculture.* Yale University Press, New Haven: 212 pp.

Shourie, A. (1973). Growth and development. Pp. 387–425 in *Rural Development and Employment* (Ed. C. Gotsch). Ford Foundation Seminar, Ibadan, Nigeria: 774 pp. (mimeogr.).

Spencer, Chukee-Dinka, R. (1977). Politics, public administration and agricultural development: a case study of the Sierra Leone Industrial Plantation Development Programme, 1964–67 *J. Developing Areas*, **12**, 69–86.

Thiesenheusen, W. C. (1969). Population growth and agricultural employment in Latin America with some US comparisons. *Am. J. Agric. Econ.*, **51**, 735–52.

Thomas, G. (1970). Climbing up from the subsistence level. *Ceres*, **3** (2), 51–3.

Thomas, J. W. (1968). *Rural Public Works and East Pakistan's Development: Rep. No.* 112. Centre for International Affairs, Harvard.

Tobias, G. (1970). *Human Resources Utilization and Development in India.* Meekakshi Prakashan, New Delhi: 95 pp.

Uhlig, C. (1967). The scope for a labour service. *Development Digest,* **5** (3), 112–114.

Velsen, van J. (1960). Labour migration as a positive factor in the continuity of Tonga tribal society. *Econ. Dev. & Cultural Change*, **8**, 265–87.

Walle, E. van de (1970). The population of tropical Africa. Pp. 247–303, in *Africa in the 70's and 80's* (Ed. F. Arkhurst). Praeger, New York: xiii + 405 pp.

CHAPTER VI

Generation of New Technology*

ROLE OF AGRICULTURAL RESEARCH IN DEVELOPING COUNTRIES

The physical environment in which tropical and subtropical agriculture must be practised has certain immutable characteristics. These can constitute constraints that make agricultural production more difficult, as well as provide potentially favourable conditions which may give these regions a competitive advantage over temperate regions. Agricultural research is essential in order to learn how to overcome, partially or entirely, the limitations imposed by the natural environment, and how to make the most effective use of the potentially favourable resources.

In analysing the factors contributing to rural progress in India, Mellor *et al*. (1968) state: 'The obvious lesson is that the first step in an agricultural development programme should be the initiation of a substantial highly integrated research programme, directly connected to farm problems at one end and to basic research and foreign efforts at the other.'

There is general agreement that the efficient application of the results of agricultural research is one of the primary means for accelerating the rate of agricultural development in developing countries. Agricultural research is therefore an acitivity in which no underdeveloped country can afford *not* to engage.

The main production problems to be solved by agricultural research are: modifications of the environment in order to enable a high level of production of plants and animals, increases in the levels of inputs having their origin off the farm, changes in the genetic constitution of plants and animals so as to make them more responsive to physical inputs, and greater protection from natural hazards.

The agricultural scientist must be aware of the many economic, human and social constraints which may impede the adoption of new practices.

The reasons for the non-acceptance or lack of effectiveness of much of the agricultural research carried out in developing countries is ascribed by

*The subject of agricultural research in developing countries has been treated in detail in a book, *Guidelines for Agricultural Research in Developing Countires* to be published shortly by the Food and Agriculture Organization of the United Nations (FAO). In this chapter, (by permission of FAO), a few topics are presented which are of direct relevance to our subject.

Bunting (1971) to the following factors:

(a) research has been fragmented among a series of specialist disciplines;
(b) the packages of innovation have not been designed as a whole;
(c) some research workers are directed from their primary purpose by the professional and intellectual attractions of pure science;
(d) inapplicability of advanced agricultural technology, developed in temperate regions, to ecologically entirely different conditions;
(e) neglect of agricultural indigenous farming systems and the concomitant neglect of food crops.

In our opinion, the most important reason for the limited adoption of research results in the developing countries is that the vast majority of farmers have no access to the new technologies proposed by research.

It has been the privilege of the author to serve as a consultant in agricultural research and development on behalf of FAO, UNDP, IBRD in many developing countries in Africa, Asia and Latin America. In general, all these countries had fairly well-established agricultural research institutions; it was however, a source of disappointment and frustration to find that in practically all cases, research had little or no impact on the agriculture practised by the small subsistence farmers who constitute the majority of the rural population of these countries. To improve their productivity and level of income should be *the* basic national aim. As long as they remain unaffected by agricultural progress, agricultural research cannot be considered to have achieved its aims and to justify its existence.

The specific needs of hundreds of millions of subsistence farmers, whose incomes, scale of operations and commodities produced are entirely different to those of the large farming units, have been largely neglected. However, the general awareness of the role of small farmers in national economic development has increased in recent years, mainly as the result of the following socio-economic trends (Stevens, 1977):

(1) The successful introduction of new technologies has mainly benefited the larger commercial farms and thereby increased the income gap between large and small farmers resulting in an increase in social tensions, a subject that will be discussed in depth in Chapters VIII and XI.
(2) Rapid population growth and the slow growth of off-farm employment opportunities have increased chronic underemployment which can only be alleviated by increasing the productivity of the rural sector in its entirety.
(3) Global scarcities of grains and rising food prices underline the need to increase the productivity of the countless small farmers throughout the world.

Improving production on small farms is therefore the only effective means of decreasing the income gap between the stronger and the weaker elements of the rural sector.

The implications for research and extension in the LDC's are clear. The focus must be shifted from the problems of large-scale commercial farming to identifying and solving the problems of the small subsistence farmers. Though the farms are small, their problems are complex and far less amenable to solution than those of larger production units. This change in emphasis requires a re-orientation of the programmes at all levels of research and a new strategy of extension and development aimed at making possible the adoption of research results by small farmers.

Generation of an Appropriate Technology

'Appropriate technology' has been defined by Norman and Hays (1979) who list four basic requirements:

(1) *Technical feasibility*: it must be capable of increasing productivity given the technical elements.
(2) *Economic feasibility; dependability and compatability* with the farming system: it must be profitable and have a risk level that the farmer can accept as well as have requirements which enable the technology to fit into the farm system.
(3) *Social acceptability*: it must be compatible with community structures, norms and beliefs.
(4) *Infrastructure compatibility*: it must have requirements which can be accommodated by the present level of infrastructure.

This definition is suitable for developed countries, in which the farmer is a man of business, who is on the look-out for innovations that may increase his income and is prepared and able to invest money in innovations. The infrastructure in which he operates (markets, transport, storage, credit, services, etc.) is not only adequate but encourages investment in improved technology.

However, the definition proposed by Norman and Hays has no relevance to the conditions in which the vast majority of farmers in the LDCs must operate, and yet, it has —consciously or unconsciously—guided most of the research work in the LDCs. This definition assumes a passive acceptance of the *status quo*. Orienting research efforts to the finding of solutions 'that are economically feasible, that have a risk level that the farmer can accept, as well as requirements which enable the technology to fit into the existing farm system', etc., effectively precludes any technology than can bring about a radical change in the traditional methods of the small farmers, until the constraints under which they operate are removed, or at least alleviated.

In the context of the LDCs, *appropriate* technology requires a determined effort aimed specifically at overcoming these constraints.

The first question is whether such a policy really discriminates against the large- and medium-scale farmers. Discrimination could prove to be costly in terms of economic development in view of the important share in agricultural production of the advanced sectors, particularly in commercial crops for export.

Schultz (1971) writes: 'Important as it is that economic policies not bypass and not discriminate against small farmers, agricultural scientists who are endeavouring to develop more 'efficient' plants (and animals too) in terms of their genetic capacities and chemists who are engaged in developing cheaper and better chemicals should *not* be placed under constraint that the fruits of their research be applicable only to small farms'.

If one considers the individual components of the most advanced technology, such as improved varieties and breeds of animals, soil management, fertilizers, plant protection chemicals, artificial insemination, etc., these are suitable for adoption by large and small farmers alike.

Conversely, the types of innovation that are inappropriate for small farmers, are those 'that require large increases in work capital, require adoption of complicated new management techniques, compete with foodcrop labour peaks, require indivisible inputs and require divisible inputs in short supply' (Harvey *et al*., 1979).

There is therefore no reason to assume that most of the results of agricultural research cannot be adopted and adapted by large farmers. There are only a few special subjects, such as breeding and management of small stock (sheep, goats, rabbits, etc.), special equipment for small farms, storage methods, mixed cropping, etc., that may be of significance exclusively to small farmers.

The reasons for the inability of the small, traditional farmer to adopt new technologies, are many and complex, and are a recurrent theme throughout this book. Research is only partly responsible for this situation, but it should carry its share of responsibility. Hence the need for a re-appraisal of how research conceives its role.

It is a widely accepted premise that the role of agricultural research is concluded when a solution to a given problem is achieved in the laboratory and the fields of the experiment station. 'Adaptation' of these results to farming practices is, according to this philosophy, the responsibility of the extension service.

The net result of such a policy is the formulation of technological proposals that frequently cannot be adopted by the vast majority of farmers and as a result, cause an alienation of the research worker from the real needs and potentials of the majority of the farming community.

The realization that the responsibility of the research worker extends right up to the stage at which a proposal can be implemented by the farmer is gradually gaining ground. It finds its expression in a new approach called 'pre-extension research' which gives research a bias towards small farmers because its basic concern is: (a) to integrate the results from individual research projects into complete systems; (b) to test these systems under the conditions under which the farmer operates; (c) to pinpoint the constraints preventing adoption; (d) to design new farming systems incorporating labour-intensive, yield-increasing technologies that can make a significant improvement in production possible; and (e) to involve the research workers in improving the development process.

Pre-extension research also has the additional function of serving as a bridge between research and extension, and this aspect will be treated in Chapter VII.

In brief, the main justifications for orienting agricultural research specifically to small farmers are the following:

(a) they constitute the majority of the rural population, and are in greatest need of improvement in their living conditions;
(b) their agricultural production is at such a low level that improved technology can have a major impact;
(c) most research findings of value to small farmers can be easily adapted and adopted by larger commercial farmers.

Integrating Research Results into Viable Farming Systems

Most research, in developed and developing countries alike, has been mainly concerned with the investigation of individual components of production—improved varieties, fertilization, soil management, irrigation, disease, pest and weed control, etc.

Agricultural production however, consists of an integration of all production factors into so-called 'farming systems'.

Whilst farmers in developed countries can easily overcome the problems involved in the introduction of a new crop or adoption of a new technology into their farming systems, this is not the case with the subsistence farmer, who faces numerous social, economic and institutional constraints impeding change.

Farming systems the world over differ according to environmental conditions, available resources, level of technology and types of commodities produced. In the developing countries, the prevailing farming systems have generally evolved over generations by trial and error. Their main advantages are relative stability of production, albeit at a low level, and the minimizing of the risk factor.

A major challenge for agricultural research is to design new farming systems adapted to the various combinations of environment, production factors and technological levels under which small farmers operate. The main objectives of such farming systems would be to incorporate the results of research on the individual components of production so as to achieve optimal conservation of resources, high productivity and better year-round utilization of labour, resulting in improved economic conditions for the majority of the rural population.

It is clear from the foregoing that research on farming systems is complex and requires an important research infrastructure. This is probably the main reason that it has so rarely been undertaken by national research organizations in developing countries. Its importance cannot, however, be overstressed and every effort should be made to initiate such research.

Because research on the farming systems is probably more location-specific than any other type of agricultural research, the principal contribution that the international institutes can make is to develop the methodology of research and to train national workers in this field.

Pre-extension Research

Experiment station research is not usually carried out under normal farming conditions, in particular if these are still of the traditional, subsistence type. Also, a dramatic improvement in the productivity of subsistence farming cannot be achieved by the piecemeal introduction of single practices.

For all these reasons, an intermediate testing stage is essential that can bridge the gap between the experiment station and the farmers' fields.

It is only in recent years, with the growing awareness of the need to orient agricultural research to the requirements and abilities of the small farmer, that attention has been focused on pre-extension studies and their necessity recognized. This approach was first applied in Senegal (*Unites Experimentales de prevulgarisation*) and has proved to be very effective.

At the pre-extension stage, research recommendations on various crops and their respective production practices (crop mix, timing of operations, planting patterns such as intercropping, plant populations, crop protection, etc.) need to be incorporated into a complete farming system, or alternative systems, and then tested and evaluated under conditions similar to those encountered by the farmer, thereby making it possible to identify the technical, economic, social and institutional constraints which would prevent implementation by the farmer. Particular attention needs to be paid to labour availability (peak seasons), availability of power and the condition of draught animals at the time they are needed, phenological and environmental factors, efficient use of soil moisture, effects of preceding crops on those following, etc.. Components of the system are then modified in the light of these findings and new research results incorporated on a sustained basis (Fig. 29). The integration process inherent in farming systems can only be successfully undertaken by interdisciplinary research on a scope unknown in conventional agricultural research.

Pre-extension studies should be carried out by a team comprising the following disciplines: agronomy, animal husbandry, agricultural engineering, extension, economics and sociology. The core of the team would consist of the team leader, an agronomist, an economist and an extension specialist.

Research on farming systems should involve the following stages. At the

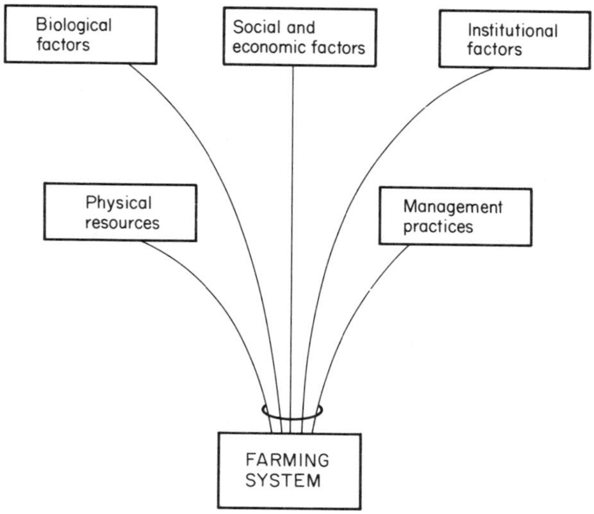

Fig. 29 Integrating all relevant factors into a farming system.

planning stage: a social-economic analysis of the existing traditional systems from which data on the use of available resources by the farm family are obtained and basic assumptions on potential improved systems can be derived. Information on farmer management practices should be incorporated as controls. Decisions are then taken on the siting of trials on the basis of special features of environment and of crop management as practised by the target population. In the *trial stage*: field experiments are planned and carried out in which selected prototypes are tested on relatively large plots. Suitable prototypes developed elsewhere (for example at the international centres) for similar conditions can be included. All the systems tested incorporate inputs from research carried out at the regional stations on mixed cropping, fertilizer use, crop protection, soil management, etc. An economic assessment of cost–benefit ratios is made for the different prototypes. In designing these systems, special attention is paid to the problems and capabilities of the small farmer.

From these experiments, one or more of the farming systems best suited to small farmers are chosen. These are then tested, each on one or more pilot farms in the actual environment in which the small farmer operates.

From the work on the pilot farms, limiting factors are identified, which can be referred back to the experiment station for further research.

This general outline can, of course, be modified in accordance with local conditions and research competence.

The actual trials would be carried out with the labour and equipment available to the farmers of the target group.

The devising and testing of farming systems does not ensure that they can be adopted by various target groups of farmers, in particular the small subsistence farmer who has no cash available for inputs and for whom any departure from traditional practice implies risking his very existence and that of his family. The next phase is therefore a study of the social, economic and institutional constraints to adoption by various target groups, *to devise ways and means to overcome these constraints*

This work should be done essentially by the same team as is engaged in the pre-extension studies, but the brunt of the work will devolve on the economist and the sociologist.

The economist on the team would be responsible for identifying economic constraints preventing or hindering adoption of the farming system proposed, and should formulate proposals for overcoming these constraints. By estimating the possible rate of adoption he will be in a position to advise on the quantities of inputs such as chemicals, seeds, etc., that will be required; the marketing and processing facilities that may need to be expanded, etc. On the basis of these data, the amounts and types of assistance that will be required in the form of long- and short-term credit, supplies of inputs, etc., can be estimated.

The economist will also monitor the effectiveness of improved farming system(s) after adoption by farmers.

The sociologist will, evidently, study the social constraints to adoption and propose how to overcome these, taking into account local political, cultural, ethnographic and social factors.

The team as a whole will study ways and means to make the proposed system attractive to the farmers comprising the target population.

A special case concerns pre-extension research under conditions of uncertainty.

Normally the farming system or systems to be tested, and subsequently modified in the light of actual farming conditions, should be based on the results of past research carried out under the general ecological conditions of the region to which it is intended to apply the system.

Unfortunately, in many developing countries there are regions for which no recommendations based on past research can be made; this applies in particular to new areas slated for agricultural development. Under these conditions policy-makers in research (and extension) are faced with a choice between two alternatives:

(a) postpone the pre-extension studies until results from research become available; or
(b) attempt to put together for testing a system or systems based on the experience and knowledge of experts, and at the same time attempt a crash programme for obtaining information on various crucial components of the system.

Alternative (a) would imply postponing improvements at the farmers' level for at least five years, if we include the pre-extension stage; in many situations this option appears to be unacceptable. Alternative (b) is then indicated, notwithstanding the lack of immediately available pertinent knowledge and the resultant risks.

Formulation of 'Technology Packages' for a Single Crop

CIMMYT has recently adopted an intermediate procedure, based on the same concept as pre-extension research, whereby teams of agronomists and economists concentrate on formulating technology packages for a single crop, instead of engaging in full-scale farming system research.

These teams do not attempt to identify 'optimum' packages of practices, but to 'forge good approximations—technologies which promise more incomes with acceptable risks to representative farmers', in the expectation that each farmer will, after adoption, 'adjust the recommended practices to fit his own particular circumstances' (Winkelmann & Moscardi, 1979).

The research itself is preceded by two related sets of activities in the environments for which the technology is to be developed: (a) an exploratory survey, which includes informal discussions with farmers, extension workers, merchants, etc., focusing on production practices and problems, markets for production and inputs, etc., and (b) a formal survey in which questionnaires, based on the information and insights resulting from the informal survey, are administered to a random sample of farmers.

Both surveys serve to identify the characteristics of representative farmers (farm size, common implements, typical rotations, critical periods, access to inputs); currently employed practices—(levels, types, and dates) and provide information on the farmer's perception of the problems he faces.

In their pre-extension research, three types of on-farm trials are carried out by the CIMMYT research teams (Winkelmann & Moscardi, 1979):

(a) *Yes–no trials* designed to study the major effects and interactions of the factors thought to be the most critical in limiting production. These are mainly in the form of factorial designs, in which two levels of inputs or practices are examined—one at current farmer's levels and the other at a significantly higher level.
(b) *'How much' trials* aimed at identifying levels at which income-seeking, risk-averting farmers might be prepared to adopt practices identified as limiting in the yes–no trials.
(c) *Verification trials*. After researchers and farmers are convinced that an appropriate strategy is available, that is consistent with the farmer's circumstances and promises a significant improvement in income at an acceptable level of risk, the strategy is verified on a number of representative sites. Recommendations are made after confirmation by the verification trials.

Justification for a Strong National Research Effort

A tremendous amount of agricultural research work is being carried out in all parts of the world. The need for research in developing countries can therefore be legitimatly questioned. In a country which is struggling to establish a sound economy, which lacks trained personnel, and in which the agricultural population is still primitive, research may appear to be a luxury which can be ill-afforded. It may well be asked whether elementary logic does not compel a developing country to concentrate on disseminating and applying knowledge already available in other countries, and which can be found in the numerous textbooks, bulletins and journals, that are devoted to agricultural progress—knowledge which is increasingly daily at such a pace that it is almost impossible to keep abreast of developments. In other words, should not the available limited personnel with adequate training be devoted to extension instead of research?

This apparent logic is, however, a dangerous fallacy. As a policy it would be self-defeating. Many problems of significance to developing countries, involving basic research, have not been sufficiently studied in developed countries. Studies of tropical soils, their fertility, their management problems, the ecology of major pests and diseases, animal physiology, soil–plant–water relationships under subtropical and tropical regions are but a few examples of scientific problems whose solutions are essential for sustained agricultural progress in developing countries.

A direct transfer of technology, developed largely in the temperate zones for temperate crops, to the subtropical and tropical regions in which most of the underdeveloped and developing countries are situated, is generally not feasible and has even been the cause of considerable damage in certain cases. Basic principles can be established anywhere in the world but their application to a specific environment requires local research teams working under local conditions.

It is increasingly recognized that improved rice production methods, for example, cannot be immediately transferred to South-East Asia from the United States and Japan. The ecology of the monsoon tropics, as well as cost and price relationships rule out the direct transfer of existing rice production technology on the basis of technological and/or economic considerations. What can be transferred is the capacity to focus scientific efforts on technical problems of economic significance, and the skill that comes from having solved similar problems, even when done in a different environment (Ruttan, 1968).

Many practical results of applied research can, nevertheless, be transferred from humid temperate zones to the tropical and subtropical regions without differences in climate being an insurmountable obstacle, provided the results are properly adapted to the new environment.

In brief, the process of technology transfer is not a simple one, and much of the effort to promote the adoption of known Western World technologies has met with little success in underdeveloped countries because the adaptation problems were not adequately solved (McPherson, 1968).

The so-called Green Revolution, might—at first sight—appear to disprove the need for 'own' research. Within a few years, varieties of wheat and rice developed in international institutes in Mexico and the Philippines respectively, have been transferred directly to large areas, mainly in Asia. The results were indisputably impressive (cf. pp. 284–293). However, it is not generally realized that the development of the new HYVs of wheat and rice is actually an example of the transformation undergone by the results of temperate zone research before they could be applied to the subtropical and tropical regions.

The basic genetic make-up of these varieties had been developed years earlier in Japan, and in other developed countries. The direct transfer of these varieties was impossible, however, because of their adaptation to the specific conditions of the countries for which they were produced. Only by developing research capacity in the regions concerned, was it possible to integrate the basic characteristics of the so-called miracle varieties (namely, their short straw and inherent ability to respond to high levels of soil fertility and management) into varieties suited to subtropical and tropical conditions.

Even after the genetic work of adaptation was completed, numerous problems have arisen following the introduction of the new varieties into different ecological situations: susceptibility to disease, lack of adaptation to flooding, problems of consumer acceptance related to quality, etc. (cf. pp. 293–298). All these problems require local adaptative research and plant breeding work.

This is also true for *all* breeding programmes. Thousands of new varieties of the principal crops are developed and released yearly in the world. One cannot, however, simply go to shopping for improved crop varieties—an introduction service must be established and the promising introductions tested for adaptability, disease resistance and technological suitability, according to the most scientific procedures. Even such a simple 'application' of scientific achievements developed elsewhere is therefore far from simple. In addition, the introduction of new varieties of crops, if carried out without the necessary scientific safeguards, may cause untold damage by introducing pests or diseases that were previously non-existent in the country.

Even the best varieties bred are practically useless unless placed under appropriate ecological conditions of nutrient and water requirements, daylength, crop sequence, and weed, pest, and disease control, while many other factors need elucidation and due consideration. These cannot be determined by reading textbooks published in foreign countries but frequently require appropriate research under local conditions.

A pertinent example is the introduction of the robusta variety of coffee in the Ivory Coast; it was widely distributed to replace the Kovilou variety which had been subject to an epidemic of tracheomycosis. Only subsequently was it found that the variety introduced, because of a greater susceptibility to drought, would not thrive in some of the places in which kovilou had grown well (Wilde & McLoughlin, 1967).

However good imported varieties may be, they will in time generally show some weakness which seriously limits their usefulness. This can usually be overcome by an appropriate breeding programme carried out by local scientists working under local conditions.

Sooner or later, emergency situations will develop: an unexpected invasion of insects, an unexplained epidemic of disease, or one of the other numerous emergency situations which normally appear in progressive agriculture. These cannot be tackled simply by consulting textbooks or journals, but require teams of research workers who are experienced in their respective professions and fully acquainted with local conditions.

Wherever the Green Revolution has been successful, this success has been the result of the interaction between the work of the international research centres in which the improved varieties originated, and research work in the individual countries themselves, aimed at adapting agronomic practices to the new varieties under the prevailing environmental conditions. The Green Revolution has both required and stimulated a considerable strengthening of national research programmes (Johnston & Kilby, 1973).

Where effective research work has been carried out in tropical regions, the results have sometimes been most impressive.

High levels of productivity have been achieved in tropical areas, with bananas, cocoa, coffee, oil palm, sugar-cane, and other export crops of developing countries. The hybrid oil palm which was developed in the Congo in the early 1940s enabled a fivefold yield increase over the traditional bush palms (Eicher, 1970). In recent years research on food crops in Mexico and the Philippines has given spectacular results. McPherson & Johnston (1967) conclude that 'the higher levels of productivity in temperate zones may very well be due to differences in research and development efforts rather than to differences in endowed resources'.

The fact that most of the research effort in the tropical regions was, in the past, directed towards export crops, to an almost complete neglect of food-crops and animal husbandry, was mainly due to the substantial backing received from industries using these commodities. It was also easier to finance research on export crops by the producers, merchants or exporters. The structure of the large holdings facilitated the adoption of new techniques and inputs and responded to clear-cut cash incentives to increase yield and quality (Oram, 1972). This has created islands of high agricultural production in areas where agricultural methods are generally primitive. Agricultural production is low, because there has been little or no government sponsored research in regard to food products. Lele (1976) states that inadequate adaptive research has been a major constraint to increasing productivity and income of subsistence farmers in many programmes in Africa, such as those of the Ujama villages in Tanzania, small farmer's credit programmes in Kenya, and others in Cameroon. The research gap has been particularly severe in the case of food-crops. As relevant information from research in export crops was more readily available, it is the schemes for promoting the latter which were the more successful, leading to substitution of cash-crops for food-crops. As a result, off-seasonal prices of foods have increased considerably, so that low-income farmers, who are unable to produce food in sufficient quantities for the whole year, are also too poor to purchase it at high off-seasonal prices.

In contrast to the neglect in the past of research on food-crops in the rainforest tropics, a considerable amount of work has been done on the problems of the *wet rice region*, particularly in Japan, Taiwan, India and following the establishment of the International Rice Research Institute of Los Banos, in the Phillipines. The range of agricultural research has been extended to other crops, in particular food-crops, in a number of recently established international institutes.

A wealth of breeding material, improved varieties and new technological inputs will in due course be made available to the developing countries as a result of the intensive research effort of the international institutes. This will only increase the need and justification for a strengthening of the national location-specific research efforts (see next section).

The impressive results already achieved with the two major cereals, wheat and rice, is an indication of what may be achieved with other important food-crops, such as root-crops, tubers, pulses and oil-crops by an interaction between international and national research programmes.

Other problems that are location-specific, or of special interest to developing countries are devising *farming systems* relevant to each region and the development of farm machinery adapted to family farms, that will increase efficiency and reduce bottlenecks without causing large-scale labour displacement.

In view of the aversion of traditional farmers to risk (cf. pp. 924–925), (a major reason for their resistance to new practices) research on the *variability* of any new technology under local conditions is of considerable importance.

Ability of a Country to Adopt Research Results from Elsewhere

It would be a mistake to conclude from the foregoing that the duties of research workers in a developing economy are limited to 'trouble-shooting'. Research workers must pioneer new developments, think ahead of farmers and planners, and spearhead agricultural progress. They have to be constantly alert to the opportunities provided by new information resulting from basic research carried out elsewhere, and by new technology advances. It would, of course, amount to inexcusable neglect not to take advantage of every step forward that is made in agricultural knowledge in any part of the world; but even the evaluation of promising leads in the scientific literature requires people trained for research, who can judge what is desirable to test and adapt to local conditions, and who are able to carry out this work. As a result of increasing specialization, only workers engaged in research in given fields are usually able to keep abreast of developments and to evaluate correctly the potential importance and applicability to their own specific conditions, of the results of agricultural research carried out elsewhere.

The remarkable progress of agriculture in Japan is attributed in great part to the appreciation by the government of the need to invest heavily in agricultural research and extension services. By establishing its own experiment stations, agricultural schools, seed propagation farms, and extension services; by sending students to study at foreign universities and inviting foreign experts, Japan was able to adapt modern Western knowledge to its own conditions and environment (Johnston, 1962).

In a far-ranging analysis of the effects of technology transfer on agricultural development, Ruttan and Hayami (1973) reach the conclusion that 'Failure of a nation to institutionalise domestic research capacity can result in serious impediments to effective international technology transfer. A major challenge for the developing countries is to develop the scientific and institutional capacity to design and adapt location-specific agricultural technology to the resource endowments and economic environments in which the new agricultural technology is to be employed.'

In an intensive study on the subject of technology transfer from developed to developing countries, Kislev and Evenson (1973) specified an international model in which productivity (in wheat and maize production) was related not only to the research programme of the country in question, but to the research programme in other countries located in similar ecological zones. The idea was to determine how much of the research discoveries of other countries could be borrowed by or transferred to the country in question. The basic relationship found is shown in Fig. 30.

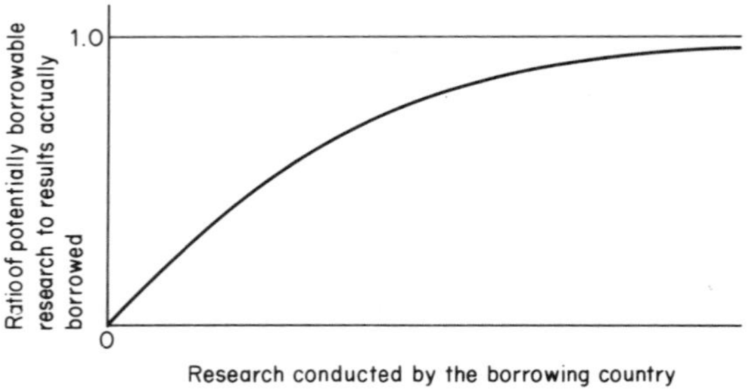

Fig. 30 Basic relationship between 'own' research and research adopted from other countries (Kislev and Everson, 1973, courtesy of the authors).

Figure 31 shows that a country does 'borrow' or benefit from the research findings of other countries, but that the extent to which it does so depends on its own indigenous research capability. The Kislev–Evenson study concluded that countries without the capacity to produce internationally significant research publications, also lacked the capacity to benefit from the research findings of other countries. Those countries which did have indigenous research capability in wheat and maize production, benefitted significantly from the research done in other similar regions, but *not* from research conducted outside those regions.

Ruttan and Hayami (1973) consider that 'the most serious constraints on the international transfer of technology are (a) limited experiment-station capacity in the case of biological technology and (b) limited industrial capacity in the case of mechanical technology.' These authors distinguish three phases of adoption by one country of technology from other countries: The first phase—*material transfer*—consists of the simple import of new varieties, animals, machines and techniques without orderly and systematic trials. The next stage is *design transfer*: new introduced crops and varieties, breeds of animals and techniques are recommended to farmers only after orderly testing. Seeds of recommended varieties are propagated through systematic multiplication. The third phase is that of *capacity transfer*, consisting mainly in the transfer of scientific knowledge and capacity. Increasingly, national research is strengthened—crop varieties and animal breeds are bred locally to adapt them to local conditions, and adaptive research on agricultural techniques is expanded. New and original approaches are developed, which are not necessarily location-specific.

To move from the first to the second stage requires an appropriate organization and infrastructure for experimentation; the transition to the third stage is mainly dependent on an adequate force of trained personnel, backed by modern research equipment.

It is essential that developing countries move as rapidly as possible from the first to the second phase. Phase three will usually be achieved gradually, as research personnel gain experience and motivation.

The benefits resulting from indigenous or national research therefore have three components: a direct contribution, a contribution through the acceleration of transfer from other countries, and a contribution to other countries.

National Structure of the Research Organization

The need for centralized organization, planning and direction of agricultural research is generally recognized as essential for ensuring effective public control of research policy in orienting and coordinating research efforts toward the needs of the community.

A review of the national structure of agricultural research organizations in a large number of countries indicates that there exist a few basic prototypes according to the bodies responsible for the implementation of research. This responsibility can be vested in a research council, the ministry of agriculture or similar ministry, or one or more autonomous bodies (universities, academies, or a number of independent institutes). Within each of these main groupings, various solutions and combinations may be found, giving rise to numerous forms of organization.

The question naturally arises as to whether it is possible to design a prototype organization that can, with more or less minor modifications, serve as a general model to most developing countries.

The main requirements for such a model structure would be:

(a) committment to solving the problems of the agricultural community;
(b) ability to carry out a balanced program of research, covering both the urgent and day-to-day problems of agriculture as well as the long-term problems;
(c) ability to make the most efficient use of research personnel, equipment and funds.

At a meeting of an international panel of experts on the organization and administration of agricultural research, called at the initiative of FAO, in 1965 (FAO, 1965), the unanimous conclusion was reached that the type of research organization that could most effectively serve as a model for most countries was that based on a central national research institute or service within, or closely linked to the ministry of agriculture, and having the following characteristics:

(1) responsibility for planning, coordination and implementation in the whole field of governmental mission-oriented agricultural research;
(2) autonomy in the implementation of the approved research programme, in order to ensure flexibility of administration with a minimum of bureaucratic interference and red tape;
(3) funds and research directives derived from a plurality of sources: (a) the ministry of agriculture for general financing and for programmes adjusted to development plans of the ministry and the most pressing needs of the farming community; (b) the national research council, for long-term and basic research; and (c) uncommitted institutional funds for exploratory research, doctorate theses, etc.; and
(4) centralized organization and administration of research, and decentralized implementation of research based on a network of regional polyvalent experiment stations.

Of course, the *size* of the country to be served is a factor to be considered. Though even small countries can have a great variety of climates and soils, this is not a hindrance to the functioning of a central research institute, provided it relies, for the execution of adaptative research, on a suitable network of regional experiment stations. However, beyond a certain size, distances within a country may become a physical factor that affects the ability of the organization to carry out its responsibilities. In very large countries, such as the USA, Australia, USSR and India, a political factor also intervenes: in addition to physical dimensions, a national government is superimposed on a number of states or provinces. In such cases, a research organization is proposed, consisting of two distinct frameworks:

(a) A national organization of research, with overall responsibility for research policy with administrative and service headquarters near the seat of government; and
(b) each political (state, province) or physical (region, watershed) entity to have its own research centre, with a corresponding network of experiment stations established according to ecological requirements.

The Headquarters would have the following responsibilities:

(a) To identify the important problems involved in agricultural development and ensure that the research programmes adopted at the regional centres conform to the overall line of national policy defined by the ministry of agriculture.
(b) To encourage and finance research of national significance by: initiating projects of national significance at one or more of the state or provincial institutions; establishing cooperative projects in which a number of states or provincial institutes are involved together, and by contracting for specific research projects at the Universities or other appropriate non-governmental institutions.

A prerequisite for the effective functioning of these headquarters is an adequate staff of scientists capable of planning, coordinating research, and evaluating research results, in the various specialized fields of research of national significance.

Duplication would be avoided by maintaining at headquarters a national register of research projects, disseminating information on the results of research achieved in the states or provinces, and organizing contacts between research directors and workers from different states.

State, provincial or *regional* research centres would engage in all aspects of agricultural research and at all levels required for the effective solution of regional problems. They would have a balanced programme of basic and applied research, research into problems of immediate significance and long-term research. The level of competence would be similar to that envisaged for a central institute in smaller countries.

There are however, research problems and services of a national character. These are generally problems that require heavy expenditure, costly and sophisticated equipment, teams of research workers from different disciplines and coordinated field experimentation in a number of regions. Duplicating work on these problems at the regional centres would be costly and ineffective. A desirable solution is that each regional centre, in addition to its work of a purely regional character, should assume responsibility for one or more fields of research on a national basis. This would prevent excessive provincialism by broadening the field of responsibility and scope of influence of each centre and would ensure effective contacts between the centres at the professional level.

Each regional centre would, therefore, also have a national role based on its interrelationships and cooperation with other regional centres. The overall guidance provided by the national headquarters would ensure that the efforts of the regional centres are coordinated into a national cohesive programme of problem-oriented research, essential for the implementation of the nation's development plans. The level of scientific competence at the regional centres would have to be of the highest possible order; the actual disciplines and commodity departments would be determined according to the specific problems of the region.

Role of Universities in Agricultural Research

The Land-Grant College Model

Of all the structural types of research organizations found in different countries, the one that at first view appears the most rational and convincing is the land-grant college model, in which responsibility for research, education and extension is vested in the colleges of agriculture, which, in turn, form part of the state universities. This approach, adopted by the individual states of the USA, has ensured close links between education, research and extension and has been exceedingly successful in transforming American agriculture. And yet, the many attempts to transfer this model to other countries have generally been unsuccessful; and in those countries in which the establishment of land-grant type colleges is in the early stages of development, serious doubts may be entertained as to their suitability.

The success of the land-grant colleges in the USA is due to a number of factors specific to that country, and which are generally lacking in developing countries:

(1) The land-grant colleges were extremely dependent on the good-will and support of the farming communities in which they were established. This made them sensitive to the needs of agriculture and responsive to the demands from farmers to solve their problems.
(2) The teaching and research staff were only a single generation removed from the pioneering farmers of the frontier days; they themselves were usually of farming stock, eager to dedicate themselves to the solution

of farmers' problems, in which they saw a worthwhile and interesting challenge. The majority were close to the land and to the farmers and had considerable practical knowledge and experience in farming. The manual labour involved in field experimentation was not considered demeaning for a scientist; on the contrary, familiarity with farm operations and ability to operate farm equipment and use tools was a source of pride.

(3) The large number of land-grant colleges ensured complete coverage of practically all the problems encountered even when no careful programming or planning of research was carried out. Research results from a single experiment station were mostly applicable to a number of neighbouring states, thereby making an 'overlapping' of research programmes possible.

(4) Wherever shortcomings in the research programmes remained, these were usually covered by the additional research structure of the federal government.

(5) This system entailed a considerable amount of duplication and repetition, but the country was sufficiently rich to afford this and sufficiently aware of the importance of research for the development of agriculture to agree to the necessary financial appropriations.

The number of countries in which a similar set of circumstances exists is small indeed! Universities are usually concerned with maintaining their academic freedom, and hence are not usually amenable to public control aimed at ensuring an oriented research programme which is a *sine qua non* of agricultural planning and development in most countries.

The Contribution of the Faculties of Agriculture to Agricultural Research in Developing Countries

The university schools of agriculture, in common with all institutes of higher learning, have two basic functions—teaching and research—both relevant to their contribution to agricultural research.

In teaching, their aim is to prepare agronomists for the functions of planning, production, technology, research and extension in the field of agriculture.

Research at institutes of higher learning has two objectives: to extend the field of human knowledge and to contribute to the training of students. Traditionally, research objectives are followed without consideration of practical implications, or restrictions on the liberty of the researcher in the choice of his problems. It is called academic, pure, fundamental, or basic research—these terms are considered interchangeable when applied to university research.

However, in certain specialized fields, of which medicine and agriculture are typical, the 'conflict' between pure and applied research has never been relevant. In these fields, applied research is closely related to the educational objectives; research cannot be divorced from the subject taught, and hence medical or agriculture research in the respective faculties cannot be 'pure'.

The nature of agricultural research carried out in faculties of agriculture is, therefore, not necessarily different from that in government agricultural research institutes. There is, however, one characteristic that distinguishes between the two: the freedom of choice of research subjects. The faculty member has the privilege of absolute discretion in the choice of the problems on which he and his students will be engaged, whilst the government researcher has a moral obligation to undertake research in subjects according to priorities that are binding on him. These are usually determined by public or governmental bodies, albeit with his participation, but in which his own desires are not decisive and may be overridden.

In common with all university research, agricultural research in a school of agriculture is not only an end in itself, but is an important part of the educational system. As such, the main objective of research is to train the student and teach him a systematic approach to the solution of problems. Problems should be chosen 'that encourage students to think imaginatively, reason scientifically and gain new understanding of principles—convert a worker into a thinker' (Bailar, 1965). These objectives can be attained equally well if the subjects chosen for the research have worthwhile practical implications, as when they are esoteric and have no relation to agricultural practice. The choice of research subjects has, moreover, a considerable effect on the shaping of the future agronomist's predilections and motivations. If the research subjects usually chosen contribute little or nothing to worthwhile knowledge or practical objectives and their only 'practical' application is to provide material for a 'paper', this will strengthen the latent careerism of the student. Research subjects that have no other objective but to satisfy scientific curiosity will tend to produce a graduate divorced from agricultural reality. However, it is possible and desirable to choose subjects that provide scientific training and, at the same time, attempt to solve problems of significance to agriculture—thereby showing the young graduate or post-graduate student that basic and applied research are interwoven and that scientific curiosity can serve agriculture. This will encourage devotion to, and involvement with agriculture.

In brief, whilst the university is generally not a suitable framework for assuming national responsibility for the planning and implementation of a mission-oriented research programme, faculty staff can make a serious contribution to such a programme. No developing country can afford the luxury of a faculty divorced from the realities of its agricultural economy and indifferent to the solution of its problems. The integration of faculty members (and their students) in the national research programme can be encouraged by: (a) involvement in the formulation of research policy and programming; (b) assigning research projects to faculty members with special competence on a contract basis; (c) inviting faculty members to join research teams; and (d) involvement in in-service training of government research and extension workers.

Training Staff for Research Work

Requirements

In most developing countries a shortage of scientific manpower is the major limiting factor in establishing and maintaining an agricultural research organization on a scale commensurate with the problems to be handled.

It is essential to have capable research personnel at all levels, and a deficiency at any level will hamper research work.

The number of research workers required in each category, the level of competence and the types of specialization required will depend on the stage of agricultural development of the country, the resources available and the priority accorded to agricultural progress. A study of the number of well-trained people needed to support a research and production programme in a given country has been carried out by Centro Internacional de Mejoramiento de Mais y Trigo (CIMMYT) staff. Using maize as an example, it was estimated that countries cultivating from 30,000 to 100,000 ha would need a national team of 11 agronomists, including breeders, entomologists, pathologists, research production agronomists, extension agronomists and protein evaluation specialists. For countries cultivating 100,000–500,000 ha, this number would increase to 26 and to 55 for countries with 500,000 to 1,000,000 ha (Violic, 1973).

The future researcher is faced with severe personal demands. He will have to work in a rural environment and understand rural people and their problems. He must not only be competent professionally but must look upon his work with a sense of mission.

Basic Training

In recent years, the capacity of educational institutions in developing countries to train their own agricultural scientists has risen steadily. The organization structures and curricula of most of these faculties of agriculture or agricultural colleges have been patterned on models from European or United States institutions which are discipline-oriented rather than production- and problem-oriented (Gray, 1974). Most of the students in the latter institutions have chosen agriculture as a vocation, either because they have a rural background and farm experience or because they are genuinely eager to serve in agriculture. For students with this background, the specialized, discipline-based curricula are entirely appropriate and give them an adequate preparation for work in a highly specialized agriculture.

By contrast, in developing countries, most students with sufficient education to qualify for university admission come from urban backgrounds, have little understanding or sympathy for rural life and have entered agricultural college with hopes for city-based jobs in a ministry of agriculture.

The curriculum content tends to be theoretical with emphasis on book-learning, while the textbooks are based on agriculture in the temperate regions. The graduates may have a fair grounding in agriculturally-related basic sciences, but are seriously lacking in many of the qualifications required for agricultural research in developing countries.

Few of the faculties of agriculture provide training beyond the BA or BSc level, but the undergraduate programme itself is generally aimed at graduate work and not at training specifically for the functions that will be undertaken, at least in the first years of their professional life, by the majority of those graduating from college with a BSc degree. For all these reasons, if the novice research worker is to perform effectively, the competence he lacks must be imparted by in-service and on-the-job training by the research organization staff.

The research worker should be enabled to keep up with advances in research in his own and neighbouring fields, by the provision of appropriate library services, regular meetings between research workers from different disciplines, colloquia on specific subjects, study groups, special courses including lectures, case studies and discussions.

Lectures should be given on a regular schedule on subjects such as the socio-economic problems of the region, farming systems, cultural characteristics of the rural population, governmental policies and development plans.

In-service training is a never-ending process. It cannot consist exclusively of what is provided by the research organization or attendance at courses, seminars, colloquia, etc.; the researcher must continue to train himself by personal effort—reading, individual learning, etc.

Young Research Workers

Research workers must learn to identify the significant problems in their field that limit production, to understand the social, economic, cultural and institutional environment in which the farmers they intend to serve operate, the problems of communication in a research organization, the constraints encountered in adopting research results and the social and economic consequences of successful adoption.

A significant deficienty of agricultural education, at university level, is lack of training in the practical skills required in farm management, crop production and animal husbandry. Graduates may know in theory how to grow a crop, but when they are in the field, they are not able to do it. At first sight, these skills do not appear to be essential for research workers. The need for such training is generally not recognized by decision-makers, or potential trainees, who frequently feel that working at farm level is degrading. Familiarity with practical aspects of farming is however of considerable importance, for all researchers engaged in field experimentation. Experimental plots must be models of good management if the results are to be meaningful, and this cannot be achieved without acquiring farming skills. This practical background will also serve the researcher in good stead in their contacts with extension workers and farmers, and in identifying practical problems requiring research.

The international centres have compiled some evidence on the knowledge, understanding and abilities of 'agricultural specialists' with respect to the production of one or more crops. Working agronomists from Asia, Africa and Latin America have consistently shown alarmingly low scores on these subjects.

In five production-training courses at the International Rice Research Institute (IRRI), involving 173 agronomists for 30 countries, the average entrance score was 31.6%.

In a practical gest at Centro Internacional de Agricultura Tropical (CIAT) of 34 agronomists in diagnostic and management skills on diversified crops, the mean score at the begining of training was 36.4%. Most animal service graduates enrolled at the CIAT trianing courses were found to be unable to handle or restrain animals for diagnosis or treatment—a clear demonstration of their lack of opportunity in learning such basic skills. The International Institute for Tropical Agriculture (IITA) reports that the beginning score for 23 persons from 11 African countries was 22%. About half the group could not identify a rice seedling from an ordinary grass!

The conclusion drawn from the above is that 'all crop research trainees should be capable of raising the crop. They should gain experience in actual production operations, training which includes commercial field-experience, and on-station and on-farm trials in which they perform all cultural operations' (Byrnes, 1974).

The negative attitude to practical farming must be changed during training, otherwise the training must be considered a failure. At CIAT, on completion of the courses, practical experience was by far the dominant choice among all trainees as the most valuable aspect of their training period. Apparently a good programme of practical training also generates an increased appreciation of the value of practical experience. Even former trainees who have risen to higher positions in their home institutions mention that though they no longer physically used the practical skills they had learned at CIAT, they had gained confidence and a better understanding to evaluate research proposals (Byrnes, 1974).

Research Techniques

In the best of cases, the young researcher will have learnt at university the rudiments of experimental design and statistical analysis. Courses and practical field work on the planning, designing and implementation of a research project and analysing the results, should be organized on a regular basis for newly appointed research workers.

Every young researcher should go through a period of apprenticeship in well-designed, problem-oriented projects in the home country or another country in their region. In particular, the international research centres are now organized to impart training in a number of specialized fields, in particular in plant breeding, farming and cropping systems, research and socio-economic research. In view of their importance, the possibilities and potentials for training of the international centres will be discussed in detail.

Training of Technicians and Extension Workers

Obviously, country-wide experimentation on farms cannot be done by researchers alone; trained technicians will be needed at all stages of implementation, including the collection of data.

On-farm experiments also provide an excellent opportunity for cooperation between research and extension (cf. p. 251). No formal framework exists in which technicians and extension workers receive training in experimental techniques; it is therefore essential that those who are involved in experimental work should be trained by the research organization. This requires the active involvement of experienced research workers in the in-service training of technical and extension personnel. Seminars and field days should be held regularly in which research techniques are taught and demonstrated. Subjects with which they should become familiarized are roughly those presented in Chapters X and XI. Research results should also be communicated and discussed with these two groups.

Mid-career Research Workers

In addition to the need to keep continuously abreast of developments in their own and related fields of research, research workers in the middle of their careers encounter specific problems.

Refresher Training

Even the best equipped researcher requires continued training. It has been estimated that it takes only four to seven years, on average, after graduation, for most of the knowledge accumulated during the formal education of a scientist to become obsolete. Young, recently recruited scientists are frequently more familiar with new techniques and theories than are their older colleagues, who completed their formal studies a few years earlier. As to still older colleagues, they are frequently hopelessly behind the times. Researchers should therefore be encouraged to keep abreast of developments in the subject matter of their disciplines. Much of this will be based on self-study, but the research organization should help by allowing attendance at university courses and symposia and by organizing special courses, especially in new techniques. With the ever-increasing rate of technical and scientific advance, even senior research workers need refresher courses and sabbaticals.

It has been estimated that at least one third of the time of a working scientist has to be devoted to keeping up with new knowledge in his field and this study must continue throughout his scientific career.

Work Towards and Advanced Degree

A constructive way of overcoming a typical mid-career feeling of stagnation is to encourage the research worker to work towards a PhD degree; and the research institute should provide the essential facilities to do so. A precondition should be that the subject of the thesis be chosen within the research area in which the researcher is working, or in which he will engage after completionof his doctorate, and which can make a significant contribution to agricultural progress.

Normally, the opportunity to work towards the PhD degree should be reserved for researchers who have several years of research experience, whose field of interest is clearly defined and whose permanence in the organization is assured.

It has been stated that 'the most expensive item in higher agricultural education was the training in depth of specialists and particularly the provision of facilities for postgraduate study and research. Furthermore, it was also the activity where costs were rising most rapidly because equipment was more sophisticated and as the demands for trained technicians to service and maintain such equipment increased' (Coombs, 1970).

In consequence of the high cost of graduate education, most candidates from developing countries were expected to get their post-graduate training at foreign universities in developed countries. Experience has shown, however, that this kind of highly specialized training, in a narrow field of basic research, using sophisticated equipment, and on subjects which are usually of little relevance to the problems of developing countries, did not usually prepare the trainees adequately for the type of work needed in their home countries. Even if production-oriented subjects are chosen by the trainees, most of these universities are situated in temperate zones, whilst the students come from the subtropical and tropical zones.

The students also become imbued with certain values that handicap them in their future work, such as the superiority of basic research over applied research, the emphasis on the writing of 'papers', etc. Other undesirable tendencies that are usually already present, such as the desire for life in town and for work that provides prestige, are thereby strengthened. Whilst in the past there were few, if any, alternatives available to the post-graduate student other than to study towards an advanced degree in a university in a developed country, this is no longer the case. Most developing countries now have their own faculties of agriculture, and the international research centres also provide post-graduate training programmes (cf. p. 201). The assumption that graduate education is too expensive for developing countries is only true if it is intended to copy slavishly the objectives and methods of graduate education practised in the rich countries. It is not true if the actual research is carried out within the framework of the department of the researcher, in his field of research and under local conditions, or in another developing country that is similar but more advanced and has better training facilities. The supervisor should preferably be a senior researcher who is accredited by a university to serve in this capacity, or a faculty member of a nearby university. The candidate for a PhD degree should be given the opportunity to spend a few months before starting research in the field at the appropriate university, discussing his problems in the light of his former experience, learning techniques, studying the literature, and taking the courses required towards the degree.

Training for Research Management

The productivity of agricultural research is greatly influenced by the quality of research management. Research workers, when accepting research management responsibilities, are making a decision that will have major consequences not only on their future lives, but also on the research unit which they are to direct. Most research workers undertaking the duties of research management are not even aware of how abysmally ignorant they are of the basic principles of management, and that learning managerial science can help them to solve the innumerable administrative problems with which they will be faced in their 'new career', such as supervision of people, decisions on scientific programmes, and the budgeting of time, money and effort. Having always considered competence as an essential for a scientific career, they should realize that they must also acquire competence in research management if they are to be successful in their new and vital role. There are directors of research units who deny the need for their fulfilling a managerial role and who insist that their main responsibility is to guide and direct the scientific work of their unit and to provide inspiration for its researchers. Such a role is possible at the lower levels of research management; at the higher levels this attitude is possible only if someone else assumes responsibility for the managerial role. If this 'someone else' is a scientist, we are simply begging the question; if he is a non-scientist, nothing remains of the axiom that the man who effectively manages research should himself be a scientist. People trained exclusively in general management, without a research background, do not understand the potentialities of research, the idiosyncrasies of the researcher, or how research has to be carried out.

In order to facilitate the transition from a vocation as a researcher to the managerial duties to be assumed, educational programmes have to be designed to help scientists acquire the specific knowledge needed 'to analyze, make decisions, take action and bear responsibility for creating national agricultural research systems and attach themselves to problems ranging from national development priorities and strategies to technological requirements, support mechanisms and research priorities' (Rigney & Thomas, 1979).

International Training Facilities

Recognition for the need for managerial training to facilitate the transition from research scientist to research management is increasing in recent years. The newness of the area and a lack of personnel qualified to accept responsibility for preparing and implementing training programmes are obstacles that have to be overcome.

There are a number of organizations or agencies that offer some type of training and/or experience in agricultural research management, including the following (IADS, 1979):

- Southeast Asian Regional Center for Graduate Study and Research in Agriculture (SEARCA);
- International agriculture research centres (CIMMYT and IRRI) (see below);
- US Department of Agriculture;
- World Bank—Agricultural Management Case Studies and the Economic Development Institute;
- Harvard University—Management Case Studies;
- In-Service Training Awards Program (INSTA), Wageningen;
- International Course for Development-Oriented Research in Agriculture (Europe);
- Indian Institute of Management (IIMA), Ahmedabad, India; and
- Central Staff College for Agriculture (National Academy for Agricultural Management), Hyderabad, India.

Training programmes for *mid-career* persons are still in the formative stage in most institutions. While they are not yet fully formulated, the above organizations and others under consideration (by FAO in Africa, for example), provide the potential for a regional approach to an accelerated training programme for research managers.

IRRI's research training centres around the research project. Trainees work in a disciplinary department, and are closely associated with an IRRI scientist on one of the latter's research projects. In some cases, this work results in a jointly authored research paper. The trainees obtain solid research experience, learning each step involved in planning, designing, executing and reporting on a research project. Swanson (1975) found that after returning to their home countries, IRRI trainees tended to emphasize experimental research (in field, laboratory and greenhouse) aimed primarily at generating new knowledge about rice production in the tropics rather than in the direct development of location-specific rice technology (in agronomic field trials or genetic crosses).

CIMMYT's training stresses a 'team' or integrated approach to wheat improvement. The emphasis is on interdisciplinary teams that are well integrated in function. Trainees learn all the essential skills and techniques during each stage of the growing season and the varietal development process, with each test first being discussed in the classroom and demonstrated in the field. After he has become reasonably proficient the trainee helps to carry out each research task within the on-going CIMMYT programme. In brief, the CIMMYT training programme concentrates on genetic technology which has wide adaptation and far less on production technology which is by nature location-specific. As a result of this training strategy, the trainees were found to have focused on wheat breeding after their return home, and have continued to work in close collaboration with CIMMYT in its international programme. By contrast, they generally did not specialize in research problems within a particular scientific discipline or engage in developing complementary packages of production technology.

The differences in the type of work in which former trainees from the two international centres engage in is also reflected in their tendency to publish. In the course of two years, 37% of the former CIMMYT trainees, working in their national wheat programmes, had produced technical papers as compared to 84% of the IRRI trainees working in their national rice programmes (Swanson, 1975).

Regional Research

Regional research aims at investigating problems that transcend the national boundaries of a number of countries within an ecological region.

Regional research can be undertaken by a number of neighbouring countries in the same ecological region, or by externally organized and financed international research institutes.

Regional Research by Neighbouring Countries

Most efforts at establishing regional research institutes to serve a number of neighbouring countries have been made in Africa.

Experience with research which cuts across natural boundaries has generally not been very successful on the African continent. Exampes are: the dissolution of the West African Institute for Oil Palm Research, originally serving Nigeria, Ghana, and Sierra Leone, and which has become the Nigerian Institute for Oil Palm Research; the West African Cocoa Research Institute, originally intended to serve all English-speaking countries in West Africa for which cocoa is an important crop, and which has become the Cocoa Research Institute of the Ghana Academy of Sciences.

Various approaches have been attempted in Africa in organizing regional research (Robinson, 1970).

(1) Establishing a regional centre, in a strategically chosen site, with appropriate laboratory facilities and administrative headquarters. One example of this approach is the East African Agriculture and Forestry Research Organization which was to serve Kenya, Uganda, and Tanzania, mainly on problems of water use efficiency and crop irrigation.

Such a regional centre, to be effective, must do field work under varying environmental conditions in national experiment stations and in cooperation with the national research workers.

(2) Establish one regional administrative centre, with multi-disciplinary research units sited in the participating countries in which the major problems are found. An example of this type of organization is the 'Agricultural Research Council of Central Africa'. The main problem of this type of organization is the difficulties of communication and control between administrative headquarters and diffuse research units, and the inconvenience of centralized library and similar facilties.

(3) The third approach is for research institutions within a region having similar environmental conditions and problems to coordinate their research programme, thereby avoiding duplication and sharing the results of research. Devred (1968) has outlined the ten main objectives of such a coordinated regional research effort:

(a) To organize certain aspects of research within a region having similar environmental conditions and problems.
(b) The employment of multi-disciplinary research effort on problems.
(c) To coordinate and rationalize the development of research projects on a regional basis, their implementation being 'farmed' out among the main research institutes already in existence.
(d) To stimulate cooperation and reduce duplication between research institutes by means of closer working relationships where there are already similar scientific interests.
(e) Research on some problems of agricultural productivity could be conducted in this way and in order of regional priority.
(f) To facilitate cooperation in research among universities and research institutions between developing and developed countries.
(g) To develop and improve intensive exchange of agricultural research information, data and results within a zone of regional research.

(h) To concentrate a portion of national resources devoted to agricultural research on key topics selected on the basis of regional priorities.
(i) To collate the specific requirements for research that are at present beyond the available financial resources at a national level, and focus the attention of the international scientific community, research institutions, universities, etc., on these requirements. This would facilitate technical aid and assistance being obtained in the most critical areas, with the added strength of a single regional approach rather than two or three individual and national approaches to the same thing.
(j) To avoid as far as possible unjustified duplication and unnecessary competition in the field of agricultural research with particular reference of course to regional research subjects.

Though these objects are laudable and appear simple to apply, this kind of regional approach has been difficult to implement (Robinson, 1970).

The first requirements for the success of a regional programme would be the existence of good national institutions, capable of implementing national programmes and participating in joint activities on common problems selected by a regional research council. The second requirement appears to be the good offices of an external agency to serve as a catalyst, and possibly supplier of funds, for a joint regional effort.

International Research Centres

Impact on National Agricultural Research

The most important recent development for agricultural research in the LDCs has been the establishment of international research centres. The first two international centres—the International Rice Research Institute (IRRI) in Los Banos (Philippines) and the International Maize and Wheat Improvement Centre (CIMMYT) in Mexico—were established by the Rockefeller Foundation (1959) and the Ford Foundation (1966) respectively.

Following the considerable achievements of these two centres, in 1971, a 'Consultative group on international agricultural research', an informal association of mainly donor agencies, was formed. The membership consists of 12 'donor' countries, a number of countries elected to represent the developing regions, the regional development banks, the Ford, Kellog and Rockefeller Foundations, and the Canadian International Development Research Centre. The group is supported by a 'technical advisory committee' of 33 scientists. The objective of the consultative group is to use the funds provided by members in a coordinating fashion, to support research in developing countries in accordance with internationally agreed priorities.

The international centres have given an entirely new dimension to the national research programmes of the LDCs, which have to contend with limited financial resources and a lack of trained personnel, and cannot possibly, on their own, afford the comprehensive research required to solve all the problems faced by their agriculture.

By contrast, the considerable resources, excellent equipment and above all the interdisciplinary teams of high-level scientists attracted to these centres, make possible work on a scale and level that cannot possibly be matched on a national scale, excepting by the richest countries.

By providing consulting services to governments of LDCs, and assigning staff members as residential specialists, the international centres help to plan, organize and implement local programmes. By taking advantage of these services, the national research systems of the LDCs can obtain valuable back-stopping for their own programmes. They get access, rapidly and without undue expense, to much of the basic information that they lack, to technology that is relevant to their needs, to seeds of improved varieties for screening at their regional stations and to genetic material for their breeding work. By orienting the national programmes to adapting the findings from the work of the international centres to their own conditions, so that they can be applied at farm level, rapid progress can be made on many problems with relatively modest resources and a modest research system. In the plant breeding programmes, for example, promising material from the international centres is made available to the relevant national research institutions at early generation stages, so that local plant-breeders can make the evaluation and selection on the basis of their local experience and judgement. Therefore, the benefits of the crop improvement programme that can be obtained by different countries will depend on the competence of the national research centres (Cummings & Kanwar, 1974). 'The more successful are the International Centres, the more productive is the national research work' (Kislev, 1977).

However, the ability to evaluate and adapt the technology provided by the international centres depends on the existence of a well-planned, effective national research system in each country, staffed by trained and committed researchers.

Besides adaptive research in varieties and crop management of the principal commodities grown in their respective countries, national programmes need to focus on integrating new technology into farming systems appropriate to their own conditions (cf. pp. 180–182).

Research Programmes of the International Centres

Planning a national research programme so as to make the most effective use of the technology developed at the international centres requires an awareness by the former of the objectives and research activities of the latter.

The following services are normally provided by these international centres to national agricultural programmes in the regions for which the respective centres are relevant (Hardin, 1973):

(1) developing of and training in modern, problem-solving research procedures;
(2) generation of complete commodity 'production packages' with suggested procedures for adaptive testing and modification to fit local conditions;
(3) provision of genetic materials;
(4) international testing of materials and practices with associated data retrieval and analysis;
(5) consulting services; seminars, workshops and direct technical assistance.

Table XII shows the various research interests of the international research centres.

Table XII Research programmes of the international research centres.

	Main crop(s)	Main research areas	Main area of responsibility
International Rice Institute (IRRI), Philippines	Rice	Breeding for high yields, and wide adaptability	All rice-growing areas
		Socio-economic and technological constraints to adoption	
		Farming systems, management	
		Small-scale machinery	
International Maize and Wheat Improvement Centre (CIMMYT), Mexico	Maize	Breeding for wide adaptation and broad genetic background	Whole world
		Multiple disease-resistant and high-lysine content	
	Wheat	Breeding dwarf varieties for high yield, resistance to stress etc. Improvement of dietary wheat. Triticale breeding	
International Institute for Tropical Agriculture (IITA), Nigeria	Major food crops of the humid tropics	Improved farming systems as alternative to shifting cultivation	Humid tropics in the world

Table XII Research programmes of the international research centres (*contd*).

	Main crop(s)	Main research areas	Main area of responsibility
Centro International de Agricultura Tropical (CIAT), Colombia	Cassava	Breeding for high protein level, low cyanide and disease resistance	
	Field beans	Breeding for high yields, resistance adaptability to hot tropics	
	Beef cattle	Improving pastures, disease and parasite control, systems for family sized units	
	Swine	Cheap nutrient sources, improved management	
International Crop Research Institute for Semi-Arid Tropics (ICRISAT), India	Sorghum, pearl, millet, chick peas, peas	Improve genetic potential for yield and nutritional quality	Semi-arid tropic in the world
		Farming systems for semi-arid tropics	
International Potato Centre (CIP), Peru	Potato	Breeding and management	Whole world
International Livestock Centre for Africa (ILCA), Addis Ababa, Ethiopia	Livestock	Studies on genetics, physiology, nutrition, adaptation to climate epidemiology. Environmental studies on pastures	Tropical Africa south of the Sahara
		Socio-economic research	

SUMMARY

The efficient application of the results of agricultural research is one of the primary means for accelerating agricultural development. And yet, whilst most developing countries have more or less well organized research organizations, the research itself has had little or no impact on the agriculture practised by small farmers who form the majority of the rural population of these countries. The generally accepted view is that the main reason for the non-acceptance of the results of research is that it is not 'appropriate' for small farmers.

Actually, most advanced technology in improved varieties, soil management, fertilizers, plant protection, etc. is suitable for adoption by small and large farmers alike. However, for many reasons that are not directly related to research, and will be discussed in the following chapters, small farmers generally have no access to new technologies.

Because there is a growing awareness of the potential role of the small farmer in national economic development, the needs of the disadvantaged sector of the rural population are receiving more attention in recent years. The major implication for research is to focus more specifically on the special problems of the small farmer. The main change required in the conventional research programme is to integrate the results of research on individual components of production into complete farming systems. The proposed systems must be tested under the conditions under which the small farmer operates, and the constraints preventing adoption must be identified. The final research product must be the design of new farming systems incorporating labour-intensive, yield-increasing technologies that ensure optimal conservation of resources, high productivity and better year-round utilization of labour, and enable a significant improvement in the economic conditions of the majority of farmers.

For this purpose, the conventional research system must be complemented by multi-disciplinary pre-extension teams, incorporating agronomic, economic and social expertise. These teams must bridge the gap between experiment station and farmer by a major 'on-farm' effort.

To achieve these aims, a strong national research organization is required. Though the results of research carried out in the entire world, and in particular by the international institutes, is freely available to the developing countries, each country's ability to benefit from international research findings depends on its own indigenous research capability.

In view of the scarce human and capital resources available in the LDCs, the organization, planning and direction of agricultural research must be centralized in order to ensure public control of research policy in orienting and coordinating research efforts towards the needs of the community.

Certain basic principles regarding the objectives of agricultural research and responsibility for research planning and execution, are common to all developing countries. A model structure, incorporating these basic principles can, therefore, be planned, and is applicable to most countries. The basic components of this model structure are a central semi-autonomous national research institute, closely linked to the ministry of agriculture, with a network of regional polyvalent research stations.

The main contribution of the universities to agricultural research is for the faculty members to be involved in the formulation of research policy and programming; joint interdisciplinary research teams and/or to accept research projects on a contract basis, and to be active in in-service training of government research and extension staff.

The most important recent development for agricultural research in developing countries has been the establishment of the international research centres. These provide valuable back-stopping to national programmes by enabling rapid access to basic information, new 'packages' of technology, genetic materials for breeding programmes and training in problem-solving research procedures.

In contrast with the considerable contributions of the international centres, efforts at organizing regional research programmes, based on cooperation between a number of countries within an ecological region, to work on common problems that transcend national boundaries, has generally not been, up to the present, very successful.

References

Bailar, J. C., Jr (1965). The evaluation of research from the viewpoint of the university professor. *Res.Mgmt*, **8**, 1333–9.

Bunting, A. H. (1971). *International Seminar on Change in Agriculture*. Duckworth & Co., London: xiv + 313 pp., illustr.

Byrnes, F. C. (1974). Agricultural production training in developing countries: critical but controversial. Pp. 215–35 in *Strategies for Agricultural Education in Developing Countries*. The Rockefeller Foundation, New York: x + 444 pp.

Coombs, P. H. (1970). *What is Educational Planning*? UNESCO, Paris: 61 pp.

Cummings, R. W. & Kanwar, J. S. (1974). Transfer of technology outreach of ICRISAT. Pp. 487–97 in *International Workshop on Farming Systems*. International Crops Research Institute for the Semi-Arid Tropics, Hyderabad, India: viii + 548 pp., illustr.

Devred, R. (1968). Agricultural research programmes on ecological bases—basic principles and general measures for strengthening cooperation. Paper AER/68/1 presented to *FAO Reg. Conf. for the Establishment of an Agricultural Research Programme on an Ecological Basis in Africa (Sudanian Zone), Rome, November* 1968: 8 pp.

Eicher, C. K. (1970). Some problems of agricultural development. Pp. 196–246 in *Africa in the 70's and 80's* (Ed. F. Arkhurst). Praeger, New York: xiii + 405 pp.

FAO (1965). *Report on First Session of the Sub-Panel of Experts on the Organization and Administration of Agricultural Research*. Food & Agricultural Organization of the United Nations, Rome: 52 pp.

Gray, C. G., III (1974). Suggestions for Land-grant College action—a personal viewpoint. Pp. 72–8 in *Transforming Knowledge into Food in a Worldwide Context* (Ed. W. F. Hueg, Jr. & C. A. Gannon) American Academy of Arts and Sciences and University of Minnesota. xiii + 202 pp.

Hardin, L. S. (1973). International agricultural research and training institutes and national programmes. Paper presented at *Round Table on Organizing and Administering Agricultural Research Systems, Beirut.*

IADS (1979). *Preparing Professional Staff for National Agricultural Research Programmes*. International Agricultural Development Service, Report of a Workshop, Bellagio, Italy: iii + 122 pp.

Johnston, B. F. (1962). Agricultural development and economic transformation. *Food Res. Inst. Stud.*, **3**, 223–76.

Johnston, B. F. & Kilby, P. (1973). *Agricultural Strategies, Rural-Urban Interactions and the Expansion of Income Opportunities*. OECD Development Centre, Paris; v + 299 pp., illustr.

Kislev, Y. (1977). A model of agricultural research. P. 265 in *Resource Allocation and Productivity in National and International Agricultural Research* (Ed. T. M. Arndt, D. G. Dalrymple & V. W. Ruttan). University of Minnesota Press, Minneapolis: x + 617 pp., illustr.

Kislev, Y. & Evenson, R. (1973). *Agricultural Research and Productivity—An International Analysis*. Yale University, New Haven, Connecticut: (mimeogr.).

Lele, U. J. (1976). Designing rural development programmes: lessons from past experience in Africa. *Econ. Dev. & Cult. Change*, **24**, 287–308.

McPherson, W. W. (1968). Status of tropical agriculture. Pp. 1–22 in *Economic Development of Tropical Agriculture* (Ed. W. W. McPherson). University of Florida Press, Gainesville, Fla: xvi + 328 pp., illustr.

McPherson, W. W. & Johnston, B. F. (1967). Distinctive features of agricultural development in the tropics. Pp. 184–230 in *Agricultural Development and Economic Growth* (Ed. H. M. Southworth & B. F. Johnston). Cornell University Press, Ithaca, NY: xv + 608 pp., illustr.

Mellor, J. W., Weaver, T. F. Lele, U. J. & Simon, S. R. (1968). *Developing Rural India*. Cornell University Press, Ithaca, NY: xxi + 411 pp., illustr.

Norman, D. W. & Hays, H. (1979). Developing a suitable technology for small farmers. *Natl. Devnt*, **21** (3), 67–75.

Oram, P. A. (1972). Accelerating agricultural research in the developing countries. Pp. 141–64 in *The State of Food and Agriculture, 1971*. Food & Agriculture Organization of the UN, Rome: x + 189 pp.

Rigney, J. A. & Thomas, D. W. (1979). Strengthening agricultural research organizations. Pp. 42–5 in *Preparing Professional Staff for National Agricultural Research Programmes*. International Agricultural Development Services, New York: iii + 122 pp.

Robinson, J. B. D. (1970). Kinds of agricultural research organizations in developing countries. Pp. 120–130 in *The Organization and Methods of Agricultural Research* (Ed. J. B. D. Robinson). Ministry of Overseas Development, London: 200 pp.

Ruttan, V. W. (1968). Strategy for increasing rice production in South-east Asia. Pp. 155–82 in *Economic Development of Tropical Agriculture* (Ed. W. W. McPherson). University of Florida Press, Gainesville, Fla: xvi + 328 pp., illustr.

Ruttan, V. W. & Hayami, Y. (1973). *Technology Transfer and Agricultural Development*: *Staff Paper* 73–1. Agricultural Development Council, New York: 32 pp.

Schultz, T. W. (1971). *Investment in Human Capital: The Role of Education and of Research*. Free Press, New York: xii + 272 pp.

Swanson, B. E. (1975). *Organizing Agricultural Technology Transfer*. PASITAM, Bloomington, Indiana: x + 76 pp., illustr.

Violic, A. (1973). Maize training. Paper prepared for *Int. Maize & Wheat Improvement Centre, Mexico*: (cited by Byrnes, 1974).

Wilde, de, J. C. & McLoughlin, P. E. M. (1967). *Experiences with Agricultural Development in Tropical Africa*. Johns Hopkins Press, Baltimore, Maryland: Vol. 1, xi + 254 pp.

Winkelmann, D. & Moscardi, E. (1979). *Aiming Agricultural Research at the Needs of Farmers*. International Maize and Wheat Improvement Center, Londres, Mexico.

CHAPTER VII

Effective Transfer of New Technology

THE GAP BETWEEN AVAILABLE KNOWLEDGE AND PREVAILING PRACTICES

In most developing countries there is an enormous gap between available knowledge of improved technology and actual practice, as exemplified by the data in Tables XIII and XIV.

Part of the reason for the gap is that maximum yields are generally obtained on experiment stations and the farmer cannot achieve the same level of management as the experiment station.

The differences in yield level between experiment station, a farm demonstration plot and an average farm in Pakistan are shown in Table XIV.

In a study on actual and potential yields of rice in the Philippines, it was found that under present conditions the maximum attainable average yield for the country as a whole (4.1 t/ha) was more than double the actual yields (1.8 t/ha) (IRRI, 1978). The factors involved in creating this gap are shown in Fig. 31.

These examples are not intended to imply that obtaining maximum yields is desirable or economically justified, a matter that will be discussed in Chapter IX. However, the enormous gap between possible and actual yields is an indicator of the economic benefits that can be expected if extension is successful in substantially narrowing this gap.

Research findings do not automatically transform themselves into

Table XIII Maximum and average agricultural yields (in tonnes/ha) in India (FAO, 1972, by permission).

Crop	Maximum yield	Average yield
Rice	10.0	1.6
Maize	11.0	1.1
Wheat	7.2	1.2
Potato	41.1	8.0
Tapioca	48.0	13.0
Yam	19.0	5.8

Table XIV Approximate yields (in units/ha) at a research station, demonstration plot, and average farm, in Pakistan (FAO, 1969, by permission).

Crop	Research station	Demonstration plot	Average farm
Rice (paddy)	125	100	38
Cotton	88	70	20
Maize	250	200	28
Wheat	150	100	23

agricultural practices; even in the best cases there is usually a considerable time lag between the development of a new technique and its application. As farmers in advanced countries differ from farmers in underdeveloped and developing countries, so the agricultural extension problems also differ. In an advanced country the problem is how to communicate effectively the results of research to a farmer who is mentally prepared to accept new practices that will give him higher returns. In contrast, in the underdeveloped and developing countries, one must not only solve the problem of how to communicate information but also that of how to motivate the farm operator to accept technological change and use improved practices to his advantage, and to ensure he has the necessary means to do so.

It is a truism to state that effective transmission of research findings to farmers is essential if research efforts are to contribute to agricultural progress. For this, a minimum base of general education, and an on-going process of adult training of the farmer is important, as well as appropriate levels of education for those serving the farming community.

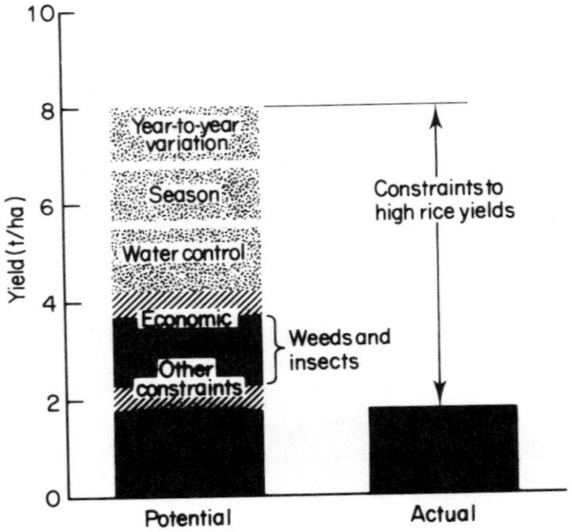

Fig. 31 Allocation among factors that constrain rice yields in the Philippines (IRRI, 1978b courtesy of the International Rice Research Institute).

There are four categories of recipients whose education will affect agricultural development (Wharton, 1965):

(a) *farmers:* including owner-operators, tenants, labourers, etc.;
(b) *those serving farmers directly*: extension agents, development planners, researchers, rural administrators, etc.;
(c) *those serving farmers indirectly*: merchants, credit agencies, suppliers of agricultural inputs, etc.;
(d) *leaders* and *policy-makers.*

We are concerned here with the needs for education of farmers and of those who serve the farmer directly.

Education and Training of the Farmer

The cornerstones on which the education of rural people in developing countries should be based are: (a) permanent change comes only from within and cannot be successfully imposed from without; (b) the changes in people's desires, feelings, tendencies, and capacities that help them to adjust to their environment usually require application of external stimuli for motivation and guidance; and (c) institutions for the promotion of change must be created and work effectively to this end (Leagans & Loomis, 1971).

General Education

In countries with a considerable proportion of illiterates in the population, the vast majority of these are usually concentrated in the rural areas.

According to United Nations data, levels of literacy (19% literate) did not prevent farmers from adopting new technology and increasing production greatly (Stevens, 1977). However, as soon as more inputs have to be integrated into agricultural systems, more sophistication is needed and lack of education may become a constraint to further progress.

Whilst it is possible to influence illiterate persons to use improved technologies and even train them in the use of agricultural machinery, it is more difficult to do so than with literate people. The efficacity of extension methods is also considerably reduced in the former case and a greater number of extensionists are required.

In Japan, education of the rural population is considered as one of the main factors responsible for the extraordinary development of agriculture in that country since the beginning of the century.

Griliches (1964) found that schooling is an important source of gains in agricultural productivity for the United States, whilst Chaudri (1968) found a statistically significant relation between schooling and farm output in a traditional setting in India. Apparently, schooling is important where there is a rapid rate of technological change.

Education and Agricultural Productivity

The main contribution of general education to increased agricultural productivity can be summarized as follows:

(a) It provides farmers with the basic skills (reading, writing, arithmetic) which facilitate the transmission of technical knowledge, make possible the keeping of farm records and making simple calculations required for deciding on the economic benefits of proposed inputs.
(b) It improves rationality—making it easier to overcome traditional social or cultural constraints which hinder progress.
(c) It increases inquisitiveness and thereby improves receptivity for new ideas, opportunities, and methods.
(d) It changes values and aspirations, and thereby strengthens the will to economize, and facilitates the adoption of new techniques (Tang, 1961).

These changes in basic skills and in outlook are of particular importance when the management of small farm units is the responsibility of many small-scale operators who must make their own decisions in the light of the specific conditions of their farms.

The basic weakness of the educational system in most developing countries is the disparity between their teaching programmes and the real needs of the rural population.

In many developing countries, more than half the pupils of primary schools drop out after the second year (UNESCO, 1964). Only a fraction of those who complete primary school are able to move up to secondary school, and a still smaller number go on to university. And yet curricula are drawn up as if the sole object of primary eduction is to prepare children for the competitive examinations leading to secondary education and the latter is viewed solely as a stepping stone to higher education.

A major reform would be for the programmes of elementary schools to be geared to preparing the majority of the children for the life they will lead and those of secondary schools should be specifically designed to train for service in the rural areas (Nyerere, 1970).

In research on the effect of literacy on the rate of modernization, it was found that people who have two or three years of schooling are not much different in their attitudes from those who have no education; the education threshold in most developing countries appears to be at least four to five years of schooling (Rogers & Svenning, 1969). The atmosphere of the traditional village discourages reading; also, little reading material is available to maintain interest and as a general result many people forget in a short time the little they have learnt. The policy generally followed in developing nations is to attempt to allocate a little education to almost everyone.

Alternative Approaches to Rural Education

Most developing countries are unable to cope with the problems of education by solutions that cost more than they are spending already. The possible alternatives are either to concentrate educational opportunities only upon children who can be carried through post-primary education (Rogers & Svenning, 1969) or to adopt new approaches that are within the means of these countries.

Self-supporting Schools

In order to make the new kind of education proposed by President Nyerere available to the majority of the rural people, he suggests that the schools should contribute to their own upkeep, by farming or other activities—'they must be economic communities as well as social and educational communities'. Each school should therefore have a farm and/or workshop which produces food and income for the school. These farms are envisaged as economic units, in which the pupils use improved labour-intensive agricultural practices. President Nyerere envisages that secondary schools at least can become reasonably self-supporting, with only teaching and supervision being provided by outside funds (Nyerere, 1970).

Opposition has been expressed to obliging schools to produce their own food for pupils and staff. It has been stated that this type of approach might be profitable as a discipline for the individual, or in education vocationally oriented towards agriculture, but given the age of the pupils, especially at primary school level, it may result in diverting the child away from the land, despite his natural inclination. At this level, agricultural education should be related to agricultural work only on a limited scale and excluding hard drudgery. 'A widespread ruralisation of subjects is most undesirable, since it would mean deliberately imposing an orientation which would be harmful to education in rural environments' (UNESCO, 1964).

The Education Corps in Iran (Blandy & Nashat, 1966)

At the time agrarian reform was instituted in Iran only 24% of the country's elementary-school teachers were serving in rural areas, so that at best only a quarter of the rural children of school age had a chance to go to school.

The Education Corps programme was launched as a means of using high-school graduates, who had been conscripted for military service, to contribute to the elimination of illiteracy in rural areas.

Corpsmen serve the normal 18 months period of conscripted service. During the first four months, they receive military training and are taught non-military subjects, such as teaching methods, sanitation, first aid, community development, etc. Those who pass an examination at the end of the training period are promoted to the rank of sergeant, equipped with special uniforms and sent to villages to serve the remaining 14 months of their term of service.

Only villages that request the services of the corpsmen, and who undertake to provide lodging and a place to teach are included in the scheme.

Classes are held at times which fit in with the needs of the rural communities. Besides reading, writing and arithmetic, the curriculum also includes hygiene, first aid and vocational skills of benefit to the children and their families. Group activities among the children, such as scouting, are encouraged. Literacy classes are also held for adults.

Besides the major contribution it has made to the solution of getting teachers for rural schools and combating illiteracy in the villages, the Education Corps has provided a number of additional benefits:

(a) Young people with secondary education, who previously might have been unable to find suitable jobs, have been given the opportunity to make a significant contribution to their country's problems.
(b) The Ministry of Education now recruits many of its regular teachers from the corps. These have to undertake to serve for an additional three years in rural villages before becoming eligible for transfer to urban schools.
(c) They have made an important contribution to the success of the rural reform and development plans.

The Ethiopian University Service

Following a proposal made by the National Union of Ethiopian Students to help overcome the acute shortage of teachers in rural areas, a compulsory year of service was established by the university as a degree requirement (David & Korten, 1966). As university education is free of charge, it was felt that the service would be a form of repayment by the students of their debt to the country. By making the break between the third and fourth years of study, the students would already have the qualifications needed for their work in the rural areas, and would benefit from the experience gained before completing their studies.

The students engage in teaching work under the aegis of the Ministry of Education and receive guidance, encouragement and academic supervision from the university staff.

Besides teaching, mostly in seventh and eighth grades, in which professional teachers are extremely scarce, the students generally carry out additional assignments such as helping in the national literacy campaign, conducting classes at night, organizing school clubs, special pest eradication and hygiene campaigns, etc.

Initially, the majority of students felt that the requirement to interrupt their studies and serve for one year in rural areas was an unfair imposition; after having become aware by their own experience of the critical need for teachers in the rural areas, most participants changed their attitude, and support for the continuation of the programme was almost unanimous.

Overall, the programme has proved to be successful, though a few students remained resentful and seriously detracted from the service they rendered. However, most performed outstandingly, even in the face of considerable obstacles, such as teaching in remote areas without basic teaching facilities. Though the programme is compulsory, good and enthusiastic work has been forthcoming (David & Korten, 1966).

Extension Education

Extension education is generally the main, if not the only, agent for farmer education in developing countries, and is a specialized form of the broader concept of adult education.

Objectives

FAO defines agricultural extension as 'an informal out-of-school educational service for training and influencing farmers (and their families) to adopt improved practices in crop and livestock production, management, conservation and marketing. Concern is not only with teaching and securing adoption of a particular improved practice, but also with changing the outlook of the farmer to the point where he will be receptive to, and on his own initiative continuously seek, means of improving his farm business and home (Chang, 1962).

Transfer of Improved Practices

Farmers in developing countries need knowledge in the following major areas (Wharton, 1965):

(a) *New inputs* which will increase output per unit area and investment of labour and capital. This involves up-to-date information on new varieties of crops or breeds of animals, agrochemicals, equipment, etc.
(b) *Techniques of production*: land use management and methods of sowing, rates and techniques of fertilization, effective disease, pest and weed control, feeding and treatment of animals, etc.
(c) *Economic factors of production*: the farmer must not only be a farm technician, he must also be a businessman. He needs to know about the choice of commodities he can produce, information on prices and marketing conditions, techniques of preparing his produce for the market, processing and storing, etc.

Much of the technological information required by the farmer must come from the research organization. Therefore an essential function of extension is to select information derived from research or from other sources that can be beneficial to the farmers they serve.

Community Development

It is generally agreed that in developing countries the objectives of extension education are not only to bring about an improvement in farming through the application of science and technology, but also to promote the social, cultural, recreational, intellectual, and spiritual life of the rural people. There is also general agreement, especially in developing countries, that whilst the advisory service has to adapt itself to the existing social framework of the farmers, it must also be active in promoting change towards a more progressive social framework as a prerequisite for technological change.

Extension programmes are therefore supposed not only to take research results to the farmers but also to attempt to substitute new values for old, overcome constraints imposed by old traditions and stimulate the development of auxiliary services.

Regulatory Functions

Many extension services in developing countries are burdened with regulatory functions, involving inspections, supervision and control, and above all, a considerable burden of paperwork.

Evolution of Objectives

Some extension services are mainly concerned with communicating technological information to farmers, others devote their main energies to community development. The emphasis generally changes from one aspect to the other in the course of development.

As a rule, the concern with community development diminishes as modernization progresses and efforts concentrate more and more on the technological aspects of extension.

In the early stages, much of the time of extension workers is taken up with regulatory services. Though there is a general awareness that these functions impair the ability of extension agencies to work for change and modernization, lack of staff and other constraints generally do not permit any other alternative. These functions are gradually transferred to other services as agriculture progresses and government services become more specialized.

Institutional Settings

An extension service can be the responsibility of government, as is the case in many countries; or it can be undertaken by universities; by farmers' associations; by banks, or by statutory bodies. There may also be different extension services working concurrently in the same country.

There are, therefore, many differences among extension systems adopted in different countries. These depend on the size of country, the important farm products, the existence of farmers' organizations and the subjects in which they are interested, the structure of the agricultural branches of governments, size of their budgets, and the types of training in the agricultural schools from which extension workers are recruited.

The Conventional Extension Service

The most prevalent type of extension in the LDCs is the government extension service.

The justification for government assuming responsibility for extension education lies in the fact that the extension programme, though aimed at improving the efficiency of agricultural production, actually contributes to general economic growth and in many cases to export earnings. Without the extension service, public funds invested in development projects would not be used as efficiently as they should be.

Without a service advising farmers on the adoption of innovations, the benefits of agricultural research financed by government may be lost, and potential economic gains foregone. Similarly, investments in rural infrastructure—roads, transport, markets, irrigation schemes, etc.—would not be effectively utilized.

Where the extension service is sponsored by government, responsibility may be assigned to a ministry of community development, ministry of education or a ministry of agriculture, the latter case being the most frequent. Even in a single ministry, extension work is frequently carried out by a large number of departments: crop production, animal production, fisheries, forestry, game, community development, water, veterinary services, home economics, etc.

Occasionally there may be a unified service responsible for extension in all fields of agricultural production and rural economy; it operates independently within the framework of national agricultural policies, objectives and priorities.

Organizational Structure

Most government extension services have a number of common features:

(1) They are usually organized as territorial hierarchies: headquarter's staff in the capital city, subordinate levels at province or region, district and sometimes division or subdistrict, and a broad base of geographically dispersed field workers in the lowest subareas.
(2) They consist of three categories of workers: (a) *extension officers*

(EOs), who are responsible for the administration of the service at national headquarters, and at the regional and subregional levels; (b) *extension advisers* (EAs), who work with the farmers at village and farm level; and (c) *subject matter specialists* (SMs), who provide technical advice and guidance to the extension officers in all matters pertaining to their respective specialities, and in particular, on special programmes.

Basically, the conventional structure of government extension services is as follows.

At headquarters, a director of extension is responsible for planning, coordinating and monitoring of the entire service. He must develop extension techniques, define regional targets for national production and realistic work programmes for extension officers; he must monitor the effectiveness of the extension service, develop training courses for extension staff and lecture at these courses. He is also responsible for effective liaison between research and extension, participates in identifying problems for research and their priorities. He is assisted by a number of heads of departments; these are generally based on commodities or groups of commodities.

Heads of departments provide guidance in administrative, technical and financial matters to the regional extension officers, and monitor their activities.

Depending on its size, the country is divided into a number of administrative units and subunits; at each level, an extension officer is responsible for the administration of the extension service in his area, providing leadership for the various subject matter specialists and coordinating their activities as well as supervising extension agents.

Down to the regional level, the function of the extension officers is basically administrative, and is differentiated mainly by the size of the areas under their jurisdiction. They are generally responsible for the implementation of the rural development programmes in their respective areas of jurisdiction and for supervising the work of the extension officers reporting to them. It is at the two lowest levels, the division and circle, that extension workers assume their distinctive roles.

The Divisional Extension Officer (DEO)

The main functions attributed to the DEO within the divisional boundaries are:

(1) To determine the most appropriate farming systems under the ecological and economic conditions prevailing in his division.
(2) To prepare agricultural development programmes.
(3) To collect information on the agriculture of the division, and social and economic changes affecting farm life.

(4) To ensure the supply of necessary equipment and materials for the effective implementation of agricultural programmes.
(5) To ensure that government schemes for rural development projects within his area are operating satisfactorily.
(6) To determine training needs of his staff.
(7) To monitor the performance of the staff under his control.
(8) To foster and maintain good public relations in the community.
(9) To assist with in-service and farmer training by giving lectures, organizing field days, etc.

From the above it is evident that the duties of the DEO entail a mix of administrative and technical functions. The technical functions are frequently submerged by the administrative responsibilities. This can only be avoided by providing adequate administrative support for the DEO.

The Extension Adviser (EA)

The EA is the key element in the extension programme, working directly with farmers and thereby translating government policy into reality at field level.

Generally, EAs are required to carry out a wide variety of functions. In Jamaica, for example, the 'job description' covers five pages of typewritten text. The following is a brief summary of his duties as outlined in this job description.

The EA must discuss with the farmers the problems relating to the development of their farms, determine the constraints that hinder the farmer, assist the farmers in developing desirable attitudes, the knowledge and skills necessary to improve their social, cultural and economic well-being, encourage them to adopt modern methods of agriculture, analyse the viability and potential of existing farms and develop farm plans. He must also identify, train and develop sound local leadership; disseminate agricultural information by maintaining demonstration plots, organize field days and lectures for farmers, demonstrate to them on their farms the proper manner to cultivate crops, organize farming groups, and organize and implement farmer training. He must advise farmers on livestock husbandry and pasture management.

In addition to the technical work outlined above, he has administrative and regulatory duties, such as preparing applications for farm houses by determining ownership of property, recommending payments for work done under a government enabling scheme and helping farmers in completing applications; ensuring that plants and seeds are delivered to farmers and that planting materials are properly used. He must also supervise government tractor operators by supervising measurement of the land and work of tractor and operator. He must forecast requirements for planting materials, determine acreage of crops and likely yields. He must also cooperate with agricultural agencies and groups engaged in youth and adult education by attending meetings, visiting schools, 4-H clubs, and other organizations, giving talks on agriculture and related methods.

Finally, he is required to keep up to date on modern agricultural methods, attend refresher courses, identify problems in which research should be undertaken and on which researchers are engaged.

And in case all this does not occupy him fully, he must 'perform such other duties as may be assigned to him from time to time'.

In India, the so-called village-level workers, besides giving advice on technical matters, may be engaged on family planning drives, collecting debt arrears or taxes, helping the farmer complete application forms, etc. (Hunter, 1970).

Subject Matter Specialists (SMSs)

Inevitably, EAs will encounter problems with which they cannot cope, and therefore require the backing of more specialized extension agents, subject matter specialists, who serve as consultants to the EAs, drawing on detailed knowledge of a specific commodity or discipline. SMSs are preferably stationed at the appropriate experimental station, so that they can maintain close contact with their counterparts in research.

Their primary role, however, is to support the EAs, who are mainly generalists, by giving them guidance and support: identifying the problems, proposing effective solutions, participating in the preparation of demonstration fields, preparing bulletins and pamphlets; they are also active in the in-service training programmes of the extension staff and cooperate in experimental work with their research colleagues (cf. pp. 250–253).

Objectives and Methods

Generally the main emphasis is on production techniques, rarely is advice given on farm planning and management as a whole. A small number of women extension workers sometimes advise housewives on cooking, food preservation, sewing, etc.

The methods used consist of farm visits and group activities, such as demonstration plots, group visits to experiment stations, radio broadcasts, printed matter and training courses.

In general, individual efforts are concentrated on the larger and more progressive farmers.

The failure of many extension programmes to affect the vast majority of farmers has been generally ascribed to a lack of extension workers, to poor organization, etc. It should, however, be clear, that structural improvements will not result in a more general and equitable rural development unless the goals of the extension service clearly reflect an unequivocal commitment to achieve this objective.

Reasons for Failure of Extension Efforts

There is no disputing the fact that extension efforts in developing countries have not always achieved their anticipated results.

'An agricultural extension service is used by almost every country as a major tool for agricultural development; yet it has proved disappointing in many ways' (Hunter, 1970).

Failures of the extension system can be attributed to:

(a) organizational defects and operational weaknesses;
(b) lack of an appropriate technology;
(c) lack of coordination between research and extension;
(d) economic and social constraints (these will be treated in Chapter X).

Organizational Defects

Though there are occasions on which organizational forms have been imported from other countries without being adapted to accommodate the cultural norms of the community to be served, or without consideration of the available human and capital resources, failure of the extension services can rarely be attributed to the form of organization adopted, but is largely due to operational weaknesses.

Communication Within the Service

One of the most important potential weaknesses of an extension service is the lack of vertical communication between workers in the field and their chiefs at regional and national headquarters and vice versa. Supervision of field staff is often erratic, or may be almost non-existent or poor. Field visits may be rare and unpredictable, instructions may then be given based on inadequate and hasty appraisals. As a result, national policies seldom reach the local extensionist, and feedback rarely reaches the top levels of administration whose decisions are then made without a full and up-to-date knowledge of the situation at the operational level (Rogers & Svenning, 1969).

In a study of the agricultural extension organization in 71 villages in eastern Nigeria, Hursch *et al.* (1968) conclude that: 'agricultural extension agents often live in villages under conditions that foster lethargy, with no meaningful communication with superiors, inadequate supervision and advice. All these factors create feelings of personal alienation and dislocation'.

In Tanzania 'field workers are mostly left to decide for themselves which crops to emphasize, what operations and improvements to stress, which farmers to concentrate on and how to organize their time' (Chambers & Belshaw, 1973). The general lack of supervision is contrasted with the tighter systems in one-crop organizations such as the Nigerian Tobacco

Company (Harrison, 1969), the Kenya Tea Development Company (Heyer *et al.*, 1971), and the paddy-growing Mwea irrigation settlement in Kenya (Chambers & Moris, 1973), with their clearly defined tasks, strict disciplines, cross-checks on performance, and close supervision of field workers.

Communication from the field staff upwards is also frequently defective. Reports are often 'routine, ritual, unusable, unused and unread' (Chambers & Belshaw, 1973). Communication within ministry headquarters is generally no less defective. At the highest level, the heads of the extension service are rarely directly involved in policy decision-making; as a result, the extension service acts as an executive arm, carrying out policies in the formulation of which they have had no part. Orders to carry out these decisions move downwards through the échelons, who have no discretion to adjust the programmes to the specific socio-economic and ecological conditions in the areas under their jurisdiction.

Straus (1953), in a study on the effectiveness of the extension service in Ceylon, came to the conclusion that notwithstanding the existence of a well-staffed extension service, they were seldom able to achieve their stated goals. Extension workers carried out 'drives' to promote a new crop, for example, on order from above, without any consideration of whether the innovation was appropriate to the individual areas involved.

Planning the Extension Programme

Chambers & Belshaw (1973) state that in African countries extension programmes are often badly designed or inappropriate.

There may be either a lack of instructions from headquarters, or an overload of programmes flowing out from the centre without any systematic appraisal of compatibility in their demands on staff time, 'the later ones burying the earlier'.

In one district in Kenya, the authors found that the official programmes for a year, when worked down to the level of location, varied in their demands on staff time, month by month, from 18% to 47% of the working days available!

The main weaknesses in planning extension work are summarized as follows:

(a) unrealistic targets, with junior field staff not involved in setting them;
(b) no systematic ordering of priorities between the competing demands of different programmes on field staff time; and
(c) no systematic work planning for field staff.

Roles not Related to Promotion of Agricultural Production

The extension service has been used to perform additional roles, other than promoting agricultural production, such as regulatory, supervisory or commercial duties, involving enforcement of regulations or laws, collection of repayments for production loans, etc. In the eyes of villagers they are therefore often seen 'as emissaries of government enforcing mysterious and even senseless regulations, who must be bribed or evaded in order that the ordinary needs of village life can be met' (Hunter, 1970).

For example, in a study of the factors impeding the diffusion of new technology in the development plan for the Pueblo Valley (Mexico), Diaz (1974) found that the change of emphasis of extension workers from demonstration and field days to loan and repayment processing had a negative influence on the diffusion of new technology.

In view of the severe lack of personnel with which developing countries are faced, it is unrealistic, and possibly harmful, to demand that extension personnel should devote themselves exclusively to providing professional advice to farmers. Supervisory and regulatory functions that are directly linked to the adoption of improved practices are not only legitimate activities for the extension worker, but may actually strengthen his hand. For example, the provision of credit for essential inputs, to be effective, would require the extension worker to supervise use of these inputs in the field and to certify that they were applied properly; regulations requiring the adherence to certain soil conservation practices, collective pest control, fire prevention measures, etc., can properly be monitored by extension workers, provided they endeavour to explain the need for these measures to the farmer and make him aware that they are intended for his benefit. On the other hand, the extension worker should have no part in regulating or other activities that have no direct relation to promoting production, even if savings in manpower can be achieved thereby. These activities will undermine the trust in the objectivity of the field worker, a prerequisite to his effectiveness. Similarly, all activities that involve much paper work must be avoided, as they inevitably cut down the time the extension worker can spend in the field. In certain extreme cases, practically *all* the time of the extension workers may be taken up with paperwork.

Number of Extension Workers and Logistic Support

A difficult problem is that of determining the minimum number of extension workers required to make a significant impact on the agriculture of a region. For extension to be effective, contacts between farmers and extension agents must be sufficiently frequent. The desirable ratio of extension workers to farms will, of course, depend on the specific conditions prevailing in each region, such as potentialities for increased production or diversification, the technical and educational level of the farmers, the density of the farming population, the mobility and dedication of the extension worker, to name just a few.

A tentative and rule-of-thumb conclusion, established at a conference on extension in East Africa, sponsored by FAO, is that the minimum objective should be a ratio between 1:350 and 1:1000 (FAO, 1962).

The actual ratio varies considerably from country to country: in India and Zambia it is over 800; in Brazil, about 5500; and in Bolivia, over 8000 (FAO, 1975). Actual contact rates may even be lower: they amount to only about 2.5% of all farms in Paraguay; and 3.3% in Ecuador (Herzberg & Antuna, 1973). In many cases in which extension programmes have been most successful (the Gezira in Sudan, Comilla in Bangladesh, and Monkara in Chad), these have been carried out with a massive investment in a limited region, often to the neglect of other parts of the country (FAO, 1976).

In most developing countries, the number of extension workers is far from adequate. This shortcoming is further compounded when the requisite logistic support is not forthcoming: visual aids, equipment, transportation, communication channels, guidance by subject matter specialists, etc.

In many African countries, for example, the extension worker who is usually less well trained and has to work under far more difficult conditions than in developed countries, with less facilities, is often supposed to work with several thousands of farm families (Nour, 1969).

In a survey of 295 Philippino farmers, it was found that 88% were aware of the extension workers presence in the village. However, when asked how often they had contact with the extension worker, the replies were as follows: never, 22%; seldom, 45%; often (once or twice a month), 18%; and very often, 15% (Lu, 1968). Insufficient numbers of extension workers have led to the 'palm tree type' of personnel structure. Because it is easier to produce and employ a limited number of graduates at the higher level than the large numbers required at intermediate level, the top échelons of the extension service may have too few mid-level technicians to work with them. As a result, little work is done in the field and contacts between the extension service and farm people are minimal (Chang, 1969).

Ineffective Extension Personnel

The village-level extension worker has the extremely difficult role of introducing new concepts and techniques to a tradition-bound community; this requires skills and technical knowledge as well as understanding of the people, their problems and attitudes. To be successful, the village-level worker requires 'the teaching ability of the schoolmaster, the skills of the farmer, the persuasive capacity of the politician, the understanding of the social worker and almost infinite patience' (FAO, UNESCO, & ILO, 1971). Yet this difficult task is generally undertaken by the lowest paid and least educated members of the government service.

Technical Inadequacy

Extension workers are frequently technically incompetent in agriculture. The little training they receive is limited to technical subjects, whilst the economic and social aspects of extension work are completely neglected. There is also little teaching in communication techniques, including audio-visual aids. Very little in-service training provided to keep the extension workers abreast of technological and other changes (FAO, 1976).

Over the past ten years, some of the international centres have been screening and testing agricultural specialists from Asia, Africa and South America and have compiled information on their knowledge, understanding and abilities with respect to one or more crops. The scores were consistently low. For example, in several thousand tests conducted by the International Rice Research Institute, Asian rice extension workers scored on the average about 25%, on a practical examination consisting of identifying common diseases, insect pests and their damage, nutritional disorders and the common chemicals used by farmers. 'In other words, they were right only one time out of four in identifying the farmers' problems' (Byrnes, 1974).

In testing 360 extension workers from Bangladesh, in 1970, the entering score was 21.3%. After receiving training, the final average was 81%, showing that it was possible for them to improve their knowledge, if given the opportunity.

The International Institute for Tropical Agriculture, in Nigeria, reported similar results. The average score for 23 extension workers from 11 African countries, who came for rice production training, was 22%. About half of the group could not distinguish between a rice seedling and ordinary grass. CIAT, on screening 36 professionals from ten countries in Latin America working in extension and teaching, found that their level of competence was similar to those described above (Byrnes, 1974).

The level of knowledge required by the extension worker depends on the professional level of the farmers he serves. As agriculture progresses, extension workers must keep ahead of farmers. When they fail to do so, they soon run out of content for their extension activities. This may happen even after a few months of contact with farmers, who quickly realize that the extension workers no longer have anything left to offer them.

A key tool of the extension worker is the demonstration plots. To be visible, demonstration plots must be local and therefore they have to be very numerous. The limited staff are generally unable to attend properly to their plots and frequently do not have the necessary skills. As a result the work is badly done, results are poor and the demonstration is counter-productive.

Many ministries of agriculture in Africa are convinced that 'the lowest levels of the agricultural extension staff are virtually careless, incapable of giving the sophisticated advice which is the only advice worth giving, and devoted not to the development of their areas but to their own farms and interests' (Chambers & Belshaw, 1973). Studies in Tanzania and western Nigeria (Harrison, 1969) have found that extension staff work only about five hours daily (including travel). The number of visits to individual farms was found to be ten per week in Tanzania (Chambers & Belshaw, 1973) and five per week in a certain area in Kenya (Leonard, 1972). In western Nigeria extension workers were found to concentrate mainly on a few relatively accessible villages (Kidd, 1968).

Poor Working Conditions

It is highly probable that much of the inefficiency and low work output of extension workers is a response to organizational defects (see above) and poor working conditions.

Chambers & Belshaw (1973) describe the problems and disincentives facing field agricultural staff in African countries due to poor terms of service, living conditions and working conditions: though they are generally secure in their jobs, their pay and allowances are less than those of their peers working for parastatal or private-sector organizations, such as tobacco extension in Nigeria, or tea extension in Kenya; housing is often not provided or is poor; promotion prospects are poor and are generally believed to be related to family or tribal connections rather than to work performance. Where work performance is evaluated, criteria are often used that emphasize distribution targets, such as number of visits made, fertilizer bags distributed, rather than actual effort or impact. As a result, the village-level worker has an additional incentive to concentrate on large farmers, who take large quantities of inputs (Jiggins, 1977). Good work is not seen to be rewarded and poor work is not seen to be penalized. in brief, there is little or no economic or professional incentive to work hard and well.

Working conditions are also hard and physically frustrating. The climate is difficult, the roads poor and transport inadequate. Small wonder therefore, 'that junior staff, even if they *start* with enthusiasm and energy, are pushed by the very situation and system within which they work, and the lack of rewards and incentives, towards an attitude of resignation, apathy or looking after "number 1" ' (Jiggins, 1977).

One of the preconditions for improving extension services is therefore undoubtedly the improvement of working conditions and the provision of incentives, such as adequate remuneration and promotion based on performance, suitable housing and transport, so as to motivate the extension workers to achieve the goals of the service.

Attitudes and Behaviour

Technical adequacy of the extension agent is a prerequisite for his effectiveness, but it is not sufficient. Failure to keep promises, laxity and negligence in the performance of their responsibilities are cited by Byrnes (1969) as major reasons for the failure of extension workers. When demonstrating a new practice or package of practices, it is not sufficient to show farmers that a better crop can be produced. The farmer must be made aware of the problems involved in adopting the new technology: the possible changes required in his farming system, details of the costs and possible benefits expected, what the rates involved are, availability of inputs, government incentives to adoption available to him, etc.

Although the extension service generally claims that there is a two-way pathway of information between extension workers and farmers, the assumption in reality is that useful knowledge flows only in one direction so that 'an aura of benevolent authoritarianism pervades many extension and training programmes' (Coombs & Mansour, 1974). In extreme cases, the recommended packages of practices are compulsory; farmers are required, under pain of penalty, to conform to instructions regarding soil conservation, planting specific crops, observing time schedules, etc. Although these rules are designed for the benefit of the farmer, they generally achieve the opposite by making both the proposed measures and the extension agent suspect in the eyes of the farmer.

In Indonesia, for example, recent government campaigns designed to increase rice production used a combination of persuation and coercion to induce rice producers to adopt a rigidly specified package of practices, and to sell their produce at government-fixed prices. The increase in yield did not offset the costs of the prescribed package of practices; the campaign embittered the farmers and weakened rural institutions (Timmer, 1975).

Many extension workers alienate farmers by showing a sense of superiority combined with a lack of understanding of the cultural values and social mores of their clients'. This may be compounded with a dislike of physical labour and 'getting their hands dirty'. These attitudes are themselves related to cultural traits where relationships between upper and lower strata are based on superiority. Straus (1953) cites the case of extension workers in Ceylon, who, even under direct orders to treat the villagers as equals, were unable to bring themselves to do so.

The main reason why Indian farmers did not readily adopt potentially profitable new practices of which they were made aware, was that they were *unable* to do so, because they lacked the necessary inputs. However, when these were made available by government 'the authoritarian behaviour of extension officials and the generally unfavourable conditions under which these inputs were made available to farmers acted as major disincentives' (Hodgdon and Singh, 1964).

A good deal of resistance to the proposals of the extension agent can be due to lack of confidence in his objectivity. The more unbiased the extension agent and the less he is suspected of making a personal profit from the adoption of his proposals, the more influential will he be.

Good results obtained from applying his proposals, proving that his suggestions were sound, will considerably increase his prestige and affect the future acceptance of his suggestions.

Farmers' Views on Extension Workers

The shortcomings of many extension agents are reflected in the comments obtained in the course of interviews with 45 farm household heads, chosen at random in five barrios in the Philippines, about the extension worker assigned to each barrio (Byrnes, 1969):

(a) He is too young compared to the majority of the farmers and so is not too experienced in matters of farming (27/45).
(b) He uses (technical) language we cannot understand very well (33/45).
(c) He goes by the book and not on what is really happening in the field (17/45).
(d) He is not very sure sometimes of what he is advising us to do (35/45).
(e) He cannot answer many of our questions.

Economic and Social Constraints

Not all the reasons for the failure of extension efforts can be imputed to deficiencies in the service. The efforts of the extension service may fail because of political, cultural or economic conditions. The forces opposing change may be too strong. In Niger, for example, when government organized its extension service in 1968, local notables, politicians and traditional chiefs opposed the new service, because they feared they might lose their traditional control over the farmers (Heck, 1979). The resources needed for the extension effort may not be available, the essential inputs may be lacking or the infrastructure inadequate. The potential benefits may be insufficient to overcome social and cultural constraints. These aspects will be discussed in detail in Chapter X.

Alternative Solutions

The long list of deficiencies in many of the existing extension services in developing countries does not imply that improvement is not possible because of the numerous constraints encountered in these countries.

Faced with the inadequacies of their extension services, many governments in developing countries have opted for one of the following solutions:

(1) To maintain the existing structure, but improve its *modus operandi.*
(2) To transfer responsibility for extension from the ministry of agriculture to other formal institutions: universities and agricultural colleges, farmers associations, cooperatives, statutory commodity boards, regional development boards, multiple organizations.
(3) To adopt unconventional methods, based on the self-help approach.

Improving the Mode of Operation of the Conventional Extension Service

Among the various efforts to improve the functioning of conventional extension services, one that has been successfully introduced in India, Turkey, Burma, Nepal, Sri Lanka and Thailand, in projects assisted by the World Bank, is the *training and visit system of extension*, developed by D. Benor.

The system, as it was conceived and subsequently adapted in the course of a decade, is described in detail by Benor & Harrison (1977), and will be briefly summarized and evaluated.

The conventional structure of the extension service as described above, remains basically unchanged: the proposed system is based on village-level workers with comparatively low standards of education, supported by subject matter specialists, working within a management structure which establishes a clear single line of responsibility.

It is in the basic extension techniques, encompassing concepts of goal definition, work planning and monitoring at each échelon of the service that the 'training and visit' system differs from conventional extension services. Essentially, it is based on a systematic, time-bound programme of visits and training. Under this system 'schedules of work, duties and responsibilities are clearly specified and closely supervised at all levels'.

Whilst the basic conventional hierarchical structure remains unchanged, each level of the service has a span of control narrow enough to permit close personal guidance and monitoring of the level immediately below. Hence the regional extension officer (REO) supervises four to eight divisional extension officers (DEOs). Each DEO is in charge of six to eight extension advisers (EAs). The REOs are supervised either directly by headquarters or by an intermediate échelon, depending on the conditions pertaining in each country.

The Extension Adviser (EA)

The EA must live in the area in which he is active. The maximum number of farm families that an EA can cover so as to make a significant impact on the agriculture of an area depends on many factors, such as the potentialities for increased production or diversification, the types of crops grown, the technical and educational level of the farmers and of the extension worker, to name just a few.

These ratios may vary from one in 300 in special cases, up to one in 1200 in areas with a low cropping density and a few dominant crops.

In any case, the EA cannot maintain personal and regular contact with all the farm families in his area. He therefore divides his area into eight groups of farms. From each group he selects, in consultation with the village leaders, about 10% of the farmers who will serve as 'contact farmers', and on whom he will concentrate his main efforts. These should be small farmers, typical in the kind of farming that they are practising, but of good standing in their community, whose opinions are likely to be respected and their example followed. They must indicate their willingness to accept the advice of the EAs, to try out new practices and to explain them to several neighbours, relatives and friends who are farming in their vicinity. The spread of the improved practices will be further enhanced by group activities. In this way, the EA can make a major impact on a relatively few farmers, but the adoption of the improved practices can spread rapidly to other farmers.

Initially, the contact farmers will be encouraged to apply the recommended practices on part of their farm only, and expand as they gain experience and confidence.

Each group is visited by the EA once every two weeks for a full day. These visits will always be on the same day of the week, so that the farmers know in advance when the EA is expected. In the morning, he visits the contact farmers, checking progress, noting any problems that have come up, discussing what needs to be done until the next visit. In the afternoon, he is available at a convenient location in the village to any farmer of the group who requires his help or advice; he can also initiate group discussions on any topical subject of interest to the group, or arrange a visit to one of the farms for demonstrating something topical.

If in the course of his visit he notes any occurrence of particular importance, such as an incipient pest attack, he notifies his DEO, who will take the necessary action.

The entire working schedule of the EA consists of:

(a) four days a week visiting his farmers (so that he covers his circle of eight groups in the course of a fortnight);
(b) one day a week in his office for meeting farmers who require his help, keeping records, writing records, etc., or for organizing group activities;
(c) one day weekly is reserved for training (see below).

The EA is required to maintain a schedule showing his fixed days of visiting each group, his office days, the programme of group activities, etc. A copy of the schedule is made available to his DEO.

A monthly report summarizing his activities, the progress achieved, constraints encountered and observations on future work is also submitted to his DEO.

A key element of the Benor system is the intensive and continuous in-service training, during which the EA learns what to recommend to his farmers during the next period of visits. He also has the opportunity to discuss with his trainers the problems he has encountered during his work. One of the two fortnightly training sessions is conducted by the team of subject matter specialists assigned to the region. These sessions are scheduled so as to provide training for groups of 30 to 40 EAs for a full day. At each of these sessions, only three or four crucial matters are treated in depth. One-third of the time, at most, is devoted to lectures, the remainder to group discussion and field demonstrations.

The other fortnightly training session is conducted by each DEA for his group of EAs. This is devoted to an informal discussion of the problems encountered by the EAs in the course of their work; the matters presented to them on the sessions with the SMSs, etc.

The Divisional Extension Officer (DEO)

The DEO spends one day a week on training sessions (alternately: the one he conducts himself and the one conducted by the SMSs) and four days in the field, monitoring and guiding the work of the EAs. He should be able to visit two EAs and their farmer groups in a day. His visits should be prescheduled and timed so that over a period of several months he sees each of his EAs and their groups.

The Regional Extension Officer (REO)

REOs should spend at least half their working time on field supervision of the extension service and the training programme. They are seconded by a team of SMSs.

Subject Matter Specialists (SMSs)

Each SMS devotes about one-third of his time to the training sessions for the EAs, one-third to field visits and field trials, and one-third on maintaining contact and cooperating with his counterparts in the research organization.

Priorities

Benor & Harrison (1977) stress that initially the extension effort should be focused on improving agricultural management practices, such as better land preparation, improved seedbed and nursery maintenance, the use of good seed, seed treatment, timely operations, weeding, proper spacing of plants, etc. The increased use of purchased inputs, such as fertilizers, should not be recommended initially. The rationale for this approach is that improved cultural practices produce sure results, involve no risks, require more work (amply available) and require little cash. Further, purchased inputs or costly investments, such as tubewells, cannot give full benefit until practices are first improved.

This approach is in contradiction with the line of thought that stresses the need for a *complete package* of improved practices. True, fertilizers, HYVs, tubewells cannot give full benefit unless management practices are improved, but it is equally true that conversely, the physical effort and time invested in improving management practices will give very meagre results unless the necessary inputs are provided which enable a worthwhile increase in yield to be achieved. The question therefore arises: how to reconcile the excellent results of the system claimed by the World Bank, and the low probability of achieving significant increases in yield from management practices only? The answer to this question is fortunately provided in the annex to the brochure describing the Benor system, in which three case studies of the application of the method are reported in detail. In every one of these cases, HYVs, fertilizers and pesticides were used in conjunction with improved management practices. For example: farmer Shri K. N. Deb Moswami, in the state of Assam, who had installed a tubewell, started growing HYV paddy-rice and wheat, using fertilizer, and found that he was losing money! On the advice of the EA (working within the re-organized training and visit system) 'he adopted line sowing, grew seedlings on raised seedbeds, provided for sufficient spacing and plant population. He also gave top dressings of fertilizers at the rate of 60 kilograms per hectare, supplemented by farmyard manure, and took care to remove weeds. Pest attacks worried him but he has now learnt to use pesticides, can talk about plant diseases in English terms, and mention the cures too'. A clear case of a classic 'package programme' and definitely in contradiction to the stated philosophy of the system. The other case histories similarly show that the excellent results obtained from the 'Training and Visit system' were based on recommendations including purchased inputs *and* improved management.

In Summary

The training and visit system is based on the following guidelines:

(a) The EA must be divested of regulatory functions and his administrative load reduced to a minimum.
(b) A well-defined realistic goal must be established, which he must be able to achieve within a predetermined period.
(c) A realistic working schedule must be established to which he must adhere strictly, under proper guidance and supervision.
(d) Effective communication throughout the system must be ensured so that he gets the technical and administrative backing needed, as well as providing the proper guidance on the policies and strategies of the ministry.
(e) He must participate in a continuous in-service training programme so that his professional competence can be gradually built up to a level commensurate with the vitally important role he is called upon to fulfil.
(f) Effective monitoring of his work is essential, providing professional guidance and encouragement in what is bound to be an extremely difficult task.

As to the actual recommendations to be made to farmers, we have seen that there is a contradiction between the declared intention of initially stressing improved management practices to the exclusion of purchased inputs, and the actual practice of recommending complete packages of improved technology.

Universities and Agricultural Colleges

The traditional land-grant college model of the USA is based on combining higher agricultural education, research and extension within the same institution. The adoption of this model in developing countries has often been advocated wherever US aid was active in promoting agricultural development.

Because the American extension system was one of the main contributors to the phenomenal increase in production achieved by American farmers, this success serves as an excellent argument that others should adopt the same system. However, the transfer of responsibility for extension from government ministries to agricultural colleges or universities has generally not given satisfactory results in developing countries.

The reasons for the inappropriateness of the US model of extension service are similar to those advanced for explaining the inappropriateness of the land-grant college approach to research in many developing countries (cf. pp. 192–193).

The United States extension service was evolved to meet the needs of a farm population at a relatively advanced stage of development and not in order to mobilize them for development, as is the primary goal in developing countries. It came in response to pressures exercised by a politically powerful farmers' sector; literacy and exposure to a well-developed system of mass media was widespread; urban jobs were available for farm labour made redundant by modern technology; *per caput* incomes were high (Barraclough, 1973). It is highly improbable that a system developed for these circumstances should be appropriate for the mass of rural people in underdeveloped countries. The family-size but large commercial owner-cultivators who are the mainstay of the American extension effort are a small minority of the farmers in developing countries, the majority consisting of small subsistence farmers unable to adopt new technologies even if they are made aware of them, unless a whole package of prerequisities is assured.

In this context it may be relevant to note that the American extension system has been largely ineffective in rehabilitating the rural population of the backward and depressed areas in the USA, until a special organization was established for this purpose in the 1930's: the *New Farm Security Administration* (Barraclough, 1971).

An Agency for International Development (AID) report states that in too many cases attempts were made to impose the US form of extension upon a foreign university for which a US university had a contract with

AID. Sometimes a new institution was established in competition with an existing one (AID, 1968).

An example of the problems encountered when establishing the US extension model in a developing country is provided by India.

Large-scale involvement in extension by the Universities in India is relatively recent. Where agricultural universities have been established (and the intention is to establish one at least to each state) they have been patterned on the model of the land-grant colleges. Confrontation with the extension services of the ministries of agriculture has already led to a re-evaluation of the function of the universities in the field of extension. The problem was whether the university should aim at direct contact with the farmers, thereby actually taking over the extension service, or should it confine itself to training extension workers, and provide the framework for farmers' courses.

In India, extension has to serve a very large number of small farmers, many of them illiterate, who need—in addition to technical advice—help in obtaining supplies, credit, equipment, etc. For these reasons, the extension staff consists of a large number of simply trained staff. Several agencies are involved, and these require close cooperation; certain programmes (soil conservation, insect control, credit recovery, etc.) may involve elements of control and even compulsion. All these functions require a government-controlled coordinated service, with statutory functions and an administrative framework for thousands of workers. It is doubtful that a university can, or should, undertake such roles. 'This would immediately turn the university into a bureacratic organisation indistinguishable from the existing government service.' (Hunter, 1970).

Conflicts have already arisen between the agricultural universities and the state ministries of agriculture. In some cases these have been resolved by the university undertaking to confine itself to the training of extension specialists and research in extension education whilst cooperating with the existing extension service and providing support for the field work of the latter.

The reservations outlined above regarding the suitability of the American extension service organization in no way lessen the importance of the basic principles underlying the US extension system and the findings of extension research in developed countries, provided they are critically examined in the light of the conditions of the individual countries, and then properly adapted to those conditions.

Farmers' Associations

In certain countries, government has encouraged the formation of farmers' associations and delegated to them the responsibility—partly or entirely—for extension education of their members.

As they are also responsible for the allocation and distribution of inputs, such as water, fertilizer and credit, among their members, as well as for the marketing of their products, they are in a position to: (a) decide on the approved practices they wish to recommend for adoption; (b) provide the inputs required at the right time; (c) provide the credit to purchase the inputs; (d) supply advice and ensure effective supervision of the correct use of the inputs through their extension agents; and (e) guarantee repayments of credit after harvest (Yudelman *et al.*, 1971).

Taiwan

In Taiwan, for example, the evolution of farmers' associations into effective extension and marketing organizations, together with increased incentives resulting from the land reform of 1949–52 played significant roles in agricultural development. They form a federated system providing multipurpose services on a cooperative basis. The associations have five objectives: (a) promotion of farmers' interests; (b) improvement of farmers professional levels; (c) increase of agricultural production; (d) improvement of farm family living conditions; and (e) development of the rural economy.

At provincial level, the major functions of the farmers' associations are supervising, auditing, training, coordinating and assisting the lower levels of all the farmers' associations in the provinces in the expansion of their service activities. At township level, which is the functional level that caters for the needs of the farmers, the association consists of six sections: credit, economics, livestock insurance, administration, accounting and agricultural extension. The farmers' associations employ their own extension workers; part of the funds for this purpose are provided by government, the remainder is derived from the profits made by the association.

The extension workers receive their information from the experiment station in their area, from the agricultural research service and occasionally from agricultural colleges. The agents specialize in certain kinds of work: some are concerned solely with disseminating farming information to individual farmers; others with general education; others with training in farm management techniques; still others work with the youth programme. There are also extension agents dealing with home life improvement (Axinn & Thorat, 1972). About 80% of the farm advisors have graduated from agricultural or home economic vocational schools, the remainder are graduates of academic high schools.

The farmers' associations place much emphasis on training programmes for their personnel—all of whom receive induction and in-service training. On entering the extension service, they learn the procedures and traditions of the agricultural extension service.

In-service training has a subject matter orientation and is given in short programmes, three to five times during the year, and in summer school.

Training in technical subjects is provided at the colleges of agriculture, agricultural vocational schools and at the local experiment stations.

Cooperatives

The adoption of new technologies by farmers can foster special activities of the cooperatives of which they are members, such as the introduction of irrigation, the supply of improved seeds, improved equipment, farm chemicals, etc., and credit. Therefore the cooperatives have a legitimate interest in promoting these innovations.

A further step forward occurs when action is taken to organize cooperative systems of production on the basis of improved methods, specialization through division of labour, joint control of insects and pests, etc.

As a prelude, irrigation systems may be developed, land holdings reallocated and consolidated. The joint adoption of improved technology makes it easier to overcome individual and social constraints; guidance provided by the cooperatives may be mainly directed towards quality control, storage and processing of farm products, whilst advice on routine production procedures is left to the government extension service.

Occasionally, cooperatives are set up for the express purpose of providing extension to their members. For example, in Argentina, the Rural Society for Agricultural Experimentation (SREA) consists of organizations of large farmers who study the results of various farm methods being used by the members. In 1966, there were 73 of these cooperative groups, each consisting of 12 to 16 large farmers, each employing an agronomist or other specialist to analyse farm records and advise members concerning improvement of methods.

In Brazil, two large and effective cooperatives provide service and advice to members, and there are a few similar organizations in Peru, Colombia, and some other countries (Hopkins, 1969).

Statutory Commodity Boards

In Jamaica, for example, a number of commodity boards have been established to promote the production of sugar, coffee, bananas, coconuts, citrus, etc., and who have sole responsibility for marketing their respective commodities. The three biggest boards for sugar, bananas and coconuts respectively, have established their own research and extension services, and have made considerable contributions to the development of their respective industries; however, their impact has been mainly confined to the large farmers and plantations.

Experience has shown that whilst the extension services provided by statutory boards can be very effective, they have a number of disadvantages when viewed from the national perspective. A commercial and monocultural attitude towards agriculture leads to a fragmentation of extension education. This not only increases expense and is an inefficient way to use scarce human resources, but their teachings and their promotion of individual enterprises often run counter to the objectives of the farmer whose interest is to develop a combination of crop and livestock enterprises that will yield him maximum net returns, rather than to confine his farm to any one crop to the exclusion of others. They may also run counter to national policies, which may wish to de-emphasize research and extension in certain commodities whilst promoting others found to be more consistent with national requirements.

Regional Development Boards or Societies

In Senegal, a number of *'Societés de Développement'* operate under the aegis of the Ministry of Rural Development (Arnon, 1977).

Each society is responsible for all aspects of development of the region assigned to it, but it also has a national responsibility for the dominant crop of the region. Hence, the extension service of each society works both on a regional basis for all agricultural commodities and on a national scale for a specific commodity.

Each of the societies has its own policies and operating procedures. There is no central extension service, capable of coordinating extension on a national scale, providing centralized training of the extension workers, giving professional support to the field staff and ensuring liaison with the research service. In addition, the system has all the disadvantages of the extension services provided by statutory boards that have been outlined above.

Multiple Organizations

Egypt is an example of a country in which agricultural extension is provided by multiple organizations. Four major agencies are directly involved: agricultural extension centres; rural social and welfare centres; cooperative societies and vocational training programmes (Axinn & Thorat, 1972).

Agricultural Extension Centres

A special form of extension service, under the auspices of the Ministry of Agriculture, is practised in Egypt. It is based on agricultural extension centres, which have been established for the diffusion of the findings of agricultural research to farmers (Sidky, 1957).

Each centre has a number of components: a general demonstration farm, a vegetable nursery, a plant for agricultural industry, a stud farm, a veterinary clinic and a slaughterhouse. There is also a small museum and library. Each centre has a resident staff; an advisory council including representatives of government agencies dealing with rural development and of the farmers which advises on policy and programmes.

Crops are grown by modern techniques on the demonstration farm. Pedigree sires are available for improving farmers livestock, chicks of improved breeds are distributed to farmers, as well as fruit and vegetable seedlings.

The centres maintain a number of demonstration plots on farmers' farms in the areas they serve.

Complementing the work of the agricultural extension centres, are the *rural social centres*, under the aegis of the Ministry of Social Affairs, and *welfare centres*, under the supervision of local administration.

Rural Social Centres

These centres provide combined programmes of health education, literacy, elementary training, and social and economic activities, including agriculture. The centres maintain a demonstration plot, distribute improved seeds, teach farmers methods of pest and disease control and livestock raising, and advise on crop diversification.

The centres are established only if requested by the community, which has to contribute land, money and labour. They are operated on a self-government basis, promoting initiative in the community.

Welfare Centres

These centres run medical clinics, boy's clubs and domestic science and needlecraft courses for girls. The personnel also organize village committees for planning and implementing agricultural development. The committees provide help to farmers in obtaining government services available to them, such as improved seeds, veterinary advice, loans of pedigree sires, etc. Other activities include poultry improvement and bee-keeping.

Cooperative Societies

Most cooperative societies organized at village level have several functions: they supply agricultural inputs, consumer goods and arrange for the sale of produce; they also provide credit.

The cooperatives have also been active in encouraging the adoption of improved practices by the farmers, including the use of selected seeds, the control of plant and animal pests and the use of improved methods of marketing (Sidky, 1957).

Commercial Firms

Generally, an innovation involves the promotion and sale of a new product. Many large firms involved in producing inputs for farming, not only engage in research but also hire agents to actively promote their products. These generally use the same techniques of extension as do the extension workers of the public bodies: personal contacts, demonstrations, mass media, etc. They are generally better paid, often more motivated (receiving a percentage of the value of their sales) and more aggressive than their official counterparts. They may therefore play a considerable role in the diffusion of new products. Their main drawbacks are: possible lack of objectivity, with resulting oversell of their products and lower credibility; their tendency to concentrate on the larger farmers and neglect the small farm units who require a larger investment in effort with a smaller expectation of returns.

When a farmer is selected by a company to serve as their agent, these drawbacks may be mitigated, because of the need to maintain credibility with one's neighbours, who regard him as a peer, rather than in his role of agent (Rogers, 1960).

The Self-help Approach

The self-help approach is based on the assumption that motivation for rural transformation must come from the people themselves, in the form of local initiatives and self-help. Outside help must come in response to the expressed needs of the communities and concentrate on fostering local institutions for cooperative self-help.

The first requirement is to initiate a broad eductional process that would alter the fatalistic attitudes, dependency and lack of self-confidence characteristic of the underprivileged sectors of the rural population, raise their aspirations and encourage individual and community action for self-improvement. This would be achieved by creating greater political awareness, fostering cooperation, strengthening democratic procedures and broadening the leadership base (Coombs & Mansour, 1974).

The basic technique adopted is the *group or cluster method* in which extension activities are concentrated on a group of farmers who are all neighbours, and membership is voluntary. The leaders must be chosen by the group members and be representative of their peers.

Among the programmes based on this concept are: '*animation rurale*' in Senegal, the 'community development programme' in India, the 'rural reconstruction movement' in the Philippines, the 'cooperative education system' in Tanzania and the 'Comilla project' in Bangladesh.

Animation Rurale (Senegal)

Subsistence farmers do not easily accept changes in their traditional practices when these are proposed by government agents, who—however well intentioned—are generally considered as aliens, with ulterior motives.

The basic concept of '*animation rurale*' is that the villagers should be stimulated by one of their own number to identify their needs for improvement, to take the necessary initiatives and to demand from the government the technical and economic assistance required to foster their efforts. In order to achieve this aim, 'activists' are sought from within the villages themselves.

In this approach, the key person is the so-called '*animateur*', a farmer selected by the villagers themselves to act as liason between them and the extension agents (called '*conseillers*'). Both '*animateurs*' and '*conseillers*' receive the same training at special training centres, the former are unpaid volunteers and the latter are recruited and paid by the extension service.

The '*animateurs*' return to their villages after training in order to stimulate the villagers to act in support of government programmes aimed at improving their lot. Periodically, the '*animateurs*' return to the training centres for short (four to five days) consultation and training courses. The scheme was initiated in 1959, and by 1967 about 7000 '*animateurs*' were in action.

The main advantage of this approach was 'to create in the village a receptivity to change, and a sense of participation' (Wilde & McLoughlin, 1967). This enabled the extension organization to work more efficiently as well as to concentrate on problems that the villagers themselves felt as being the most urgent.

The '*animation rurale*' initially achieved considerable success in creating among villagers a greater awareness of their interests, needs and capacities for development. Enthusiasm and *esprit de corps* were stimulated in long-dormant rural areas.

Unfortunately, the necessary backing by the technical services and support organizations that were to provide material help was inadequate. There was no willingness on the part of different departments to cooperate in a concerted action to promote *animation rurale* on a sustained basis. Apparently, Senegal did not have sufficient human talent and material resources to devote to the project and to translate an excellent, but elaborate idea into successful action. Powerful traditional leaders also frustrated the efforts of the *animateurs* (Coombs & Mansour, 1974).

Media-forums

A similar solution, adopted in India, is the so-called media-forums, which are patterned on the informal discussion groups which are part of most peasant cultures. These media-forums were first established in 1959; by 1965 there were already 12,000 forums, in which a quarter of a million villagers were enrolled.

Extension officers, who could previously devote themselves in the traditional way to three or four villages were now able to serve 50 to 100 villages by using the village forum approach.

Very briefly, the basic elements of these forums are as follows: each village included in the programme chooses two representatives, called 'conveners'—one male, the other female; after a period of training at special centres, each convener receives a transistor radio to be used on their return to their respective villages; Once or twice a week, the village forum members meet, men and women separately, to hear special, regularly scheduled, radio programmes; the conveners initiate group discussions and report to the broadcasters on questions raised for clarification and on decisions taken by the group.

In an investigation carried out to ascertain the effectiveness of the media-forum system, three groups of villages were compared: (a) those with radio forums; (b) those which had radios but no forums; and (c) those without radios or forums (Neurath, 1960).

The forum members showed an impressive gain in knowledge of innovations, far greater than the villagers of the controls. Of particular interest was the finding that illiterate members of the group gained as much knowledge as did the literate members. The effectiveness of the forums is attributed to the regular attendance of the members, due to group pressures, and to the participation of the individual in group decisions. The effectiveness of the forums is however dependent on a number of factors: the relevance of the programme content to the problems of the participants; emphasizing in the discussion that follows the programme, the local application of the ideas presented; feedback from the audience, indicating their reactions, interests and questions for clarification; and to careful organization, operation and maintenance procedures (Rogers & Svenning, 1969).

The Comilla Project

Probably the best known of the efforts to provide an unconventional solution to the problem of professional training for farmers combined with community development is the Comilla project, initiated in 1959 by the Academy of Rural Development at Comilla in what was formerly East Pakistan. The academy undertook to 'develop and test patterns and procedures that might be suitable for developing East Pakistan agriculture' (Rahim, 1969) and the Comilla Thana was made available by government as a pilot project for this purpose.

The Comilla experiment consisted of a number of closely interrelated pilot projects, based on initial studies made by the academy staff before the action programme was launched. The core of the project was to develop an integrated system of primary and central cooperatives, directed by a 'central cooperative association'. The other projects that followed were rural administration, irrigation and rural electrification, education, woman's education and family planning.

All the agencies involved in rural development in the Thana were located in a single physical location, and their operations coordinated by the Thana council, in close cooperation with the central association. The central cooperative association provided loans to the primary societies, maintained a mechanized unit including tractors, equipment and power pumps, organized joint marketing, and was responsible for the extension and training services. It is with the latter that we are concerned here.

The Comilla Extension Education System

Earlier experiences in the area with village-level workers had been disappointing.

These workers were mostly educated persons, who could not live in the villages which did not provide minimal educational and social services for their families. Hence they remained aliens and were distrusted by the villagers.

Because of the need for intensive supervision of the credit programme and the cooperatives, a regular communication channel between the association and the village-level societies was essential. This was achieved by regular weekly meetings of the cooperative members and meetings of the managers at the central association.

The extension programme was fitted into this system of communication. Special training programmes were organized for the managers, at which they learnt new management practices, including the use of modern inputs, such as fertilizers, insecticides, etc., and were shown in demonstrations at a small farm near headquarters. Equiped with visual aids and written instructions, they returned to their villages and instructed the farmers enrolled in the project in accordance with what they had learnt themselves. They in turn, were asked questions which they presented to their trainers at the following training session. The extension programme was further strengthened by demonstration plots in the villages (Rahim, 1969).

Achievements: The Comilla Project, in the course of slightly more than a decade, had a record of impressive achievements, but also showed a number of shortcomings. It demonstrated the possibility of involving villagers in rural development, increasing the role of local-level public administration and combining the services of distant government agencies into a coordinated effort. After its success in a single locality, the government of Bangladesh is attempting to apply the system on a nationwide basis.

However, the project has not been an unqualified success. Amongst its shortcomings should be noted: (a) only about half the farmers have benefitted from the training programmes; (b) the main beneficiaries have been the better-off farmers; (c) whilst extension work directed to increased agricultural production was effective, other educational efforts, such as in adult literacy and family planning, were completely inadequate (Coombs & Mansour, 1974).

VOCATIONAL TRAINING

Conventional extension methods can be supplemented and strengthened by systematic farmer training courses, provided either in special centres or in the villages themselves.

The courses can be long-term, for one or two years for selected farmers, short-term intensive courses on farming production methods or new practices, or special programmes for certain groups, such as courses on home-economics and poultry-raising for farmers' wives.

Many of these centres are residential, but in other cases, the farmer may come for a single day of instruction.

An interesting experiment has been carried out in Comilla (Bangladesh) to provide a system of continuous on-the-job training for cooperative organizers, cooperative accountants, model farmers, tractor drivers, irrigation pump operators and women's programme organizers. These groups come to the training centre for a few hours in the morning, on certain days of each week, throughout the year, so that the training should cause minimum interference in their normal work schedule (FAO, UNESCO, & ILO, 1971).

Training courses in the villages are carried out jointly by two or three extension workers, who are first carefully prepared for their task. The teaching is based, as far as possible, on discussion, with emphasis on clarifying the advantages of proposed innovations, the techniques for proper application, the constraints that may be encountered during application and how to overcome them. It is important to treat the participants as experienced farmers, whose opinions should be considered seriously. There is generally no point in introducing an innovation before the farmers are convinced that it is worth a trial.

The main problems encountered in providing training for farmers is the selection and design of courses appropriate for each area and the difficulties of maintaining a competent staff (Coombs & Mansour, 1974).

Training for Non-farm Occupations

Vocational training for non-farm occupations is of importance to the rural population as a whole, in that it opens possibilities for gainful employment of redundant farm labour and equips villagers with skills that keep them in the village. It is important for farmers too, by making possible the provision of services that are required following the modernization of their production methods, at an adequate level of competence. In traditional agriculture, the farmer who needs the help of an artisan, usually relies on a village jack-of-all-trades. As modern technologies are adopted, there is a concomitant increase in skill requirements to repair farm equipment, trucks, electric or diesel pumps, and even transitor radios or sewing machines.

Subsequently, the need arises for managers for operating cooperatives, clerks to handle credit programmes and even small manufacturing plants; hence the need for special training programmes for artisans, craftsmen, small entrepreneurs and managers as an essential component of the development process. The main problem is of course the great variety of skills for which training must be provided.

Coombs & Mansour (1974) mention three categories of non-agricultural training programmes:

(a) technical skills to prepare young people for gainful employment;
(b) technical upgrading of existing artisans and craftsmen;
(c) training for managers and employees for cooperatives, small industries, credit schemes, etc.

Notwithstanding their importance, there are still very few training programmes in the developing countries devoted to non-farm occupations. Some countries that have pioneered in this subject are India, Senegal and Nigeria.

Extension Work with Youth

The potentialities of youth movements and organizations as an educational factor can be exploited for extension purposes.

Experience in many countries has shown that rural youth is very receptive to new ideas and can make a considerable contribution in influencing the attitudes of the older generation (Summers, 1962).

Youth movements can give rural youth a sense of mission, dedication, and values which can give their lives new content. This is the one great advantage that underdeveloped countries have over developed countries; namely, they have a plenitude of challenges, objectives, and undertakings to offer their youth; the developed countries are relatively poor in this respect, and for this reason are burdened with the problems of a disillusioned youth with all its attendant evils.

In *Japan*, an important feature of the extension service is the *farm youth training programme* carried out on special farms, of which over 50 have been established (Axinn & Thorat, 1972).

First conceived during the depression period of the 1930s, when the need was felt to counteract the effects of the depression on farm youth, the programme has since been expanded and has proven its value in normal periods. Training is given to farm youths with at least nine years of education; the courses are of one to two years duration, during which the youths live, learn and work together. The programme emphasizes the improvement of both agriculture and home life. The number of participants is about 5000 boys and 1000 girls every year.

Training of rural youth within the framework of their *military service* has already been discussed (cf. pp. 169–170).

Many youth training schemes for farming have not been successful. The majority of the trainees do not wish to go back to the family subsistence farm or do not have land and capital for farming on their own. Most of them view the training schemes as an opening to employment outside agriculture.

Where facilities for farming are favourable, as in the Sudan Gezira scheme, the results have been far better, with over 70% of the youth trained for farming returning to the farms (Nour, 1969).

Extension Training for Women

In many parts of Africa, men generally have little interest in agricultural work and woman is still the real farmer. In parts of Asia, women often do most of the rice planting and help with the harvesting. In many cases, the growing of family-subsistence crops is the sole responsibility of women, whilst cash-crops are grown by men. Marketing of the crops is also frequently handled by the women.

It will be readily understood that unless women are made aware of new techniques, it is extremely difficult to modernize farming (Jurion & Henry, 1967).

Women as a target group are generally neglected by the extension services in developing countries; with the possible exception of advice preferred on home-economics. Though this neglect is largely the result of deeply-rooted attitudes to the role of women in traditional societies, these prejudices should not be allowed to perpetuate a situation that is an obstacle to progress.

Availability of an Appropriate Technology

Efforts to modernize agriculture cannot be successful unless innovations are available that have a high benefit–cost ratio and have been shown to be adapted to a given environment by adequate location-specific testing. This is of course a function of the existence of an effective research organization with an appropriate research programme, etc.—subjects already discussed in the previous chapter.

Two illuminating examples of the crucial role of the availability of an appropriate technology, carefully tested under local conditions, are provided by the experience of the Papaloapan basin project in the tropical region of Mexico (Poleman, 1964), and of a joint Government of India/Ford Foundation rural development project carried out in a number of districts (Brown, 1971).

The Mexican Project

Superficially, at least, the planning and implementation of the Mexican project was in accordance with all the rules. Government had undertaken to 'anticipate virtually all of the colonist's operational requirements and to supply him with something approaching a complete "farm in being" and enough follow-up aid and guidance to enable him to operate it efficiently'.

In addition to providing such overhead facilities as roads, schools, and drinking water, all necessary efforts were made to introduce efficient commercial-scale agriculture. The settlers were sold farmsteads on long-term credit and installed in ready-built houses. Land was cleared by government machinery and hired labour, and the settlers were supplied with enough equipment to permit immediate exploitation, so that they could begin to farm at once on a commercial scale and at near optimum capability to repay their debts. Short-term loans were provided to cover the running expenses of crop production. A central machinery station was established which carried out contract ploughing and other mechanical services for the settlers, who paid for them with funds received as loans. Technical aid and training was combined with the credit scheme. This of course entailed close supervision and a large degree of control over the operations of the settlers.

Full-time field advisers were assigned to work closely with the settlers; and in practice assumed almost complete charge of their operations. Calendars of operations were prepared, detailing what crops would be grown, when planting would take place, what implements and cultural practices would be applied. The granting of credit was made conditional upon satisfactory compliance with these instructions.

Notwithstanding all the planning and investments, the follow-up with technical and financial aid, and what at first sight appears to be a solid foundation for successful settlement, the Papaloapan project ended as a complete failure ten years after it was started.

Though the reasons are not simple, the principal cause of the scheme's failure was *the lack of reliable information on crop production methods under local conditions.* In this scheme, a new system of production was based on the assumption that technical practices which have proved suitable in one environment could be transported to another without preliminary and follow-up experimentation. This inevitably resulted in errors in crop selection and poor judgment in the choice of techniques by those who were supposed to provide guidance. The supervisors and instructors lacked experience in suitable cultivation practices under the prevailing conditions, so that their instructions to the farmers were formulated on conjecture; unexpected complications and failures kept cropping up, and most settlers were unable to earn enough to cover their production costs, let alone their long-term obligations. The net result was an erosion of the settlers' confidence in the scheme and in the instructors, leading to a mass exodus from the project.

Poleman (1964) concludes that the success of a land settlement scheme depends on a sound research programme carried out in three stages:

(a) an experimental stage, during which location-specific experimentation provides answers to the main problems, such as the optimum size of

holding, the local suitability of crops and varieties, cultivation practices, etc.;
(b) a pilot stage, during which the effectiveness of the programme is tested on a modest scale, providing at the same time training for the personnel; and finally
(c) full-scale implementation with on-going experimentation.

The Indian–Ford Foundation Project

In 1960, the Government of India, with the support of the Ford Foundation, adopted a ten-point rural development programme. In each state, a district would be chosen in which:

(a) credit, supply and marketing cooperatives were functioning successfully;
(b) rainfall was adequate and/or irrigation facilities were available;
(c) there were no drainage problems; and
(d) the local agrarian leadership was receptive to change.

The scheme encompassed 2.9 million farmers; 9000 multi-purpose extension workers were enrolled, enabling a ratio of one extension worker to 825 farmers. These workers received intensive on-the-job training; learning how to prepare crop plans, discuss the proposed package of inputs and demonstrate its effectiveness, and how to use mass media.

Service resources were installed, including soil testing laboratories, seed processing facilities, agricultural engineering workshops, communication centres and evaluation units.

The main on-farm strategy consisted of preparing individual farm plans with each farmer, who agreed formally to follow its provisions. The detailed inputs, costs and expected returns served as a basis for the provision of credit. This was extended in kind for seeds and fertilizers.

In brief, a textbook project, carefully planned and implemented, in highly-favoured regions and with selected farmers.

By 1967 more than half of the farmers were participating on 39% of the cropped area, which is considered a unique achievement for India.

Less satisfactory were the results achieved by the participants. No significant increase in output or yield per hectare over the five years previous to the programme was recorded in 12 out of 15 districts. Many of the cultivators who followed the recommended practices and rates of application suffered a net loss in income.

Though there were weaknesses in the administrative structure for agricultural development, the main reason for the disappointing results was the *lack of an adequate yield-increasing technology with satisfactory benefit–cost relationships* (Brown, 1971).

Liaison Between Research and Extension

Reasons for Alienation Between Research and Extension

The stereotype concept of the channels of communication between research, extension and the farmer is that the extension worker communicates research findings to the farmer and transmits the latter's problems to the researcher who then incorporates these problems into his work programme:

Research
↓ ↑
Extension
↓ ↑
Farmer

The simplicity and logic of this model is misleading and it rarely functions as described. It does however, ensure that the research worker is effectively insulated from direct contact with farmers.

While the need for close cooperation between research and extension appears to be axiomatic, its achievement in most countries is the exception rather than the rule. Extension work and research are usually organized in different services; there exists a general tendency towards separation of interests and even alienation between the two services.

Extension workers complain that research workers isolate themselves in their laboratories and experimental stations, do not pay attention to economic factors or the real problems of the farmers, delay publishing their findings and are not prepared to commit themselves to a firm opinion.

On the other hand, research workers complain that extension workers do not really trust or accept their findings; do not ask research workers for information when they need it; and do not make it clear to research workers just how important or far-reaching are the problems which they meet out in the field.

In short, the opposing claims are that research workers are not adequately informed about actual practical questions and that they do not properly communicate their findings to those who would take advantage of them, and vice versa.

These attitudes easily lead to a situation in which the extension worker, instead of serving as a link between research and the farmer, becomes, on the contrary, an obstacle between them. Lack of contact between research and extension easily leads to conflicting conceptions and opposing instructions to farmers, to the detriment and confusion of the latter.

Practically everybody is agreed that such a situation is incompatible with the needs of the agricultural community and must be avoided, nowhere more so than in developing countries. Lip-service to this idea is, however, not sufficient, nor can the problems be solved only by contacts at the individual level.

The two services, though they have a common goal, have basic differences in their methods of work and objectives, and therefore require distinct and separate administrative machinery in order to ensure their efficient functioning. This can, however, strengthen the centrifugal tendencies. These tendencies have to be counteracted and there are a number of means which make this possible.

Improving Liaison Between Research and Extension

(1) *A Common Decision-making Framework*

A common decision-making framework for the two separate administrative units provides the possibility for extension workers to participate in establishing research programmes, determining priorities, etc., thereby ensuring maximum cooperation between them. This will be discussed in detail below.

(2) *Verification Trials*

There is one area in which the activities of the research worker and extension worker converge in the field. Innovations resulting from agricultural research must be tested for their adaptability and relevance to the farmers' environment (in so-called 'verification trials'). Experiments under actual farming conditions should be the final essential stage of any research project aiming at introducing a new farming technique. Typically, new varieties, recommended fertilizer rates, plant production methods, etc., should be validated by on-farm testing. Ideally, local extension workers should be involved in this work. Another type of research activity, in which extension workers should be co-opted, is pre-extension research (Chapter VI).

(3) *Organizing, in Common, Certain Units Which Serve Both Services*

Examples include libraries, central laboratories for the diagnosis of fertilizer requirements and others for disease identification, etc.

(4) *Senior Research Workers Should be Actively Involved in the In-service Training of the Extension Personnel*

Field-days and seminars should be held regularly in which the results of research should be communicated to, and discussed with, the extension personnel.

The Planning of Research and Extension Work

Research–extension–farmer cooperation should begin at the planning stage. Apart from the commodity committees that function at national level (cf. Chapter VI), each regional station should have an advisory committee in which research, extension, local government and farmers are represented and which makes proposals defining general goals and priorities and the allocation of the resources available for regional research to the various problem areas. A number of subcommittees need to be established for each of the main commodities or groups of commodities of the region. These subcommittees consist of the research and extension specialists in the area of competence of the subcommittee, and one or more representative farmers. The subcommittees evaluate the results of current research, decide on what research findings are ripe for adoption, discuss the extension programme for the transfer of their findings to farmers, and indicate priorities for future research (Fig. 32).

These institutionalized contacts between research, extension and farmers ensure that research is not divorced from farming realities and that innovations are adopted by consensus. The farmer is assured that his problems are brought to the attention of the research worker and that the necessary attempts are being made to provide a solution according to priorities determined in consultation with his own representatives. The extension worker feels that he is a full partner in this planning process, and has a say in on what innovations should be adopted, and on when and how this should be done.

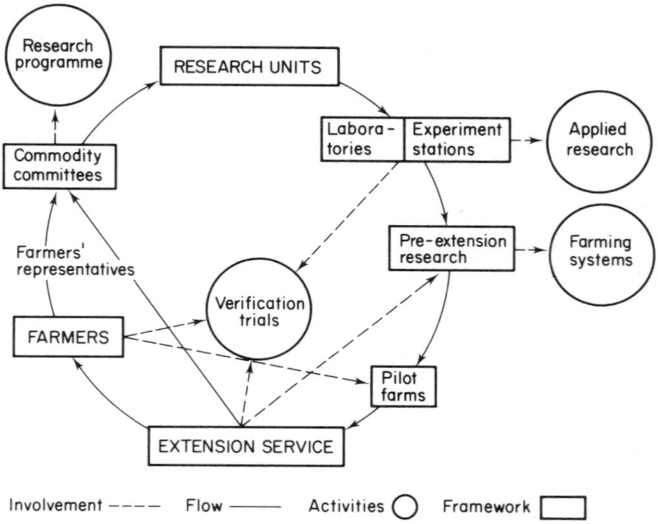

Fig. 32 The agricultural research process: involvement of extension workers and farmers in planning and implementation.

The psychological importance of this form of contact and reciprocal influence of research worker, farmer, extension worker and local government cannot be overstressed.

However, farmers' representation in the planning process involves certain problems in most developing countries, which will be discussed in more detail.

Participation of farmers does not signify that proposals for research themes must come from them only, though their initiatives in this respect should be welcomed. The research workers have a primary role in the formulation of the research programme; but their proposals must be communicated to their 'clients' so that there is opportunity for dialogue and the possibility of influencing the final outcome.

Discussion and decision-making on the planning of research and extension work at two different levels—national and regional respectively, may appear to be cumbersome procedure and a possible source of conflicting interests. It is indeed more difficult to implement than decision-making from the top. However, recognition of the need to accomodate location-specific problems, requirements and potentials within the framework of national priorities and available resources should be the overriding considerations, and this requires grassroots participation, if it is to be effective.

Farmers' Participation

The participation of farmers in research and extension programmes, which includes formulating policies and requirements, planning, as well as implementation (i.e. pilot farms, on-farm experimentation, etc.) is of great importance.

When local people play a role in guiding research and extension efforts, the acceptance and effectiveness of research and extension work at grass-root level are considerably enhanced. Therefore, the contributions of well chosen local leaders can be considerable.

There is generally no great difficulty in finding suitable farmers who are prepared to cooperate with research on pilot-farms or in the carrying out of experiments on their farms. The major problem is the lack of more or less formal local organizations of rural people that would enable them to make their needs and wishes known to the authorities in general, and to research and extension in particular, as well as to chose their representatives on decision-making bodies.

In some countries, government has been successful in encouraging the formation of farmers' associations which play a significant role in agricultural development.

As these associations are responsible for the allocation and distribution of inputs such as fertilizers, pesticides and credit among their members, as well as for the marketing of their produce, they are in an excellent position to know the nature of the problems facing the farmer and the approved practices they wish to recommend for adoption. The participation of their representatives in the work of the commodity committees can therefore have a significant beneficial impact on the orientation of research and the adoption of research findings.

However, the number of developing countries in which such farmers' associations are active is still restricted. In the majority of cases, rural life is still dominated either by rich landlords, or by a gerontocracy, or some other 'elite' group.

The importance of government action in fostering rural associations that are truly representative and that will promote the interests of the majority of the farmers cannot be overstressed. The problems encountered in these endeavours and the means used to overcome them are treated in Chapter X.

This brings us to the question of how should the representative farmers to serve on the commodity committees be chosen in the absence of formal organizations. It is frequently stated that attempts to replace traditional leaders, or approach farmers directly rather than through the traditional leaders, can be counterproductive and should be made very carefully.

There is no doubt that the readiness to adopt new ideas and practices is influenced by the nature of the leadership and the degree of control it exercises over the community. Reliance on 'traditional leaders' may be justified as a short-term, tactical approach, but normally a strengthening of their traditional stranglehold—economic, moral and social—on their people will be self-defeating. It will reinforce the conservative, anti-change tendencies—which are very strong in any case—and will stifle the initiative of more progressively minded members of the clan and village. Bose (1961) has a point when he states 'that it is necessary to change the entire cultural pattern of an underdeveloped society before we can hope for permanent technological changes.' Dumont (1962) also stresses the evil influence of the gerontocracy of the villages. By making the strongest economic power the preserve of the chiefs of large families, many primitive societies entrust the levers of progress to the oldest, who are often the least receptive to modern techniques.

The absence of formal organizations that would be able to nominate their representatives for the commodity committees should not discourage attempts at ensuring the farmer's participation in research and extension planning.

The only possibility is to coopt informal leaders, or those who, with proper encouragement, can develop into such leaders.

Where difficulties are encountered in identifying farmers who are truly representative and who can be effective members of the committees, a useful preliminary step is for teams consisting of a research worker and an extension officer to meet with the groups of farmers and discuss with them their needs and problems. Field days and group visits to demonstration plots are excellent opportunities for such meetings. At these meetings, it will generally be possible to identify the most alert, involved, and articulate among the farmers, who could be obvious choices for membership in the research and extension commodities.

In every rural community can be found farmers who have commonsense and native intelligence, a modicum of initiative and desire to better their situation, and who are otherwise representative of the majority of the farmers of the region. It is from among these that members of the regional commodity committees and subcommittees should be chosen. Incidently, the same guidelines can be used for the choice of farmer cooperators for managing pilot-farms and for on-farm experimentation.

Possibly, the involvement of small farmers in decision-making bodies of relevance to the improvement of their lot, will help to increase the awareness of their peers of the potential advantages of organizing themselves into local farmers' associations.

Direct Contact Between Researchers and Farmers

Meeting the Farmers

The existence of joint committees and of a pre-extension research unit, and the two-way communication channel of the extension service does not absolve the individual research worker from the need to acquaint himself personally and directly with the problems of the farming community he is supposed to serve. Journeys through the area served by the experiment station, talks to farmers and their advisers will help him to orient his research work to the real needs of farming. The greater the success of the research and extension efforts, the more dynamic agriculture will become, and the greater will be the need for the research worker to remain aware of the changes that are occurring in farming practice.

Conversely, visits by the farmers and extension personnel to the experiment station should be encouraged by open-days, field-days and organized visits, lectures, etc.

Learning from Farmers

A related principle is that the effectiveness of agricultural research and extension workers will be greatly increased if they attempt to learn from the traditional farmers at the same time as they try to teach them. Many failures in introducing innovations into traditional agriculture occurred because research and extension workers did not understand the rationale of the traditional farmers' methods.

Therefore, the research worker and the extension worker must learn to know local conditions, and the habits and traditions of the people they are to serve. A practice that may appear at first sight to be irrational, may actually be a necessary adaptation to local conditions. The classic example is the former attitude of agronomists trained in the European tradition to the long fallows which are characteristic of shifting cultivation in the humid tropics. The traditional methods were considered a terrible waste of resources; however, the many efforts made to replace the 'long fallow' by more modern methods—before learning the reasons that had led to the adoption of the traditional practice—were costly failures.

One country in which close contacts between research workers, educators and farmers, and the concept of 'learning from farmers' have been institutionalized is the People's Republic of China.

In 1976, a seven-man team from the International Rice Research Institute visited the People's Republic of China. The accomplishment which made the greatest impression on the team was the narrow knowledge–practice gap in Chinese agriculture. As soon as research and field tests had demonstrated the validity of a new technology, farmers quickly put it into practice. The IRRI team identified the 'remarkably close association between scientists and educators, and the workers at the *grassroots* of production' as a main factor accounting for the prompt adoption by farmers of new research findings. Research scientists from the experiment station spend time each year working with farmers. Work on farms is a required experience for all faculty members and students at agricultural colleges. These close associations ensure that the researchers and educators are kept aware of the problems faced by the farmers. Research scientists also work closely with production-team leaders in identifying research problems, and the leaders are actively involved in research implementation on the farms (IRRI, 1978).

In order to 'learn from farmers' many members of the staff of the Chinese Academy of Agricultural Science have lived and worked with farmers in order to learn the farmer's beliefs and production methods. On the basis of information collected from farmers, publications dealing with rice, cotton, wheat, fruits, animal diseases, plant protection, and other subjects have been prepared. Ancient agricultural publications, including a work on agricultural techniques, published in the first century BC, have also been reprinted and published. Applying, and where possible improving, traditional techniques and tools are encouraged, simultaneously with the introduction of modern methods. Acupuncture and other methods of animal treatment are used by veterinarians, as well as modern scientific methods; native agricultural drugs as well as chemical pesticides are used in plant protection, equal emphasis is given to the use of organic manures, including night soil, as to chemical fertilizers, etc. (Kuo, 1972).

Adoption of New Practices

The Diffusion Process in a Farming Community

The generally accepted pattern of diffusion of a new practice within a farming community is that described by Lionberger (1960) (Fig. 33).

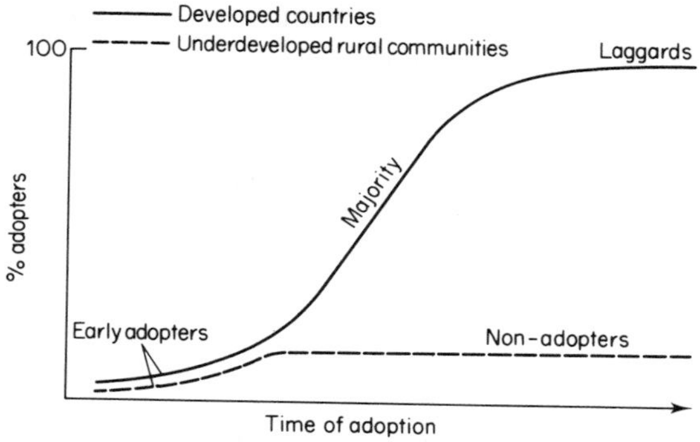

Fig. 33 Contrasting patterns of adoption of agricultural innovations in developed countries and in LDCs.

According to this concept, an innovation is first adopted by a very small group of people who are sufficiently educated to become aware of the potential for profit of the innovation; who are sufficiently rich to invest in the inputs required, and can afford to take the risks involved; and who by their position in society can more easily withstand the social pressures against change (if they exist). After the innovation proves to be a success, the early adopters are followed by those who are less prone to take risks, but can afford the inputs required. The innovation spreads at an accelerated pace; the last to adopt are the laggards—the poorest, eldest, most conservative and most averse to risks.

From this general pattern, two practical conclusions have been drawn: (a) it is useless for the extension worker to devote himself to the laggards, who will be the last to adopt an innovation; and (b) extension efforts should be concentrated on the progressive farmers, who are most open to new ideas and have the means to implement them. Once these have adopted the new practice, the others will follow in any case.

This 'practical' approach has the greatest appeal in those countries which are poorest in extension resources—human and material.

It gives an appearance of justification and legitimacy to the tendency of extension workers to spend more of their time working with the large farmers, with whom they feel socially more compatible and with whom it is far easier to achieve visible results and recognition. In several Latin American countries, for example, where 1–2% of farming families control over two-thirds of the good farmland, they receive an even higher proportion of the extension service's efforts (Barraclough, 1971).

Unfortunately, it is in poor countries that the diffusion of innovations does not, as a rule, follow the pattern described above. In these countries, the village community generally consists of a minority who are able to adopt an innovation and need only to be made aware of its potential benefits, and *a vast majority of small farmers who lack the strength and/or the means needed to make adoption possible.* Even after the richer and progressive farmers in their area have adopted a new practice, they do not follow suit. Instead of a minority of laggards, there may be a majority who are *non-adopters*, not by choice, but by force of circumstance.

A survey in three Indian sample villages, chosen at random, showed that almost all the families had heard of the Japanese method of rice culture, and knew of its potential benefits. However, the major reasons given for not adopting a practice that produced almost twice as much per unit area as the traditional method was that it required much greater physical effort, and the villagers stated they did not have the strength needed (Deutsch, 1971).

More frequently, the main cause for non-adoption is the inability of subsistence farmers to take risks (cf. pp. 309–310) and devote any part of their meagre resources to the purchase of the inputs required for innovation. Unless special measures are taken, the adopters become richer and the non-adopters remain poor. If prices fall as a result of increased production due to the innovation, they become still poorer. We will discuss the implications of this situation further in Chapter XI.

An additional factor that prevents the diffusion of new practices to follow the model described by Rogers is that in a highly hierarchical rural society—dominated by a small clique of large farmers, such as prevail in many developing countries—the small farmer's access to information is limited.

In a study in five villages in Uttar Pradesh (India) cited by Dasgupta (1977), it was found that most of the information on new technology is transmitted by extension agents who prefer contacts with the richer and larger farmers, and the further diffusion of this knowledge is limited to relatives and friends. Even mass media, which are not controlled by the village elite, are not really accessible to the underpriviledged; in the case of written material because they are illiterate and in the case of mass media because they are too poor to own a radio. As a result, large sections of the villagers are not only ignorant of new technologies and their potential advantages, but are also unaware of the availability of government grants for the boring of tubewells or even of the availability of credit, seeds or fertilizers through the cooperatives.

Developing countries cannot afford to allow the non-adopters to stagnate, both for economic and social reasons. A two-pronged extension strategy must therefore be adopted; conventional methods can be followed for those farmers who are *able* to adopt new techniques, whilst special efforts must be designed to make it possible for the most disadvantaged sector to follow suit.

Leonard (1977) from his study on the extension service in Kenya is convinced that the substantial bias of agricultural extension services in favour of the wealthier and more progressive farmers can be lessened if Ministry of Agriculture decision-makers show the necessary political will. The measures he proposed are: (a) a carefully redefined extension strategy and (b) guidelines for working with the middle and bottom rungs of farmers. One possible way would be 'to define extension goals by the number of farmers who progressed beyond a defined threshold of production levels'. Another possibility is a greater emphasis on group activities.

Adoption of Innovations by the Individual

Stages of Adoption

Five stages in the adoption process by the individual innovator have been recognized (Rogers, 1962). The individual first becomes *aware* of the innovation, but is not yet motivated to seek further information. If he feels that the innovation may be relevant to his needs he becomes *interested* and actively seeks additional information on the subject. He then attempts to *evaluate* the possible costs and benefits to himself of the innovation. If his evaluation is favourable, he may decide to give the innovation a *trial*, by applying it on a small scale to determine its utility under his conditions. Even farmers surrounded by neighbours who have previously adopted the practice and proven its value will generally insist upon personal experimentation before adopting the innovation completely. This highlights the problem of the small farmer, who cannot allocate part of his holding for 'experimental' purposes.

In the light of his experience during the trial stage, the individual may decide to apply the innovation fully. Rogers (1962) has named the five stages: awareness, interest, evaluation, trial, and adoption. Different names have been assigned to these stages by various authors.

Factors Influencing Receptivity to Change

Types of Innovation

The characteristics of the innovation itself, as perceived by the farmers, affect its rate of adoption.

Rogers (1962) mentions five such characteristics: *relative advantage* in relation to the procedure that it is to supersede (profitability is one criterion); *compatibility* with existing values and past experiences of the adopters; *complexity*, the degree to which an innovation is relatively difficult to understand or apply; *divisibility*, the possibility of trying the innovation on a limited basis (this may be most important for the early adopters); *communicability*, the ease of communicating the innovation to others. Characteristics which favour rapid adoption are rapid and strikingly visible results, relatively low cost, ease of application, possibility of use on small areas and high returns. Innovations that have these characteristics are called *lead practices* not only because they are relatively easily applied, but because their use predisposes the farmer to adopt other improved practices. The use of fertilizers, for example, is such a lead practice.

Striking visual characteristics of the HYVs of rice and wheat undoubtedly contributed to the rapidity with which they spread. Because of their short stature, they are easily distinguished from the traditional varieties; the heavy load of grain on the erect, short straws is also strikingly visible. Their ability to remain erect, when traditional varieties around them are completely lodged, is also an important attribute (Fig. 34).

In the adoption of pesticides, the time-lag is generally significantly greater than for fertilizers; the main reasons for this are: greater complexity in identifying the agents and the right control method; the greater danger in handling pesticides; and the ineffectiveness of control on small areas, unless a common control programme is adopted over a larger area.

Certain innovations depend for their success on simultaneous acceptance by all farmers in a given community. A progressive farmer who introduces an early-maturing variety on his own, will expose his crop to the concentrated attack of birds and rodents. A farmer who sprays his crop against an insect pest will find his field invaded by insects from the untreated fields of his neighbours.

Castillo (1975) describes the procedure followed by a technician who wished farmers to change the planting time of rice in a Philippino village in

Fig. 34 Striking difference between a new, semi-dwarf variety of wheat (centre) and surrounding tall-strawed variety. The latter lodged badly at a later stage.

order to be able to take advantage of periods of more abundant water supply. The proposal had to be adopted by all the farmers or not at all. The nature of the problem and the prospect of a solution was explained to the farmers individually; commitment to the change was then sought in a group meeting, and social pressure to conform was brought to bear on the reluctant farmers by their peers.

Environmental Conditions

The profitability of new technologies, and the amount of risk involved depend to a considerable degree on the environmental conditions under which they are applied. Differences in adoption rates of HYVs and fertilizer use by Turkish farmers, according to region and topography are shown in Table XV.

In the Mediterranean region, the HYVs have proved to be well adapted, and they yield well above the traditional varieties; most farmers have adopted them. The favourable environmental conditions are also reflected in the high rates of fertilizers used.

In the hills of the Aegan and South Marmara regions, the cold weather presents hazards to which the traditional varieties of wheat are better adapted, hence the rate of adoption of HYVs is low. The amounts of fertilizers applied to HYVs are about half those used in the Mediterranean region, and still lower amounts are used on the traditional varieties.

In the plains of South Marmara, the Mexican varieties proved to be very susceptible to Septoria disease and less so in the Aegean flatlands; these differences are again reflected in the adoption rates of the new varieties.

It was also found that differences in climate between villages within a given region may be sufficiently great to account for sharp differences in adoption rates (Winkelmann, 1976).

Table XV Adoption of HYVs and fertilizers in wheat production by farmers[a] in different regions in Turkey, 1972 (Demir, 1976, by permission of CIMMYT).

			Ferlizers (in kg/ha, N + P)	
Region	Topography	HYVs (in % of farmers)	on HYVs	on traditional varieties
Mediterranean	Hills	90	153	—
	Plains	97	124	—
Aegean	Hills	23	64	30
	Plains	77	60	27
South Marmara	Hills	32	64	42
	Plains	43	65	48

[a]In order to avoid obscuring the data by economic constraints, the data relate only to larger farmers. Effect of farm size will be discussed separately.

Table XVI Adoption of maize hybrids and fertilizers by farmers in three agroclimatic regions of Kenya, 1973 (Gerhart, 1975, by permission of CIMMYT).

			Per cent of adopters	
Region	Altitude	Rainfall	New hybrids	Fertilizers
1	>1500 m	Good	96	75
2	>1500 m	Good	95	93
3	<1500 m	Variable	17	6

In a similar study in Kenya (Gerhart, 1975), the adoption rates by farmers of new maize hybrids and fertilizer used in three agroclimatic regions were determined (Table XVI).

The new hybrids, which incorporate high-altitude germ plasm from the Andean region of Latin America, were developed at an experiment station in region 2, and give good performance at high altitudes, hence the high adoption rate of the new hybrids in regions 1 and 2, with concomitant fertilizer use. In the lower-altitude region, with less reliable rainfall, the hybrids showed little or no yield advantage over the traditional ones. After 25% of the farmers in this region had initially adopted the new hybrids, their number fell to 17% by 1973*.

The Complementary Nature of Technological Factors

Agricultural progress cannot, as a rule, be piecemeal. There is little point in introducing improved varieties if they are unable to develop their potentials owing to lack of nutrients; there is no justification in adopting practices aimed at producing what might become a bumper crop, if disease prevention and pest control are not carried out and a large part of the crop is thereby lost. Living standards will hardly rise if weed control is ineffective because it has to be carried out by back-breaking manual labour.

For these reasons, single-practice programmes, such as introducing irrigation, applying fertilizers, adopting high-yielding varieties, using good seeds, controlling pests, etc., usually give poor results.

An illustrative example of the inefficiency of fertilizers, when used on varieties that are not adapted to high levels of fertility, is provided by the Mexican experience (Wellhausen, 1967). In Mexico, native varieties of wheat were susceptible to black stem rust. When fertilizers were applied to these varieties, the profuse tillering and heavy growth of foliage created a favourable micro-climate for the germination of rust spores. As a result, heavy infections and epidemic spread of rust caused complete crop failures.

*The proportion of farmers using fertilizers in this region was still lower, presumably because of the lesser response of the traditional varieties.

By contrast, in unfertilized stands, the sun penetrated the thin stands, causing dew to dry up early in the morning, spore germination under these conditions was limited and a build-up of the disease to epidemic proportions was thereby avoided. Yields under these conditions were low—around 750 kg/ha—but fairly constant. Conversely, an example of the need for fertilizer application before improved varieties can give highly increased yields is described by Kellogg (1962). In fertilizer trials on maize in India, the increases obtained over unfertilized local varieties were of 1290 kg/ha from hybrid seed alone, 1110 hg/ha from the application of fertilizer alone, and 3480 kg/ha from the combination of the two.

Another aspect that adds to the argument against single practices, even when effective, is that at best they provide only small increases in yield over traditional farming and therefore have very little impact. Increases of 10 or even 20% over an average grain yield of 500–600 kg/ha that is usual for traditional farming, are smaller than the normal fluctuations in yield due to climate, and will therefore not even be attributed by the farmer to the improved practice. It is also doubtful whether the increase is sufficient to justify the cost of the input required. In the early stages of development, it is almost essential that spectacular increases of at least double or triple the normal yield be obtained. This is usually not possible with a single practice, but may be reasonably expected with an appropriate 'package' of practices.

Instead of the major increases in food production due to fertilizers postulated by the Freedom from Hunger Campaign of FAO, research in Africa has consistently shown that only nominal increases were obtained when fertilizers were applied in an otherwise unchanged subsistence agriculture. Some increases were of the order of 50%, but in view of the very low base, these were not very impressive in absolute figures. Most of the experiments showed that fertilizers alone could decrease the rate of yield decline, or at best maintain the low yield levels characteristics of traditional agriculture (Cline, 1971).

Certain techniques, when applied in combination, can give spectacular results in a very short time. A combination of improved variety, appropriate fertilization, adjusted plant population, and efficient weed control and plant protection, can give increases in yield that range up to several hundred per cent. The expenditure required from the farmer for inputs is apt to be low in relation to the additional yield produced, provided only that the prices for fertilizers and pesticides are not inflated as a result of deliberate policy, dependence on unscrupulous middle men, unrealistic distribution costs, or other man-made factors that disrupt the cost-ratio of crop and input factors.

In over 500 demonstrations carried out in Africa by the FAO fertilizer programme on a number of crops, the yields obtained with improved practices, with and without fertilizers, were compared with those resulting from the traditional farmers practices (Fig. 35). Overall, the incomplete package

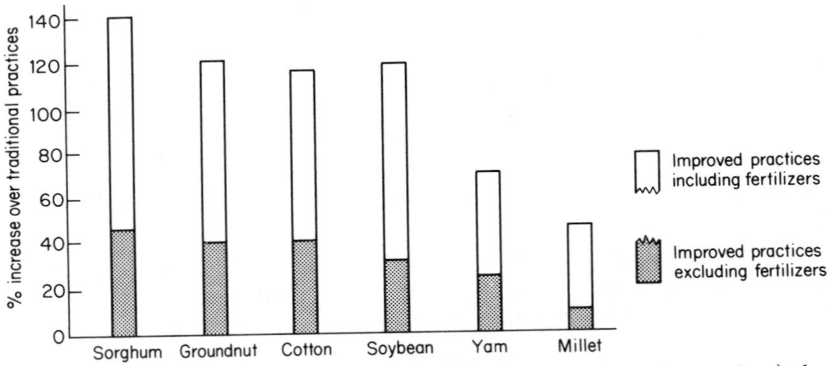

Fig. 35 Yield increases (in percentage of traditional farmer's practices) due to improved practices, with and without fertilizers (based on data from Mathieu, 1979).

of improved practices gave yield increases that did not exceed 30–40%, whilst yields were more than doubled in most crops when the improved package included fertilizers.

The relative lack of response of yam and millet to improved practices, in all probability reflects the lack of breeding effort that has been devoted to these crops.

Detailed studies by Allan (1971) at Kitale, Kenya, have shown the effects on yields of maize of six factors: planting time, plant population, weeding, variety, phosphate and nitrogen application. The effects of various combinations of factors on yields expressed as a percentage of traditional methods are shown in Fig. 36.

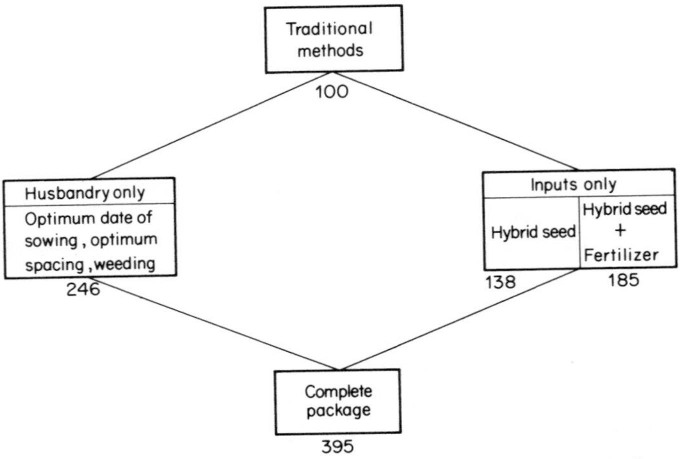

Fig. 36 Effect of single and combined factors of production on yields of maize in Kenya (traditional methods = 100) (based on data from Allan, 1971).

Another example of the synergistic effect on yield of combined factors is shown by an experiment at ICRISAT, conducted on sorghum, in which the effects of variety, fertilization and management, singly and in combination, were tested. The yield increases from improved variety, fertilization and improved management together over the traditional practices was about 3000 kg/ha, double that of the sum of the increases due to the same three factors applied singly. When two factors were combined, there was only a slight synergistic effect (Krantz & Kampen, 1977).

'Package' Programmes

An interesting example of the large-scale introduction of a combination of improved practices into traditional farming is the 'package programme' carried out in a number of districts in India (Malone, 1965). More than one million farmers were already taking part in the plan, comprising about 2% of India's farming population. In the fourth five-year plan period, India intended to extend the 'package' principle to more than 100 of the country's 325 agricultural districts.

The basic concept underlying the programme is that agricultural progress will be more rapid, and the adoption of improving practices will be more effective if:

(1) A 'package' of complementary, improved production practices, locally adjusted to climate, soil and irrigation conditions, is established by specialists.

(2) The 'package' deal is applied across an entire farming community, by helping whole groups of farmers in each village to make a break with tradition by adopting the package of improved practices. These include clean seeds treated against seed -borne diseases, equipment for improving seedbed preparation, fertilizers, plant protection measures and effective irrigation practices. The 'package' is adapted to each locality, but always consists of a combination of interacting practices.

(3) The technical supplies (seeds, fertilizers, pesticides, and implements) needed for the execution of the programme, and the necessary credits to finance the plans, are made available to the villages in time and in sufficient quantity, together with a number of supporting services that are required for this purpose. These include suitable transport and marketing arrangements, adequate storage facilities, seed and soil testing laboratories, workshops, and credit institutions.

(4) A general educational programme clarifying the benefits to be derived from the package plan, including demonstration plots, is executed. Each demonstration is carried out on two plots in a farmer's field, on one of these plots the 'package' of improved practices is applied, whereas the other is farmed in the traditional way.

(5) A simple farm plan is worked out with each farmer participant, indicating the crops he intends to grow and the supplies he will require.

The 'package programme' in India was started in 1960, in seven selected districts, with support from the Ford Foundation, which provided financial help and technical guidance. Malone (1965) studied the effect of the programme in the Tanjore District in which almost 63,000 farmers participated. The programme in the district was carried out by three extension officers and 20 community workers for each extension area, consisting of about 90 villages and 83,000 farms. Energetic extension workers achieved an enrolment in the plan of 50–70% of farmers within two years.

Participation included all sizes of farms, with tenants as well as owners participating. In the three-year period following the introduction of the plan, the tonnage of fertilizers used in the Tanjore District increased by 90%; the application of fertilizers on package programme farms was double the average for the Tanjore District. The input of 'new' practices (fertilizers, plant protection materials, etc.) increased by 250%. More livestock feed was purchased than before. Labour input increased by about 20% on the same acreage, as a result of employing improved practices and diversification of crops. Rice yields increased by 19% (about 450 kg/ha); this modest increase was due mainly to the small response to fertilizer application of the rice varieties used at the time. More minor field crops: vegetables, fruit, and milk, were produced. On the average, there was one dollar of net gain for every dollar spent.

The implementation of the 'package programme' in the Tanjone District has shown that farmers take the first steps away from traditions more quickly and more successfully if they do it as a group, together with their neighbours.

The 'Pueblo project' in the central plateau of Mexico, is another example of the 'package programme' approach to introducing improved practices into traditional subsistence farming. This project was aimed at rapidly increasing maize yields among 50,000 smallholders (Sanchez, 1970).

After an initial exploration aimed at obtaining information on the agricultural, social, economic and political factors operating in the area, the following 'package' was adopted:

(1) a 'package' of desirable techniques was determined on the basis of experimental work conducted with the participation of farmers on their own land;
(2) easy credits for inputs were provided;
(3) an acceptable relationship between cost of inputs and prices paid to the farmers for maize, was established;
(4) a minimum price and assured market for the produce was guaranteed;
(5) crop insurance was provided.

In the course of four years, the number of participant farmers increased from 30 to about 5000; over 26,000 inhabitants derived a direct benefit from the scheme in the form of increased yields and income, which averaged about $100/ha more than from traditional farming.

A useful device for popularizing the 'package programme' was initiated in El Salvador, and later spread to the Philippines, Vietnam, India and elsewhere. A 'packet' containing all the required basic inputs—seed of the improved variety, fertiliser, chemicals for plant protection—with explicit instructions on their use is given to the individual farmer. The inputs are sufficient for an area of about 100 m^2. The farmer can use the 'packet' in a corner of his field and compare the results with his traditional crop. Fertilizer companies are now selling such packets commercially (Brown, 1970).

After adopting an input package, its very success may lead to a change in growing conditions, so that a new package has to be designed. For example, when the traditional varieties of wheat are replaced by HYV's, the latter may at first respond only to heavy rates of nitrogen fertilizer. After a few seasons, soil reserves of phosphorus and potassium become exhausted, and yields decline steeply unless P and K fertilizers are added to the input package. Changes in farming systems may cause a complete change in the type of weeds infesting the crops. If conditions are created that are favourable to the proliferation of perennial weeds, quite serious problems may arise, requiring control methods that were previously unnecessary.

The Time-lag in the Adoption of New Practices

After a new technique has been developed by agricultural research, its widespread adoption by farmers may be delayed for a more or less lengthy period.

This time-lag is affected by a number of factors:

(a) the time required by research and extension before they are prepared to recommend the adoption of a new practice;
(b) the time interval between awareness of the individual farmer of the new technique and its adoption;
(c) the length of the diffusion process in the farming community;
(d) the type of innovation and the interactions between new inputs;
(e) environmental factors;
(f) social, economic and institutional factors.

When Should an Innovation be Recommended?

A dilemma with which researchers are frequently faced is when to recommend the adoption of a new practice. The natural tendency is, of course, to play safe by testing a new variety or new practice for a number of seasons before recommending its adoption. However, delaying the adoption of a new variety or technique capable of increasing productivity in agriculture can be the cause of potential economic loss as great as that incurred in taking the risks of premature adoption. The responsibility for recommending adoption or delay lies squarely with the research workers, and cannot be avoided.

However, even if a research worker is prepared to recommend a new practice, the extension worker may be an obstacle.

Even in a highly developed country, such as the USA, a considerable time-lag may occur between 'awareness' by the extension agent of an innovation and his readiness to urge adoption. The reason may be the same as that which prompts the research worker to delay recommending the practice—the wish to be very sure that the farmer is not exposed to risk and, it goes without saying, fear of risking one's professional reputation. An example of such a delay is given by Rogers and Yost (1960): extension agents in Ohio, after having been made aware of the effects of stilbestrol, a growth-promoting sex-hormone feed adjunct for beef cattle, required about two years before recommending the new practice to their farmers.

The question arises whether a procedure can be devised which shortens the time-lag in large-scale adoption whilst minimizing risk due to insufficient testing under local conditions.

The following can serve as general guidelines for such a procedure:

(a) If a practice or new variety has proved to be highly successful in one or more countries with similar ecological conditions there is a *prima facie* justification for a shortened procedure of testing in the country contemplating adoption.
(b) The tests should be carried out for one year under as wide a variety of conditions as possible, in experiment stations and on farmers' fields.
(c) If the results are favourable, a decision should be taken on the extent of adoption for the following season: area, number of farmers, region or regions, etc. It should be stressed that whilst a recommendation based on the results of these preliminary tests is the responsibility of research, decisions in this respect should be taken jointly by research, extension and representatives of farmers.
(d) Wherever possible, government should insure farmers participating in the adoption programme against excessive risks; this can be done by guarantees: (1) against loss of costs of production; (2) against lower yields as compared to the standard practice or variety; and (3) of a minimum level of return per unit area (cf. pp. 435–436).

At the same time as the programme of adoption is being executed, systematic testing and research on the new practice or variety must be initiated aimed at refining the practice or improving the variety according to local needs and conditions.

An example of the taking of a calculated risk in the adoption of a new variety that showed unusual potential, is that followed by the International Rice Research Institute (Los Baños, Philippines) for the now famous IR8 strain of rice. Normally, a selection is made from three to four seasons of preliminary yield tests, two to three seasons of general yield trials and two to three seasons of regional adaptability tests in different parts of the country (Castillo, 1975).

The IR8 strain began to look very promising by late 1965. In spring 1966, the few kilograms of seed available was space-planted and heavily fertilized, producing an increase in yield of about 800 times. This nucleus seed was distributed to a number of seed-producers in the Philippines and other countries in the summer of 1966. Its performance was sufficiently impressive to justify its official release in the autumn of 1966.

In 1967, over half the rice grown under controlled irrigation in central Luzon (Philippines) was planted to IR8. In India, 500 kg of seed was imported in summer 1966; the same year 20 tons were obtained. By 1967, farmers had planted over 60,00 hectares (Cummings, 1971).

The further phenomenal spread of the IRRI type rice varieties is already well known!

Education of Those Serving the Farmers

Professional education at the secondary and university levels must be aimed at ensuring the cadres of trained personnel on whom the development of agriculture depends. These cadres include extension agents, research workers, teachers for agricultural schools and colleges, and economists, sociologists, etc., to serve in the various governmental and private institutions concerned with agricultural planning, production and marketing.

A prerequisite for the modernization of agriculture is to make it an attractive challenge to the bright, innovative and ambitious members of rural society, who at present are apt to view agriculture as a backward occupation characterized by drudgery and as an obstacle to self-improvement (Hapgood, 1965). This relates to future farmers as well as to those who serve farmers.

It is the declared main aim of many agricultural schools to train rural youth to become farmers able to use modern methods. However, most young people from farm families who need training, are rarely able to come to agricultural schools because they are needed on the farms or because they lack the necessary qualifications. Also, it is unrealistic to expect graduates to go back to subsistence smallholdings.

The minority of farming families who can afford that their children continue to study beyond elementary school, will generally not encourage them to study agriculture. Castillo (1975) writes:

> 'The dreams and expectation of parents were definitely away from farming, in spite of some improvement in material level of living and increased income attributed to increased yields by a majority of the respondents and despite their modernising outlook in farming. Apparently the changes in farming have not made it attractive enough to make them want their own children to be in it. Furthermore, college education which they aspire for is not associated with farming. As a matter of fact, they could not understand why persons with college degrees, such as agricultural extension workers, would work in the villages and get their feet muddy'.

Education and training cannot take place outside the context of economic and social development. If agriculture is not modernized, it will not provide occupational opportunities for trained personnel. not only will the efforts expended on education and training be wasted, but they will become a source of resentment and frustration and will be counterproductive if they increase the selective exodus from agriculture (Prebisch, 1971).

Education of the Extension Worker

There is need for an extension service with a sufficient number of trained people, who understand the new technologies developed by research and who are in a position to demonstrate these new practices and their applicability to a reasonable number of farmers.

Areas of Competence Required

Ideally, the extension worker must be competent in at least six areas (Byrnes & Byrnes, 1971):

(1) *Technical.* Basically, he should have the ability to apply principles and generalizations to specific problem-situations; this requires a knowledge of the desirable production methods for the commodities produced in the area for which he is responsible.
(2) *Economics.* He should have the ability to weigh and recommend alternate strategies, based on the calculation of cost–benefit ratios, interests, the knowledge of market outlooks, agricultural policies, availability of credit, etc.
(3) *Science.* The ability to read and understand professional literature and the ability to carry out field experiments.
(4) *Farming.* The extension agent should be willing and able to

demonstrate new techniques to the farmer, even if this entails physical work and practice.

(5) *Communication.* He should be able to convey innovations to different categories of farmers and to encourage them in their adoption.

(6) *Social.* The extension worker must be familiar with the customs, values and ways of thinking of the farming population he serves. He must realize that traditional farmers are suspicious of innovations and of the motives of innovators because of their long historical experience of being socially exploited. He must realize that in many villages, any stranger is viewed with distrust, increasingly so if he is a government employee. The general success of village workers in India and Pakistan has been ascribed to the fact that after their period of training, they return to their respective villages, where they are not considered strangers. However, a communication gap may occur because they are government officials (Rogers, 1960). Therefore, in all cases, the extension worker must learn to gain the respect and confidence of the villagers and this he can only do by understanding them.

Level of Education Required

An important problem is the desirable level of training and education required by the extension agents. The level depends directly on the level of education, technical competence and economic acumen of the farmers themselves, and will therefore change in the course of development.

In the early stages of development, both the lack of highly qualified personnel and the objective needs of the farming community do not enable or justify aspiring to a high level of education of the extension agents. In a study carried out by the Allahabad Agricultural Institute (1958) in India, it was found that village-level extension agents with only an elementary education were more effective in reaching Indian villagers than were extension agents with high-school or university educations. On the other hand, if the agent is not sufficiently ahead of the farmer in agricultural knowledge and general educational level, he will not achieve the prestige and creditability required to influence his 'client'. In all probability, the former type of extension worker is effective only during an initial period of village development. During this period, able farmers, technicians with secondary school training and agricultural teachers with incomplete training have to undertake the tasks of village-level workers.

In research on the effectiveness of the extension service in Kenya, Leonard (1977) found that 'secondary education has a uniformly detrimental effect on the work performance of agricultural extension workers in the current Kenyan context, one similar to that of many developing countries'. Leonard ascribes the negative impact of education on the extension worker to the creation of expectations that cannot be fulfilled, leading to a low level of job satisfaction and poor commitment to an extension career. He was however able to show that in those fields in which good work by the extension agent led to substantial promotion, the better educated agents performed better than primary school leavers.

His conclusion is that improved training of extension personnel will only cause frustration unless proper incentives for good work are provided. The improved capacity of the extension agent can however only be used effectively if the other prerequisites for adoption of improved technology by the majority of the farmers exist or are also provided.

Special Training Programmes

In order to prepare these people for their tasks, special training programmes are required. In India, for example, when the community development programme was started, training centres were established throughout the country to train over 100,000 field workers in extension work and community development (Axinn & Thorat, 1972).

The minimum qualification to the six-month training programme was a high-school certificate. The main subjects taught were agricultural production and extension methods; in addition, aspects of public health, cooperatives, education, rural industries and animal husbandry were included. 60% of the time was devoted to lectures and the remainder to training in the villages.

This background, though fairly superficial, provided the minimum working knowledge necessary, to be supplemented on the job by guidance from extension officers and subject matter specialists.

As the general level of farming improves, the professional level of the extension worker must be raised. This is generally accompanied by increasing specialization in his work.

Intermediate Agricultural Education

The role of intermediate agricultural education is to bridge the gap between the university graduate and the farmer. In countries with a predominantly agricultural economy, intermediate agricultural education and training are major factors in the modernization of agriculture, by providing technicians for work in the field, the laboratory, the workshop, in processing plants and in community services (FAO, UNESCO, & ILO, 1971).

Intermediate agricultural education is generally divided into higher level (diploma courses) and lower level (certificate courses). However, because of the wide range of entrance requirements and of standards of training, what may be termed a certificate in one country, may be equivalent to a diploma in another country (Nour, 1969).

In developing countries, government is almost the sole employer of trained agricultural technicians; as agriculture is modernized the number of technicians employed by commercial interests increases (FAO, UNESCO & ILO, 1971).

Types of Courses

Diploma courses: These courses, of two to three years duration, are more or less equivalent to junior colleges in the USA. Variously named technical institutes of agriculture (Guatemala), schools of agriculture (Jamaica), colleges of agriculture (Malaysia), junior colleges of agriculture (Japan, Taiwan, Korea), their main purpose is to train competent mid-level agricultural technicians to serve as extension workers, technical assistants, cooperative organizers, teachers in agriculture for lower-level schools, etc. It is estimated that developing countries need from five to ten times as many agricultural technicians as agricultural scientists (Wilson, 1968).

In countries, such as Japan, the modernization of agriculture and the increased level of education of the farmers has led to the raising of the standards of training of extension workers. Graduates of the agricultural colleges are now used for extension (FAO, UNESCO & ILO, 1971).

Certificate courses: There is a wide variation among these courses, which offer from one- to four-year courses and are variously known as agricultural high schools (South Korea), practical farm schools (Ceylon), schools of agriculture (Malaysia), senior vocational schools (Taiwan), etc.

Their main objectives are to train students for farming, to develop local leadership, to train farm managers and village-level extension workers (Chang, 1963).

In most Latin American countries, vocational courses in agriculture at secondary level are offered in specialized residential schools (Becerra & Samper, 1969).

Shortcomings

In principle, the need to give practical training to students is generally emphasized, and most schools have farms at their disposal for this purpose. In practice, the farms are generally poor planned and managed, and field work consists mainly of drudgery, causing resentment and dislike for manual labour. An important reason for this situation may be the difficulty of securing instructor–teachers who combine practical expertise with teaching competence (Farouky & Skilbeck, 1969).

One of the main aims of the field work is to introduce the students to modern, labour-intensive methods of farming. Simple field tests, comparing new and traditional techniques, carried out by the students who should also be involved in the measuring, recording, analysis and evaluation of the results, could be a powerful tool in making farm work interesting and stimulating.

In many countries the institutions for intermediate education were set up before independence, and their curricula have not been adapted to the new requirements of the countries (FAO, UNESCO, & ILO, 1971). New institutions, established with the help of foreign sponsors, have frequently adopted programmes used in the donor countries. Many of the teaching programmes are therefore unrelated to the work of the future graduates, with too much science being taught and too little attention paid to farm management, extension techniques, social aspects, etc. Much of the learning consists of textbook teaching of material taken from totally different environments, with little or no relevance to the development problems of the home region of the students. Teaching, laboratory, and workshop equipment is usually inadequate.

One area of intermediate agricultural education that has been largely neglected in many developing countries is that of women, who can be powerful agents for change in agriculture by improving their home and family life and who, in many countries, do much of the farm work (cf. pp. 154–156).

Probably the main shortcoming of these institutions is their failure to motivate the students to the 'grassroot' work they will be called on to undertake, and to understand the vital importance of their work in the development of their country (FAO, UNESCO, & ILO, 1971).

Improving Intermediate Education

In spite of the shortcomings of many agricultural schools, they provide the main manpower for government regulatory and extension services, village-level schools, statutory boards, etc. In Taiwan, the farmers' associations (cf. p. 237) employed in 1960 over 8000 people on their staff, of whom 98.2% were graduates of intermediate agricultural education.

There is, therefore, every justification for improving intermediate agricultural education so as to ensure that it is relevant to the development needs of the area it serves.

The major proposals made by FAO, UNESCO, and ILO for intermediate agricultural education to serve its objectives properly and economically are:

(a) Systematic *planning* of the national structure of agricultural education in accordance with the present and future needs in trained manpower and the range of functions those who are trained will be required to perform.
(b) Effective *coordination* between the authorities and institutions involved in agricultural education. This can be achieved by a national council for agricultural education.

(c) The *curriculum should be reformed* so that subject matter, concepts and teaching methods are relevant to local development needs.
(d) Strengthening the *infrastructure* of the institution by providing adequate and up-to-date teaching and training facilities.
(e) Improving the *selection and training* of the teaching staff and develop a satisfactory career structure for them.
(f) Increasing the *participation of women* in intermediate agricultural education, and open up to them additional vocations besides home economics.
(g) A major emphasis on *motivation* for 'grassroot' work and adequate training for the practical problems to be encountered in the future work in the field of the graduates.

Higher Agricultural Education

Objectives

Higher agricultural education presents the future agronomist with severe personal demands. He has to work with rural people and live in a rural environment. He must be able to adjust himself to this environment and be prepared to look upon his job with a sense of mission. He must develop an ability to mix with rural people. The agronomist must not only be competent professionally but also have a wider education; he must also have high personal qualities.

A great difficulty encountered in developing countries is that the number of students and professors who have actually lived or worked on farms is extremely low.

The profession of agronomist evidently does not appeal strongly to many young men in developing countries. In Latin America, for example, out of 926 university students whose fathers were farmers, or livestock men or were occupied in similar pursuits, only 3% had enrolled in agronomy courses and 12% in veterinary medicine. Graduates of the agricultural colleges, their professors, and the men who have previously graduated from these institutions occupy positions in the ministries of agriculture, carry on research and extension work, and constitute the scientific and official elite of Latin American agriculture. The fact that a majority of them have no farm background goes a long way toward explaining some of the shortcomings of the agriculture of the region and particularly the communication gap between agricultural officials and actual farmers (Hopkins, 1969).

On conclusion of his studies the agronomist will be called upon to undertake the functions of an agricultural adviser, teacher, administrator in an agricultural organization, or to be a research worker. Each of these occupations requires additional training in the appropriate sciences. An agronomist planning to become an *adviser* must be trained in instruction methods, know something of psychology and sociology and must be familiar with the way of life of rural people, of farming conditions and traditions. He must acquire proficiency in farm work, such as in ploughing and in driving a tractor, and learn to listen to the experienced farmer.

In higher agricultural education there must be equal emphasis on theory and practice. Practical knowledge is extremely important, and practice and theory should be interwoven. The student aspiring to a higher agricultural education should have one to two years of practical experience. On the other hand, basic sciences must not be ignored. The agricultural sciences are based on natural and other exact sciences. Therefore, a good grounding in botany, zoology, chemistry, mathematics and physics is essential.

Extension Workers

Subject matter specialists and extension officers need to be fully qualified agronomists with at least a BSc degree or its equivalent.

Many colleges of agriculture have established departments of agricultural extension with the aim of training agricultural extension workers. In Taiwan, for example, the Department of Agricultural Extension at the College of agriculture of the National Taiwan University has a four-year programme in agricultural extension that offers training in five areas: all-university courses; agricultural subject matter; basic sciences—natural, life and rural sociology; teaching methods—including adult education and psychology; and extension methods—including programming, evaluation, philosophy, and comparative extension (Axinn & Thorat, 1972).

Those who have not taken the college course on extension will require an in-service special 'training' course teaching these subjects. Formal training should in any case be supplemented by regular seminars, including such subjects as communication, organization and administration; social systems, farming systems, recent advances in agricultural techniques, etc.

Generalists or Specialists?

A question fundamental to the efficiency of the future extension worker is: what should the university aim at producing—generalists or specialists? In big countries that are highly developed there may be justification in training top-rank specialists, but the problem changes when one has to consider low-level and backward agricultures with diversified farming. In such a case the university should mainly aim at turning out agonomists with wide backgrounds, who will be fairly competent in all agricultural branches. A more limited number would be trained as specialists in different branches of production and disciplines, to serve as subject matter specialists.

In-service Training

Whatever the background of the extension agent, suitable introductory and in-service training must be provided. Even for the university graduate, learning cannot end on completion of his formal studies. The objectives of in-service training should be:

(a) to keep up with research: by regular meetings between researchers and extension workers, joint colloquia, etc.;
(b) to impart basic knowledge, not only in the fields directly related to agriculture, but also in sociology, economics, psychology, etc.;
(c) to improve extension methods, by constant evaluation of methods, the joint study of research findings and extension methods, the exchange of experience.

In-service training is a never-ending process. It cannot consist exclusively of what is provided by the services; the people involved must themselves continue training by personal effort—reading, individual learning, etc.

Summary

Education of Farmers

General education contributes to increased agricultural productivity by facilitating the transmission of technological knowledge, improving rationality and receptiveness for new ideas and by changing the values and aspirations of farmers.

However, the basic weakness of the educational systems in most developing countries is that they are not geared to preparing children for the life the majority of them will lead; hence the need for major reforms in their educational programmes.

Because most developing countries lack the means to maintain an adequate conventional rural education system, alternative approaches have been sought and applied, such as self-supporting schools (Tanzania); the education corps (Iran); the university service (Ethiopia); etc.

Extension Education

Extension education is an informal out-of-school educational service for training and influencing farmers to adopt improved practices.

The most prevalent institutional setting is the government extension service. Though an agricultural extension service is the major tool for agricultural development in most LDCs, it has proved disappointing in many ways. The *failures of the conventional extension services* can be imputed to organizational defects and operational weaknesses; lack of appropriate locally tested improved methods, making possible dependable and substantial increases in production; lack of coordination between research and extension; and economic and social constraints.

Amongst the main organizational and operational defects are: poor communication between workers in the field and the administrative hierarchy; badly designed or inappropriate extension programmes; the performance of regulatory, supervisory and commercial duties not related to extension work; an insufficient number of field workers, ineffective extension personnel and poor working conditions; unsatisfactory attitudes and behaviour towards farmers.

Faced with the inadequacies of their extension services, many governments in the LDCs have opted for one of the following solutions: improve the existing structures; transfer responsibility from the ministry of agriculture to other institutions (universities and agricultural colleges, farmers' associations, cooperatives, commodity boards, etc.); adopt unconventional methods, based on the 'self-help' approach.

The conventional extension efforts need to be supplemented and strengthened by systematic farmer training courses, either in special centres or in the villages themselves. Vocational training for non-farm occupations has generally been neglected, despite its importance for providing gainful employment and improved services in the rural areas. Other important areas that are relatively neglected are extension work with youth and with women.

Whilst the need for close *cooperation between research and extension* appears to be axiomatic, its achievement in most countries is the exception rather than the rule. Cooperation between the two services can be improved by extension workers participating in the formulation of research programmes, including the determination of priorities; by participating in experimental field work; by organizing common service units (libraries, diagnosis laboratories, etc.); and by involving research workers in the in-service training of extension workers.

The importance of the *farmer's participation* in research and extension programmes cannot be overstressed. This should include the formulation of policies as well as implementation (on-farm trials, pilot-farms, demonstration plots, etc.). Government action is required to foster rural associations that are truly representative and devoted to promoting the interests of the *majority* of farmers. Direct contacts between farmers and research workers, without by-passing extension personnel, should be encouraged.

Research and extension personnel should learn to know local conditions and the habits, traditions and methods of work of the people they are required to serve; to learn from farmers at the same time as they try to teach them.

The Adoption of New Practices

The generally accepted pattern of diffusion of a new practice in a farming community, whereby an innovation is first adopted by a few pioneers and then rapidly spreads to the great majority of the farmers, is rarely true in the LDCs. In the latter, the village community usually consists of a small minority who are able to adopt an innovation and need only to be made aware of its potential benefits, and a vast majority of small farmers who lack the means needed to make adoption possible. This fact has implications for extention work in the LDCs which are entirely different to those generally accepted.

The Complementary Nature of Technological Factors

Single-practice programmes, such as introducing irrigation, applying fertilizers or adopting new varieties, etc., usually give poor results, because no single factor can be fully effective if other essential conditions are lacking.

By contrast, certain techniques, when applied in combination, can give spectacular results in a very short time. Hence the justification for 'package programmes'.

Education of Extension Workers

The extension worker must be technically competent and have a basic knowledge of economics and science.

He must have practical knowledge of farming and know how to demonstrate new practices to farmers. Also, he must be familiar with the customs, values and ways of thinking of the rural society in which he works.

To prepare extension workers for their tasks, special training programmes are therefore required. Many of the intermediate-level schools that train pupils for farming in general and for extension work in particular, have teaching programmes that are unrelated to the work of the future graduates, with too little attention paid to farm management, extension techniques, social problems and practical farm work. The major improvements proposed for intermediate education are: a reform of the curriculum, so that subject matter and concepts should be relevant to local development needs; adequate teaching and training facilities; selection and training of the teaching staff; participation of women in intermediate agricultural education; a major emphasis on motivation for agricultural work and adequate practical training.

Finally, *in-service training* should be considered a never-ending process. It should be provided by the service and supplemented by personal effort in individual learning.

References

AID (1968). *Building Institutions to Serve Agriculture: A Summary Report of the Committee on Institutional Cooperation.* Agency for International Development, Purdue University.

Allahabad Agricultural Institute (1958). *Extension Evaluation.* The Leader Press, Allahabad, India.

Allan, A. Y. (1971). Fertilizer use on maize in Kenya. Pp. 10–25 in *Improving Soil Fertility in Africa: FAO Soils Bull. No.* 14.

Arnon, I. (1977). *Organization et Programmation de la Recherche Agronomique dans la Republique du Senegal.* Food and Agriculture Organization of the UN, Rome: 56 pp.

Axinn, G. H. & Thorat, S. T. (1972). *Modernizing Agriculture—A Comparative Study of Agricultural Extension Education Systems.* Praeger, New York: xv + 216 pp., illustr.

Barraclough, S. L. (1971). Pp. 51–62 in *Behavioural Change in Agriculture* (Ed. J. P. Leagans & C. P. Loomis). Cornell University Press, Ithaca, NY: xii + 506 pp.

Barraclough, S. L. (Ed.) (1973). *Agrarian Structure in Latin America.* Lexington Books, Lexington: xxvi + 351 pp., illustr.

Becerra, J. & Samper, A. (1969). The situation, problems and trends in agricultural education and training in Latin American countries. Regional paper, *World Conf. on Agricultural Education and Training in the Latin American Region.* FAO, UNESCO & ILO, Copenhagen, Denmark.

Benor, D. & Harrison, J. Q. (1977). *Agricultural Extension—The Training and Visit System.* World Bank, Washington, DC: ix + 55 pp., illustr.

Blandy, R. & Nashat, M. (1966). The education corps in Iran. *Int. Lab. Rev.*, **XCIII**, 521–9.

Bose, S. P. (1961). Characteristics of farmers who adopt agricultural practices in Indian villages. *Rural Sociol.*, **26**, 138–46.

Brasseur, R. E. (1976). Constraints in the transfer of knowledge. *Focus,* **18** (3), 12–19.

Brown, Dorris (1971). *Agricultural Development in India's Districts*, Harvard University Press, Cambridge, Ma.: xvi + 169 pp., illustr.

Brown, L. R. (1970) The green revolution, rural employment and the urban crisis. Paper presented at *Columbia University Conf.* on *International Economic Development*, New York.

Byrnes, F. C. (1969). Farmers' resistance—to what? *Development Digest*, **7** (2), 29–37.

Byrnes, F. C. (1974), Agricultural production training in developing countries: critical cut controversial. Pp. 215–35 in *Strategies for Agricultural Education in Developing Countries.* The Rockefeller Foundation, New York: x + 444 pp.

Byrnes, F. C. & Byrnes, K. J. (1971). Agricultural extension and education in developing countries. Pp. 326–51 in *Rural Development in a Changing World* (ed. R. Weitz). MIT Press, Cambridge, Mass.: 587 pp.

Castillo, Gelia T. (1975). *All in a Grain of Rice.* Southeast Regional Center for Graduate Study and Research in Agriculture, College, Laguna, Philippines: xii + 416 pp.

Chambers, R. & Belshaw, D. (1973). *Managing Rural Development.* Institute of Development Studies, University of Sussex, Brighton, England.

Chambers, R. & Moris, J. (Ed.) (1973), *Mwea, an Irrigated Rice Settlement in Kenya.* Weltforum Verlag, Munich, FRG: 539 pp., illustr.

Chang, C. W. (1962). *Increasing Food Production Through Education, Research and Extension: FFHC Basic Study No.* 9: FAO, Rome: vi + 78 pp., illustr.

Chang, C. W. (1963). *Extension Education for Agricultural and Rural Development.* FAO Seminar, Bangkok.

Chang, C. W. (1969). The situation, problems and trends in agricultural education and training in Asia and the far east region. Regional paper, *World Conf. on Agricultural Education and Training.* FAO, UNESCO & ILO, Copenhagen, Denmark.

Chaudri, D. P. (1968). *Education and Agricultural Productivity in India*, PhD Dissertation, University of Delhi.

Cline, M. G. (1971). Agricultural research and technology; Response.

Pp. 94–100 in *Behavioural Change in Agriculture* (Ed. J. P. Leagans and C. P. Loomis). Cornell University Press, Ithaca, New York: xii + 506 pp.
Coombs, P. H. & Mansour, A. (1974). *Attacking Rural Poverty*. Johns Hopkins University Press: xvi + 292 pp.
Cummings, R. W. (1971). Agricultural research and technology. Pp. 79–83 in *Behavioural Change in Agriculture* (Ed. J. P. Leagans and C. P. Loomis). Cornell University Press, Ithaca, NEW York: xii + 506 pp.
Dasgupta, B. (1977). *Agrarian Change and the New Technology in India.* UNRISD, United Nations, Geneva: xxvii + 408 pp.
David, C. & Korten, F. F. (1966). Ethiopia's use of national university students in a year of rural service. *Comparative Education*, **10** (3), 482–92.
Demir, N. (1976). *The Adoption of New Bread Wheat Technology in Selected Regions of Turkey.* CIMMYT, Mexico City.
Deutsch, K. W. (1971). Developmental change: some political aspects. Pp. 27–50 in *Behavioural Change in Agriculture* (Ed. J. P. Leagans and C. P. Loomis). Cornell University Press, Ithaca, New York: xii + 506 pp.
Diaz, H. (1974). *An Institutional Analysis of a Rural Development Project: The Case of the Pueblo Project in Mexico.* PhD Thesis, University of Wisconsin.
Dumont, R. (1962). Accelerating African agricultural development. *Impact of Science on Society,* **12**, 231–53.
FAO (1962). *Development Centre for East, Central and Southern Africa: ETAP Report No.* 1566. ETAP, Rome.
FAO (1969). *Bigger Crops and Better Storage.* FAO, Rome: vii + 49 pp.
FAO, UNESCO & ILO (1971). Intermediate agricultural education and training. Commission paper, *World Conf. on Agricultural Education and Training.* Copenhagen, Denmark.
FAO (1972). *The State of Food and Agriculture, 1972.* Food and Agriculture Organization of the UN, Rome: x + 189 pp.
FAO (1975). *Agricultural Extension and Training.* Food and Agriculture Organization of the UN, Rome.
FAO (1976). *The State of Food and Agriculture, 1975: Agriculture Series No.* 1. Food and Agriculture Organization of the UN, Rome: vi + 150 pp.
Farouky, S. T. & Skilbeck, D. (1969). The situation, problems and trends in agricultural education and training in the near east region. Regional paper, *World Conf. on Agricultural Education and Training.* FAO, UNESCO, & ILO, Copenhagen, Denmark.
Gerhart, J. (1975). *The Diffusion of Hybrid Maize in Western Kenya.* CIMMYT, Mexico City.
Griliches, Z. (1964). Research expenditures, education and the aggregate agricultural production functions. *Agr. Econ. Rev.*, **54**, 961–74.
Hapgood, D. (Ed.) (1965). Policies for promoting agricultural development. Pp. xiii, 321 in *Report of a Conference on Productivity and Innovation in Agriculture in the Underdeveloped Countries. Massachusetts Institute of Technology, Center for International Studies*. MIT Press, Cambridge, Mass.
Harrison, R. K. (1969). *Work and Motivation; A Study of Village-Level Agricultural Extension Workers in the Western State of Nigeria.* Nigerian Institute of Social and Economic Research, Ibadan: (mimeogr.).
Heck, van, B. (1979). *Participation of the Poor in Rural Organization*, Food and Agriculture Organization of the UN, Rome: xiii + 98 pp.
Herzberg, J. & Antuna, S. (1973). *Analytical Study of the Extension Services for Ecuador and Paraguay*. Food and Agriculture Organization of the UN, Rome.
Heyer, J., Ireri, D. & Moris, J. (1971). *Rural Development in Kenya.* East African Publishing House, Nairobi.
Hodgdon, L. L. & Singh, H. (1964). *Adoption of Agricultural Practices in Madhya*

Pradesh. National Institute of Community Development, Government of India, Rajendranagar, Hyderabad 30, Andhra Pradesh.

Hopkins, J. (1969). *The Latin American Farmer.* US Department of Agriculture, Washington, DC: pp. 137.

Hunter, G. (1970). *The Administration of Agricultural Development: Lessons from India.* Oxford University Press, London: 160 pp.

Hursh, G. D., Niels, R. & Kerr, G. B. (1968). *Innovation in Eastern Nigeria; Success and Failure of Agricultural Programs in 71 Villages of Eastern Nigeria.* Michigan State University, East Lansing, Mich.

IRRI (1978a). *Rice Research and Production in China: An IRRI Team's View.* International Rice Research Institute, Los Baños, Philippines: 119 pp., illustr.

IRRI (1978b). *Economic Consequences of the New Rice Technology.* International Rice Research Institute, Los Baños, Philippines: v + 402 pp., illustr.

Jiggins, Janice (1977). Motivation and performance of extension field staff. Pp. 1–19 in *Extension Planning and the Poor.* Overseas Development Institute, Agricultural Adminstration Unit, London: vi + 57 pp.

Jurion, F. & Henry, J. (1967). *De l'Agriculture Itinerante a l'Agriculture Intensifiée.* Institut National pour l'Etude Agronomique du Congo (INEAC), Bruxelles: 498 pp., illustr.

Kellogg, C. E. (1962). Interactions in agricultural development. United States paper prepared for *the United Nations Conf. on the Application of Science and Technology for the Benefit of the Less Developed Areas.* US Government Printing Office, Washington, DC.

Kidd, D. W. (1968). *Factors Affecting Farmer's Response to Extension in Western Nigeria.* CSNRD, Michigan State University, East Lansing, Mich.

Krantz, B. A. & Kampen, J. (1977). Water management for increased crop production in the semi-arid tropics. Paper presented at the *National Symp. on Water Resources in India and Their Utilization in Agriculture.* Water Technology Centre, IARI, New Delhi.

Kuo, L. T. C. (1972). *The Technical Transformation of Agriculture in Communist China.* Praeger, New York: xx + 206 pp., illustr.

Leagans, J. P. and Loomis, C. P. (Ed.) (1971). *Behavioural Change in Agriculture: Concepts and Strategies for Influencing Transition.* Cornell University Press, Ithaca, New York: xii + 506 pp.

Leonard, D. K. (1972). *Organizational Structures for Productivity in Kenyan Agricultural Extension: Working Paper No.* 20. Institute for Development Studies, University of Nairobi.

Leonard, D. K. (1977). *Reaching the Peasant Farmer. Organization Theory and Practice in Kenya.* University of Chicago Press, Chicago and London. xxi + 297 pp., illustr.

Lionberger, H. F. (1960). *Adoption of New Ideas and Practices.* Iowa State University Press, Ames, Iowa, 164 pp.

Lu, H. (1968). *Some Socio-Economic Factors Affecting the Implementation at the Farm Level of a Rice Production Program in the Philippines.* PhD Thesis, University of the Philippines.

Malone, C. C. (1965). Some responses of rice farmers to the package program in Tanjore District, India. *J. Farm Econ.*, **47**, 256–68.

Mathieu, M. (1979). *Progress Report: 4th Consultation on the FAO Fertilizer Programme.* FAO, Rome.

Neurath, P. M. (1960). *Radio Farm Forums in India.* Government of India Press, Delhi.

Nour, M. A. (1969). The Situation, Problems and Trends in Agricultural Education and Training in the African Region. Regional paper, *World Conf. on Agricultural Education and Training.* FAO, UNESCO & ILO, Copenhagen, Denmark.

Nyerere, J. (1970). Education for self-reliance. *Development Digest*, **8** (4), 3–13.
Poleman, T. T. (1964). *The Papaloapon Project.* Agricultural Development in the Mexican Tropics, Stanford University Press: 167 pp.
Prebisch, R. (1971). *Change and Development: Latin America's Great Task.* Praeger, New York: xxxi + 293 pp., illustr.
Rahim, S. A. (1969). The Comilla program in East Pakistan. Pp. 415–24 *Subsistence Agriculture and Economic Development* (Ed. C. R. Wharton, Jr). Aldine Publishing Company, Chicago, Ill.; xiii + 481 pp., illustr.
Rogers, E. M. (1960). *Social Change in Rural Society.* Appleton-Century-Crofts, New York; xi + 490 pp., illustr.
Rogers, E. M. (1962). *Diffusion of Innovations.* The Free Press, New York; xiii + 376 pp., illustr.
Rogers, E. M. & Svenning, L. (1969). *Modernization Among Peasants.* Holt, Rinehart & Winston, Inc., New York.
Rogers, E. M. & Yost, M. D. (1960). *Communication Behaviour of County Extension Agents.* Ohio Exp. Sta., Wooster, Bul. 850.
Sanchez, R. (1970). *The Pueblo Project.* Agricultural Development Council, New York: 39 pp., illustr.
Sidky, A. R. (1957). *Progress of Egyptian Agriculture in Five Years, The Agricultural Policy—Application and Achievements.* Ministry of Agriculture, Cairo.
Stevens, R. P. (1977). Tradition and Dynamics in Small Farm Agriculture. The Iowa State University Press, Ames: xiii + 266 pp., illustr.
Straus, M. A. (1953). Cultural factors in the functioning of agricultural extension in Ceylon. *Rural Sociology*, **18**, 249–56.
Summers, E. A. (1962). The Role of Agricultural Extension in the Application of Science and Technology. Paper presented at *the UN Conf. in the Application of Science and Technology, Geneva.*
Tang, A. (1961). Research and Education in Japanese Agricultural Development. Paper presented at *Annual Meeting of the Econometric Society, Stillwater, Oklahoma.*
Timmer, C. P. (1975). The political economy of rice in Asia: Indonesia. *Food Res. Inst. Stud.*, **14** (3), 196–231.
UNESCO (1964). *Education and Agricultural Development.* Freedom from Hunger Campaign Basic Study No. 15, Paris: 62 pp.
Wellhausen, E. J. (1967). Opportunities for crop improvement. Pp. 237–55 in *Rural Development in Tropical Latin America* (Ed. K. L. Task & L. V. Crowder), Vol. 2. New York State College of Agriculture, Ithaca.
Wharton, C. R. (1965). Education and agricultural growth; The role of education in early stage agriculture. In *Education and Economic Development*, (Ed. C. A. Anderson & M. J. Bowman). Aldine Press, Chicago: x + 436 pp., illustr.
Wilde de, J. C. & McLoughlin, P. E. M. (1967). *Experiences with Agricultural Development in Tropical Africa.* Johns Hopkins Press, Baltimore Maryland: Vol. 2, xii + 446 pp., illustr.
Wilson, F. B. (1968). *Education and Training for Rural Development in Africa: A working paper.* FAO, Rome.
Winkelmann, D. L. (1976). Promoting the Adoption of New Plant Technology. Paper presented to *the World Food Conf., Iowa State University, Ames, Iowa.*
Yudelman, M., Butler, G. & Banerji, R. (1971). *Technological Change in Agriculture and Employment in Developing Countries. Development Centre Studies, Employment Series, No.* 4. Development Centre of OECD, Paris: 204 pp.

CHAPTER VIII

The Green Revolution

High-Yielding Varieties (HYVs)

The sowing of improved varieties of the traditional crops is, for the individual farmer, probably the cheapest of all the means available to him for increasing production. As cereals are the main source of nutrients for the populations of the developing countries, increased yields will depend to a large extent on the results of efforts to produce and propagate high-yielding adapted varieties of wheat, rice, maize, sorghum, and millet.

The core of the so-called 'green revolution' are new high-yielding varieties (HYVs), the most prominent of which at present are wheat and rice. The fact that the new HYVs are not sensitive to differences in day-length, makes them adaptable to a wide range of environmental conditions and has thereby greatly increased the possibility of their wide adoption in many regions of the world. Considerable effort is also being invested in breeding for high yields in the other cereal crops mentioned above, such as maize, sorghum and millet, as well as many other crops.

The internation institutes, IITA and CIAT, now have major programmes devoted to the development and improvement of root crops which originated and have been continuously cultivated in the tropical regions: sweet potato (*Ipomoea batatas*), cassava *(Manihot esculenta)*, yams *(Dioscorea spp.)* and the edible aroids—taro *(Colocasia esculenta)* and tannia *(Xanthosoma sagittifolium)*. This work could well result in the 'green revolution' of these areas, to which the main cereal crops are not adapted.

Impact on Production

Many of the countries in which the 'green revolution' has taken place were originally importers of grain and produced part of their own supplies at high cost (Dalrymple & Jones, 1973). HYVs have made a fundamental contribution to the economic development of these countries.

By the early 1960s, new HYVs of wheat, developed at the International Maize and Wheat Improvement Centre in Mexico, had already been taken up by Mexican farmers and planted on 90% of the country's wheat land, doubling the average yield per hectare as compared to the traditional varieties (Ojala, 1972).

In Indonesia, modern varieties of rice were first introduced in 1967. Their use was required as a condition for obtaining maximum institutional credit. They covered 190,000 ha in the 1969 dry season and 306,000 ha in the 1970–71 wet season. By that time HYVs accounted for 10% of the total rice area in Indonesia.

In Pakistan, the first of the Mexican wheat varieties were introduced in 1965. By 1971, 3,037,000 ha, or about 50% of the wheat area, were planted to the new varieties (IRRI; 1975).

Altogether the area planted to new HYVs of wheat and rice rose on an approximately straight-line trend, from practically zero in 1965–66 to 43.9 million tons in 1976–77 with a slight slowing-down in 1973–75, partly as the result of a fertilizer shortage (FAO, 1979).

The proportion of the total areas sown to HYVs of wheat and rice in the developing countries in 1976–77 is shown in Table XVII.

Wellhausen (1970) gives a few examples of the impact of HYVs of wheat on the economies of a number of countries. With seed and production technology imported from Mexico, Pakistan raised its total wheat production in five years from 4.6 million to 8.4 million tons—an increase of 83%. Similarly, India, with the same varieties and production technology imported from Mexico, raised production from a high of 12 million tons in 1965 to over 20 million tons in 1970—an increase of 64%.

The development and distribution of the new HYVs along with new production techniques perfected by the International Rice Research Institute at Los Banos, have caused considerable increases in yields of rice in many parts of South and South-East Asia. The Philippines attained self-sufficiency in five years with extensive plantings of the semi-dwarf variety IR8*. In 1971, a rice shortage again occurred which was attributed to typhoons, the Muslim–Christian conflict which had seriously hampered rice and maize production, lack of agricultural credit, inability of the Rice and Corn Administration to guarantee a floor support price for rice, and damage brought about by tungro disease (Castillo, 1975).

It is naturally impossible to separate the effect of the adoption of the HYVs (with attendant inputs) from those of other factors, such as variations in climatic conditions. According to Efferson (1972), favourable

Table XVII Share of HYVs as a percentage of total area sown of wheat and rice in 1976–77 (FAO, 1979, by permission).

	Wheat	Rice
Asia	62	30
Africa	22	3
Latin America	41	13
Near East	17	4

weather occurred in Asia during the period the new varieties were introduced. However, the increases in production have frequently been so spectacular as to be attributable largely, even though not entirely, to new technology.

The spread of hybrid maize in developing countries has generally been slow, the main reason being the complex problems of seed production and the need to replace the hybrid seed every year. Emphasis in maize-breeding in developing countries has shifted mainly to open-pollinated synthetic varieties whose seed can be produced simply, cheaply and needs to be replaced only every three or four years. Though potentially not as high-yielding as the best adapted hybrids, synthetics are capable of producing yields twice as high as that of local open-pollinated main varieties (Ojala, 1972).

In Thailand, during the past decade, production of maize has increased at the rate of about 52,000 hectares and 100,000 tons per year, using new HYVs and cultural practices. West Pakistan reported a 30% increase in yield over the past year. India has boosted the area planted to maize by 40–50% since 1960, with yields per unit area steadily climbing. The new hybrid dwarf varieties of sorghum are moving almost as rapidly, especially in India and Pakistan (Fig. 37).

Fig. 37 Field of hybrid semi-dwarf sorghum. Yields were 5 t/ha, rain-fed and 8 t/ha, irrigated. Photograph by courtesy of 'Hazera' Seed Growers' Cooperative.

Influence on Farm Income

Initially, innovative farmers profit from the increased production per unit of land *and* the relatively high prices prevailing in countries in which demand, in the past, has exceeded supply. However, as more and more farmers adopt the new HYVs, output is increased and the price of the commodities produced generally goes down.

For innovative farmers, lower prices will generally mean the end of exceptionally high profits, but income from HYVs will remain greater because of high production levels and the freeing of land for other high-income crops.

However, farmers who have not yet adopted HYVs, are caught on a 'technological treadmill'. As prices go down, the laggards are forced to adopt the new technology, if only to remain where they are (Dalrymple, 1969). The hardest hit by the fall in prices is the small farmer who cannot afford to purchase the additional inputs involved (cf. pp. 293–294) and who will receive less and less for any small surplus over immediate needs he may be fortunate in producing.

In a survey carried out by the International Rice Research Institute, in 32 villages in selected rice-growing areas in Asia, with a high potential for production, the following relationships between farm ownership and farm size on the one hand adoption of HYVs and increased income on the other were found (Table XVIII).

The advantage derived from HYVs patently accrued more to owners than to tenants and to owners of larger farms as compared to owners of smaller farms. The inequality is however built into existing structures, as evidenced by the figures for farmers growing traditional varieties only. Under the circumstances any innovation is bound to have different impacts on different groups or individuals.

Short-term Effects on Employment

In a survey of 32 villages carried out over a broad geographic rice-growing area in Asia, and reflecting a wide range of conditions, it was

Table XVIII Percentage of rice farmers reporting increase in rice income since 1964–65 in relation to tenure and size of holding owned (Castillo, 1975, by permission).

	Type of tenure		Farm size	
	Owners	Tenants	Less than 4 ha	4 ha and over
HYVs only	59	33	50	67
Local varieties only	20	1	26	66

found that a large proportion of rice farmers reported an increase in the use of family or hired labour following the adoption of HYVs of rice. The increase in preharvest labour requirement was related to the concurrent adoption of other preharvest improved practices. Villages with a high percentage of farms reporting a large increase in labour requirements were those in which a high percentage of farmers had also adopted fertilizer use, insecticides, and straight-row planting (Barker & Anden, 1975).

The adoption of HYVs may have no effect on the labour requirements of such operations as land preparation, seeding or transplanting; but harvesting, threshing, and handling of the higher yields, if carried out in the traditional manner, are bound to require more labour. The application of fertilizers may also make weeding more frequent and intensive.

It is therefore generally agreed that, with no change in the degree of mechanization, the 'green revolution' has increased labour requirements per unit of land, but has decreased labour per unit of output. Increases per unit of land have been larger and more consistent for wheat than for rice (Dalrymple & Jones, 1973).

Though there were wide differences in expenditure on hired labour by wheat farmers in different districts in India, for example, those who had adopted HYVs and fertilizer use had close to a 40% increase in labour expenditure per hectare compared with farmers using traditional methods. The largest increases occurred on farms in the 4–5 hectare group using improved varieties giving the highest yields, as compared to those using traditional varieties. In this size-group, neither category of farmer used machinery, so that the increase in labour requirements can be ascribed to the use of yield-increasing inputs (Yudelman *et al.*, 1971).

A more detailed analysis of the impact of improved varieties on labour requirements was made by the Punjab Agricultural University, as shown in Table XIX.

In certain crops, when harvesting requires a large amount of manual labour, and the yield differences between improved and traditional variety

Table XIX Labour requirements (man-hours) per cropped acre (0.405 ha) in Punjab at optimum level of technology (Punjab Agricultural University, 1973).

Crop	Labour requirements		Increase over local variety (%)
	Local variety	Improved variety	
Wheat	239	313	31
Cotton	391	490	25
Paddy	555	695	25
Sugar-cane	1231	1403	14
Groundnut	374	420	12
Bajra millet	252	267	6
Maize	389	408	5

is great, the increase in labour requirements may be quite considerable. In a study in a zone in Gujarat (India), for example, it was found that improved varieties of cotton required 91% more human labour than did native varieties (Desai *et al.*, 1970).

Early-maturing photoperiod-insensitive HYVs of rice have enabled multiple-cropping and thereby provided an additional source of labour (cf. pp. 290–292).

Increased labour requirements may be beneficial where additional productive work for abundant labour is sought; it may create a bottleneck where and when seasonal labour is in short supply. In the survey of farmers in India mentioned above, it was found that the greatest absolute increase in expenditure for labour, incurred by those farmers who have adopted HYVs of wheat, was for casual labour which is mainly seasonal (Yudelman *et al.*, 1971).

Although many factors influence rural wage rates, the general indication is that the introduction of HYVs has increased wage levels as a result of the increased demand for labour, thereby allowing rural labout to share somewhat in the economic benefits due to the new varieties. In a study in Indian Punjab, it was shown that whereas the share of the new wheat varieties increased in the course of three years from 3.5 to 57.9% of the whole wheat acreage, real wages of the agricultural and non-agricultural rural labour had also increased significantly (Soni, 1970). Aggarwal (1973) reports that wage rates more than doubled in the five years following the introduction of HYVs in the Ludhiana district, Punjab. Because HYVs require much more intensive and timely care than traditional varieties, competition between farmers for labour during peak requirement periods can be a major factor in increasing wage levels.

Effects on Nutrition

There is no doubt that HYVs have made *more* food available in countries in which undernutrition or malnutrition are prevalent. The effect on food *quality* is however not so clearcut. It may be positive, if the adoption of HYVs frees land for crops with high nutritive value, such as fruit, vegetables and dairy products; it may be negative if they displace high-protein pulse crops or if the new varieties have lower protein levels, and/or the protein is of lower biological value. It is hoped that these latter problems can be partly or entirely overcome by appropriate plant-breeding work.

Already, new HYVs are available that have had a higher protein content bred into them. The normal level of traditional varieties is around 9–12%; whereas some of the new varieties have as much as 16–17%. However, even cereals that have a relatively high protein content, cannot resolve protein hunger more than marginally. The human stomach is not capable of ingesting the levels of carbohydrate that are necessary to acquire sufficient protein from cereals alone (Palmer, 1972a).

Progress has also been made in breeding cereal varieties with protein of higher biological value. Mention need only be made of the maize mutants with much higher lysine and tryptophane contents and a barley strain with higher lysine levels.

The danger of displacement of pulse crops is real, mainly because of their low yields and low prices.

Palmer (1972a) lists the following factors influencing a change in food production away from pulses towards cereals in consequence of the introduction of new HYVs: (1) deliberate discriminatory intervention in market factors favouring cereal production; (2) inegalitarian shifts in mass purchasing power (see below); and (3) the increasing share of household budgets going to cereals (or tubers) as income falls and as the relative price of cereals *vis-à-vis* pulses is reduced because of lower costs of production of the former.

Increased Productivity per Unit of Water

Because of high yields, the efficiency of water use by HYVs is much greater than that of traditional varieties. In a comparative study of well irrigation in the Aligarh district (India), it was found that the response to increments of water and associated inputs (mainly fertilizers) was four times as large for Mexican wheat as for the native variety, and nearly five times as large for hybrid millet as for the traditional variety (Moorti, 1971).

Multiple-cropping

Two-crop rice culture has long been practised in parts of southern China where climate and other factors permitted. However, attempts to extend the practice to other areas have generally not been successful, mainly because the more productive varieties were generally late-maturing, so that harvesting of the first crop and planting of the second crop came too late in the season (Kuo, 1972).

Many of the new HYVs are early-maturing and therefore their growing season can be shortened by as much as one-third. They are also relatively insensitive to length of day. For these reasons, two or more crops can be planted and harvested in the course of a 12-month period in tropical and subtropical regions. In the research stations of Los Baños, Philippines, three crops of rice are grown per year, with a yield totalling 15 tons/ha.

By using irrigation in the dry season, the full potential of these regions can be achieved. Because of the sunny weather, dry season yields of productive rice varieties may average 50% more than those obtained during the wet season.

Where sufficient water is lacking for producing a rice crop during the dry season, other crops with lower water requirements, such as wheat or sorghum, can be grown. Land planted with an early variety of rice followed by sorghum produced 20 tons/ha of grain at Los Baños. This should be compared to the 12 tons/ha that is considered a normal yield in some parts of Asia (Brown, 1970).

In Taiwan, multiple-cropping has expanded from 18% of the total crop area in 1946 to 89% in 1966.

The breeding of early-maturing rice varieties enabled the farmers of central Taiwan to move from double- to quadruple-cropping by adding a winter catch-crop and a summer catch-crop between the first and second rice crops (Ho, 1966).

In general, the regions of double rice cropping get the highest total crop production per unit area of cultivated land. Maximum returns are obtained where double cropping of rice is practised on relatively fertile alluvial soils with adequate irrigation water, and no excessive rainfall or cloudiness in winter. Multiple-cropping increases labour requirements and can help to level off seasonal peaks in labour requirements.

Giles (1975) made a study of two hypothetical 12-ha farms in Ludhiana district, India; the one typically bullock-powered, growing a single crop per year, the other multiple-cropped and using selective mechanization.

The multiple-cropped farm was found to have a greater labour requirement per hectare, despite the relatively high degree of mechanization, and a more even distribution of labour requirements throughout the year (Fig. 38) (cf. p.).

Besides the advantage of higher yields, multiple-cropping has justified investments in irrigation facilities and makes possible more efficient utilization of labour, draught animals and farm equipment which would otherwise remain idle during the dry season. The increased intensity of cropping will

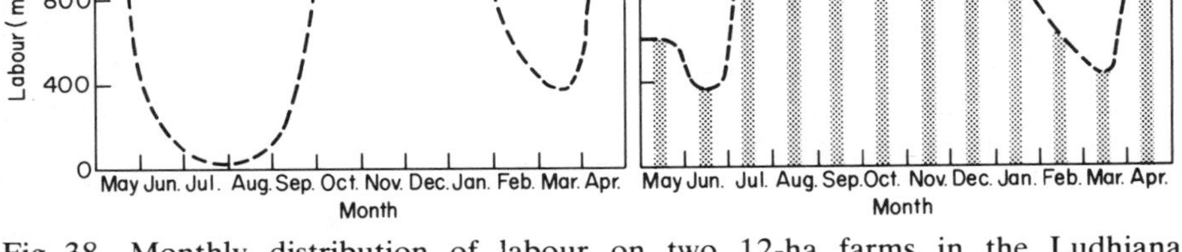

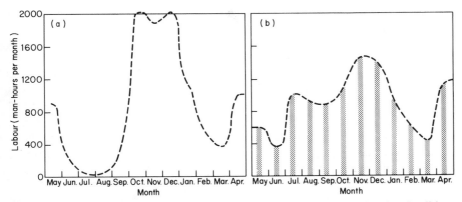

Fig. 38 Monthly distribution of labour on two 12-ha farms in the Ludhiana district: (a) single crop (100% crop intensity), animal-powered, 10,500 man-hours total; (b) multi-crop (225% crop intensity), selective mechanization, 11,000 man-hours total (Giles, 1975). By courtesy of the author.

therefore, in due course require and justify selective mechanization, such as equipment for pumping, tractors and implements adapted to smallholdings, seeding, and threshing machines, etc. (cf. p. 363).

Catalysis Effects

The adoption of high-yielding and fertilizer-responsive varieties of wheat and rice, has also proved to be the main catalyst of changes in agriculture.

High returns from HYVs make possible rural savings which are invested in inputs such as fertilizers and pesticides which increase production in the short run, as well as land improvement and expansion of tubewells and irrigation which are effective over the long run.

A few examples of the effects of this large-scale introduction of the new wheat and rice varieties on the development of rural industries have been described in Chapter I.

The profitability and possibility of replacing human and animal power by mechanical power are leading to increased use of tractors (cf. pp. 298–299).

Large investments in production inputs such as industries supplying fertilizers, chemicals for plant protection as well as substantial improvement in distribution and storage facilities become necessary. Other supporting services, such as credit and extension, have to be expanded, and modification of the land tenure system may be needed in some countries.

In brief, the dependence of agriculture on the non-farm sector increases considerably, and this dependence has a considerable impact on development of rural industries, services and infrastructure.

The adoption of improved varieties is generally the spearhead of modernization of agriculture.

In the Philippines, for example, farmers who adopted new rice HYVs also accepted other improved rice production practices (Table XX), such as careful seedbed preparation, fertilization, spraying against disease and pests, straight-row planting, weeding, etc. (Castillo, 1975).

Not only is agriculture modernized, but rural life is also improved. With the adoption of new practices in growing rice (high-yielding varieties with heavy applications of fertilizer) by the Filipino farmer, 'came a modernizing outlook on agriculture, new attitudes, aspirations, perceptions, and amenities in life' (Castillo, 1973). In the course of seven years (from 1965 to 1970) the introduction of new rice technology into a Filipino village had resulted in an extension of the irrigation system and double-cropping. New schools and new chapels were established, roads improved, transistor radios and hand tractors became commonplace (Lewis, 1971).

Ladejinsky (1970) reports a similar change in the mental attitudes of Indian farmers adopting the combination of varieties and fertilizers in wheat. The importance of the resulting psychological change, as expressed in a desire for better farming methods and a better standard of living, cannot be overemphasized.

Table XX Comparative adoption patterns of improved practices by two groups of rice farmers in Laguna, Philippines[a] (1967 in % of 1966, when all farmers grew only traditional varieties).

	Farmers who changed to IR8 in 1967	Farmers who did not change to IR8
Nitrogen (kg/ha)	469	133
Expenditure per hectare on		
Fertilizers	353	125
Herbicides[b]	160	225
Insecticides	633	250
Labour (man-days/ha)	141	100
Yield	178	92

[a]Based on data from *IRRI Agricultural Economics Annual Report*, 1967. The figures in Table XX give an indication of the dramatic increases in the use of fertilizers and insecticides, and to a lesser degree weedicides, that occurred in the first year of sowing of IR8. A comparative group of farmers, who did not sow IR8 also showed a certain increase in the use of these inputs, possibly influenced by their neighbours.
[b]Though the percentage increases in the use of weedicides is markedly greater for the farmers who did not sow IR8, the actual amounts used by the two groups were practically identical, indicating that this input was already in general use before the introduction of HYVs.

Problems Arising from the Adoption of HYVs

The promise of the HYVs is not an unqualified one. They have brought in their wake great advantages, but have also created new problems.

After more than a decade of experience in countries in which the seed-fertilizer revolution was initiated in the mid-1960s, it is now possible to assess its long-term effects, the so-called second generation problems, many of which were not anticipated.

'For a strategy of technological improvement to succeed on a sustained basis, it must include plans to cope with the consequences of its success' (FAO, 1968).

Much further effort, extensive financing and comprehensive planning will be essential if the fullest possible benefit is to be attained from the new technology. These problems are reviewed briefly here; several aspects will be treated in more detail in the following chapters.

An assessment of the impact of the adoption of the HYVs must take into account not only the anticipated increase in yields, but also the need for additional inputs and its socio-economic implications, ecological restrictions that may lead to failure, problems of consumer acceptance, short- and long-term effects on employment, and above all the ability of the small farmer to participate in the Green Revolution and enjoy its benefits.

Need for Additional Inputs

Under adverse growing conditions, or without a combination of essential inputs, HYVs generally have little or no yield advantage over old varieties.

For this reason, Palmer (1972b) considers the term 'high-yielding' varieties a misnomer. We will return to this point later (cf. p. 310). High-yielding varieties therefore demand, besides substantial supplies of fertilizers and good water management, a considerable improvement in traditional methods of cultivation and plant protection.

The high rates of fertilizers used in conjunction with HYVs, in particular nitrogen, increase the vulnerability of the crops to a large variety of pests and diseases at all stages of growth. This in turn requires larger inputs of chemicals for plant protection.

Because the potential for increased yields of HYVs depends on additional inputs, the costs of production per unit of land are considerably greater for new varieties than for traditional ones. Ladejinsky (1970) states that in the Indian Punjab, for example, it takes 10,000–12,000 rupees for a farmer with 7–10 acres to switch to one of the HYVs of wheat.

In one study in the Phillipines, growing HYVs instead of native varieties under traditional practices, caused an increase in production costs per hectare of 50–75%.

The costs of the additional inputs required may be a major constraint for large numbers of subsistence farmers. Two surveys of farmers in the Philippines who decided not to grow improved rice varieties for another season, indicated that over 50% did so because of low price of the grain or added expenses (Barker & Quentana, 1967).

Many farmers in India are reported to be alarmed at the expense of growing HYVs, and often compromise by halving the fertilizer dose. In a survey by Delhi University, it was found that only 8% of farmers used the recommended dose of fertilizers on hybrid millet (Agricultural Economics Research Centre, 1968). As a result, yields of HYVs were not always markedly better than the yields of local varieties. For small farmers, who form the vast majority, the high cost of inputs, compounded by the risks of growing new varieties, is frequently too great a burden.

Ecological Restrictions to Adoption

Traditional varieties have been selected over generations for their suitability to local growing conditions. It is not surprising that new varieties are generally much less resistant to drought, temperature extremes, flood, pests and diseases than are traditional varieties.

In the Philippines, for example, the first 'miracle' rice IR8 was abandoned soon after it reached its peak of adoption in 1968, mainly because of serious outbreaks of disease. It was replaced by IR5, which hit its peak in 1969 and disappeared by 1972 because of its susceptibility to *tungro* virus disease, and was replaced by IR20 (Castillo, 1975).

District-by-district analysis in India and Pakistan show a very high correlation between the increase in crop production and controlled water supplies (Falcoln, 1970). 80% of fertilizers used in India in 1968 was concentrated in 25% of the districts, most of the latter having an irrigation system; these are also the districts in the vanguard of adoption of HYVs. In India, none of the HYVs of rice introduced to date are fully adapted to the monsoon conditions (Agricultural Attaché, 1973). A similar district-by-district study in Turkey on the adoption of HYVs of wheat in relation to ecological conditions has been described in the previous chapter.

For rice, the main problem is that most Asian regions do not have the ideal conditions required for the new HYVs. Even short periods of deep flooding can be disastrous, and unless water supply is carefully regulated, the yields obtained are far below their potential levels (Dalrymple & Jones, 1973). A water depth in the range of 25–30 cm is considered suitable, and depth in excess of 50–60 cm as too deep for modern varieties (Sriswasdilek *et al.*, 1975).

HYVs of rice are better adapted to the dry season than to the wet season. The sunny dry-season skies enable modern varieties to realize their high yield potential, and the incidence of diseases and pests is much lower. Possibly the latter factor is even more important to farmers, because it reduces the risks involved in growing new varieties which require much larger investments, provided that a regular source of water is available.

Though the HYVs of rice can realize their full potential only during the dry season, in certain areas, such as the Laguna province in the Philippines, an increasing number of farmers are growing the modern varieties in the wet season and the traditional varieties in the dry season. The explanation given by Barker (1970) is that because dry-season water resources are uncertain or limited, many farmers cannot afford the risks involved in planting the modern varieties, with the required high inputs of fertilizers, etc., during the dry season; whereas the traditional varieties do well on medium to low levels of fertilization and the cash investments required are therefore far lower. When the supply of water is more assured there is a preference for sowing HYVs in the dry season (Tagamupay-Castillo, 1970).

The low adoption rates of HYVs in Afghanistan, Algeria, Iran, Iraq, Lebanon and Morocco, as well as the African countries south of the Sahara, also reflect to a large extent the adverse growing conditions of these countries (Palmer, 1972b).

In Turkey, new wheat varieties have been adopted in the higher-rainfall coastal areas, but not in the great wheat producing area–the Anatolian plateau.

In brief, many areas are not suitable for HYVs. Even within Asia, where the rate of adoption of the new varieties was very rapid, only about 9% of rice land and about 23% of the wheat land in 1968–69 was in improved varieties. These were presumably the areas which had the more favourable prerequisites for adoption (mainly an adequate and/or controllable water supply) and therefore the prospects for further increase in areas sown to the new varieties is likely to be much slower (Falcoln, 1970).

In some cases, the use of HYVs in rainfed areas may actually result in a lower income for farmers. In Gapan (Philippines) for example, between 1965 and 1970, the area under HYVs rose from 0 to 31.6% of the total, yet yields rose only from 1.7 to 1.8 tons/ha. Concurrently, the amount of fertilizer increased from 9.2 to 20.5 kg/ha, and the number of farmers using herbicides, insecticides and tractors more than doubled (Griffin, 1973).

Whilst the investment in inputs increased steeply, yields increased only marginally and farmers inevitably became poorer.

Decline in Yields and 'Spillover' Effects

A frequently observed occurrence is that the yields of HYVs tend to decrease as the number of farmers and the extent of the areas sown increase. The early adopters are generally farmers with better resources; as the number of adopters increases, more farmers with poorer water resources and less cash for essential inputs sow the new varieties, with a resultant decline in yields. Nearly 100% of the land of the early adopters was fully irrigated; this level dropped to 50% for the later adopters. Poor irrigation facilities are associated with lower inputs—an additional factor—in lowering yields (IRRI, 1967) (Fig. 39).

It will be observed from Figure 40 that in contrast to the decline in

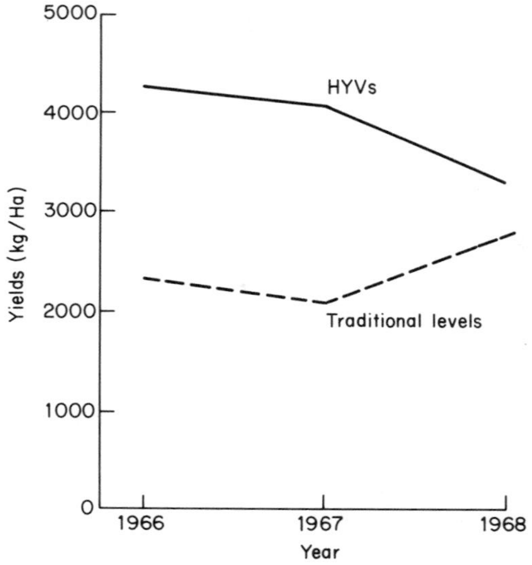

Fig. 39 Contrasting yield trends during the wet season of HYVs and traditional varieties on 155 farms in the Laguna district (Philippines) (based on data of Barker, 1970).

yields of modern varieties in the years following their introduction, yields of traditional varieties tend to rise.

With the adoption of HYVs in an area, farmers not only begin using liberal amounts of fertilizers, but the distribution system for fertilizers improves and credit for farm inputs may become more available. In the case of the Laguna farmers (Figure 40) the supply of credit extended by the landowners to their tenants increased approximately fivefold after the adoption of HYVs in the area. This credit was used not only for fertilizers, but also for additional hired labour in weeding, land preparation, and transplanting. The traditional varieties also benefitted from the higher level of inputs and care, and hence the trend for their yields to increase (Barker, 1970). It will be observed that because of these opposite trends, by 1968 the yield differences between modern and traditional varieties had become very small. It must however be pointed out that the comparisons have been made during the wet season, in which the yield advantage of the modern varieties is potentially much lower than in the dry season.

Problems of Consumer Acceptance

The primary goal of the wheat and rice breeders at the international centres was to achieve a rapid increase in yields of varieties with a wide range of adaptability. Quality problems were considered of secondary importance.

Difficulties may arise if the produce of newly introduced varieties is less acceptable to people than that of traditional varieties to which they are accustomed. For example, the cooking quality of the famous IR8 rice is less popular than the local varieties; as a result, the price paid for the new rice has been 20% or more below that paid for traditional varieties in a number of Asian markets (Barker, 1970). Many of the 'new' varieties with undesirable characteristics have been replaced, and in many cases the price differential has disappeared. For example, IR8 has been replaced by IR26 which is free of the defects of the earlier variety.

In Israel, the use of the new high-yielding dwarf wheats was limited because of their poor baking quality; this difficulty has been overcome lately by an intensive breeding programme.

A new hybrid maize was introduced to Spanish American farmers in the Rio Grande valley (New Mexico). After it was shown that the new hybrid produced a threefold higher yield than the traditional variety, it was adopted by the majority of the farmers. Four years later, most of the farmers reverted to the former variety, because the farmers' wives complained that the texture of the dough, the taste and the colour of the tortillas made from the new grain were unacceptable (Apodaca, 1952).

The problem of consumer acceptance may have special significance when it relates to foreign markets to which the grain is exported. For example, in Pakistan, fine-quality rice of the Basmati variety is exported and provides substantial foreign exchange earnings. For this reason, the government has found it profitable to raise the price paid to farmers for Basmati rice to as much as twice the price level of the coarse-grain modern varieties, such as IR6 and IR8 (Barker & Anden, 1975). This price advantage of Basmati over HYVs more than compensated for the yield advantage of the latter, especially if the added cost of inputs required by HYVs is taken into account.

Long-term Effects on Employment

The positive impact of the 'green revolution' upon employment (pp. 287–289) is likely to be temporary. In the long run, rising real wages and labour unrest inevitably lead to an increase in mechanization, in particular on the big farms. Higher profits also make mechanization possible. After mechanization is adopted on the larger farms, it spreads to the medium and smaller farms through contract work.

In a survey (1971–72) of 32 villages, in a broad geographic area representative of a wide variety of rice-growing conditions in Asia and in which about 90% of the farmers have adopted HYVs, the results of Table XXI were obtained.

There is a consistent increase in labour-saving practices with increasing size of farms. In particular the high proportion of farmers with less than 3 ha using mechanized threshers (79%) is a reflection of the work pressures during peak periods of labour requirement, such as harvesting. Of interest is that the need for labour-saving practices on larger farms is also evident in the use of herbicides, even though no economy of scale is involved in this case. In Mexico, the profits generated by the green revolution have made it possible for large farmers to invest heavily in labour-saving machinery, which has become a symbol of status and progress; government has actively promoted this trend. As a result, by 1971, many of the unskilled field hands, in Sonara, who had participated in the clearing of new land, could count on less than six month's work a year. The introduction of multiple-cropping only mitigated the situation temporarily, because wheat production by then required only about 3 man-days/ha, and alterna-

Table XXI Percentage of rice farmers using labour-saving practices in relation to farm size (Barker & Anden, 1975, with permission).

	Farm size		
Practice	Less than 1 ha	1–3 ha	Over 3 ha
Herbicides	6	20	29
Rotary weeding	3	20	37
Tractors	13	41	57
Mechanical thresher	36	43	63

tive crops such as soya and sorghum were also highly mechanized (Alcantara, 1976).

Changes also occurred in the type of machinery used. A typical wheat harvesting combine originally required eight to nine men to operate it and to handle the sacks of grain. As the cost of labour increased, the old machinery was replaced by modern combines, operated by a single worker, that poured the grain unsacked into waiting trucks. It is true that more truck drivers were required because of the considerable increase in yields, but the increase of employment in transport could not compensate for the loss of employment for field labour.

Labour-saving combines were not the only innovation to be adopted; larger tractors replaced small tractors and new labour-saving attachments were increasingly introduced (Alcantara, 1976).

In Thailand, rice-farmers found it more economical to hire tractors for ploughing, instead of using buffaloes which have to be maintained all year and are used during a few weeks at ploughing time. In 1968, some 20,000–25,000 tractors ploughed an estimated quarter of the rice-acreage, mostly on a custom-hire basis (Brown, 1968).

Surprisingly, there is no clear consensus regarding the overall long-term effect of mechanization on farm employment. Johl (1975) for example, argues that new technology is essentially bullock-displacing, rather than labour-displacing. The new jobs created by the adoption of HYVs, for operating tractors, pumps, the intensive application of modern inputs, double-cropping etc., would more than compensate for the loss of work associated with the use of bullocks. Furthermore, the increased productivity of agriculture improves the overall national economy, thereby increasing employment opportunities. This was confirmed by the findings of a survey of 32 villages over a broad geographic rice-growing area in Asia; it was found that even where the use of tractors and herbicides was widespread, labour requirements had increased with the adoption of HYVs. The labour-saving effects of tractors and herbicides had apparently been more than offset by the increased preharvest labour requirements of HYVs (Barker & Anden, 1975).

This reasoning probably holds true as long as mechanization is strictly selective; when indiscriminate, mechanization may cause considerable displacement of labour. In the Punjab area of Pakistan, for example, mechanization reduced the labour force by about 50% (Bose & Clark, 1968). Mechanization of wheat production may cause a reduction in labour requirements from 30–40 man-days/ton, using exclusively animal traction, to less than 1-man-day/ton (Dalrymple & Jones, 1973).

Problems Deriving from Multiple-cropping

Multiple-cropping, made possible by the use of HYVs, is not without its problems. In certain cases it has been traditional practice for villagers to graze the land after removal of the crop. This becomes either impossible or at best greatly restricted when growing HYVs. The loss of grazing rights may have important social and economic implications.

Because of the almost continuous period of plant growth, insect pests, diseases, and weeds become more troublesome and more difficult to control.

Fertilizer applications must be increased to make possible the higher levels of production and, in particular, deficiencies in micro-nutrients may arise.

As more crops are grown within a given season, less time is left between the harvest of one crop and the planting of the next, so that a large number of operations have to be carried out in a relatively short time—this may easily lead to slipshod cultivation methods.

There may also be difficulties of harvesting and drying the grain during the monsoon season, so that drying equipment may become necessary (Fig. 40).

In brief, multiple-cropping requires, besides appropriate varieties: a relatively high degree of technical knowledge and managerial skill; an assured water supply throughout the year; considerable increases in inputs such as fertilizers and pesticides; improved methods for handling and marketing the produce (Dalrymple, 1971).

Fig. 40 Portable IRRI batch drier built at low cost in the Philippines from locally available materials. Needs little maintenance and easy to operate. Dries one ton of paddy in four to six hours. Operated by 3 HP gasoline engine or 2 HP electric motor. By courtesy of the Agricultural Engineering Department, International Rice Research Institute. Los Baños.

Need for Diversification of Crop Production

By increasing the profitability of farming and by reducing the areas of land required to meet the demand for staple foods, both the need and the possibility for diversification of agricultural production have increased.

In countries where the promotion of high-yielding varieties has been successful, provision has to be made for crop diversification and alternative uses of the land taken out of cereal production. The crops to be considered for these purposes will depend to a large extend on the food habits of the people.

In view of its importance, this subject will be treated in more detail (cf. pp. 334–341).

Need for More Effective Extension Work

New technology involves new inputs and management practices that are extremely complex for the traditional farmer. This is compounded by the fact that HYVs generally have a high obsolescence rate, as they are regularly replaced by new, improved varieties. These in turn frequently have different requirements to those of their predecessors. Their successful adoption is therefore largely dependent on effective extension work.

Distribution of Assets

The distribution of assets among various farm-size categories is extremely uneven in quantity and quality. The introduction of new technologies is generally followed by asset distribution moving further in favour of large farms. An example of such a trend is provided by a study in Ferozapur district (Punjab) in which the ratio of assets between large and small farms moved from 3.25 in 1967–68 to 4.21 in 1971–72 (Kahlon & Singh, 1973).

Distribution of Land

Given the limitations of land supply and increasing population pressure, land prices tend to increase over time. However, because new technology increases the productivity of land, price increases have generally been greater in areas in which new technology was adopted (Dasgupta, 1977).

The higher the price of land the more difficult it is for small farmers to increase their land area and the greater the temptation to sell off land. With increasing productivity, the tendency of large farmers to repossess land for self-cultivation increases, transforming tenants into landless labourers. Hence, the skewness of land ownership has further worsened under the HYV technology.

Land, major implements and irrigation are the major assets of the larger farms—all these are factors of production capable of generating more profits and assets; by contrast, on small farms, residential and farm buildings and animals constitute a high share of the total, so that the skewness of income has further worsened with the introduction of new technologies (Dasgupta, 1977).

A surprisingly similar situation developed in Mexico after the green revolution. The value of land within irrigation districts tripled and quadrupled within a few years.

Land resources were increasingly transferred from smaller to larger farmers. Plots of small private farmers (*colonos*) unprotected by *ejido* law, were bought up at such a rapid rate that many *colono* settlements in the largest irrigation districts of the country 'became little more than conglomerates of daily labourers or city workers'. In the case of the *egidatarios* whose land is communal property and cannot be sold, the law was bypassed by renting. In the Yaqui valley alone, 80% of the beneficiaries of land reform had alienated their plots in favour of the better organized and more politically powerful large farmers of the area (Alcantara, 1976).

Uncertainty

The adoption of HYVs increases uncertainty and risk for the following reasons (Dasgupta, 1977):

(a) The new varieties are, at least initially, more prone to location-specific diseases and pests. This has been more evident in the case of rice than wheat and is reflected by a higher level of participation by small farmers in the growing of HYVs of wheat, especially in the Punjab.
(b) Inability to apply the necessary inputs in the right amounts, either because of unavailability at the right time, or lack of money.
(c) Uncertainty regarding price and market conditions. Because the production of HYVs is more market-oriented than that of traditional varieties, their production is more sensitive to market fluctuations and price differentials.

Uncertainty resulting from the adoption of HYVs is high for farmers of all sizes; it is greatest, however, for small farmers, because of inadequate information on the new technology, the lower price he gets for his produce (because of his lack of storage), his pressing need for cash, and above all his inability to obtain the necessary inputs. At the same time, he has the lowest risk-bearing capacity.

Problems of Inadequate Infrastructure

The transition from subsistence farming to commercial agriculture is inevitably dependent on the development of an appropriate and efficient infrastructure, which provides the facilities needed to reproduce, distribute, and supervise the use of the materials required for improved agricultural production.

The role of these components of the infrastructure relating to the so-called 'agro-distribution system' and consisting mainly of marketing outlets, a transport network, storage and processing facilities, will be discussed in detail in Chapter X.

Suffice to state that in those areas in which the 'seed-fertilizer' technology has been adopted, a phenomenal marketed surplus has often occured. As a result, acute problems have arisen, because of transportation, storage and marketing bottlenecks. For example, in the southern rice area in West Pakistan, in 1969, storage facilities at rice mills were filled to capacity, but rice could not be moved because the railway system was completely swamped. Large uncovered piles of rice accumulated at railheads and prices to farmers fell considerably (Efferson, 1969).

Similar situations in milling, grading, storage and transport occurred in other countries (Falcoln, 1970). In the spring of 1966, scores of village schools had to be closed in northern India, in order to provide emergency storage space for a bumper harvest. Even so a large proportion of the harvest had to be stored in the open with resultant losses (Brown, 1970).

The availability of an adequate infrastructure is not only an essential precondition to change in agriculture, but also provides farmers with a stimulus to effect such change.

Problems of an Inadequate Institutional Framework

Many of the developing nations, even when they are rich in natural resources, remain poor because their social institutions are either inadequate or not oriented to meet new economic and social needs (Deutsch, 1971). Certain authors argue that a major constraint to modernization of agriculture is inadequate service from these institutions responsible for supporting agriculture.

A major reason for the increasing disparity between large and small farms in India as a result of the green revolution, in terms of asset accumulation and profitability, is the lack of institutional support: government officials at all levels, discriminate against the small farmer who is generally illiterate, poor and belongs to a low caste (Dasgupta, 1977).

Towards the end of the 1960s, the Indian government sponsored two special programmes aimed at giving institutional support to small and marginal farmers amd agricultural labourers: The Small Farmers' Development Agency (SFDA) and the Marginal Farmer and Agricultural Labour Agency (MFAL).

SFDAs activities were strictly confined to providing a subsidy for credits to small farmers and covering the risks undertaken by banks and cooperatives when advancing loans to small farmers.

The sister agency MFAL was instituted to encourage marginal farmers to introduce dairy-farming; poultry; pig, goat and sheep rearing; and to adopt improved agricultural practices–soil conservation, minor irrigation, horticulture, etc. For agricultural labourers, the intention was to provide work through rural works programmes, homestead land and housing.

In both cases staff was inadequate, there was lack of coordination between departments so that the physical target areas were unfulfilled, and most of the benefits went to large farmers who were able to manipulate the bureaucracy in their favour.

The experiences of both MFAL and SFDA show that good intentions are not sufficient when setting up organizations with the laudable aim of helping small farmers and agricultural labourers. Unless the ability of the rural elite to sabotage such enterprises is neutralized, and the programmes themselves are adequately planned, staffed and financed, the results are bound to be disappointing (Dasgupta, 1977).

A similar situation is reported for Mexico, where large farmers adopted the new technology, which had proven to be more profitable, but they did so within the framework of a federal investment programme that poured billions of pesos into irrigation works, roads, storage facilities, electricity, railroads, long-term agricultural credit, and ultimately, into a guaranteed price for wheat so high that it involved a national subsidy to wheat farmers of about 250 million pesos annually. The adherance of the most progressive large landowners to the 'green revolution' in wheat was thus bought with public funds at a very high price (Alcantara, 1976).

In a study on the factors impeding the diffusion of new technology in the development plan for the Pueblo valley (Mexico), Diaz (1974) states that 'the very roots of backwardness are found in the dysfunctional institutional structure which serves agriculture in Mexico'. Farmers who wish to acquire credit from Banco Agricola and Banco Ejidal encounter cumbersome and time-consuming procedures for obtaining (and even for repaying!) loans; the promised inputs arrive late and they receive little technical assistance from the banks in their use. This means that obtaining credit is too costly in terms of time required, and returns to inputs are too low because of late delivery and/or inadequate advice.

Socio-economic Effects

Though the 'green revolution' has relatively few economies of scale, it has led to differential effects at farm levels, leading to an increase in the disparity between different levels of the rural population of each country. Almost everywhere it has favoured the rural elite for the following reasons:

(a) It increased the profits and assets of the commercial farmers, and therefore their economic power.
(b) Dependence of the small farmer on the large farmer increased, because most means of production are owned by the latter.

(c) Inputs supplied by government are delivered to the villages through the intermediary of the dominant sector, who own most retail shops for fertilizers, seeds and other inputs.
(d) The extension village-level workers are generally matriculates and members of the richer families, with whom they tend to maintain contact, to the exclusion of the disadvantaged farmers (Dasgupta, 1977).

As a result, the green revolution has strengthened the existing social structures and the present dominant rural classes.

In Mexico for example, the birthplace of the new HYVs of wheat, their introduction has raised wheat production *per caput* considerably, but the profits have accrued mainly to owners of large mechanized farms while the produce itself has been exported (Palmer, 1972b).

In the Punjab, available evidence shows that the 'green revolution' was of benefit to all categories of farmers and labourers. Employment opportunities were increased for local and out-of-state workers. Working hours decreased, and wage rates increased in real terms. All farmers gained, but as gains are proportional to marketed surpluses, bigger farmers gained considerably more than the others (Johl, 1975).

A similar trend was found in Rajasthan. In a village that had adopted the new technology, including irrigation, farmers' incomes increased by 9.3% on small farms of less than 3.75 ha, by 11.9% on medium farms of 3.75 to 10 ha, and by 50.1% (!) on farms larger than 10 h (Bapna, 1973).

In brief, the adoption of HYVS can be a profitable enterprise if moderately reliable irrigation is available. Even small farms can make a profit, provided credit is available to enable them to meet the high costs for seeds, fertilizers, plant protection chemicals and other inputs.

However, the larger the farms, the greater the potential benefit, both in relative and in absolute terms. Beyond a certain size level, the net gains become sufficiently great to enable reinvestment in the farm; this, in turn, increases profits besides making the larger farmer increasingly independent of high interest credit. Labour saving machinery can be purchased and additional land rented or purchased to make more efficient use of the new equipment. The larger farmer thereby enters 'a path of dynamic growth' (Pearse, 1980).

Pearse further points out that this path of growth is unlikely to be open to the small farmer; on the contrary—the progress achieved by the larger landowner may effectively block that of the small farmer. He cannot compete with the large farmer in renting or purchasing land to increase his holdings; on the contrary, the temptation to sell his land will increase with rising prices. In spite of the apparent scale-neutrality of the new technology, small farmers who attempt to adopt it are definitely at a disadvantage.

As to the immediate benefits that accrue to the farm labourers as a result of the 'green revolution', these are very small as compared to the benefits of the landowners.

It has been calculated that on wheat farms in the Aligarh district in India, in 1967–68 farm labour received only about 10% of the increased returns due to new technology; the net return to landowners increased about 80% and 10 % went for other factors of production (Lele & Mellor, 1972).

A survey made by the Special Committee on Agrarian Reform confirms the already widespread impression that increased productivity alone has not improved the position of the great mass of rural workers, but, on the contrary, frequently makes it worse (FAO, 1971).

The social and political effects of the green revolution are of major importance and will therefore be discussed in detail in Chapter XI. It should, however, be emphasized here that it was not the green revolution *per se* that created the socio-economic and political problems, many of them unanticipated, that have arisen. The social and institutional structures of many developing countries do not allow the benefits of any new technology to spread equitably throughout the society and consequently, development has always tended to be confined to limited sectors. The green revolution, by its very success, has accentuated the disparities and highlighted the need to reduce, if not eliminate them.

The problems outlined above should therefore *not be* construed as an indictment of the green revolution, without which agricultures in the countries in which it has been a success could not have broken out of their stagnation. They do, however, highlight the need to adopt policies that will improve the ability of the small impecunious farmer also to benefit from the new opportunities provided by the green revolution (cf. pp. 518–519).

SEED TECHNOLOGY

Importance

In many countries, a major constraint to the introduction of HYVs has been the absence of an organized industry for the production of quality seed. Furthermore, much of the available seed has deteriorated as a result of mixing with other varieties (FAO, 1976). An FAO survey indicates that of the 25 countries using HYVs of cereals on a large scale, only about one-third have adequate seed multiplication and related services (United Nations, 1974). For an in-depth study of the problems involved in the establishment of a seed production programme, the reader is referred to Feistnitzer & Kelley (1978) and Douglas (1980).

The introduction and early propagation of new HYVs was first based on the purchase of seeds in other countries. However, expansion of the areas sown to HYVs cannot remain dependent on foreign supplies. The cost in foreign exchange to the country and the high cost of the seeds themselves to the farmer make it essential to produce locally seeds of proper quality.

High-quality seed multiplication requires an appropriate organization with the necessary technical facilities and high ethical standards.

The provision of good seeds has two main objectives: to ensure that the farmer will sow the variety that is most suitable for his conditions, and, so far as possible, to guarantee that he will get a full and uniform stand of plants.

The individual farmer, even in advanced countries, is usually not well-equipped to maintain the purity and genetic identity of the varieties which he grows. With hybrid varieties, whose importance in a number of widely grown crops is increasing, it is of course out of the question for the individual farmer to produce his own seed.

In addition to the need for maintaining the genetic identity of improved varieties, a certified source of seed will ensure that the farmer obtains sound seed with a high percentage of germination and emergence and free from seed-borne diseases and noxious seeds. A further advantage will accrue if the seed is already treated against seed and soil-borne diseases and soil insects. A striking example of what centralized control of seed disinfection can achieve is the seed-treatment campaign against the bacterial blight of cotton in northern Nigeria, which was extremely effective in controlling the disease at very low cost to the farmer.

In general, it is assumed that at a certain stage seed production will be taken over by cooperative or private enterprise.

This transfer has not always been to the farmer's advantage. Dasgupta (1977) reports that with the growth in private trade in seeds of HYVs there was a steady deterioration in quality. Many seed producers did not adhere to quality requirements and did not treat the seeds against seed-borne diseases as required by official regulations. Adulteration became widespread and is one of the reasons for the uneven record of supposedly HYVs over time and space. The mixing of HYVs with traditional varieties not only reduced yields, but also created difficulties at harvesting time because of the different maturing dates of the varieties in the mixture.

With the decline in the quality of seeds marketed as HYVs the faith of the farmers in them has declined.

Even after a seed-producing industry has developed, government intervention must be maintained. The research organization must continue to maintain pure seed-stocks and supply seed-growers with registered seeds; the extension service must organize adequate field supervision of the production of certified seeds; legislative measures against adulteration must be enforced.

Requirements

Requirements for the commercial production of good quality seeds are:

(1) facilities for drying, cleaning, grading, treating and bagging the seed;
(2) facilities for storage and an adequate distribution service;
(3) the enactment of seed-laws and the setting up of the necessary agencies for supervision, control and enforcement.

The supply of certified seed in advanced countries is usually assured by private firms or cooperative enterprises. In developing countries this may be an important function to be carried out by an appropriate service in the ministry of agriculture.

The seed to be certified is generally grown under supervision by authorized seed-multipliers, who use 'registered' seed provided by the official seed organization. These farmers should have at least 10 hectares, in a contiguous area, available for growing the seed, as well as temporary storage facilities. The seed farmer is generally required to sign an agreement with the seed-producing organization stipulating that: (a) he will grow the seeds in accordance with recommended cultural practices for seed multiplication, including stringent measures to prevent contamination by seed from other sources; (b) he will allow regular inspections by the seed-certification service; and (c) that the seed-producing organization has exclusive rights to purchase the seeds.

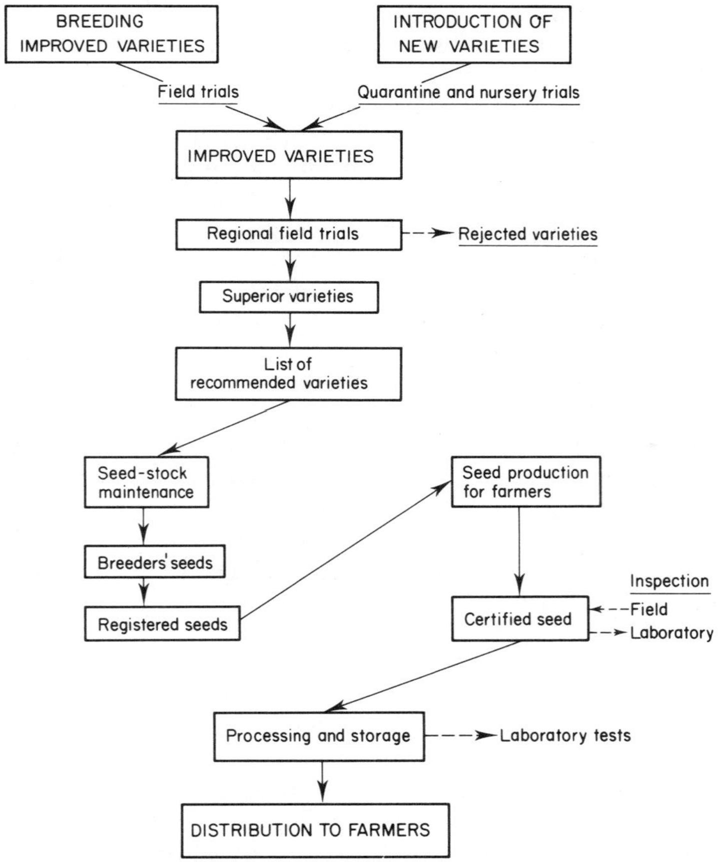

Fig. 41 The stages of an efficient seed-production scheme.

Stages of Production

The stages of an efficient seed-production scheme are shown in the flow chart of Fig. 41.

In order to ensure a regular seed supply of locally suitable varieties and to maintain reserves for drought years, FAO (1979) recommends the establishment of a seed security system, preferably with widely decentralized seed-production centres at provincial and district levels, rather than at national and regional levels. Appropriate policies to encourage farmers to use improved certified seed are also called for.

FERTILIZERS

Fertilizers as a Lead-practice

Increased use of fertilizers is important even in the early stages of agricultural development, for practical and psychological reasons: practical because returns are quick and little capital is required; the use of fertilizers is probably the single most responsive factor for increased yields per hectare or per unit of water.

Fertilizers also have the great practical advantage that they can be successfully applied by the individual farmer who has the necessary initiative without him being dependent on his neighbours. By contrast, many other practices, such as insect and disease control, are almost certainly doomed to failure, for obvious reasons, unless organized collectively and carried out in a planned fashion over fairly large areas. The *psychological* reasons are that few inputs have such strikingly visible effects on crops. The fertilizer itself is a tangible input, so that the relation between cause and effect is most evident. For these reasons, fertilizer application is considered a 'lead'-practice which predisposes the farmer to adopt other improved practices. Every improvement in varieties and management practices that increases yields, also increases fertilizer requirements. Increased amounts are needed to make possible potential yield increases due to improved practices, and to replace additional nutrients removed from the soil.

Fertilizers bear such a highly complementary relationship to other yield-increasing practices, that the amounts of fertilizers used per hectare of land have been found to be a reliable index of progress in the adoption of yield-increasing technologies in general. Williams and Couston (1962) report a 0.87 coefficient of correlation between fertilizer consumption and grain yields in 40 countries. In Japan, fertilizer is a major expense item, which accounted for 30% of all cash expenditures for agricultural production (Christensen & Stevens, 1962). Generally, countries with low levels of fertilizer use are relatively underdeveloped countries; conversely a high level of fertilizer use is the hall mark of countries with a modern, highly productive agriculture and an efficient industrial sector (Fig. 42).

Fertilizer use is generally adopted by farmers of all size-groups and

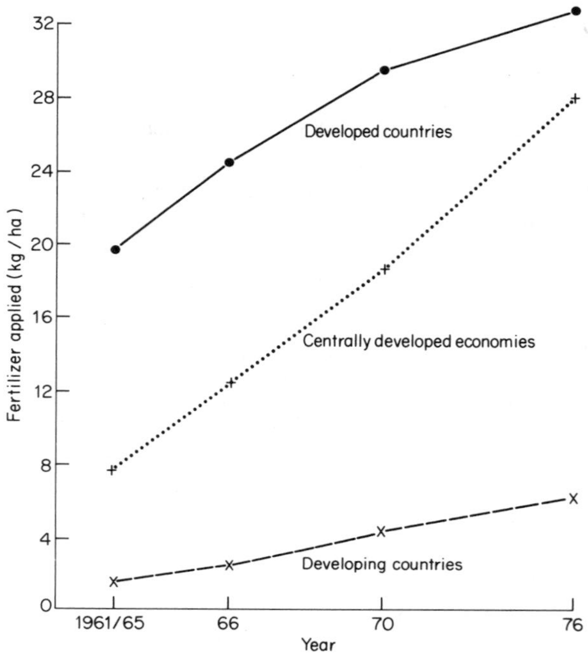

Fig. 42 Consumption of fertilizer per hectare of agricultural area 1961–5 to 1976 (FAO, 1977).

under all types of tenure relationships, provided they have the means to do so. There may be a delay of two or three years as smaller farmers wait to see the effects on more affluent farms. There seems to be some evidence that small farmers tend to use smaller amounts of fertilizers, but because of the greater input in labour (more careful weeding, for example) yield increases are substantially the same as on large farms (Gotsch, 1973).

The Seed–Fertilizer Revolution

Possibly the most striking and widespread interaction between agricultural production factors is that found to exist between high-yielding varieties (HYVs) and fertilizers. It is the ability to respond markedly to fertilizers which is the main characteristic of the modern varieties and the basic reason for their high productivity.

The term 'green revolution' is generally linked to the introduction of the new so-called 'miracle' high-yielding varieties of wheat and rice. However, it is a fact that, without fertilizers, the impact of the new varieties of these crops is at best small, and frequently imperceptible.

On the other hand, the response of traditional varieties of food crops to fertilization on many African soils in many trials over significant periods of time were found to be nominal. Though the increases obtained may appear to be high on a percentage basis, e.g. a 50% yield increase, absolute increase itself is very low, equivalent to 500 kg/ha perhaps of maize. Most experiments showed that chemical fertilizers could decrease the rate of yield decline under continued cropping; sometimes it was even found possible to maintain yield levels through fertilization, but very rarely were major yield increases recorded. The conclusion from these experiments was that the use of fertilizer by itself is not a decisive factor for increasing production in that environment (Cline, 1971). From the foregoing it is clear that a combination of fertilizer application *and* improved varieties is a *sine qua non* of the green revolution. For this reason, the green revolution is also called the seed–fertilizer revolution (Johnston & Cownie, 1969).

Figure 43 illustrates the difference in response to fertilizers between new HYVs of rice and a traditional variety.

The main points to be noted are: (a) that there are little (wet-season) or no (dry-season) yield differences between HYVs and Peta when no fertilizers are applied; (b) the yields of HYVs increase with increasing rates of nitrogen application, more steeply in the dry season than in the wet season, whilst the yields of Peta actually decline with increasing rates of nitrogen application; (c) both high yields and greater response to fertilizers in the

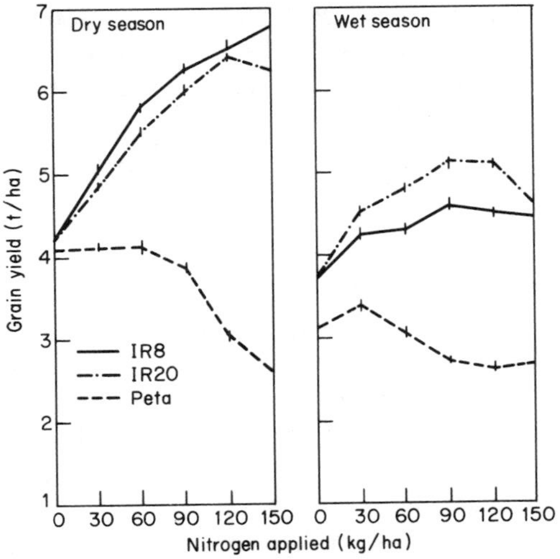

Fig. 43 Differential response of HYVs of rice (IR8, IR20) and of a traditional variety (Peta) to increasing rates of nitrogen application. Data are averages for 1968–73 cropping seasons (from Brady *et al.*, 1975). By courtesy of the International Rice Research Institute, Los Baños, Philippines.

Table XXII Comparative yields (kg/ha) of improved and traditional varieties of wheat and rice in India, with and without fertilizers (Griffin, 1973, by permission).

	Wheat		Rice	
	Traditional variety (C-306)	Improved variety (Sonora-GY)	Traditional variety	Improved variety (IR-8)
No fertilizer	1350	2230	2500	3000
With fertilizer	3690	4600	3500	4570

dry season, reflect the effect of intensive sunlight enabling HYVs to achieve their yield potentials.

An additional example of this relationship is given in Table XXII.

The high-yield potential and resistance to lodging of the new rice and wheat varieties permits them to maintain highly favourable response ratios to fertilizers up to levels three or four times as high as the level at which traditional varieties begin to lodge and/or show a decrease in yields. This enables yield increases of 200% and more to be achieved (Johnston and Cownie 1969). In the Punjab, replacing 80% of the traditional wheat varieties by the 'miracle' wheats, resulted in a tripling of fertilizer consumption within four years. The rate of application increased from a mere 5–8 kg/ha on average, to as high as 100–150 kg/ha (Ladejinsky, 1970).

Not only do the new varieties of rice respond to much higher rates of fertilizer application, but they are far more efficient in its use. Traditional varieties yield about 10 additional kilograms of grain for each kilogram of nitrogen applied: for the new, improved varieties, the ratio is 20:1 (Hopper, 1968).

A Precedent for the Seed–Fertilizer Revolution

The green revolution, or more properly, the seed–fertilizer revolution, is popularly considered to be a recent event in developing countries. Actually, this is not so. In Japan, at the beginning of this century, the productivity of agricultural labour was practically doubled following the adoption of two highly complementary factors of overwhelming importance: the development of varieties that gave a strong response to increasingly heavy applications of fertilizer, and corresponding increases in fertilizer use (Johnston, 1962).

Fertilizer Requirements

The use of fertilizers has increased dramatically in the course of the last thirty years, from a total consumption of NPK of 15 million tonnes in 1950

Table XXIII Fertilizer consumption (millions of tonnes) in developed and developing countries (FAO, 1979, by permission).

	1966–7	1976–7	Annual growth 1966–7 to 1973–4
Developed	31.6	44.9	4.6
Developing	5.1	15.4	13.2
Centrally planned	14.4	34.4	10.1
World	51.0	94.6	7.3

to about 100 million tonnes in 1978 (Peter, 1978), whilst cultivated acreage increased by less than 20% (Table XXIII).

Most of the world's fertilizers are used in the already developed countries. In the early 1960s, Europe, North America and Japan used nearly 90% of the world's supply of all fertilizers on less than 40% of the world's arable land (OECD, 1967). However, some of the most striking increases in fertilizer use are occurring in the developing countries. In 1950, the developing countries, with three-quarters of the world's population, used only 731,000 tonnes NPK, this amount was used by Pakistan alone in 1978 (Peter, 1978). By 1978, the developing countries used 29% of the nitrogen consumption, 21% of the phosphorus consumption and 11% of the total world consumption of these fertilizers.

The rapid adoption of fertilizers in agriculture in the process of modernization, and the considerable potentials for increased use are illustrated by the example of Mexico. Between 1948–52 and 1966, nitrogen consumption increased twofold, phosphate consumption tenfold, and potassium fivefold. The area of crops receiving fertilizers increased from 5% to 15% of the harvested area. Though this is a significant increase, it underlines the fact that the vast potential of fertilizers was still unexploited on most farmland (Venezian & Gamble, 1969).

With the current rapid increase in fertilizer use, it is expected that by 1985 the purchase of fertilizers by farmers in developing countries will exceed that of all other manufactured inputs combined (FAO, 1969).

Effects on Yields

Much of the rise in crop yields, both in industrialized and developing countries, can be attributed to increased use of fertilizers. For example, it has been estimated that half of the rise in crop production per unit area in the United States from 1930 to 1960 was due to additional fertilizers (Duroset & Barton, 1960). In India, increased fertilizer use was estimated to be responsible for 4.6 million tonnes out of the 11.2 million tonnes increase in food production during the second five-year plan (FAO, 1967).

Table XXIV Factors limiting fertilizer use in developing countries (FAO, 1975, by permission).

Category	Lack of basic information	Deficiency of foreign exchange	Inadequate infra-structure	Number of countries	Distribution
A	×	×	×	25	Mostly in Africa
B		×	×	39	Equally distributed in Asia, Africa, Latin America
C		×		22	Mainly Asia, Latin America

It is, however, difficult to separate out the effects of other improved practices which, in addition to making their own specific contribution to yield increases, greatly enhance the response of the crop to fertilizer applications (cf. pp. 262–263).

Over all the regions and countries of the fertilizer programme of the Freedom From Hunger Campaign of the FAO, nitrogen had positive effects in 97% of the locations, phosphorus in 90%, and potassium in 85% (Couston, 1967). The most successful fertilizer treatment increased yields by an average of 60% but was profitable in only about 30% of all locations. If the other production inputs had also been introduced, the overall effect on productivity would have been greater. The value/cost ratio averaged from 4 to 8, indicating that the investment in properly used fertilizers is highly profitable (Olson, 1968).

Constraints to Fertilizer Use

The main constraints to a necessary increase in fertilizer use on a national scale by developing countries are: (a) lack of basic information on fertilizer use; (b) a deficiency of foreign exchange; and (c) an inadequate infrastructure for marketing, distribution, advice and credit (FAO, 1975).

FAO (1975) has grouped the developing countries into three main categories, in accordance with the major constraints encountered in increasing fertilizer use (Table XXIV). Category A requires, in the first stage, mainly experimental and demonstration work, to be followed by pilot schemes for fertilizer distribution and use.

Lack of Information and Improper Use

Limited knowledge of the most effective way of using fertilizers reduces their impact. Factors such as optimal rates, the best timing of application, the most suitable nutrient carriers, and combinations of fertilizers in the right proportions, determine their effectiveness. Small farmers lack equipment for fertilizer application; the fertilizers are generally broadcast by hand on the surface of the soil, without being incorporated in the soil. Efficiency is thereby reduced and, in the case of urea, losses by volatization are increased.

In many cases the yield potential of the HYVs–fertilizer combination is not yet being fully exploited, either because of *insufficient* or *unbalanced* fertilizer application, or both. In a field study in West Pakistan, for example, in 1968–69 the average yield for the sample of farmers studied was 2360 kg/ha for dwarf wheat and 1370 kg/ha for the local varieties. This difference was associated with an average of just under 55 kg/ha of nitrogen applied to the dwarf varieties, compared to 41 kg/ha used on the local varieties. The grain/nitrogen response ratio averaged 17 kg of grain per kilogram of nitrogen for the dwarf varieties, compared to just over 8 kg for the local varieties. The average amounts of nitrogen actually applied to the dwarf varieties were considerably less than half of the optimum level, and this is one of the major factors responsible for the gap between potential yield levels and the yields actually obtained by farmers (Eckert, 1971).

In a 1970 government survey in West Pakistan, it was found that 80% of farmers were applying nitrogen to dwarf wheat, and only a quarter were applying phosphate, and then at considerably less than half the recommended rate. More than half the farmers had not even heard of phosphate (Government of the Punjab, 1970). Potash is not even mentioned in this study. There is no doubt that because of the heavier yields resulting from the growing of improved varieties, soils reserves of phosphorus and potassium are being rapidly depleted at an increasing rate, and that a deficiency in one or both these elements is probably already reducing the effectiveness of nitrogen applications. An illustrative example of this situation is provided from fertilizer trials on dwarf wheat in West Pakistan. In these trials, it was found that whereas the average yield of grain was 2860 kg/ha with an application of 113 kg/ha of nitrogen, the yield increased to 3800 kg/ha when 56 kg/ha P_2O_5 was given in addition to the nitrogen application (Eckert, 1971).

Foreign Exchange and Infrastructure Deficiencies

Increases in fertilizer use pose considerable problems for countries that are dependent on fertilizer imports, and often already have big deficits in their balance of payments and a shortage of hard currency. These problems have been compounded by the steep increases in the price of fertilizers that have occurred in recent years. World market prices of fertilizers had tripled and in some cases even quadrupled in 1974 over 1971–72 levels (FAO, 1976). It is estimated that developing countries had to spend an additional $1200 million in foreign exchange in 1974 in order to import the same quantity of fertilizer as in 1973 (United Nations, 1974).

In many Latin American countries, the importance of fertilizers for the development of agriculture is generally recognized, information on rational use is available, and there is already a sizeable demand among farmers. Fertilizer consumption is limited mainly by the lack of foreign exchange and institutional deficiencies.

The lack of foreign exchange leads to shortages and a black market, in which fertilizer is sold at higher than government controlled prices. Such black markets may develop even when the overall amounts imported by government are sufficient but they become available only after the right time for their use in the field. The importance of proper timing is vital because of the lack of storage facilities in the villages. Dasgupta (1977) reports that in some places in India, private traders used their superior bargaining position to oblige their customers to buy additional goods from them as a condition for obtaining fertilizers.

Adulteration is also a frequent feature of the private trade, and a major cause of the discontinuation of fertilizer use by farmers after one or two crops. Even when farmers can obtain their fertilizers on credit from village cooperatives, bureaucratic operational methods discourage many farmers from taking this opportunity.

This situation is not unique to India. On the other side of the world, Alcantara (1976) describes a similar situation in Mexico.

It is highly improbable that the developing countries can solve the problems of lack of foreign exchange without outside help; to overcome the institutional deficiencies is a major challenge to the countries themselves.

The International Fertilizer Scheme initiated by the Food and Agriculture Organization of the United Nations and incorporating the efforts of the World Bank and the main fertilizer producers' association, is evidence that the need to increase the availability of fertilizers to developing countries has been internationally recognized. The scheme includes the following elements: the establishment of a fertilizer pool, the mobilization of financial and technical assistance for the purchase of fertilizers, and increasing the ability of the developing countries to produce fertilizers. Major elements of a world fertilizer policy aimed at facilitating the increased use of fertilizers in developing countries include: (a) stabilization of fertilizer prices at reasonable levels, by ensuring a balanced expansion of production in line with food production objectives and the avoidance of cyclical imbalances between supply and demand; (b) adoption of appropriate policies and programmes for the promotion of the effective use of fertilizers; (c) the organization of efficient marketing and credit systems; and (d) effective control of the quality of the fertilizers marketed.

The special problems deriving from the energy crisis will be discussed in Chapter IX.

Fertilizer Production by Developing Countries

An important contribution to the reduction of fertilizer import require-

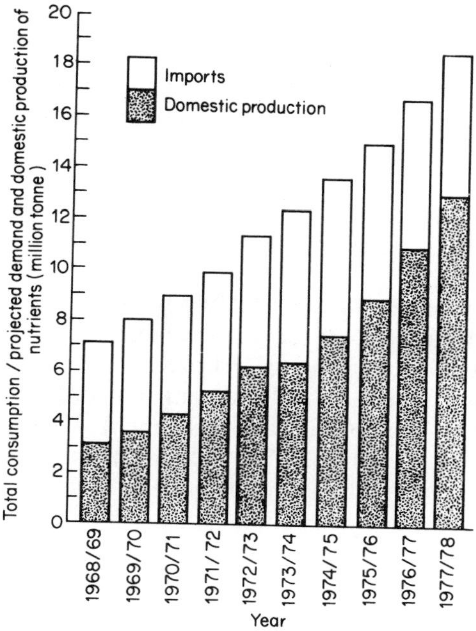

Fig. 44 Fertilizer consumption/projected demand, domestic production, and import requirements in developing countries, 1968–69 to 1977–78 (FAO, 1976). By courtesy of the Food and Agriculture Organization of the UN, Rome. Notes: figures exclude the Asian centrally planned economies; 1974–75 estimated; 1977–78 projected.

ments of developing countries is the expansion of domestic production. Trends in this respect are very favourable as shown by Fig. 44.

It is estimated the expansion in nitrogen and phosphorus fertilizer capacity in the LDCs will raise their share in world capacity from about 12% in 1976–77 to 20% by 1982–83 (FAO, 1979).

The Use of Fertilizers in Dry Regions

From the point of view of fertilizer utilization, farmers in dry regions are faced with two contrasting situations, according to whether the crops are rainfed or irrigated.

Rain-fed Agriculture

In the semi-arid regions in which water for irrigation is not available, the limited rainfall allows the production of a limited number of crops only, provided adapted varieties are sown and special cultural techniques are used.

The efficiency of fertilizers under these conditions is in direct relationship to the varying amounts of rainfall: if too little precipitation occurs, response to fertilizer will be nil; it may even be negative, if excessive vegetative growth is promoted by fertilizers, so that water reserves are depleted at an early stage and too little moisture remains in the soil at the time of grain formation; when rainfall is sufficient, response to fertilizer can be spectacular.

The farmer must decide on the kind and quantities of fertilizers to apply, without knowing what will be the amount and distribution of rainfall on which the response of his crops to fertilizer will depend. If he decides to apply fertilizers and there is too little rain, not only will he have a poor crop, but he will lose part of his cash investment. If he does not apply fertilizers, and rainfall is favourable, his yields will remain low and he will be unable to re-coup losses sustained in bad years.

For these reasons, a very cautious approach to fertilizer use is the rule in semi-arid regions—even in countries with modern agriculture. However, even under these conditions, progress in agriculture is dependent on a rational use of fertilizers. In particular, improved techniques of water conservation that improve the moisture régime of the soil, and are therefore an essential factor in augmenting yields, are also able to increase the response of crops to fertilizers.

Irrigated Agriculture

Because of high amounts of sunlight throughout the year, the yield potential of the dry regions is far higher than that of either temperate regions or the humid tropics. Under irrigation, this potential can be achieved only with very high rates of fertilization. The whole attitude to fertilizer use of the farmer who previously grew rainfed crops must be re-adjusted when he starts irrigating his crops; moving from a very cautious approach to the use of very high applications. Levels of fertilization have to be far in excess even of what is used in the intensive agriculture of temperature regions.

The Use of Fertilizers in the Humid Tropics

In the humid tropics, during the early stages of development of traditional farming, plant nutrients are the major limiting factor. We have already mentioned (cf. p. 263) that numerous trials of the FAO Fertilizer Programme in West Africa have shown that a modest application of fertilizers was able to achieve the same yield increases as a 7–10 years fallow, and to maintain these yield levels for several years, whilst yields on fallow declined rapidly. Even so, yield increases due to fertilizers are usually limited. Marginal productivity resulting from increments in fertilizer use on most food-crops is low because of the absence of other essential production factors that need to interact with fertilizers—hence the need to improve crop husbandry concomitantly with the adoption of fertilizer use.

For plantation-crops the situation is different. These are mostly grown under favourable rainfall regimes; they are generally well-managed and high-yielding varieties are planted. Fertilizers applied under these conditions have proven able to produce striking increases in yield. For example, in Ghana, with a combination of high-yielding varieties, reduced shade and efficient disease and pest control measures, fertilizers have been found to increase yields of cocoa more than tenfold, to 3000 kg/ha (FAO, 1962). Fertilizers applied to improved varieties of oilpalm have made earlier bearing possible (at the age of three to four years, instead of the usual seven to ten years) and have increased the yields of palm kernels threefold and of palm oil sixfold (Ministry of Economic Planning, 1961).

The problems of fertilizers used under tropical conditions have been reviewed by Engelstad & Russel (1975). The main problem encountered for nitrogen and potassium is loss by leaching, in particular in soils with a low cation exchange capacity. Whilst losses by leaching in soils under natural forest are small, because of the re-cycling of nutrients, in cultivated tropical soils they may exceed 50% of the amounts applied.

The opposite problem occurs with phosphorus, as many tropical soils (oxisols, ultisols, and andosols, in particular) fix large amounts of added phosphorus. For this reason, there is frequently an apparent lack of response to phosphorus fertilizer until sufficient soluble phosphorus has been applied to satisfy fixation capacity. This poses a big problem, because the amounts are generally in excess of what small farmers can afford without financial assistance.

The more fertile alluvial soils of the tropics are generally used for the production of paddy-rice which has its own specific problems, entirely different to those encountered in upland crops.

Another important problem is soil acidity. Under shifting cultivation, acidity is counteracted by the alkaline plant ash, which incidently also contains micronutrients. With the transition to permanent cultivation the need to counteract acidity arises. This can be done by the use of ash or mulches of deep-rooted species. Liming is of course the most effective procedure; however, lime is generally not available and transport costs may be prohibitive. Also, raising the pH of the soil may induce zinc, manganese, and iron deficiencies (Greenland, 1975).

Control of Diseases and Pests

The direct loss of agricultural products as a result of the combined effects of pests, diseases and weeds cannot be estimated accurately. For the United States, in spite of advanced farm technology, it is estimated that the annual loss from these causes is at least 30% (Ennis *et al.*, 1967). The overall loss in developing countries in the tropical regions is certainly greater, not only because of lack of control methods, but also because the continuous warmth and high humidity favour many kinds of diseases and pests. Rodents and birds also multiply rapidly.

Importance

The following examples will suffice to show the importance of the problem. In India, it has been estimated that rats consume 21 million tons of food annually, one-third of the country's production. In Taiwan a campaign against rats saved an estimated 140,000 tonnes of food a year. In West Africa it is estimated that insects destroy 25% of the food in storage (Millikan & Hapgood, 1967).

Risk of loss from pests and diseases must be reduced to tolerable limits if farmers are to adopt new methods for increasing production.

Certain cash-crops, such as cotton, cannot be grown on a commercial scale unless satisfactory insect control methods can be applied. The use of insecticides has been a key factor in the expansion of cotton production in Central America, where the cost of insect control may amount to 20–45% of the total costs of production (Leurquin, 1966). In the case of bananas, the control of sigatoka, an important disease, may require spraying 15 to 30 times a year and is a major item in the costs of production of this crop (McPherson & Johnston, 1967).

An important problem that has been largely neglected and frequently unnoticed is nematode infestation, which affects nearly all tropical export-crops and many food-crops, causing severe reductions in yield. The most effective control method, soil disinfection by nematicides such as ethyl-dibromide, is extremely expensive. Palliative measures are the use of nematode-free planting material, crop rotation, etc. Trees and shrubs can serve as reservoirs for extremely harmful plant nematodes that can thereby survive the dry season and high temperatures. For example, the boabab (*Andasonia digitate*), a widely distributed tree in the Sudano-Sahelian zone, serves as a host to the extremely harmful nematodes *Meloidogyne* spp. and *Rotylenchulus reniformis* (Gundy & Luc, 1979).

Varietal Resistance

Pest and disease problems are generally greatly intensified with the adoption of pure varieties and as the level of fertilizers, water and other production inputs increases.

New varieties may be far more susceptible to diseases or pests than the traditional varieties they replace. Disease pathogens adapt themselves easily to genetically pure strains of different crops and the spread of an epidemic is far more rapid in a uniform stand than in a mixed population.

Furthermore, in order to achieve high yields with a new variety, such as the HYV rice IR8, increased planting density and heavy fertilization are essential. The resultant lush vegetative growth creates ideal conditions for disease epidemics, in particular blast in rice and rust in wheat, and for infestations by insects.

Conversely, the use of disease-resistant varieties is without doubt the most effective and economical method of disease control, especially for those diseases for which no effective chemical control methods have yet been developed. In many cases, the continued cultivation of a crop in a certain area became possible only after resistant varieties had been developed.

Breeding of plants for insect resistance has received far less attention in the past than has breeding for disease resistance. However, a few varieties that are resistant to insects have been developed: varieties of maize resistant to the corn-borer (*Oscinia nubilalis*), varieties of wheat resistant to the Hessian fly (*Mayetiola destructor*) (Painter, 1958), and jassid-resistant varieties of cotton, are a few examples.

Effects of Irrigation

In general irrigation favours the incidence and spread of many crop diseases by its influence on plant physiology, microclimate and soil moisture régime. It affects both the host and the pathogen. The rapid succession of crops (especially if monoculture is practised), the longer periods that the stomata are widely open, the increase succulence of the tissues as a result of favourable moisture supply and higher rates of nitrogen fertilization, the increased plant populations necessary to make efficient use of the favourable growing conditions, are all contributory factors that increase the susceptibility of the crop and create more favourable conditions for infection.

When desert land is first developed for irrigation, a period of relative freedom from plant diseases is usually experienced. However, certain diseases, when once introduced, spread rapidly in these newly irrigated areas, especially if no proper crop rotation is practised. In the Columbia basin, in north-western North America, Fusarium root-rot of beans increased from a trace in the first year of irrigation, to serious damage in the third year, when beans were grown continuously (Menzies, 1967). Many pathogens are introduced with infected seeds into regions in which the disease that they produce was previously non-existent.

Limitations of Insecticides

For a few years, great successes were achieved with the new insecticides; but gradually their shortcomings revealed themselves. Optimistic forecasts, predicting the complete extermination of major pests, proved to be premature. Certain insects were effectively controlled, to be replaced by others which had previously been minor pests. For example, the chemical control of jassids (*Empoasca* spp.) on cotton in the Sudan, resulted in an increase of the whitefly (*Bemisia gossypiperda*); when the whitefly was controlled by an insecticide, damage due to cotton bollworm (*Heliothis armicera*) increased markedly (Joyce, 1955). It was shown that three consecutive sprayings with one of the chlorinated hydrocarbons were sufficient to annihilate all the natural enemies of these pests, and this was the main reason for the epidemic increase in pests which were not directly affected by the insecticides used.

In Peru and Nicaragua, chemical control had become largely ineffective in cotton, even with more than 30 insecticide applications during the growing season. Not only was pest control unsatisfactory, but the high costs made cotton production economically unfeasible. A tremendous reversal was achieved by applying the method of 'integrated control' based on the ecological and biological principles described below (Apple & Smith, 1973).

Biological Control

Biological control is defined as 'the direct and indirect use of living organisms to control pests and to reduce damage below economic levels' (Chant, 1966). The living organisms employed are insects or pathogens.

A large percentage of insect species prey on other insects, and there is no doubt that the main factor preventing a 'population explosion' of an insect species is its natural enemies. These fall into two main groups: the *predators*, which are usually larger and stronger than their victims, and the *parasites*, which live on or in the bodies of insects that are much larger and stronger than themselves.

Not all natural enemies of insects are beneficial to Man. The natural enemies of harmful insects have their own predators and parasites which are therefore, *ipso facto*, harmful from Man's point of view. These, in turn, also have their natural enemies, which are therefore beneficial.

The relationships between insects, their natural enemies, and the hyperparasites, are extremely complicated, and it is easy to understand how Man's intervention can readily upset the very delicate system of checks and balances occurring in Nature between these components of a biological system.

Threshold Treatments

Insecticide recommendations are either preventive or based on pest thresholds. Preventive treatments are justified for the protection of seeds and seedlings against endemic pests and major pests that appear in most years and for which pesticide timing is critical. In most other cases, threshold treatments, based on numbers of insects/m^2 or number of infested plants/m^2 are to be preferred.

It is well known that the natural enemies of an insect species are not capable of controlling it completely (this could lead to their own extinction) or even continuously. Normally, they are able to maintain the numbers of the harmful insect within certain limits, so that it does not cause appreciable economic damage (Billiotti *et al.*, 1962). Under circumstances which favour a very rapid increase of the harmful insect, however, its natural enemies are no longer able to cope with the population explosion, and the intervention of Man, using appropriate control techniques, becomes necessary.

The fundamental premise on which control by threshold treatmens is based, is that complete eradication of a pest is neither necessary nor even desirable. The development of an integrated control system requires the continued existence of best populations, albeit at a low density. This is known as the 'dirty-field principle', and is an essential condition for the permanence of a balanced system of pests and natural enemies. The presence of the insect pest, when maintained at a certain level, does not cause economic damage (Chant, 1966). This is achieved by 'adjusting the spray programme, so as to cause minimum damage to the natural enemies of the pest that is being controlled' (Ordish, 1967).

Integrated Control

Integrated control is based on a combination of all available methods: resistant or tolerant varieties, threshold treatments, alternating methods of application and insecticides, crop rotation, mixed cropping, biological control and trapping (Francis, 1979).

Problems of Adoption

Generally pest control is a 'late' input, that is fairly easily adopted by farmers after other innovations, such as HYVs and fertilizers, are in use, mainly because the efforts and costs already invested are otherwise lost. The disappearance of insects and other pests after treatment is usually spectacular and the good results are immediately evident. However, the initiative for pest and disease control cannot, as a rule, be left to the individual farmer—especially if he is still lacking in experience of modern technology. Firstly, he may fail to identify the pest or disease and be unaware of the appropriate control methods. Secondly, even if he does effectively control an outbreak in his own field, he will frequently have wasted his efforts and money if his neighbours neglect to do the same. Thirdly, there appears to be a rather significant difference in the speed of adoption of pesticides between larger and smaller farms. This is apparently due to the greater complexity in the use of these chemicals, as compared to new varieties or fertilizers, and to the highly toxic nature of many of the chemicals used (Gotsch, 1973). And, fourthly, after relatively short periods, simple chemical control treatments frequently become ineffective and must be replaced by more sophisticated approaches.

Accordingly there is a need for the creation and maintenance of an efficient crop protection service as one of the first essentials of agricultural development. This service will have to forecast insect invasions and disease epidemics whenever possible, advise the individual farmer on appropriate measures, and organize joint large-scale control measures whenever the need is indicated. Mobile groups of specially trained monitors will have to carry out a constant surveillance of the pest and disease situation within defined areas of jurisdiction.

Special equipment is also generally needed for the efficient application of fungicides and insecticides, this cannot be provided by individual small farmers.

The government of the Ivory Coast, for example, has provided free control of the cocoa capsid pest, but only on well-maintained plantations. This has encouraged farmers to improve their overall management practices and has increased the country's productive potential by 50% (Dumont, 1966).

Problems of Supply

The developing countries are almost entirely dependent on imports of pesticides. Those most in use are the broad-spectrum pesticides (e.g. aldrin, dieldrin, DDT) which, besides being extremely effective for a large number of pests, are also by far the cheapest to produce. However, in recent years, legal restrictions on the production and use of these pesticides in the developed countries (because of their negative effects on the environment) have caused supply shortages and have pushed prices up drastically (FAO, 1976).

Future production in exporting countries will be geared to the needs of developed countries, namely, target-specific and non-persistent pesticides which are expensive and require sophisticated methods of application. Developing countries will therefore be doubly affected: not only by the shortage and high cost of the customary pesticides, but also by the neglect in developing new plant protection chemicals suitable for crops and conditions in developing countries, because of the small and uncertain markets (FAO, 1976).

Relationship Between Insecticide Use and the Adoption of HYVs

The use of insecticides by rice farmers, both before and after adoption of HYVs, was investigated by the International Rice Research Institute, in 32 selected villages in various regions of Asia with a high potential for rice production. The results of this survey are shown in Fig. 45. The percentage if farmers who were already using insecticides before adopting the modern varieties is surprisingly high (47% for all the villages). Use of insecticides

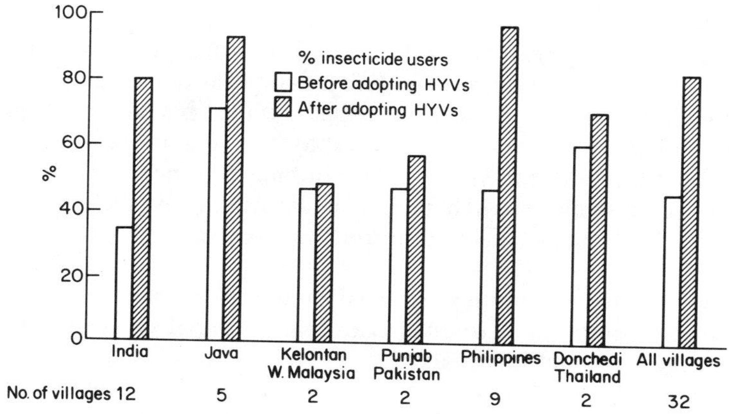

Fig. 45 Use of insecticides by rice farmers before and after adoption of HYVs, in 32 selected villages in various regions of Asia with a high rice-producing potential (based on data from Barker & Anden, 1975).

generally increased after adoption of HYVs—excepting in Western Malaysia where input subsidies were discontinued (Tamin & Mustapha, 1975); the average increase for all villages was 70.6%. In IRRI insecticide experiments on rice in 1971 and 1972 yields were 36% higher when an insecticide was applied (IRRI, 1978).

Insecticides were the most popular modern input in Filipino villages because of the high incidence of pests. Even the more backward farmers in the region use insecticides (Tan, 1975).

Vertebrate Pest Control

Isom & Worker (1979) estimate that the depredations caused by vertebrate pests may equal or exceed losses from diseases, insects, nematodes and weeds.

Species of the genus *Rattus* are among the principal crop pests in the tropical and subtropical areas of the world. Chronic losses to crops are serious and sporadic. Outbreaks can cause severe food shortages such as those which have occurred in large areas in several Asian countries following rat population eruptions. Traditional methods of control are generally ineffective. Post-harvest losses, direct and indirect are also considerable. The damage caused by mice, though considerable, is less conspicuous. Rodents also cause heavy damage in livestock and poultry production, and are frequently involved in human and animal disease transfer. Sustained baiting with chronic toxicants appears to be the most promising method of control for rodents.

Vampire and fruit-eating bats and grain-eating birds have also caused considerable damage.

The activities of Quelea birds in East Africa are a well-known example of the extensive damage that can be caused to crops by grain-eating birds. The departments of agriculture in Kenya and Tanzania have established special Quelea control units which combat the bird plague by aerial spraying of breeding colonies, the use of explosives and the cutting down of branches bearing nests and young birds (Bohnet & Reichert, 1972).

FAO is conducting research aimed at controlling weavers and Quelea grain-eating birds as a regional project in Africa involving 13 nations (Sanchez, 1975).

Effective control for vertebrate pests requires regional coordination in programming campaigns, training extension specialists and developing strategies, methods and materials.

One difficulty encountered in control is that the measures taken frequently disrupt the food chain of many carnivores and upsets ecological balance. New methods for repelling pests need to be developed.

Improving the Adoption of Pesticides

FAO recommendations for improving the adoption of pesticides in the developing countries include (FAO, 1977):

(a) better formulation of pesticides suited to tropical conditions;
(b) more adequate standards and uniform methods for controlling the chemical, physical and biological properties of the products sold to farmers;
(c) establishment of local formulation facilities where raw materials are available; and
(d) promotion of more efficient and safer use of pesticides at farm level, in particular simple, easy to operate by hand units suitable for small farms.

Many countries have developed, or are developing legislation relating to these matters. However, legislation cannot be effective until inspection services are organized, technical facilities established, and personnel trained and appointed to enforce regulations.

WEED CONTROL

Importance

It is common knowledge that weeds compete with crop plants for light, air, moisture and nutrients; but the full extent of the economic harm which they cause, and their direct effects on crop yields, are not generally realized. In arid climates the deprivation of water to the cultivated crop is frequently the most damaging aspect of weed competition—both in the growing crop and in the fallow period between crops.

Weeds grow very fast in the subtropics and the tropics, and constitute a major handicap in agricultural production. The need to keep crops weed-free is the dominant factor limiting the amount of land a family can cultivate. For this reason, the majority of shifting cultivators have tiny holdings.

Hoeing and hand-weeding are probably the oldest agricultural occupations, and the back-breaking work involved still forms a considerable and inevitable part of the human effort involved in primitive agriculture. It has been estimated that in Kenya, for example, 388 man-hours/ha are required for weed control in maize production and 428 for cassava (Clayton, 1960). With traditional methods, a man can cultivate only 3 to 4–5 ha per year (Watters, 1971).

In Tanganyika, research proved conclusively that groundnut yields were increased considerably by weeding within a few days of crop germination. Efforts to persuade farmers to undertake timely weeding were, however, largely ineffective because this operation would have competed for scarce labour at a time when land preparation had first priority in order to enable planting before the rains petered out (Joy, 1969).

Weeds can also be a major constraint when mechanization is introduced, if more land is prepared for sowing than can be effectively weeded during the growing seasons.

The effects of weed competition are mainly felt in the young crop; the much more rapid rate of growth of both the aerial parts and the roots of many weeds gives them a considerable advantage in depressing and even crowding out crop plants among which they are growing. Weeds also increase harvesting costs, require costly cleaning of seeds, reduce water flow in irrigation and drainage channels, and increase fire hazards.

In shifting cultivation, weed infestation generally increases until it becomes unmanageable, and the only practical solution in the past was to abandon the land for longer periods, until the regeneration of the forest depressed and finally controlled weed growth. This has been found to be the most effective form of weed control for traditional farming.

The problem is further aggravated by the generally accepted dictum that weeding is degrading work for men, so that only women are available for this back-breaking chore. The introduction of herbicides could, therefore, play an extremely important role in increasing the efficiency of farming.

Herbicides

The development of selective organic herbicides is probably the most significant advance in agriculture of the last two decades. A whole array of new techniques and chemicals—low-volume spraying, improved surfactants, the precision application of herbicides to the soil and granular formations; new approaches such as pre-sowing and pre-emergence weed control; the discovery of new herbicides that kill grasses selectively, and of chemicals that are converted into herbicides by the plant or in the soil—have supplied the farmer with means of selective control of weeds in most crops, even including those that were considered to be most susceptible to hormonal herbicides.

The main constraints to the widespread use of herbicides, as a means of replacing back-breaking labour and effectively ensuring weed-free crops are:

(a) The lack of exact information on the kinds and amounts of herbicides to be applied, and the methods of application, appropriate for each crop under given ecological conditions. The use of these chemicals also requires sophisticated equipment and extremely meticulous application according to the instructions received. It is difficult to expect that these conditions will be met by farmers who have little or no experience of modern agriculture. These problems can be overcome by effective training of farmers and supervision of the treatments; but a high level of knowledge and skill of the extension workers is required.

(b) *The relatively high costs of the herbicides:* though extremely small amounts of herbicides are required per unit area, many are extremely expensive. Use of herbicides was found to save 270 man-hours per hectare, about one quarter of the total labour input in rice production in Taiwan. However, on small farms, the cost of herbicides was greater than that of weeding by family labour (Lai, 1972). On the other hand, it is reported that in trials in the Minshan area in the People's Republic of China, it was found that the cost of controlling weeds by 2,4-D was much lower than weeding by hand (Kuo, 1972).

  The problem of cost of weed control can best be overcome by combining a suitable cropping system, a sequence of cultural operations in coordination with rainfall season, and judicious use of herbicides (Francis, 1979).

(c) *Persistence of herbicides in the soil*: herbicides in the soil may be destroyed by microorganisms, may decompose chemically or by photodecomposition, may be inactivated by adsorption on soil colloids, or may be lost by leaching or volatilization (Klingman, 1961). The principal factor involved in the degradation of herbicides is *biological*. Decomposition of herbicides to subtoxic levels by soil microorganisms may require a few days or up to a year or more.

  In general, well-managed irrigated soils are warm, moist, fertile and well-aerated—conditions that are conducive to the rapid decomposition of organic herbicides. By contrast, in unirrigated agriculture, during the extended periods when the soil is dry, microbiological decomposition may cease completely.

  The problem of the persistence of herbicides in the soil has two contradictory aspects: persistence is desirable in order to ensure a weed-free environment for as long as possible, but it is undesirable if it interferes with the sowing of a susceptible crop following the crop for which it had been used. In addition, the build-up of residues may reach toxic levels for most, if not all, crops.

The detrimental effects of herbicide residues are usually more serious in arid than in humid regions, and in rain-fed agriculture more than under irrigation. The erratic and unpredictable rainfall of semi-arid regions makes it difficult to forecast the time that will be required for the breakdown of a herbicide to innocuous levels. Long dry periods may permit persistence of herbicidal activity, which may then interfere with cropping schedules.

The build-up of herbicide residues to a toxic level can be at least partly overcome by dilution (tillage, irrigation), accelerating breakdown (adding organic manures or maintaining a favourable soil moisture régime), or sowing a non-susceptible crop, and, mainly, by careful use of the herbicides themselves through accurate dosage and calibration, appropriate choice of herbicides, avoiding the continuous use of a single type of herbicide, etc.

(d) *Development of resistance to herbicides*: the use of 2,4-D, atrazine and paraquat has freqnetly led to a rapid increase of weeds that are tolerant to these herbicides or have developed tolerance. Examples are *Phalaris* in wheat treated by 2,4-D and *Parthenium hysterophorus* in banana, citrus and coffee following application of paraquat (Labrada, 1975).

Complementarity Between Fertilizer Use and Weed Control

The dwarf upright stature of modern varieties and the heavy rates of fertilizers used with these varieties contribute to luxuriant weed growth. Therefore, a high degree of complementary effects between weed control and fertilizer use for obtaining high yields with HYVs can be expected (Table XXV).

In a study in three rice-growing villages in Thailand, it was found that using high levels of fertilizers or weeding intensively alone resulted in

Table XXV Complementary effects between weed control and fertilizer use on rice in 59 farms in Donchedi district, Thailand (Sriswasdilek *et al.*, 1975).

Levels of fertilizer use	Weeding	No. of farms	Fertilizer use (kg/ha N)	Average weeding (man/d/ha)	Yield (t/ha)	Net return[a] ($/ha)
low	low	25	24	0.4	2.4	79
low	high	8	6	15.1	2.5	76
high	low	13	20	0.9	3.0	87
high	high	13	27	15.2	3.9	127

[a]Return over variable cost of inputs.

insignificant increases in yield, whilst a combination of both gave substantial increases in yield and net income (Sriswasdilek *et al.*, 1975).

Implications

The following basic aspects of weed control need to be taken into account in formulating any recommendations:

(a) Without mechanization, more energy is expended on weeding than in any other agricultural operation.
(b) The major adverse effect of weeds occurs during the first three to four weeks of crop growth.
(c) Weeding by hand or hoe is a slow business. By the time the farmer has completed his first weeding, the future yields of his crops have probably been depressed by between one-third and one-half.
(d) The dominant system of food-crop production is based on mixed cropping, which has manifold advantages, but makes selective weed control more complicated.

Farmers cannot cope effectively with weeds using traditional methods. Improved tools and sowing in rows may alleviate the problem somewhat, but do not provide an adequate solution.

Not only is the farmer unable to cope effectively with weeds in subsistence farming, but there is no possibility of adopting yield-increasing inputs, such as fertilizers, as long as weeds proliferate. A solution to the weed problem is therefore of the utmost urgency.

In view of the foregoing, the most promising solution appears to be the use of a general-purpose herbicide, applied before planting, that can provide a weed-free field for three to four weeks after planting. The search for such a reasonably priced herbicide, and the determination of the most effective method of application, should have the highest research priority. In view of the vast array of known herbicides, a successful quest should not present great difficulties.

A preplanting herbicide has the advantage that it does not require the same degree of sophistication and does not involve the difficulties encountered in the use of selective post-planting herbicides. The use of the latter could be confined to commercial crops.

It is worth noting that the demand for herbicides in developing countries increased more rapidly in recent years than that for insecticides and fungicides (FAO, 1979).

IMPROVED HUSBANDRY

Importance

The foregoing pages have been devoted largely to the role of relatively sophisticated and expansive inputs as essential ingredients of the green revolution. However, one must not overlook the importance of improving conventional cultural practices.

These may contribute considerably to the overall increase in yield made possible by HYVs, the use of fertilizers, effective crop protection, etc. Without careful tillage, better stands of plants, timely sowing and weeding, more costly inputs will remain largely ineffective.

Certain desirable agronomic practices, such as timely sowing, may conflict with the peak labour requirements of traditional subsistence crops and therefore be unfeasible in practice. This may account for slipshod and hurried methods of soil preparation, sowing and weeding.

As long as mechanization is not a suitable solution, other approaches to overcome these difficulties need to be developed.

However important in the context of modernizing agriculture, improved husbandry practices by themselves have limited effects on yield, and cannot bring about a marked expansion of production. Even an improved crop rotation without fertilizer for the legume, will be largely ineffective. And yet, in the early stages of development, improvements in husbandry are extremely important, as they build up traditional agriculture to a point where a qualitative change becomes possible. They increase the farmer's ability to buy more costly inputs, such as fertilizers and insecticides, and enhance the effects of the latter.

Simple Agricultural Practices

In rain-fed crops, simple practices for conserving soil moisture may increase yields in semi-arid regions and may even prevent crop failures. In northern Nigeria, yields of rain-fed cotton, grown in traditional agriculture, are around 165–220 kg/ha of cotton seed. Experimental work indicated that by using improved varieties and fertilization, yields could be increased to a maximum of 660 kg/ha. More recent research has shown that mulching increases the percolation rate of water into the soil, reduces soil temperatures and evaporation and thereby further increases yields considerably. These relatively high levels were practically independent of variations in rainfall; thus in the course of five years with distinctly different rainfall patterns, yields per annum of 2400 kg/ha or more were obtained. Tie-ridging* can also increase cotton yields significantly by conserving and storing water from the early rains; as a result, the cotton plants are better able to survive the dry period which usually occurs after the early rains (Wilde & McLoughlin, 1967).

*Tie-ridging consists of forming cross-ridges between the main ridges, at distances of 2 metres or more. The ties prevent run-off of water during heavy downpours and thereby improve infiltration and reduce erosion. An attachment for cross-tying can be fitted to the ridges, which is quickly dropped at the desired intervals during ridging. (Tschiersch, 1978).

Improved Crop Rotation

The merits of integrating crop production and animal husbandry have been discussed, and the spectacular effects on productivity of including in the rotation leguminous crops for pasture or forage in Israel and Australia have been described in Chapter III. Other improved rotation practices are also possible as a first step in transforming traditional agriculture, and as a framework within which the response of crops to other factors affecting production is increased.

Intercropping and Relay Cropping

The advantages and drawbacks of mixed cropping in traditional farming systems have been discussed in Chapter III. Research workers are generally reluctant to undertake experiments on mixed cropping, partly because of the great number of possible crop combinations, but mainly because of preconceived ideas that no benefits are to be expected at advanced levels of farming (Andrews, 1972).

It is mainly the international agricultural research stations that have undertaken this type of research and have developed improved methods of mixed cropping, such as intercropping, relay cropping, sequential cropping, etc., which maximize the advantages and avoid most of the drawbacks of traditional methods.

Intercropping consists of a systematic planting together in alternate rows of two or more crops. *Relay cropping* consists of planting a second crop between rows of the first crop, when it is nearing maturity. Competition is therefore negligible.

Intercropping and relay cropping have been shown to be more efficient in the use of available resources than single-crop stands:

Available Light

By a proper choice of cultivars, planting time and spatial arrangements, it is possible to extend the time of full leaf spread, with maximum leaf area occurring for each crop when high solar energy is avilable (Sooksathan & Harwood, 1976).

Nutrient Uptake and Utilization

These have been found to be more efficient in maize-rice and maize–soybean intercrops than in monocultures of the respective crops (Suryatna & Harwood, 1976).

Insect and Disease Damage

These were reduced in the maize–rice and maize–soybean intercrops (Suryatna & Harwood, 1976).

Fig. 46 Intercropping experiments at ICRISAT on deep black soil. The plot on the left shows a three-crop intercrop system with pigeon pea, sorghum, and mung beans interplanted. The mung bean is harvested in about 65 days, the sorghum at around 100 days and the pigeon peas at 190 days. By courtesy of the International Crops Research Institute for the Semi-Arid Tropics (Hyderabad).

Drought Risks

These were lower in maize–sorghum intercrops than in single-crop stands (Cutie, 1975).

Labour Requirements

These are higher for intercropping, but labour demand is better distributed throughout the year (Norman, 1970).

A combination of the effects mentioned above enables well-designed intercrops to yield 30–60% more than the individual crops (Syarifuddin *et al.*, 1974). Still better results may be expected from varieties that will have been bred specifically for inter- and relay cropping. Shallow root-systems and ability to tolerate some shading are desirable traits in developing varieties for this purpose.

In bean–maize intercropping trials in CIAT, consistent yields of 1500–2000 kg/ha of beans in association with a 5000 kg/ha maize crop have been obtained. Average yields of bean and maize crops in tropical Latin America are about 600 and 1300 kg/ha, respectively (Francis, 1979).

In experiments on intercropping with sorghum in Nigeria, in an area with a growing season for rain-fed annual crops of 180 to 200 days, sole cropping, relay cropping and intercropping of long-season sorghum, early millet or maize and late-planted cowpeas were evaluated for the most effective use of the growing period; all crops were cultivated mechanically.

These experiments have shown that intercropping, at high levels of productivity, can give better yields than relay cropping, which in turn was markedly better than a single, late-maturing sorghum crop, spanning the whole growing season. Higher returns per hectare for relay cropping amounted to 59%, and for intercropping to 80%.

Improved stability of yields appears to be a major advantage of intercropping. In analysing the data from 94 experiments on sorghum/pigeonpea intercropping, Rao & Willey (1980) found that for a particular 'disaster' level quoted, sole pigeonpea would fail in one year in five, sole sorghum in one year in eight, but intercropping would fail in only one year in 36.

Intercropping gave yield advantages over a wide range of environmental conditions.

DIVERSIFICATION

It has been pointed out earlier that by increasing the profitability of farming and by reducing the areas of land required to meet the demand for staple crops, both the need for, and the possibility of diversification of agricultural production have increased.

Motives

Judging from recent development plans, the main motives for the tendency to introduce the production of new commodities are: (a) to lessen dependence on a single crop; (b) to switch from a crop with weak market prospects to one for which the demand is likely to increase; (c) to encourage the production of crops that have an assured local market (actual or potential import substitution); (d) to favour labour-intensive crops; and (e) to encourage crops that can be processed by local industries (United Nations, 1973).

Crop Diversification

By contrast with irrigated and semi-arid regions in which many temperate and subtropical crops can be successfully grown, crop diversification in the humid tropics is faced with certain environmental constraints. The much-attenuated solar radiation, the persistently high night temperatures, the lack of seasonality and the excessive rainfall limit potential yields and the possibilities of crop diversification (Chang, 1962).

Crops, such as coffee, cocoa, cotton, and tobacco require short dry periods for ripening. Under constantly high rainfall coffee for instance, has no defined blooming season, and the trees usually carry a few berries of poor quality at all times; cocoa has few pods per tree and shows a high incidence of blackpod disease (Smith, 1963).

Commercial Crops

A simple way of increasing farm income is by shifting to crops producing a high value per unit of land. Changes in the types of crops grown, in particular changes from food-crops for home consumption to cash-crops, also makes it possible to utilize labour more intensively.

In Kenya, many farmers were able to increase their cash income quite rapidly by growing coffee, tea, pyrethrum and other export-crops, thus overcoming the constraints due to the relatively small size of the domestic market for food (Otieno *et al.*, 1975).

Annual crops that are labour-intensive and at the same time sufficiently remunerative, should be the first candidates for adoption.

An example of a crop that is labour-intensive, quick to flower, easy to handle and requires no pesticides and can considerably increase the income of farmers, is pyrethrum.

In Kenya, the crop was adopted quickly and with minimal government extension effort; by 1972 some 90,000 families were engaged in production (Gerhart, 1973). Pyrethrum was also recently introduced into the higher parts of Ecuador. If world demand for this insecticide is sustained, it may become an important export crop (Duckham & Masefield, 1970).

As income rises in the developing countries, the market for *vegetables* and *fruits* generally expands. Vegetable-crops can be issued in very intensive cropping systems because they have high cash value, can be raised as seedlings and transplanted, thereby shortening their growing time in the field and minimizing competition with the principal crop; they can be grown between the rows of plantation crops; can be intercropped or relay cropped with field-crops, such as maize and sugar-cane.

Whilst neighbouring cities may provide attractive markets, the greatest and most remunerative opportunities are usually those provided by export, in particular to richer countries. The range of vegetables that can be grown in the subtropical and tropical regions is much greater than that of the temperate regions; crops can also be produced at periods of the year when local produce in temperate regions is not available, or only at very high cost.

There are a number of examples of successful large-scale replacement of traditional crops by fruits and vegetables for export. In Taiwan, in the past, sugar accounted for 80% of the country's exports and was practically the only agricultural export. By 1967, sugar accounted for only 7% of exported agricultural products and had been largely replaced by bananas and canned fruits and vegetables (mainly mushrooms, asparagus and pineapples) (Shaw, 1971).

In Haiti, the out-of-season production of tomatoes for the US market, grown as a replacement for the traditional sugar-cane, has great potential for development. In Israel, the traditional wheat-crop has been largely displaced in Arab rain-fed farming by onions for export, tomatoes for canning and sugar-beets. It has been shown that by replacing part of the traditional crops by more labour-intensive and profitable crops, and by using modern inputs, the net income of the Arab farmer from the same cultivated area was almost three times as high as formerly. Under irrigation, out-of-season strawberries and vegetables, grown under plastic cover, gave even greater returns.

Another example of the potential importance of export-crops, is the experience of the Ivory Coast, where through the development of intensive market-gardening techniques, pineapple production for export was increased from virtually nothing in 1950, to 50% of the total French market in 1966 (Jones, 1971).

There are greater risks involved in production of perishable crops for export, mainly because of price fluctuations outside the control of producers; local canning plants can contribute to a lessening of the risks and can also absorb lower-grade produce.

Industrial Crops

Industrial crops can provide opportunities not only for export but also for local processing without which much of the potential economic benefit is lost to the producing country. The increased extraction and processing of agricultural production can add considerably to the value of exports. The by-products of processing, such as oil-cake from oil-crops, can provide valuable high-protein food for livestock industries.

Agricultural processing industries are among the least sophisticated and therefore constitute an important beginning in industrialization. They help prepare manpower with the skills, aptitudes and experience required for industrial society.

Plantation Crops

Tropical countries have a practical monopoly in the production of certain plantation crops, such as bananas, coffee, cocoa, tea, rubber, oil palm, and coconuts. These are particularly well adapted to rainforest areas, because they provide adequate cover and protection against soil erosion and have deep-rooting systems which enable a re-cycling of plant nutrients that tend to be leached as a result of the high rainfall (cf. p. 81).

The main disadvantages are the quite considerable time-lag between planting and the first economic returns to the farmer, and the relative inability to adjust to changing marketing conditions, once plantations have been established. For example, rubber (*Hevea brasiliensis*) takes six to seven years to reach the production stage, and 12 to 13 years until the full bearing stage is reached; hence investment decisions must be taken on the basis of a world market outlook at least 10 years in the future; it is notoriously and extremely hazardous to predict such an outlook.

Farmers can be persuaded to adopt perennial crops provided they are sufficiently attractive, even if they have to wait a number of years for the first fruits. In Tanganyika, for example, there was heavy planting of coffee under the stimulus of high prices in the early 1950s, and within a decade the country's coffee exports had doubled (FAO, 1962).

Animal Husbandry

The improvement of animal husbandry faces many difficulties, because it involves complex interactions of technical, economic and socio-cultural factors (Lele, 1975).

In the pastoral areas, in which the limited rainfall precludes arable farming, the improvement of livestock production is critical for raising the standard of living of the rural population.

The first and most obvious step is evidently to adjust stock numbers to the carrying capacity of the range. This has however proved extremely difficult to achieve.

The tradition, in many pastoral areas, of private ownership of cattle and communal land ownership, encourages overstocking, thereby causing overgrazing and deterioration of the range.

Improvement of the local breeds by appropriate upgrading programmes, better veterinary services, and an adequate water supply, could more than compensate for the reduced stock numbers.

Improvements in livestock management can also be an important complement to crop production in non-pastoral areas.

The potential advantages of integrating arable cropping with animal husbandry in drier regions, and the replacement of shifting cultivation by improved pastures—as one of several possible solutions—have been discussed in Chapter III.

In most types of subsistence agriculture, some livestock are kept mainly as scavengers on crop wastes (such as straw, stalks, stubble, etc.), on village communal grazing lands or wastelands unsuitable for arable cropping (too shallow, stony, steep, etc.).

In the present context, we are referring to improved methods of livestock management, capable of producing three- or fourfold increases in production.

Tropical and subtropical climates have the advantage of being relatively warm to hot throughout the year, so that simple buildings for livestock are sufficient, and forage production is possible throughout the year.

The expansion of the livestock industry in the humid tropics has been promoted in a number of developing countries, actively supported by loans from the World Bank. The justifications for this policy are: the strong market for beef; the need to close the 'protein gap'; the suitability of many tropical areas for cattle; and the availability of improved breeds adapted to a tropical environment (Nelson, 1973).

The introduction of dairy cattle, especially in the vicinity of larger towns, can be a major factor in developing commercial agriculture, improving the standard of living of farmers and improving nutritional standards of the population as a whole.

In former years, large-scale animal husbandry was not considered appropriate for Japan because of its hot and humid climate. Maintaining a livestock industry on grass and clovers that did not thrive under high temperatures and strong sunlight in a country with a high density of population and with a highly intensive agricultural production, was not considered justified. In recent years, livestock production has developed considerably because it is now based on intensively cultivated and highyielding forage crops, such as sweet potatoes, of which tubers and leaves are fed to the animals. This crop, grown with heavy applications of fertilizers, yields about 250 tonnes of green matter per hectare.

Animal husbandry has become one of the most promising possibilities for utilizing agricultural land that has been made available as a result of the more intensive multiple cropping adopted in recent years. As a result of these developments consumption of milk and dairy products by urban consumers in Japan has increased fivefold and meat consumption has doubled (Johnston, 1962).

When marketing problems arise, as a result of increased grain production due to the adoption of improved practices, a possible solution is to feed the surplus grain to poultry, pigs or cattle.

A branch of production that can be very profitable and is relatively easy to introduce or expand is *poultry*.

Modern techniques of broiler production, for example, can be adopted almost anywhere, with minor adaptations to local conditions. For poor countries, poultry husbandry has a number of advantages: feed requirements per kilogram of meat produced are much lower than for animal husbandry, the pay-out period is very short, and investments are relatively modest.

The *pig* is the main type of animal production in Chinese agriculture because—unlike poultry—it can be raised without becoming an active competitor for grain required for human consumption (FAO Study Mission, 1978). With its multiple uses and easy management, pig raising is ideally adapted to the small family farm.

For certain tropical savanna areas the economic harvesting of *game animals*, together with the tourist value of wildlife, may be a most efficient use of national resources.

Tribe (1970) mentions three ways in which game can be managed for the production of meat and by-products (skins, horn, bone-meal, etc.).

(a) *Game culling* or the selective reduction of wild populations.
(b) *Game ranching* in which wild animals are encouraged to graze in an area operated as a commercial ranching enterprise. These ranches are generally located in areas with bush savanna vegetation and no less than 600 mm rainfall. The game may graze in association with domestic cattle.
(c) *Game farming* involves the domestication or semi-domestication of the species involved. The animals (usually gazelles and buck) are grazed in fenced paddocks and may receive additional feeding. The eland has shown considerable promise: it grows rapidly; produces high milk yields; has a low water requirement; and is amenable to routine handling.

All three forms could be potentially important, though game culling and ranching appear to be the most promising. The principal advantages of game animals over domestick stock mentioned are that: they are physiologically and ecologically better adapted to the savanna rangelands; their production of meat per hectare is greater; they are less damaging to their habitats; they are less susceptible to disease and can be maintained in tsetse areas; they require less development and operating capital.

The state of knowledge on wildlife management and utilization is still limited and the advantages mentioned above have not been conclusively proven, hence the need to proceed with caution in advocating their large-scale economic use. Whilst the idea of game ranching is biologically sound and technologically possible, it requires a high degree of expertise in biology and management of wild animals. Mismanagement and inadequate control may result in the destruction of an important natural resource.

The absence of suitable techniques of harvesting, processing and marketing may also be a considerable handicap to the use of game animals. However, many enterprises in the Transvaal, Zimbabwe, and Australia have been commercial successes.

Problems and Limitations of Diversification

Changes in the types of commodities produced are not easily accepted by subsistence farmers, and will certainly not occur spontaneously. They require a number of preconditions and in particular, government initiative in extension, marketing, financing, etc. A further difficulty lies in the fact that high-value crops for export can, as a rule, be developed only when local markets are capable of absorbing part of the produce.

It should be pointed out that, even where cash-cropping has been introduced, farmers in many traditional cultures still insist on food self-sufficiency. Only land beyond that needed to produce the food requirements of the family, will be devoted to cash-crops, provided that sufficient manpower is also available for this purpose. The security provided by self-sufficiency in food is generally a more important consideration than the higher income provided by cash-crops.

In certain cases, non-economic factors militate against the replacement of traditional crops by others more suited to the natural environment, or the production of new commodities whereby agriculture could be more diversified.

A most striking example of the influence of eating habits on growing a crop, which is not adapted to the environment, is the predominance of wheat production in the plains of the Indus and higher Ganges, an area providing food for some 100 million inhabitants. Because wheat can only be sown during the cool season, which is also the dry season, its production in this region is possible only under irrigation, a rather wasteful use of water, a scarce resource. The monsoon rains are of no benefit to the main crop and are used for the production of crops of secondary importance, such as millets, sorgo, etc., which do not form part of the staple foods of the population. This is a clear case of an agricultural production out-of-step with environmental conditions because of traditional food habits (Delvert, 1972). Religious beliefs can be a major constraint to the introduction of beef production in certain regions.

On the other hand, the resistance of traditional farmers to adopt new crops should not be exaggerated. As the productivity of the land and of the labour engaged in subsistence production improves, the farmer will show more readiness to devote some of the resources available to him to the production of cash-crops. In tropical regions, in particular, there is a long tradition of growing new crops. Because of the wide variety of soil and climates in the tropics, it is possible to grow many crops and farmers are accustomed to doing so. Hence, they are generally willing to try new ones, provided market conditions suggest that sufficiently profitable returns may be expected (Behrman, 1969).

An interesting observation is that when a new crop is introduced, it is generally easier to overcome the resistance to new technology shown by farmers than is the case when traditional crops are grown (Wharton, 1969).

Once export agriculture becomes firmly established, lack of diversification may become a major problem. Many Latin American countries that are overly dependent on one or a few export-crops, have attempted diversification programmes. An example is the coffee diversification programme of the Instituto Mexicano de Cafe, whose objective is to reduce the coffee area by promoting the production of avocado, citrus, rubber and livestock (Instituto Mexicano de Cafe, 1965).

The production of new branches of animal husbandry also involves a number of problems. The development of beef production in tropical regions has been faced with certain difficulties, such as the prevalence of animal diseases and inadequate feed supplies. Breeding programmes have relied too heavily on temperate breeds, and have neglected the potentials of certain breeds adapted to tropical consitions; there has been serious lack of knowledge on forage and pasture production under tropical conditions and their utilization; problems of disease control, nutrition and management have been neglected. Cultural practices and taboos have inhibited efficient practices in animal husbandry: religious prohibitions against slaughter of cattle in India and the role of cattle as a symbol of wealth and prestige in many African societies are examples.

Benett & Schork (1979) make the interesting observation that in those areas in Ghana that are served by well established credit unions, the habit of functioning in a money economy has been acquired by pastoralists, at least to some extent. Cattle are drawn upon as capital to be used when required. Besides biological, technical and cultural problems, there are also economic and institutional limitations. Most important is the lack of a physical infrastructure and adequate marketing facilities. Beef production is relatively capital-intensive and has a limited social and employment impact. It has been estimated that investment costs per family (clearing the area from forest, sowing, fencing and other installations, purchase of a breeding unit) may amount to $30,000 per family. Even though it may be an economically attractive investment, it hardly appears to be feasible on a large scale for settling campesinos (Nelson, 1973). One difficulty attending the introduction of modern *poultry production* is that subsistence farmers are accustomed to poultry as unattended backyard scavengers, subject to many diseases. Because of inferior management practices, productivity is very low. It may therefore be difficult to convince them of the economic viability of a poultry enterprise based on modern principles of production. Once established, skillful management and modern methods of nutrition, sanitation and disease control are essential.

The same problems apply to intensive *pig raising*.

SUMMARY

The introduction of the new seed–fertilizer technology has had a number of consequences that occur fairly consistently in the adopting countries.

Benefits

(a) Supplies of rice and/or wheat have increased, sometimes considerably.
(b) The income of innovative farmers has increased markedly.
(c) Prices of the two cereals (rice and wheat) have generally declined.
(d) Employment opportunities and wages have increased, but this is frequently a short-term benefit only.
(e) Productivity per unit of water has increased markedly.

(f) Multiple cropping has expanded.
(g) The adoption of HYVs has catalysed other changes: better management practices, land improvement, expansion of irrigation, increased mechanization, development of rural industries, expansion of infrastructure and improvements in services.

Problems

(a) The introduction of new HYVs predicates a complete change in technology, involving improved seed production, the use of fertilizers, effective plant protection, diversification, and frequently, a controlled water supply.
(b) The new technology requires a substantial increase in investment in irrigation, in land development and equipment, as well as working capital.
(c) Whilst most of the inputs required are neutral to economies of scale, there are considerable differences between different groups in access to the supply of these inputs. As a result, inequality between farming sectors has increased steeply.
(d) The new technology makes agriculture far more dependent on the non-farm sector for inputs, credits, marketing, etc. It also becomes increasingly monetarized.
(e) The early HYVs were generally less resistant to drought, temperature extremes, floods, pests and diseases than traditional varieties. Therefore they are restricted to favourable growing conditions, thereby increasing regional disparities.
(f) Risks are greater than with traditional varieties.
(g) The yields of HYVs tend to decrease as the number of farmers and the extent of the areas sown increases.
(h) The increase income that accrues to large farmers is an incentive to adopt full mechanization.
(i) This in turn aggravates unemployment and causes social unrest.
(j) Mechanization also increases the motivation to consolidate larger holdings and reduce tenancies.
(k) The infrastructure and institutional framework in most developing countries are generally inadequate to promote the adoption of new technologies by all farmers, and /or to cope with the increased production that does occur.

In Brief

Besides the undisputable benefits, the main problem engendered by the green revolution is the increased inequality between rural sectors and between regions, the implications of which will be discussed in detail in Chapter XI. However, these problems are not the direct and unavoidable result of the adoption of new technology, but are mainly due to the inadequacies in policies and strategies in the adoption process.

Essential Components of the Green Revolution (other than HYVs)

In many countries, a major obstacle to the introduction of HYVs is the absence of an organized industry for the *production of quality seed*, with the necessary technical facilities and high ethical standards; also required are the enactment of seed-laws and agencies for supervision, control and enforcement.

It is the ability of HYVs to respond markedly to *fertilizers* that is the main characteristic of the modern varieties. Without fertilizers, their impact is either small or imperceptible, hence the term 'seed–fertilizer revolution'. The main constraints to a necessary increase in fertilizer use on a national scale by the LDCs are: a lack of basic information on fertilizer use, the high cost, a deficiency of foreign exchange, an inadequate infrastructure for marketing, distribution, advice and credit.

Pest and disease problems are generally greatly intensified with the adoption of pure varieties in general, and of HYVs in particular. Whilst a major effort is being made to breed newer varieties resistant to diseases and pests, the use of fungicides, and still more of insecticides is frequently unavoidable.

The creation and maintenance of an efficient crop-protection service is essential.

Rodents and birds are also important crop pests in the LDCs, and their effective control requires regional coordination.

Weeds are a major handicap to agricultural production in the LDCs. There is no possibility to adopting yield-increasing inputs unless weeds are effectively controlled. Because of the effects of weeds are mainly felt in the young crop, at a time when the farmer cannot possibly cope effectively with the problem, the most promising solution appears to be the use of a general purpose herbicide, that can provide a weed-free field for about three to four weeks after planting.

In addition to the specific inputs mentioned above, *improved husbandry* in general, is a prerequisite for the successful adoption of HYVs, including soil and moisture conserving practices, improved crop rotation, inter- and relay cropping.

Diversification of production becomes possible as increased yields make a reduction in the areas devoted to subsistence crops possible. Of particular interest are industrial-crops, export-crops (out-of-season vegetables and plantation crops) and other speciality crops. The integration of arable cropping with animal husbandry, wherever possible, is one of the promising innovations.

REFERENCES

Aggarwal, P. C. (1973). Green revolution and employment in Ludhiana. Pp. 40–63 in *Rural Development and Employment* (Ed. C. Gotsch). Ford Foundation Seminar, Ibadan, Nigeria: 774 pp.

Agricultural Attaché (1973). *Brief on Indian Agriculture*. American Embassy, New Delhi.

Agricultural Economics Research Centre (1968). *Evaluation of the HYV Programme*. University of Delhi.

Alcantara, Cynthia H. de (1976). *Modernizing Mexican Agriculture: Socio-economic Implications of Technological Change 1940—1970*. UN Research Institute for Social Development, Geneva: xvii + 350 pp.

Andrews, D. J. (1972). Intercropping with soybean in Nigeria. *Exp. Agric.*, **8**, 139–50.

Apodaca, A. (1952). Corn and Custom: the introduction of hybrid corn to Spanish American farmers in New Mexico. Pp. 35–9 in *Human Problems in Technological Change* (Ed. E. H. Spicer). Russell Sage Foundation, New York.

Apple, J. L. & Smith, R. F. (1973). Crop protection problems in Latin America. *Development Digest*, **XI** (1), 96–106.

Bapna, S. (1973). *Social and Economic Implications of the Green Revolution: A Case Study of the Kota District*. Global 2 Report, Agro-Economic Research Centre, Vallabh Vidyanagar, India (mimeogr.)

Barker, R. (1970). Economic aspects of new high-yielding varieties of rice: IRRI report. Pp. 29–53 in *Agricultural Revolution in Southeast Asia*, Vol. I (Ed. A. Russell), The Asia Society, New York.

Barker, R. & Anden, T. (1975), Factors influencing the use of modern rice technology in the study areas. Pp. 17–40 in *Changes in Rice Farming in Selected Areas of Asia*. International Rice Research Institute, Los Banos: 377 pp., illustr.

Barker, R. & Quentana, E. U. (1967). Farm management studies of costs and returns in rice production. Paper presented at *Sem. Workshop on the Economics of Rice Production*. International Rice Research Institute, Los Banos, Philippines.

Behrman, J. R. (1969). Supply in response and the modernization of peasant agriculture. Pp. 232–42 in *Subsistence Agriculture in Economic Development* (Ed. C. R. Wharton, Jr). Aldine Publishing Co., Chicago: xiii + 481 pp.

Benett, Alice & Schork, W. (1979). *Studies toward a Sustainable Agriculture in Northern Ghana*. Research Centre for International Agrarian Development, Heidelberg. 125 pp. + 12 appendices, illustr.

Biliotti, E., Grison, P. & Milaire, H. (1962). La lutte biologique. *Bull. Tech. Inf. Inq. Serv. Agric.*, **1967**, 143–62.

Bohnet, M. & Reichert, H. (1972). *Applied Research and Its Impact in Economic Development*, Welforum Verlag, Munich, FRG: 210 pp.

Bose, R. R. & Clark, E. H. (1968). *Some Basic Considerations on Agricultural Mechanization in West Pakistan*. Williams College (cited by Dalrymple, 1969).

Brady, N. C., Yoshida, T., & De Datta, S. K. (1975). *A Summary of Research to Increase the Efficiency of Chemical Fertilizers*. International Rice Research Institute, Los Baños: 23 pp.

Brown, L. R. (1968). The agricultural revolution in Asia. *Foreign Affairs*, **46**, 688–98.

Brown, L. R. (1970). *Seeds of Change*. Praeger, New York: xv + 205 pp., illustr.

Castillo, Gelia, T. (1973). The changing Filipino rice farmer. Pp. 87–171 in *Rural Development and Employment* (Ed. C. Gotsch). Ford Foundation Seminar, Ibadan, Nigeria: 774 pp.

Castillo, Gelia, T. (1975). *All in a Grain of Rice*. Southeast Regional Center for Graduate Study and Research in Agriculture: xii + 410 pp.

Chang, C. W. (1962). *Increasing Food Production Through Education, Research and Extension*. Food and Agriculture Organization of the UN, Rome: vi + 78 pp., illustr.

Chant, D. A. (1966). Integrated control systems. Pp. 193–218 in *Scientific Aspects of Pest Control: J. Wash. Acad. Sci.*: National Academy of Science, Washington, xi + 470 pp., illustr.

Christensen, R. P. & Stevens, R. D. (1962). Putting science to work to improve world agriculture. Pp. 158–76 in *Food: One Tool in International Economic Development* (Ed. E. O. Hardolsen). Iowa State University Press, Ames, Iowa; x + 419 pp., illustr.

Clayton, E. S. (1960). Labour use and farm planning in Kenya. *Emp. J. Exp. Agric*., **28**, 83–93.

Cline, M. G. (1971). Agricultural research and technology response. Pp. 90–100 in *Behavioural Change in Agriculture* (Ed. J. P. Leagans & C. P. Loomis). Cornell University Press, Ithala, N.Y.: xii + 506 pp.

Couston, J. W. (1967). Physical and economic summary of trial and demonstration results. *FFHC Fertilizer Program 1961/62–1964/65*. Food and Agriculture Organization of the UN, Rome.

Cutie, J. T. (1975). *Diffusion of Hybrid Corn Technology: The Case of El Salvador*. CIMMYT, Mexico City.

Dalrymple, D. G. (1969). *Technological Change in Agriculture: Effects and Implication for the Development Nations*. USDA Foreign Agricultural Service, Washington, DC: v + 82 pp.

Dalrymple, D. G. (1971). *Survey of Multiple Cropping in Less Developed Nations*. US Department of Agriculture, Foreign Economic Development Services: 108 pp.

Dalrymple, D. G. & Jones, W. I. (1973). Evaluating the green revolution. Paper presented at *a joint meeting of the American Association for the Advancement of Science and the Consejo Nacional de Ciencia y Technologia, Mexico City*: 86 pp., illustr. (mimeo gr.).

Dasgupta, B. (1977). *Agrarian Change and the New Technology in India*. UNRISD, United Nations, Geneva: xxvii + 408 pp.

Delvert, J. (1972). Remarques sur l'agriculture de l'Asie de la Mousson. Pp. 127–37 in *Etudes de Geographie Tropicales*. Mouton, Paris.

Desai, D. R., Patel, G. A. & Patel, R. J. (1970). Impact of modern farming technology on rural employment in Saurashtra. *Ind. J. Agric. Econ.,* **25** (3), 33–9.

Deutsch, K. W. (1971). Developmental change: some political aspects. Pp. 27–50 in *Behavioural Change in Agriculture* (Ed. J. P. Leagans & C. P. Loomis). Cornell University Press, Ithaca, New York: xii + 506 pp., illustr.

Diaz, H. (1974). An institutional analysis of a rural development project: the case of the Pueblo project in Mexico. *Ph.D. Thesis*, University of Wisconsin.

Douglas, E. (ed.) (1980). *Successful Seed Programs: A Planning and Management Guide*, International Agricultural Development Service, Westview Press, Boulder, Colorado.

Duckham, A. M. & Masefield, G. B. (1970). *Farming Systems of the World*. Chatto & Windus, London: xviii + 542 pp., illustr.

Dumont, R. (1966). *African Agricultural Development*. Food and Agriculture Organization of the UN, Rome.

Duroset, D. D. & Barton, G. T. (1960). Changing sources of farm output. Department of Agriculture, Agricultural Research Service, Washington, DC. *Product Research Report*, **2**, 3–12.

Eckert, J. B. (1971). *The Economies of Fertilizing Dwarf Wheats in Pakistan, Punjab,* Lahore (cited by Johnston & Kilby, 1971).

Efferson, J. M. (1969). *Observations on Current Developments on Rice Marketing in West Pakistan*. Department of Agriculture, Lahore (mimeogr.).

Efferson, N. (1972). Outlook for world rice production and trade. Pp. 127–42 in *Rice, Science and Man.* IRRI, Los Baños, Philippines.

Engelstad, O. P. & Russel, D. H. (1975). Fertilizers for use under tropical conditions. *Adv. Agron.* **27**, 175–208.

Ennis, W. B., Janson, L. L., Ellis, I. T. & Newson, C. D. (1967). Inputs for insecticides. Pp. 130–75 in *US President's Science Advisory Committee*, Washington, DC: Vol. III, xxii + 332 pp., illustr.

Falcoln, W. P. (1970). The green revolution: generations of problems. *Am. J. Agric. Econ.*, **52**, 698–710.

FAO (1962). *FAO Africa Survey Report on the Possibilities of African Rural Development in Relation to Economic and Social Growth*. Food and Agriculture Organization of the UN, Rome: 168 pp.

FAO (1967). *Fertilizers: Annual Revue of World Production, Consumption and Trade*. Food and Agricultural Organization of the UN, Rome: 205 pp., (mimeogr.)

FAO (1968). *The State of Food and Agriculture*. Food and Agriculture Organization of the UN, Rome: vii + 202 pp.

FAO (1969). *Indicative World Plan*. Food and Agriculture Organization of the UN, Rome: 672 pp.

FAO (1971). *Report of the Special Committee on Agrarian Reform*: *FAO Docum.* 71/22 (*mimeogr.*). Food and Agriculture Organization of the UN, Rome.

FAO (1975). *The State of Food and Agriculture, 1974*. Food and Agriculture Organization of the UN, Rome.

FAO (1976). *The State of Food and Agriculture, 1975: Agriculture Series No*. 1. Food and Agriculture Organization of the UN, Rome: vi + 150 pp.

FAO (1977). *The State of Food and Agriculture, 1976: Agriculture Series No*. 4. Food and Agriculture Organization of the UN, Rome: vi + 157 pp.

FAO Study Mission (1978). *Learning from China; A Report on Agriculture* and the Chinese People's Communes. Food & Agriculture Organization of the UN, Rome, viii + 112 pp.

FAO (1979). *The State of Food and Agriculture, 1978. Agriculture Series No*. 9. Food and Agriculture Organization of the UN, Rome.

Feistnitzer, W. P. & Kelley, A. F. (editors) (1978): *Improved Seed Production*, Food and Agriculture Organization of the UN, Rome.

Francis, C. A. (1979). Small farm cropping systems in the tropics. Pp. 318–48 in *Soil, Water and Crop Production* (Ed. D. W. Thomas & M. D. Thomas). Avi Publishing Co. Inc., Westport, Conn.: ix + 353 pp., illustr.

Gerhart, J. D. (1973). Management problems and rural development in Kenya. Pp. 196–217 in *Rural Development and Employment* (Ed. C. Gotsch). Ford Foundation Seminar, Ibadan, Nigeria: 774 pp., (mimeogr.).

Giles, G. W. (1975). *The Reorientation of Agricultural Mechanization for the Developing Countries*, Shin-Noriasha Co., Tokyo.

Gotsch, C. (1973). Economics, institutions and employment generation in rural areas. Pp. 5–38 in *Rural Development and Employment* (Ed. C. Gotsch). Ford Foundation Seminar, Ibadan, Nigeria: 774 pp., (mimeogr.).

Government of the Punjab (1970). *Fertilizer and Mexican Wheat Survey Report: Fourth Plan Economic Research Project, Punjab*. Government of the Punjab.

Greenland, D. T. (1975). Bringing the green revolution to the shifting cultivator. *Science*, **190**, 841–4.

Griffin, K. (1973). Policy options for rural development, Pp. 18–79 in *Rural Development and Employment* (Ed. C. Gotsch). Ford Foundation Seminar, Ibadan, Nigeria: 774 pp., (mimeogr.).

Gundy, S. D. Van & Luc, M. (1979). Diseases and nematode pests in semi-arid West Africa. Pp. 257–265 in: *Agriculture in Semi-Arid Environments* (Eds. A. E.

Hall, G. H. Cannell, & H. W. Lawton). Springer Verlag, Berlin, Heidelberg, New York. xvi + 340 pp., illustr.

Ho, Yi-Min (1966). *Agricultural Development of Taiwan, 1930—1968*. Vanderbilt University Press.

Hopper, W. D. (1968). *Strategy for the Conquest of Hunger*. Rockefeller Foundation, New York: 131 pp.

Instituto Mexicano de Cafe (1965). *Tecnificacion de la Caficultura y Diversificacion de Cultivos*. Instituto Mexicano de Cafe, Mexico.

IRRI (1967). *Annual Report*. International Rice Institute, Los Baños, Philippines.

IRRI (1975). *Changes in Rice Farming in Selected Areas of Asia*. International Rice Research Institute, Los Baños, Philippines: 377 pp., illustr.

IRRI (1978). *Economic Consequences of the New Rice Technology*. International Rice Research Institute, Los Baños, Philippines: 402 pp., illustr.

Isom, W. H. & Worker, G. F. (1979). Crop management in semi-arid environments. Pp. 200–233 in: *Agriculture in Semi-Arid Environments* (Eds. A. E. Hall, G. H. Cannell, & H. W. Lawton), Springer Verlag, Berlin, Heidelberg, New York. xvi + 340 pp., illustr.

Johl, S. S. (1975). Gains of the green revolution: how they have been shared in the Punjab. *J. Develop. Stud.* **11**, 178–89.

Johnston, B. F. (1962). Agricultural development and economic transformation. *Food Res. Inst. Stud.* **3**, 223–76.

Johnston, B. F. & Cownie, J. (1969). The seed fertilizer revolution and labor force absorption. *Am. Econom. Rev.*, **59**, 569–81.

Johnston, B. F. & Kilby, P. (1973). *Agricultural Strategies, Rural-Urban Interaction and the Expansion of Income Opportunities*. OECD Development Center, Paris: v + 279 pp., illustr.

Jones, G. (1971). *The Role of Science and Technology in Developing Countries*. Oxford University Press, London: xvii + 174 pp., illustr.

Joy, L. J. (1969). Diagnosis, prediction and policy formulation. Pp. 376–81 in *Subsistence Agriculture and Economic Development* (Ed. C. R. Wharton, Jr). Aldine Publishing Co., Chicago, Ill.: xiii + 481 pp.

Joyce, R. J. (1955). Cotton spraying in the Sudan Gezira. II. Entomological problems arising from spraying. *Pl. Prot. Bull., FAO*, **3**, 97–103.

Kahlon, A. S. & Singh, G. (1973). *Social and Economic Implications of Large Scale Introduction of High-Yielding Varieties of Wheat in the Punjab with Special Reference to the Ferozepur District*. Punjab Agricultural University, Department of Economics and Sociology, Ludhiana, India.

Klingman, C. (1961). *Weed Control as a Science*. Wiley, New York: viii + 421 pp.

Kuo, L. T. C. (1972). *The Technical Transformation of Agriculture in Communist China*. Praeger, New York: xx + 266 pp., illustr.

Labrada, R. (1975). Some aspects of the incidence of weeds in Cuba. *Pest. Artic. News Summ.*, **21**, 308–12.

Ladejinsky, W. (1970). Ironies of India's green revolution. *Foreign Affairs*, **38**, 758–68.

Lai, Wong-Choch (1972). Current problems of farm management on mechanized farms. Pp. 162–90 in *Farm Mechanization in East Asia* (Ed. H. Southworth). The Agricultural Development Council, New York: 433 pp., illustr.

Lele, Uma K. (1975). *The Design of Rural Development—Lessons from Africa*, The Johns Hopkins Press, Baltimore and London. xiii + 246 pp.

Lele, Uma J. & Mellor, J. W. (1972). *Jobs, Poverty and the 'Green Revolution'*. A/D/C Reprint, The Agricultural Development Council, New York: 7 pp.

Leurquin, P. (1966). Cotton growing in Colombia: achievements and uncertainties. *Food Res. Inst. Studies*, No. 2.

Lewis, H. T. (1971). *Elocano Rice Farmers: A Comparative Study of Two Philippino Barrios*. University of Hawaii Press, Honolulu.

McPherson, W. W. & Johnston, B. F. (1967). Distinctive features of agricultural development in the tropics. Pp. 184–230 in *Agricultural Development and Economic Growth* (Ed. H.M. Southworth & B. F. Johnston). Cornell University Press, Ithaca, New York: 608 pp., illustr.

Menzies, J. D. (1967). Plant diseases related to irrigation. Pp. 1058–64 in *Irrigation of Agricultural Lands* (Ed. R. M. Hagan, R. H. Haise & T. W. Edminster). *American Society of Agronomy, Madison, Wisconsin: xxi* + 1180 pp., illustr.

Millikan, M. F. & Hapgood, D. (1967). *No Easy Harvest*, Little, Brown & Co., Boston, Mass.: xiv + 178 pp., illustr.

Ministry of Economic Planning (1961). *Notes on Eastern Nigeria Oil Palm*. Grove Rehabilitation Scheme, Enuga, Nigeria (mimeogr.).

Moorti, T. V. (1971). A comparative study of well irrigation in Aligarh District, India. *Cornell International Agricultural Development, Bull*. 19.

Nelson, M. (1973), *The Development of Tropical Lands: Policy Issues in Latin America*. Johns Hopkins University Press, Baltimore, Md: xvii + 306 pp., illustr.

Norman, D. W. (1970). Traditional Agricultural Systems and Their Improvement. Paper presented at a *Sem. on Agronomie Research in West Africa, University of Ibadan, Nigeria.*

OECD (1967). *Supply and Demand Prospects for Chemical Fertilizers in the Developing Countries*. OECD, Paris: 206 pp., illustr.

Ojala, E. M. (1972). International trade in agricultural products. *Food Res. Inst. Stud. in Agric. Economics, Trade and Development,* **11**, 111–28.

Olson, R. A. (1968). The fertilizer programme of the Freedom from Hunger Campaign, Case Study 13a. *Int. Semi. on Change in Agriculture, Reading*: (mimeogr.), 6 pp.

Ordish, O. (1967). *Biological Methods in Crop Pest Control*. Constable, London: 242 pp., illustr.

Otieno, J., Muchiri, G. & Johnston, B. F. (1975). The implications of Kenya's present economic structure and the rates of growth of the population and labour force. Pp. 90–103 in *Farming Equipment Innovation for Agricultural Development and Rural Industrialization* (Ed. S. B. Westley, & B. F. Johnston). Institute for Development Studies, University of Nairobi, Kenya: 238 pp.

Painter, R. H. (1958). Resistance of plants to insects. *A. Rev. Ent.*, **3**, 267–90.

Palmer, Ingrid (1972a). Food and the new agricultural technology. *United Nations Research Institute for Social Development, Geneva, Report No*. 72.9: 85 pp.

Palmer, Ingrid (1972b). Science and Agricultural Production. *United Nations Research Institute for Social Development, Geneva, Report No*. 72.8: 100 pp.

Pearse, A. (1980). *Seeds of Plenty, Seeds of Want. Social and Economic Implications of the Green Revolution*. Clarendon Press, Oxford. xi + 262 pp.

Peter, von. A. (1978). The economics of fertilizer use and fertilizer resources. Pp. 479–99 in *Potassium Research—Review and Trends*. International Potash Institute, Berne: 499 pp., illustr.

Punjab Agricultural University (1973). *Enterprise Budgets*. Ludhiana, Punjab, India.

Rao, M. R. & Willey, R. W. (1980). Evaluation of yield stability in intercropping: studies on sorghum/pigeonpea. *Expl. Agric*., **16**, 105–116.

Sanchez, F. F. (1975). Rodents affecting food supplies in developing countries: problems and needs. *FAO Plant Protection Bull. No*. 23 (3/4), 96–102.

Shaw, R. D'A. (1971). *Jobs and Agricultural Development—A Study of the Effects of a New Agricultural Technology on Employment in Poor Nations*. Overseas Development Council, Washington, DC: 74 pp.

Smith, G. W. (1963). The relation between soil moisture and incidence of diseases in cacao. *J. Appl. Meteor*., **2**, 614–18.

Soni, R. N. (1970). The recent agricultural revolution and agricultural labour. *Ind. J. Agric. Econ.*, **25** (3), 23–9.

Sooksathan, I. & Harwood, R. R. (1976). A Comparative Growth Analysis of Intercrop and Monoculture Planting of Rice and Corn. Paper presented at *Sem. International Rice Research Institute, Los Baños, Philippines*: (mimeogr.).

Sriswasdilek, J., Adulandhaya, K. & Isvilanonda, S. (1975). Don Chedi, Suphan Buri. Pp. 243–63 in *Changes in Rice Farming in Selected Areas of Asia*. International Rice Research Institute, Los Baños, Philippines: 377 pp., illustr.

Suryatna, E. S. & Harwood, R. R. (1976). Nutrient uptake of two traditional intercrop combinations and insect and disease incidence in three intercrop combinations. Paper presented at *Sem., International Rice Research Institute, Los Baños, Philippines*: (mimeogr).

Syarifuddin, A., Suryatna, E. S., Ismail, I. G. & McIntosh, J. L. (1974). Performance of corn, peanut, mung bean and soybean in monoculture and intercrop combinations of corn and legumes in dry season 1973. *Contrib. Cent. Res. Inst. Agric. Bogor, Indonesia*, **12**, 1–13.

Tagamupay-Castillo, G. (1970). Impact of agricultural innovation on patterns of rural life (focus on The Philippines). Pp. 13–52 in *Agricultural Revolution in Southeast Asia,* Vol. II.: *Rep. 2nd SEADAG Int. Conf. on Development in Southeast Asia* (Ed. A. Russell), The Asia Society, New York.

Tamin, M, bin., & Mustapha, N. H. (1975). Kelantan, West Malaysia. Pp. 210–23 in *Changes in Rice Farming in Selected Areas of Asia*. International Rice Research Institute, Los Baños, Philippines: 377 pp., illustr.

Tan, E.K. (1975). Pigcawayan, Cotabato. Pp. 325–45 in *Changes in Rice Farming in Selected Areas of Asia*. International Rice Research Institute, Los Baños, Philippines: 377 pp., illustr.

Tribe, D. (1970). Animal ecology, animal husbandry and effective wildlife management. Pp. 123–41 in *Use and Conservation of the Biosphere*. UNESCO, Paris: 272 pp.

Tschiersch, J. E. (1978). *Appropriate Mechanization for Small Farmers in Developing Countries*. Publications of the Research Centre for International Agrarian Development. Verlag Breitenbach, Saarbrücken, Germany, iii + 106 pp.

United Nations (1973). *Implementation of the International Development Strategy*, Vol. I. United Nations, New York.

United Nations (1974). *The World Food Problem: Proposals for National and International Action*. UN World Food Conference, Rome.

Venezian, E. L. & Gamble, W. (1969). *The Agricultural Development of Mexico: Its Structure and Growth Since 1950*. Praeger, New York: xxii + 281 pp., illustr.

Watters, R. F. (1971). *Shifting Cultivation in Latin America*. Food and Agriculture Organization of the UN, Rome: 305 pp., illustr.

Wellhausen, E. J. (1970). The urgency of accelerating production on small farms. Pp. 5–9 in *Strategies for Increasing Agricultural Production on Small Holdings* (Ed. D. T. Myren). Pueblo, Mexico: 86 pp., illustr.

Wharton, C. R., Jr (1969). Risk, uncertainty and the subsistence farmer. *Development Digest*, **7** (2), 3–10.

Wilde, J. C. de, & McLoughlin, P. E. M. (1967). *Experience with Agricultural Development in Tropical Africa*. Johns Hopkins Press, Baltimore, Md: Vol. 1, xi + 184 pp.

Williams, M. S. & Couston, J. W. (1962). *Crop. Production Levels and Fertilizer Use*. FAO, Rome: 48 pp., illustr.

Yudelman, M, Butler, G., & Banerji, R. (1971). *Technological Change in Agriculture and Employment in Developing Countries: Development Centre Studies, Employment Series No*. 4. Development Centre of OECD, Paris: 204 pp.

CHAPTER IX

Human Labour, Animal Traction, and Mechanization

Distribution of Agricultural Power

The power invested in agriculture derives from three main sources: human, animal, and mechanical. The relative importance of these sources depends on environmental factors, tradition and the stage of modernization of agriculture (Table XXVI).

Whilst Asia derives about half its power from animals, Latin America is dependent on mechanization as its main power source, and Africa occupies an intermediary position. On the whole, tractors are already an important source of power in all the three continents in which most of the developing countries are found. Continental averages, however, may conceal real differences between countries and even within countries, as exemplified by a comparison of various districts in India (Table XXVII).

Table XXVI Distribution of agricultural power (Giles, 1975, by permission).

Region	Total HP/ha	Percentage of available power/ha		
		Human	Animal	Mechanical
Asia[a]	0.22	26	51	23
Africa	0.10	35	7	58
Latin America	0.25	9	20	71
Aggregate percentages		24	26	50

[a]Excluding the People's Republic of China.

The Ludhiana district already uses almost four times more power per hectare in agricultural production than the Indian overall average; of this amount only 4% is derived from labour whilst mechanization already provides a surprising 74%. The West Godavari district, on the other hand, though it has a somewhat higher power consumption than the overall Indian average, has almost the same power distribution pattern.

Table XXVII Distribution of agricultural power in contrasting districts in India (Giles, 1975, by permission).

Area	Total HP/ha	Percentage of available power/ha		
		Human	Animal	Mechanical
Ludhiana district	0.82	4	22	74
West Godavari district	0.40	20	60	20
India	0.23	26	62	12

Human Power and Equipment

It is estimated that more than half the cultivated area in the world is still tilled by hand tools and animal draught equipment (Hülst, 1975). In many tropical regions, the main agricultural tools to this day are the hoe and the machete.

Hand-digging consumes human energy at the rate of about 750 kcal per hour. It is estimated that the energy output of a man working in a tropical climate is not more than 1500 kcal per day. Therefore, active digging cannot be carried out for more than two hours every day.

With only a hand hoe the size of a holding that can be prepared for planting by a man is about 0.5 hectares, a figure that has been found to be remarkably consistent across the continent of Africa (Boshoff & Minto, 1975) (Fig. 47).

Fig. 47 The hoe: basic hand-tool of traditional agriculture. The two contrasting types, long-handled and short-handled, used by different peoples. Drawn by D. Arnon

In monsoon Asia, the bulk of farmers use equipment that has remained unchanged for centuries. The principal instrument for ploughing, harrowing, puddling, sowing and intercultivation is the wooden 'stick plough' of antiquity. Other implements are the hoe, levelling plank, cart, Persian wheel, sickle, wooden threshing board, etc.; all tools produced by the farmers themselves or by local artisans. The lack of suitable implements is not only a source of back-breaking labour, and reduces the areas which could be effectively cultivated; it also adversely affects yields, because essential agricultural operations are either delayed or not well executed.

In order to prepare a seedbed using a hoe the farmer must wait until the sun-baked ground has been softened by rain; however, the onset of the rains is the most favourable time for sowing, so that the hand-hoe farmer invariably misses the optimum time for planting. Lacking suitable equipment, he also cannot cope with workloads at other peak periods and as a result seedbed preparation, sowing and weeding are generally slipshod. To a lesser extent, this is also true for animal draught with primitive equipment. There is little place in this system for machinery, which can only be introduced after a complete change in the land-use system. Moreover, inefficient tools and equipment cause a waste of effort, result in poor quality of work and make agricultural work more tedious and strenuous than it has to be. Therefore, at first, much can be done to increase the productivity of labour by the improvement of simple traditional tools and implements and the partial replacement of manual labour by animal draught. For some regions this alone may constitute an important technological improvement.

Replacing the scrap metal of which many hand-tools are made at present by properly tempered high-carbon steel, which holds a cutting edge, makes possible a more effective use of human labour, by reducing fatigue and increasing output. Similar results can be obtained by improving balance, manipulation and weight of hand-tools, (Kline *et al*., 1969).

To prepare a reasonably weed-free seed-bed with the native ox-drawn wooden plough, the Ethiopian farmer must plough the land three to four times. A considerable amount of time can be saved by replacing the wooden tip by a steel share, which is far more efficient in preparing the seed-bed.

Replacement of steel-rimmed by rubber-tyred wheels, of the traditional Indonesian rice harvesting knife by the sickle, and of sickles by scythes, for example, are major technical improvements.

Simple, inexpensive, hand-operated planters for sowing maize, beans, sorghum, etc., have been developed and are already widely used. They are suitable for land that has not been cleared of stumps or stones; they eliminate stooping and reduce fatigue. The development of simple drills in order to change from broadcast to row planting in certain crops has manifest advantages. Sowing machines make possible economies of seed and can increase yield by ensuring a full plant population as a result of uniform distribution and placement of the seed. (Fig. 48) In India, drill sowing of Ragi millet increased yield by approximately 20% as compared with broadcast sowing (Patil, 1963). Equipment for plant protection may be the only possible way of large-scale control of diseases and pests (cf. p. 324).

At the Allahabad Agricultural Institute, the output of a pair of bullocks was increased by 60% by the use of an improved harness, replacing the conventional wooden yoke (Giles, 1963). Multipurpose tool-frames or tool-bars have been developed for ox-drawn equipment, to which attachments for tillage, weeding, seeding, cultivating, ridging, and transport can be added (Figs. 49 and 50).

Groundnut lifters have been developed, either as single-purpose tool-frames or tool-bars.

Improved types of equipment are available for post-harvest operations: improved hand-operated or small-engine powered threshers can replace slow and tedious hand threshing (Fig. 51). A simple hand or foot operated winnowing fan can produce a cleaner product than wind-winnowing in a fraction of the time. Simple hand implements can increase efficiency of weed-control (Fig. 52) and fertilizer application (Fig. 53).

Tschiersch (1978) presents detailed information on the availability of

Fig. 48 The IRRI multi-hopper seeder. A light-weight, compact machine, made entirely of local materials, pulled by the operator, and capable of sowing pregerminated seeds at the rate of 7 man-hours per ha. By courtesy of the Agricultural Engineering Department, the International Rice Research Institute, Los Baños.

Fig. 49 Bullock-drawn tool carrier with ridger and floats used in establishing broad ridges and furrows 150 cm apart. The graded furrows increase infiltration, and reduce erosion and also provide drainage during prolonged rainy periods.

Fig. 50 The carrier of Figure 49 with attachments for cultivating crop (sorghum) planted on 75 cm ridges. By courtesy of the International Crops Research Institute for the Semi-Arid Tropic (Hyderabad).

Fig. 51 IRRI multi-crop axial flow thresher: powered by a 7 HP engine, with three men needed to feed, thresh, and bag the grain. Threshes paddy, sorghum, soybeans, and other small grains; output: one ton of paddy per hour. Easily moved behind small hand-tractor or jeep. By courtesy of the Agricultural Engineering Department, International Rice Research Institute, Los Baños.

improved tools, equipment, and agricultural machinery; their advantages and drawbacks and their suitability for small farms, and their manufacture in developing countries.

The main obstacle to the wide-spread use of even the simplest types of improved equipment is that the cost is still too high for the majority of small farmers, who have neither the cash nor access to credit to buy such equipment.

Animal Traction

Animal draught power is widely used in Asia and Latin America and in recent years has increased considerably in the savanna zones of tropical Africa. Factors that have inhibited the spread of animal traction in other parts of Africa are the prevalence of the tse-tse fly in wide areas and the general separation of arable cropping and animal husbandry, so that many farmers are not familiar with tending farm animals.

Fig. 52 IRRI push-type hand weeder: when the implement is pushed between the rows, the spokes press the weeds under the soil. Output: 35 to 75 man-hours per ha. The implement is already widely used in many Asian countries. By courtesy of the Agricultural Engineering Department of the International Rice Research Institute, Los Baños.

In the latter case, the basic problem facing planners is whether to adopt animal traction as a first step to mechanization, or whether to pass directly from the most primitive equipment to modern mechanization.

Advantages

Before considering this problem, it is necessary to point out that there are circumstances under which hand cultivation will continue to be necessary: where topography is too steep, holdings too small, or complete clear-

Fig. 53 A granular fertilizer and insecticide applicator, which applies the chemicals at a depth of 5–7 cm. Its use can double fertilizer use efficiency, compared to conventional broadcast methods. By courtesy of the Agricultural Engineering Department, the International Rice Research Institute, Los Baños, Philippines.

ance of forest too expensive, to justify the use of animal or tractor-drawn equipment. The need for mechanization will also be less in evidence for perennial crops, such as coffee, or where ample labour is available. The introduction of animal traction as a first stage in development has a number of important advantages:

(a) It is more appropriate to the level of understanding of the majority of farmers in developing areas.
(b) No foreign exchange is required for the purchase of animals or for most of the necessary equipment
(c) The capital investment required is not considerable, and within the reach of many farmers. Running costs are relatively low, and repairs are not expensive.
(d) Draught animals can be sold without loss, after a number of years of service.
(e) Animal-drawn equipment is adapted to relatively small holdings.
(f) Indirectly, crop production may benefit as a result of the manure produced, the need to include forage crops in the rotation, better and more timely tillage, etc.

Replacing hand labour in land preparation by animal power can result in a quite considerable extension of the cultivated area; this has been known to be the case in Gambia, in which farmers trained in the use of oxen were able to expand groundnut production by 20–40% (Peacock, 1967), and in certain districts in Tanzania where the area under cultivation was increased by 80% (Collinson, 1965). The increased areas under cultivation provide more employment for weeding and harvesting. The local manufacture of farm implements, such as carts, ploughs, harrows, etc., also generate employment.

Disadvantages

However, there are also a number of disadvantages:

(a) In most parts of the world, the area a draught animal can till with a primitive plough is much smaller than the area needed to feed it on natural grazing. Therefore, a considerable area of land must be devoted to grazing, and forage must be grown for feeding during the dry season.
(b) The animals are generally half-starved on overgrazed areas during the periods of the year when they are doing little or no work. Because of low feeding standards, the animals have a low work output, and mortality is high especially when they are subjected to great effort during the peak season.
(c) The African bush savanna of moderately high rainfall is infested by the tse-tse fly, which makes the keeping of cattle impossible or limits it to dwarf breeds, unless extensive bush clearing and other control methods are resorted to.
(d) Labour bottlenecks may be created, in particular, in weeding. This can at best be overcome partly by adapting appropriate animal-drawn weeding implements.
(e) The animals cannot plough heavy or hard soils, in advance of the rains, so that sowing may be considerably delayed.

In relation point (e), Giles (1963) gives the following typical example. In the Raipur district in India all the draught animals, employed in pulling country ploughs, would require 56 days to prepare the rain-fed paddy fields of the district properly. The time available for satisfactorily ploughing these fields is, however, only about 10 to 15 days. The result, therefore, is late and poorly prepared paddy fields with an adverse effect on yields.

Problems of Adoption

The introduction of draught animals, mainly oxen, in Africa, has been successful in some regions, and a disappointment in others, even in regions in which the tse-tse is not found. Conditions for the introduction of draught animals are most favourable in the tropical uplands and the zones of transition between rainforest and dry savanna. The supply of forage in these regions is adequate and cropping patterns enable animal draught power to be used relatively efficiently.

In regions with a very short wet season, such as are found in certain monsoonal climates, it may be essential to use draught animals, whatever the costs, to get the land prepared in time. However, this also implies that it will not be possible to employ the draught animals usefully throughout the year. Because of the relatively short periods of peak requirements, the capacity of the draught animals may be utilized at no more than 25–40%.

In the drier areas, pastures are very sparse, and the period during which the land must be prepared for sowing is very short (Shaeffer–Kehnert, 1973). Therefore, as pointed out above, the animals are at the lowest state of nutrition at the start of the rainy season, when they are most needed for relatively heavy work. Improvements in their nutritional level would entail relatively considerable increases in production costs.

Animal traction has been successful in particular with farmers who have some livestock tradition, who have sufficient grazing land, and where there are no basic limiting factors, such as the tse-tse fly.

The successful promotion of animal traction to replace hand-tools has generally entailed an intensive and sustained government programme including: ox-training, farmer training, demonstration and provision of animal-drawn equipment, and supply of credit (Kline *et al*., 1969).

General programmes for rural vocational training of village artisans in the servicing and repair of animal-drawn equipment, and training courses for extension and sales personnel, have also been adopted in a number of African countries promoting the use of draught animals (Tschiersch, 1978).

MECHANIZATION

Justification

The tractor is frequently considered to be the symbol of agricultural development, and yet agricultural progress does not necessarily depend on immediate mechanization.

It is true that agriculture based on abundant cheap labour doing back-breaking work perpetuates a low standard of living. It has been estimated that a power level of approximately 0.5 horsepower per hectare is needed for an efficient agriculture (Giles, 1967) (Fig. 54). Africa, for example, averages only 0.05 HP/ha, or only one-tenth of the minimum requirement for efficient agricultural production (Kline *et al*., 1969).

Sooner or later, therefore, mechanization is inevitable if productivity of labour in agriculture is to keep pace with other sectors, and production per hectare is to keep pace with population increase. Premature and unselec-

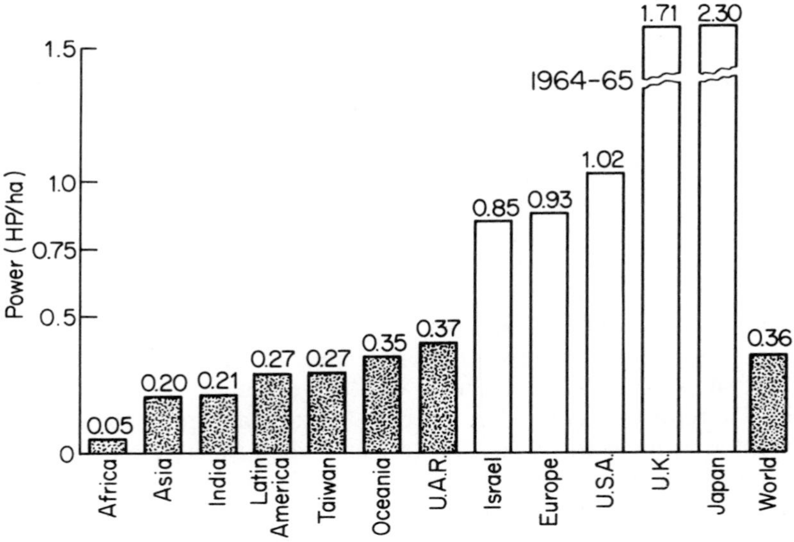

Fig. 54 Power available for agricultural field production in world regions and selected countries (Giles, 1967). By courtesy of the author.

tive mechanization, however, is not only socially wrong but also can cause considerable damage when practised indiscriminately. Introduced too rapidly, without proper planning, it has frequently caused severe losses both to individual farmers and to the national economy. The most spectacular failures on certain large mechanization schemes are well known.

Agricultural machinery requires relatively large capital investments. The main justification for mechanization in developed countries is to save labour. Outlays on farm equipment represent the second largest expenditure for purchased inputs by farmers modernizing their agriculture (FAO, 1969). In most underdeveloped countries, capital is the scarce resource and labour is in ample supply, at least in their early stages of development.

However, many developing countries are faced with the tragic paradox that despite the conditions of chronic underemployment generally characteristic of subsistence agriculture, there is considerable economic justification for mechanizing agriculture on large farms, for the following reasons (Dandekar, 1969):

(a) mechanization provides a channel for productive investment for farmers' savings in this sector;
(b) poor and unskilled hand labour available with low production efficiency because of malnutrition, is an unproductive substitute for mechanization;
(c) it reduces the dependence of this sector on workers who are dissatisfied with their lot and whose latent hostility may be a source of trouble.

These tendencies have been strengthened by the green revolution. Though new productive varieties increase labour requirements markedly on small farms, the increased income may serve as an incentive to replace labour by machinery, and large landowners usually have sufficient capital to invest in mechanization, if this is in their interest.

Notwithstanding factors mitigating against indiscriminate mechanization, the latter has been encouraged by governments in many developing countries for reasons of prestige, mistaken policies (cf. p. 438) and tied-aid policies of certain 'donor' nations (Eicher *et al.*, 1970). If distortions due to government subsidized capital and foreign exchange are eliminated, the incentive to mechanize indiscriminately would be considerably reduced.

Selective Mechanization

In certain cases, selective mechanization may be fully justified, even at an early stage of development, and family farms may remain basically labour-intensive, even when one or more operations are mechanized.

(a) *As a Means of Rapidly Increasing the Area under Cultivation*

Mechanization may be an essential complement to irrigation for developing large tracts of arid land.

An example is the rapid expansion of cotton and wheat production in northern Mexico since World War II by large-scale irrigation development, which has accounted for a substantial part of the phenomenal growth of agricultural production in Mexico, and was made possible by mechanized large-scale clearing of land, levelling, etc.

Selective mechanization may be an important component in the success of large-scale settlement plans.

One of the reasons for the success of the Mwea settlement project in Kenya is imputed to the replacement of animal-draught by selective mechanization on a large scale. 'The introduction of machinery has made possible a degree of planning, discipline and extension unthought of in the past. Since mechanization started, it has been possible to draw up a detailed timetable of operations for each area and to adhere to it throughout the cropping season. The probability of large proportions of the acreage slipping out of season with adverse effects on yields has been obviated. The orderly progression of cultural operations, made possible by mechanization of puddling, has produced an atmosphere in which strict discipline can be enforced without opposition' (Giglioli, 1965). It should be stressed that only land preparation was mechanized—preparing nurseries, transplanting, weeding, crop protection and harvesting were carried out by traditional methods.

Where heavy investments are made in infrastructure for the development of land under forests, the economic justification of the project may depend on bringing the land in question into production as rapidly as possible. This can be achieved by replacing the present systems of cutting and burning by mechanized clearing. The use of chainsaws, tractors fitted with bulldozer plates to clear second growth, and tree-crushing machines are some of the possible alternatives to burning that can reduce the time required, lower costs and enable the exploitation of the timber (Nelson, 1973).

These advantages of mechanical clearing must be weighed against its drawbacks.

Apart from the high cost of operating and maintaining machinery in tropical rainforests, the clearing of large areas within a short time involves considerable danger of erosion and makes it almost impossible for farmers to cope with vigorous jungle re-growth and other management problems. By contrast, clearing small areas at a time, by traditional slash-and-burn methods with hand labour, produces higher yields (the ash providing plant nutrients) at a lower cost. The transition from shifting to permanent cultivation is gradual and involves less risks (Sanchez & Buol, 1975).

Mechanization may be essential as a means to change over from an extensive livestock grazing system to intensive crop production on medium-sized and large farms. Such a change has taken place in the Rio Gran de Sul (Brazil) on farms ranging in size from 100 to 1000 hectares. Despite mechanization, the change from livestock to crop production resulted in a 50% increase in labour utilization. Capital investment was more than doubled; gross output on mechanized crop farms was six times higher than on traditional livestock farms and net income was four times higher. The transition was made possible by substantial amounts of credit made available to farmers by government (Rask, 1969).

(b) *Timeliness of Operations*

At first sight, the statement: 'It is not necessarily inefficient for the cultivator in India to plough his field half a dozen times with a "stick-plough" while a moldboard plough would do it in one operation and save labour, because the farmer would have no alternative use for his labour' (Heady, 1962) may appear quite logical. This, however, does not always hold true. High yields may depend on the timeliness with which certain operations are carried out. In particular, sowing dates, enabling the most effective use to be made of rainfall, may be critical for crops in certain regions.

Experimental results at Samaru in northern Nigeria show that timely planting of food- and cash-crops is not only the biggest single factor in determining yields, but is also essential if the full benefits of other improvements, such as better varieties or fertilizer care, are to be realized (FAO, 1966).

Land preparation for sowing in time may be possible on large areas only if tillage can be mechanized, because only tractor-drawn equipment is capable of tilling heavy, sun-baked soils before the onset of the rains and completing the sowing of large areas within relatively short periods.

Since with animal power only 0.2–0.3 ha per day can be ploughed, this would severely curtail the amount of land to be cultivated. With the introduction of new HYVs and the concomitant use of expensive inputs such as improved seeds, fertilizers, pesticides, etc., timeliness of operation acquires increased importance in view of the high investments involved.

Finally, timeliness in land preparation and sowing may be a decisive factor enabling multiple-cropping.

(c) *Quality of Work*

A quality of work may be achieved by mechanized tillage that is not possible with animal-draught, and certainly not with manual labour. For example, in the Near East, ploughing with the traditional plough only scratches the soil surface. As a result, the wheatfly (*Syringopais temperatella*) causes enormous damage and reduces yields considerably. Mechanical tillage, which turns the soil to a somewhat greater depth, is fully effective in destroying the larvae of the pest. Only tractor-drawn ploughs can break heavy sods effectively and turn under large amounts of vegetation.

(d) *Increasing Work Opportunities*

The assumption that mechanization of agriculture always displaces labour is not correct. In those cases in which the use of machinery increased yields, there was also a need for additional labour. Where mechanization by speeding land preparation and other operations, facilitates double-cropping, the effect is to increase the demand for labour. Double-cropping in rice production has increased labour input more than any other single factor (Castillo, 1975). Mechanized pumping can make water so much cheaper that intensive cropping becomes economically justified.

In a study in India, the replacement of the Persian wheel (powered by draught animals, with its low efficiency and capacity) by diesel and electric-powered pumps was found to reduce the cost of pumping in a given situation by over 80% (Balis, 1968). This has led to a tremendous increase in the use of ground-water for irrigation in India and Pakistan; the reduction in water costs made growing certain labour-intensive crops economically justified.

The use of tractors for transporting farm output may make relatively distant markets accessible and thereby encourage the growing of labour-intensive truck-crops or the establishment of dairy farming, etc., The hiring of a tractor for seedbed preparation instead of by animal draught can be well justified under certain circumstances. For example, mechanized seedbed preparation for rice can obviate the necessity of maintaining a team of bullocks throughout the year, thereby enabling a substantial saving and free land for crop production.

Mechanization, if adopted selectively, need not reduce employment opportunities in agriculture, as shown by the following example. In a study on the effects of the green revolution in the Ludhiana district in Punjab (Aggarwal, 1973), it was found that notwithstanding increased use of labour-saving machinery, the demand for agricultural labour remained practically constant. Tractors had replaced bullocks for seedbed preparation, planting, and other operations. Persian wheels were replaced by power-driven centrifugal pumps, threshing was being done by threshing machines, and yet the reduction in labour requirements was less than 1%. The reasons for this apparent paradox were: (a) the use of machinery had made possible an expansion of cultivated land by nearly 20%; (b) almost all the land was being double-cropped; and (c) because of the high cost of improved inputs (quality seed of HYVs, fertilizers, pesticides, etc.) the crops required and received more intensive care, such as more frequent irrigation, fertilizer applications, weeding and spraying (cf. p. 292).

(e) *Coping with Peak Periods of Labour Requirements*

Much of farming in the subtropical and tropical regions is associated with seasonal rainfall, and involves a peak requirement of labour of two to three month's duration. This labour bottleneck is one of the major constraints limiting the area farmed by a family unit, in regions in which land is not the limiting factor (Norman, 1970).

One of the main functions of selective mechanization is to reduce labour requirements during peak periods. As the family farm generally has to maintain a labour force adequate to cope with peak labour loads, the reduction in labour requirements is not just the amount saved in the peak period but that saved throughout the year. The labour so released may make it possible for the farmer to diversify his production. In Taiwan, for example, mechanization has made possible an increase in the area of such commercial crops as melons and vegetables grown after the first rice crop and before transplanting the second rice crop (Lai, 1972). On farms using hired labour, selective mechanization may alter the ratio of family labour to hired labour. In Taiwan, this ratio was found to change from 3:1 to 2:1 (Wu, 1972).

(f) *Social Factors*

Farmers' motivations for introducing machinery need not always be economic. Mechanization also has certain important social implications which should not be overlooked. In most developing countries, agriculture has low social status, and it is usually young people with initiative and some education who leave farming in favour of urban occupations. Through selective mechanization and other technological inputs, farming, instead of remaining an archaic occupation, can become a modern enterprise and this may influence young, capable people to stay in farming.

This, in turn, encourages the takeover of the management of the farm by the younger generation who are more amenable to adopting new practices.

In a survey of 150 farmers in Laguna district (Philippines) over 60% indicated that avoiding the physical and other problems involved in the care and feeding of the Carabou took precedence over economic returns in their decision to buy tractors (Barker *et al.*, 1969).

Similar results were reported in a study in Sri Lanka (Carr, 1975), in which many farmers reported that the increased leisure and better working conditions compensated for the higher costs involved in using a hired tractor rather than their own animal power.

Even when tractor-hire services are economically available to farmers, the latter may prefer to buy tractors or power tillers, either for reasons of prestige or because farm operations can be timely, and the tractor can be used for transport.

Mechanization can result in a considerable reduction in the drudgery involved in farm work. In north-eastern Japan, it is estimated that farmers become unable to operate animal-driven ploughs and weeders in the field by the time they are 44; mechanization can therefore actually lengthen their working lifespan (Dalrymple, 1969).

The status of part-time farmers also changes as a result of mechanization. Normally, farmers whose holdings are so small that they have to depend on non-farm employment to supplement their incomes, cannot cope efficiently with the work-load on the farm when using human or animal labour. In Japan, the adoption of power cultivators resulted in: (a) a considerable increase in the number of part-time farmers, because mechanization enabled them to devote less time to their farms and hence devote part of their time to off-farm work; and (b) an improvement in the quality and timeliness of their work on the farm. As a result the economic situation and the social status of part-time farmers were improved (Kawamoto, 1972).

(g) *As a Means of Intensification of Production*

The possibility of replacing one-crop-a-year production patterns by multiple-cropping is dependent on a shortening of the time involved in freeing the field from one crop and preparing it for the next one. This will usually depend on the possibility of mechanizing certain operations, such as harvesting, threshing, land preparation, etc. (see below).

The International Rice Research Institute has devised an intensive cropping system aimed at providing a diversity of food for a balanced diet and to increase production and income. It is based on growing five crops a year, but predicates the use of machinery suitable for small farms (Bradfield, 1971).

In Taiwan, which has a particularly acute shortage of land, mechanization of small farms has promoted multiple-cropping and diversification into pig and poultry farming. As a result, not only has farm output increased, but labour requirements are spread more equally over the year (FAO, 1969).

The Transition from Human and Animal Power to Mechanization

Because of selective mechanization, there is no clear-cut demarcation between an era of hand labour and animal traction, and an era of mechanization. In one development scheme in the Central African Republic, for example, tractors are used for heavy breaking, clearing and levelling of land; ox-drawn ploughs for planting and cultivating; and hand-work is employed for the remaining tasks (Kline *et al.*, 1969). As the economy invests more and more in industrial growth, which draws off farm labour, the proportion of farm workers to total population decreases, and the need to increase the productivity of the individual farm worker, and to raise the speed and precision of operations becomes progressively more acute. As a result, the need and justification for mechanization increases.

Japan is an excellent example of a gradual mechanization of farming which kept pace with a declining farm labour force. First, machinery was introduced for threshing, land levelling and polishing of rice. Then followed mechanization of tillage operations and of disease and pest control. Later, fertilizer applications and sowing were mechanized, and finally harvesting and transplanting machinery was adopted on a wide scale. As a result of this progressive mechanization, labour input in paddy-rice cultivation, which in 1958 amounted to 456 h/ha, declined in 7 years by about 22% (Morita, 1968).

The need for farming families to supplement their income by outside work has generally been considered as an indication of farmers' poverty (Kim, 1972). And indeed, as long as wages for non-farm work are low, the additional income from outside work can do no more than alleviate the harsh living conditions of the farming family. It is only when non-farm employment constitutes a significant part of the family's income—and this predicates an acceptable level of wages—that it provides the means to strengthen the economy of the family farm.

In Japan, for example, non-farm employment brings twice as large earnings as work on the farm. Therefore, the younger members of the farm family take jobs whenever possible in the rural non-agricultural sector. Because of the instability of non-farm employment, uncertain social security and inherent attachment to their farm land, they do not however abandon the family farm. Under these circumstances the introduction of farm machinery becomes necessary to make up for the outflow of farm labour, and the non-farm income makes the purchase of the machinery possible.

Because of the simultaneous occurrence of an increase in mechanization and a decrease in the labour force employed in agriculture, the causality involved in the two trends may be obscured (Yudelman *et al.*, 1971).

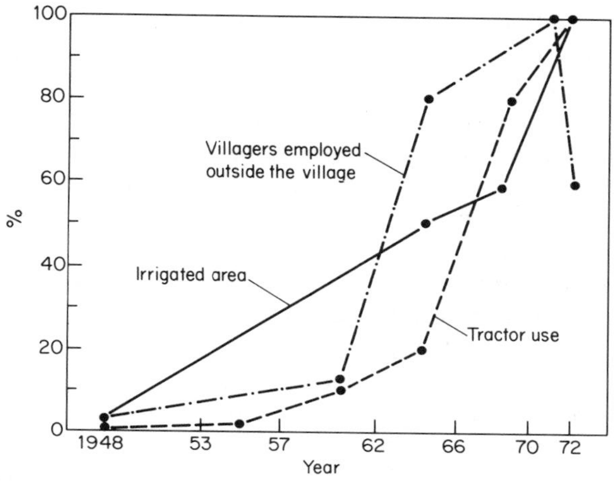

Fig. 55 Relationship between intensification of agriculture, employment outside the village, and use of tractors (Arnon *et al*., 1975). Note that when the irrigated areas reached their peak, the percentage of villagers seeking outside employment fell. All data in per cent of the situation in 1972.

However, it can be said that as a general rule, where employment opportunities outside agriculture become available to the rural population, mechanization is not labour-displacing but labour-replacing (Fig. 55).

Mechanization, in turn, becomes a catalyst in modernizing the agriculture of the farm. The use of farm machinery is relatively expensive and is only justified if productivity, and hence income, is improved by the use of inputs such as fertilizers, pesticides, etc. Therefore, non-farm income has played an important role in the modernization of the family farm in Japan.

The importance of integrating agricultural and industrial development within the rural areas that has already been stressed in relation to employment problems (see p. 150) is also relevant to farm mechanization opportunities. If farm equipment can be manufactured in the developing country itself, using locally available materials, mechanization can make an important contribution to rural industrial development (Tschiersch, 1978). This is exemplified by the contrasting approaches to industrialization adopted by Korea and Japan (Kim, 1972).

In Korea, industries have been mainly centralized in a few big cities and little development of the economy of the rural areas has occurred. The drawing-off of labour from the farms and the concomitant rise in rural wages has created the need for mechanization without farmers having the means to invest in machinery.

By contrast, in Japan, non-farm jobs have been made available mainly through de-centralization of industry and the establishment of factories in rural areas. In addition, an excellent transportation system makes it possible for members of farming families to commute daily to and from their working places, without having to migrate. As a result, in Japan as in Korea, the need for mechanization to replace workers of farm families engaged in non-farm occupations has arisen, but only in the former have the means necessary for purchasing the machinery become available.

Giles (1975) has proposed as an indicator for meaningful planning of mechanization the concept of 'horsepower per cultivated hectare'. The horsepower figures used for this purpose are an expression of the effective capacity of the three power sources used in agriculture: human, animal, and mechanical.

A programme of mechanization should be based 'on the kinds and amounts of power within economically available limits, consistent with maximizing crop production and labour utilization'.

With this concept in mind, Giles studied the relation between the effective horsepower per hectare (HP/ha) and the average aggregate yield of the major crops in 23 regions, countries, districts and farms, during the period 1964 to 1971. The results of this study are shown in Fig. 56.

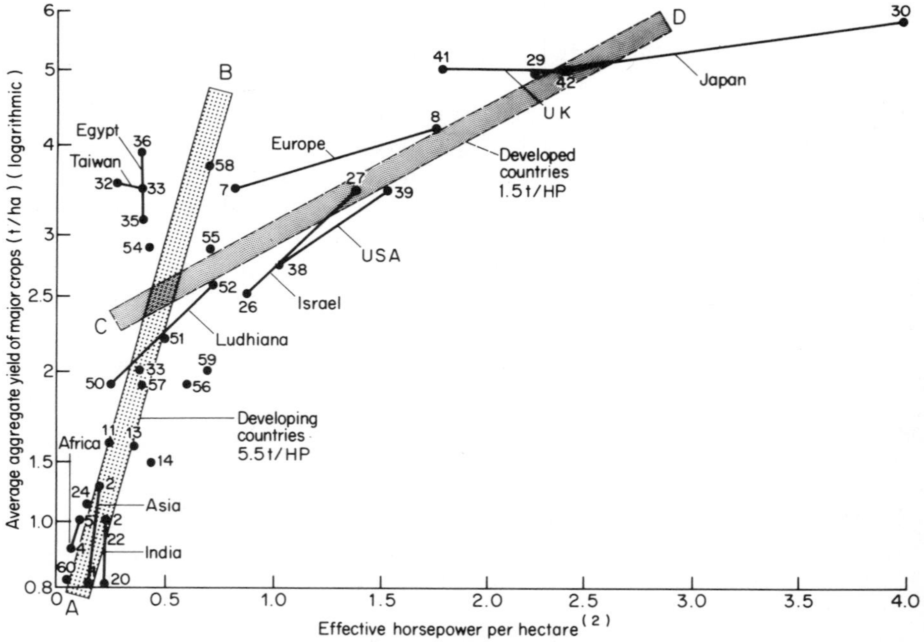

Fig. 56 Relationship between power-level and major food-crops (1964–71) (Giles, 1975). Agricultural areas have generally developed initially along lines A–B (in particular, in areas of high population density); at a later stage, along C–D. By courtesy of the author.

Agriculture has generally developed initially along the line AB, later along CD. The steep slope of AB is mainly due to the effects of inputs in addition to power, such as irrigation, fertilizers, etc. At around 2.5 ton/ha mechanization appears to become economically viable, and then proceeds rapidly along CD. The rate of replacement of human and animal power by machinery along CD is of course dependent on the rate of absorption of labour by economic sectors other than agriculture.

Whilst India as a whole progressed along the AB line during the period under review (points 20 to 21), the Ludhiana district (points 50 to 52), after reaching the hypothetical 2.5 ton/ha turning point, is advancing rapidly along CD.

Japan advanced and increased its HP/ha by rapid mechanization during a period of relatively cheap fossil fuel. Its extremely high input of machinery—4 HP/ha as compared to 1.55 HP/ha for the USA—is probably a reflection of the smallness of its farms and the type of specialized machinery developed for smallholders.

Relationship Between Use of Tractors and the Adoption of HYVs

The use of tractors by rice farmers both before and after adoption of HYVs, was investigated by the International Rice Research Institute, in a survey of 32 selected villages in various regions of Asia with a high potential for rice production. The results of this survey are shown in Fig. 57.

In the Punjab (Pakistan) about 70% of the farmers who subsequently adopted HYVs, were already using tractors when they were still planting traditional varieties; and their number remained unchanged after adoption of modern varieties. This probably reflects the government's policy of encouraging the import of tractors.

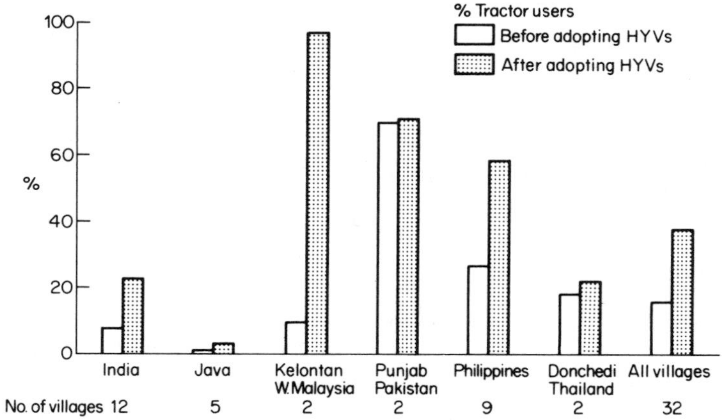

Fig. 57 Use of tractors by rice farmers before and after adopting HYVs (1971–72) (based on data from Barker & Anden, 1975).

Table XXVIII Percentage of rice farmers in Pedapullera village using tractors before and after adoption of HYVs (Pasthasarathy, 1975).

	Before adoption (1965–66)	After adoption (1971–72)
Wet season	58	84
Dry season	62	92

By contrast, in Java, the use of tractors is very limited, both before and after adoption of HYVs—because of the predominance of very small, garden-like farms and the prevalence of highly labour-intensive methods of production.

In brief, the use of tractors generally increased quite markedly after the adoption of HYVs, with the two exceptions mentioned above. The most dramatic increase occurred in western Malaysia, from 10% tractor users to 96%, even though the farms in the two villages investigated averaged only one hectare in size. The main reason is apparently the general changeover to double-cropping after adoption of HYVs. Both four-wheeled tractors and two-wheeled power tillers are used; the former are mainly owned by contractors (Tamin & Mustapha, 1975).

It must be stressed that the villages included in the survey are *not* representative of the countries in which they are located; they are selected villages situated in areas with a high potential for rice production. Also, the figures do not relate to the villagers in general, but only to those who were willing and *able* to adopt new practices, including HYVs.

The average figures presented above, whilst indicating general trends, may conceal major differences between villages, even in a given area. For example, the data for Pedapullera, a prosperous, progressive village in Andhra Pradesh (India), show that the use of tractors by the farmers was far greater than the average figures for all the Indian villages included in the survey, both before and after adoption of modern varieties (Table XXVIII).

The use of the tractor for ploughing and threshing was fairly widespread in this village before the adoption of HYVs, but became almost general after adoption.

Problems of Mechanization

The transition from hand labour and animal traction to mechanization usually involves a number of problems; technical, economical, and social.

The economic justification for mechanization will depend on the efficiency with which tractors and machinery can be operated, and the extent to which output can be increased by increasing yield and/or by extending cultivated areas for the efficiency of operation.

In developing countries, a number of factors operate which tend to make mechanization costs excessive. These factors include the high cost of imported machinery and fuel, difficulties with maintenance and repair, the need for thorough land clearance, and inefficient operation and utilization of the equipment.

The need for foreign exchange for the import of tractors and equipment, and for recurrent expenditures for fuel, oil and spare parts, can constitute a heavy burden on the national economy.

Operating Problems

One of the principal reasons for the lack of success of mechanization is the low efficiency in its use due to inadequate training of operators and of mechanics, lack of maintenance and repair facilities and, very often, unavailability of spare parts. A very common situation in many developing countries is for half the farm tractors to be standing idle at any given time, because of lack of spare parts. In 1966, for example, an estimated 40% of the tractors in India were idle (Brown, 1970). A similar estimate is made for Kenya (Otieno *et al.*, 1975). The problem is generally aggravated by importing equipment from a number of different sources, usually because of political or fiscal considerations.

The difficulties experienced with maintenance and repairs need not, however, be considered as a permanent handicap. Training of drivers and mechanics and the establishment of adequate repair facilities make it possible in due course to overcome these difficulties. The losses involved during the initial 'breaking-in' period can be considered as a kind of subsidy to facilitate mechanization.

Efficiency of operation can be improved by providing contract services to farmers, by pooling individually-owned land into larger holdings, by organized settlement schemes (in particular, irrigation schemes), and by large state or collective farms.

Experience has shown that groups of farms organized for maintaining tractor services have advantages over government hire services. The latter are notoriously inefficient, presumably because of the lack of personal incentive. Farm-groups can consolidate stretches for cultivation and therefore less travelling time to and from work is involved and fuller use of tractors throughout the year is possible (Mettrick, 1968).

Whether farmers should maintain the services cooperatively, or should encourage private *entrepreneurs* to provide the services, depends very much on local conditions, mainly on the possibility of cooperation and the technical know-how available. Private tractor contracting can be profitable to the entrepreneur and advantageous to farmers providing there is sufficient demand, sufficient experience in operating the machinery, and the number of tractors justifies the existence of adequate repair and servicing facilities. FAO (1968) reports that private contracting services have been successful in Argentina, Ceylon, Chile, Kenya, Malaysia, Sudan, Thailand and other countries, whilst all government contracting services have failed to cover costs and have either been continuously subsidized or abandoned.

Whilst entrepreneurs may provide mechanical services to a consolidated block of land, the individual farmer may remain responsible for all other operations on his particular plot.

One of the principal factors influencing the efficiency of operations and hence the cost per unit area is the extent to which the machinery is utilized. It has been estimated that the minimum working time required to make the use of a tractor economically justified is 500 hours per year and the minimum area is 50–80 ha (Wilde & McLoughlin, 1967).

Need for Increased Productivity

The introduction of mechanized equipment cannot be justified economically unless production is intensified to pay for increased costs. Expenditure cannot become modern so long as income remains primitive. Therefore, mechanization must be accompanied by improved cultural practices, which raise yields per unit area.

There have been quite a number of failures of large-scale mechanization schemes, of which probably the best known is the notorious groundnuts scheme in what was once Tanganyika. Mechanization has usually been most successful on irrigation schemes, mainly because control of the moisture supply makes possible the achievement of high yields and multiple-cropping, which in turn ensures the use of mechanical equipment over extended periods. Conversely, the timeliness and quality of mechanical operations may be a precondition to the achievement of high yields. However, not all mechanized irrigation schemes have been successful. Wilde & McLoughlin (1967) cite as an example the Richard–Toll irrigation scheme in Senegal, where some 3500 hectares were developed for mechanized rice production, and which never proved profitable. Though all operations—land preparation, harvesting and threshing—were mechanised, yields were too low to justify the high investment costs of land development and the recurrent costs of mechanization.

Bottlenecks

There is a danger inherent in partial mechanization of agriculture, in that labour bottlenecks may be created. For example, in a mechanized project that was started in northern Nigeria after World War II, it was expected that mechanical land preparation would make it possible for each farmer to crop about 10 ha. However, this proved to be impracticable because the farmer could not possibly cope with the manual task of weeding, which required on this area over 400 man-days of work during a six-week period (Baldwin, 1957). Another main bottleneck which may be created is that of harvesting, particularly important for crops in which delayed harvesting may involve considerable losses.

By increasing the expansion of cultivated land, mechanization tends to decrease the period of fallow, and hence the period during which soil fertility can be restored. Land clearing, involving removal of stumps and stones must also be exceptionally thorough if mechanization is to be adopted. The costliness of these operations also increases the incentive for intensive cultivation and shortening or abolishing the fallow period.

In hand-cultivation, the smallness of the fields, the remaining stumps and roots and the practice of mixed cropping somewhat reduce the dangers of erosion. These are considerably increased following mechanized cultivation, and proper safeguards must be adopted.

Social

The main social implications of mechanization are (a) the benefits resulting from the changeover to mechanization are realized by the larger landowners, businessmen, and entrepreneurs (see below); and (b) the displacement of labour, even when no alternative sources of employment are available.

We have seen that certain types of machinery, including tractors, can actually increase employment opportunities under certain circumstances; others are always labour-displacing.

In Israel, increased mechanization reduced labour requirements drastically in the course of ten years. The number of workdays involved in the production of groundnuts decreased from 80 to 30 per ha, in sugar beets from 60 to 20 per ha, in cotton from 120 to 17 per ha (Arnon, 1969).

It is in the mechanization of harvesting (grain combine-harvesters, cotton pickers, sugar cane harvesting equipment, etc.) that the most drastic reduction in labour requirements occur. A grain combine-harvester can replace 90 labourers; since harvest-work is frequently more highly paid because of the seasonal peak labour demand, any generalized use of mechanized harvesting in countries with no alternative labour employment opportunities, can have 'catastrophic results for rural livelihoods' (Pearse, 1980).

Increased Risks

An aspect that is frequently overlooked is that factors beyond the control of the farmer, such as the risks of economic setbacks due to unfavourable climatic conditions, disease epidemics, etc., can increase considerably following capital investment in general and mechanization in particular. In a mechanization study in West Africa, for example, it was found that whereas a 10% decline in yields caused by adverse weather conditions reduced farm incomes by 12% on farms depending entirely on manual labour, the reduction in income increased to 23% at an intermediate level of mechanization, and reached 30% on farms using tractors (Tschiersch, 1978).

Mechanization of Small Farms

Appropriate mechanization for the small farmer should have the following characteristics: (a) leads to increased production; (b) does not displace labour; (c) is economically justified, and (d) is accessible to the majority of small farmers (Tschiersch, 1978).

Whilst economies of scale are not critical for many improved cultural practices, such as use of better varieties, fertilizers, weeding, etc., this does not generally hold true for machinery.

According to FAO estimates, economic justification for the use of a 40 hp tractor would be 15–25 ha on an intensive irrigated farm, and 40–100 ha for a dryland savanna farm (FAO, 1968).

Several factors tend to restrict mechanization and its benefits to large farms: (a) the availability of credit at a reasonable rate for the purchase of equipment; (b) most equipment has been developed in western countries and is designed for relatively large farms; (c) special, non-traditional skills are required for operating the machinery; and (d) operating costs are high in relation to output because of high maintenance costs. The financing of after-sales services is almost the same for a small tiller as for a conventional farm tractor and therefore considerably increases the purchase price.

Where labour scarcity at peak periods, and/or the high cost of labour, make the purchase of machinery unavoidable, the small-scale farmer is frequently faced with the problem that even machinery designed for his special needs is beyond his financial capacity.

In addition to the various organizational forms discussed above, which by pooling mechanized equipment make it economically available to relatively small farm units, much progress has been made in recent years in adapting modern farm machinery to the needs of the small family farm. Japan pioneered in this field, in particular with small mechanical tillers and threshers, but west European countries have also made important contributions.

The rotary foot-pedal thresher developed in Japan facilitated expansion of multiple-cropping by reducing labour requirements during the critical harvesting season (McPherson & Johnston, 1967).

There are also certain types of machinery that are eminently suitable for small farms, such as motors and pumps for small tubewells, seed drills, two-wheeled tractors and small, motor-driven hoes (Fig. 58).

The 'ultra-low-volume' knapsack sprayer is an item of advanced technology that can be used by the small farmer (Greenland, 1975).

Small-machine mechanization has a number of weak points. Kline *et al.* (1969) point out that many of the small tillers and tractors made in Europe, the United States, and Japan are not designed for hard, continu-

Fig. 58 Japanese three-row power weeder: capacity of one hectare for one man working an eight-hour day; powered by a two-stroke petrol engine. By courtesy of the Agricultural Engineering Department, International Rice Research Institute, Los Baños.

ous work in tough soils. They are generally too light for most farm operations, require too much attention and are underpowered. Large tractors provide multi-purpose services, whilst of the machines developed in Japan for small farms—power tiller, transplanter, binder, etc.—each serves a specific purpose and requires a separate power source, so that the total cost of the machinery and power requirements per hectare is relatively high (Kim, 1972) (see Table XXIX).

Though the average tractor horsepower per farm is the lowest for Japan, among the countries listed, the power per hectare is the highest, indicating that Japanese-style mechanization is the most expensive in terms of power requirements.

Khan, (1974) states that Japanese machines are far too complex and uneconomic for tropical Asia. Japanese combine harvesters and paddy transplanters are 'examples of functionally suited but economically unacceptable machines for the tropical regions'.

Up to the present, none of the major tractor manufacturers in industrialized countries have adapted a specifically designed tractor for use in developing countries, which has gone into mass production (Tschiersch, 1978). For these reasons, increased efforts are being made to develop and

Table XXIX Tractor horsepower requirements in selected countries, 1970.[a]

	Average tractor horsepower	
Country	Per farm	Per hectare
USA	79.1	0.7
United Kingdom	41.2	0.8
France	18.7	1.3
Germany	17.2	2.1
Italy	4.5	1.0
Japan	3.6	2.8

[a]Source: Farm Machinery Almanac (Sin No Lin Syo) Japan, 1971 cited by Kim (1972).

test new types of farm equipment that can be manufactured in the developing countries themselves; however, because development of new equipment requirements is a specialized field, needing a high degree of technical expertise, the commercial development of machinery designed specifically to meet the needs of small farmers in the tropics is still extremely limited.

The Agricultural Engineering Department of the International Rice Research Institute started a programme to develop low-cost, small, power-operated machines for manufacture in developing countries.

The main target group of this programme are the medium-sized family farms which are too large to work economically with draught animals and too small to use imported, large-sized machinery. Under this programme, a broad range of equipment has already been developed and is being produced in the region and has been widely accepted by farmers.

Initially, the efforts of IRRI to encourage and assist manufacturers to produce the equipment was confined to the Philippines. Subsequently, they have broadened the scope of their work to other developing countries.

Engineering drawings and other technical assistance is provided to interested manufacturers in developing countries. An interesting example is a relatively cheap, easy-to-maintain, power tiller. This small tractor can be readily made in small shops with standard shop equipment and sells in the Philippines for about half the price of comparable imported tillers (Fig. 59).

The employment generated in the production, marketing and servicing of locally produced equipment can substantially alleviate the problems of farm labour displaced by mechanization.

In an up-to-date and concise review of the problems of mechanization for small farmers in developing countries, Tschiersch, 1978) concludes that 'there is no single, technologically uniform solution to the problem of mechanization of small scale agriculture. Instead, there is a wide range of solutions of different levels and combinations of technology'.

Fig. 59 IRRI single-axle 5–7 HP tractor. This lightweight tiller, designed for small farmers is easily operated for a wide range of farming operations. Produced in the Philippines, primarily from locally available materials; it costs less than half the price of comparable imported tillers. Operating and maintenance costs are low: (a) ploughing-under stubble; (b) puddling. By courtesy of the Agricultural Engineering Department, International Rice Research Institute, Los Baños.

It is impossible to draw general and unequivocal conclusions on the merits of mechanization for the family farm. As a rule, premature and indiscriminate mechanization may bring economic benefits to the few at high social costs to developing societies as a whole. However, with development, mechanization generally becomes unavoidable; whether the process of mechanization will be disruptive or not, will largely depend on the strategies adopted in its implementation. The more selective is the process, the more favourable its potential impact on the rural economy.

Energy Requirements of Agricultural Production

Farming systems have generally been appraised according to their efficiency in producing foods, their potential for productive employment and their economic returns. Until the recent past, fuels have been relatively plentiful and cheap, and in most countries constituted a relatively minor cost factor in agricultural production.

Fossil fuel requirements for the mechanization of agriculture amount to about 2.4% of the world's yearly energy requirement. Although the requirement for the developing countries amounts to only 0.2% of the total world energy requirements (Westley & Johnston 1975), the steep increase in cost of fuels since 1973 has become a serious financial burden for many countries. Therefore, the present energy supply problems make a reconsideration of different farming systems and agricultural practices in the light of their energy requirements essential, and new guidelines for planning agriculture need to be developed.

Farming Systems

In traditional agriculture, the sun is the main source of energy, supplemented by human labour and animal power.

In the early nomadic food-gathering and hunting stage, man relied solely on his own energy and expended approximately 70% of this in his search for food (Pimental, 1973).

In 'cut-and-burn' agriculture, the human energy expended on producing an adequate food supply is somewhat reduced in comparison with the earlier nomadic stage. It is estimated that the energy required for this purpose amounts to about 875 hours of manpower per hectare, the equivalent of about 132,000 kcal/ha (Haswell, 1953). Human energy output in tropical Africa is estimated as no more than 1500 kcal per day (Boshoff & Minto, 1975).

A dramatic increase in available energy occurred with the harnessing of domestic animals for agricultural production, and until the 17th century, agricultural output in western agriculture was mainly determined by the energy of domestic animals supplemented by human energy.

The fuels used were mainly firewood, straw and dried animal dung; addi-

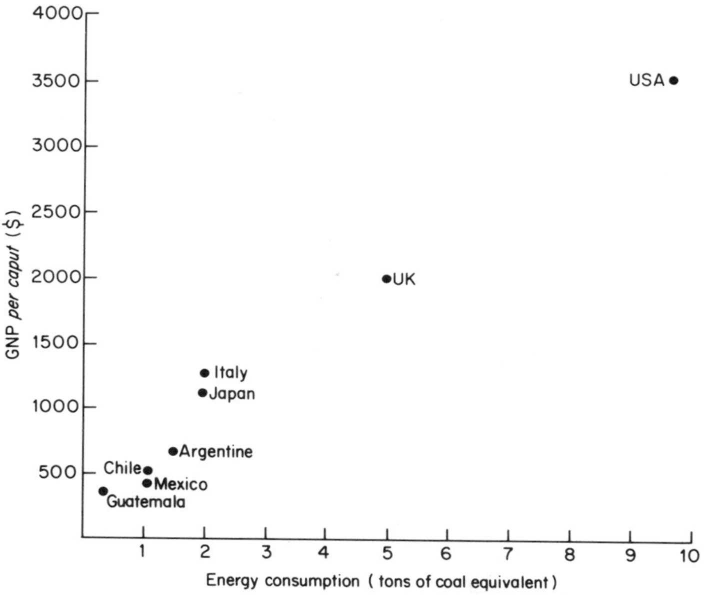

Fig. 60 *Per caput* energy consumption in relation to GNP *per caput* in a number of selected countries (based on data from Darmstadter, 1972).

tional sources of energy were obtained from water, wind and sunshine—all of which were renewable. A similar situation exists in present-day rural India, where about 90% of the energy used is supplied by humans, animals, firewood, dung, and crop residues (Crosson & Frederick, 1977).

The steam engine provided a great increase in available power; its direct effect on agriculture was marginal, however, in comparison with its impact on industry.

A further 'quantum jump' occurred with the use of liquid fuels and development of the internal combustion engine, major factors in the modernization of agriculture.

Some idea of the relative energy values of different sources of power are given in Table XXX.

In the transition to modern agriculture, fewer farmers produce more food, but both the reduced input of human labour and the impressive increases in yields depend heavily on energy inputs derived from fossil non-renewable sources. This fuel is used directly for the operation of combustion engines for tractors, pumps and other machines and for drying crops; indirectly in the manufacture of machinery, fertilizers, pesticides, and other chemicals; and in transport, processing, etc. The energy requirements of mechanized agriculture are nearly 60 times higher than those of 'cut-and-burn' agriculture.

Table XXX Relative energy values of different power sources in agriculture.

	Horsepower	Manpower
Men working at maximum output	1/12	1
Bullock	1/4	3
Walking tractor	3–10	36–120
Wheel tractor (medium sized)	45–50	420–600
Track-laying tractor (medium sized)	90	1080
Wheel tractor (large)	180	2160
Combine harvester	140	1680

The amounts of power—human, animal, and mechanical—expressed in horsepower, used for agricultural operations in different continents are shown in the following data from Hülst (1975):

USA	1.65
Europe	0.9
Latin America	0.2
Asia	0.2
Africa	0.09

On a world scale, countries in temperate regions account for nearly 90% of world consumption of inanimate energy, whereas tropical and subtropical regions, in which most of the developing countries are found, account for only about 8% of the total (McPherson & Johnston, 1967).

The Energy Ratio in Different Farming Systems

Leach (1976) compared the energy inputs and returns per unit of land in different types of food production systems. Subsistence farming has low inputs and low production levels, but the energy ratio (ratio of output to input) is very high and usually exceeds 10. In more intensive traditional systems, both inputs and production increase somewhat, but the ratio remains relatively unchanged. In exceptional cases, such as cassava production in tropical Africa, the amount of energy produced may be equivalent to that of some modern farming systems.

As agriculture is intensified by the use of agrochemicals, farm machinery and irrigation, productivity increases markedly, but energy inputs generally rise still more rapidly, so that the energy ratio declines. The ratio is lower in animal production than in crop production, and lowest in intensive poultry production, in which it is reduced to 0.1.

An example of the relative efficiency in energy rate of traditional agriculture as compared to modern agriculture is provided in a study by Heichel (1974) showing that in the relatively primitive paddy-rice culture of the Philippines 16 calories of digestible energy are harvested for each calorie of cultural energy invested, whilst in modern maize production, the comparable ratio is 5:1.

Another example is provided by a comparison between maize production in the USA and in Ghana. In the USA, the 2.4-fold increase in maize yields achieved by improved technology during the period from 1945 to 1970 required a 3.1-fold increase in mean energy inputs. Yield of calories from maize per one fuel kilo calorie input decreased by 24%: from 3.7 kcal to 2.8 kcal (Pimental, 1973). This compares with a caloric yield of 27.9 kcal for maize produced in the hand-labour system, as used in Ghana (Black, 1971). In the latter system average yields of maize were only about one-sixth of those obtained in the highly mechanized agriculture of the USA, but caloric return per caloric input was 10 times higher.

In general, in many modern cropping systems a 10- to 50-fold increase in expenditure of cultural energy only doubles or triples the digestible energy yield compared with more primitive systems, indicating that progressively larger expenditure of cultural energy is used less efficiently in crop production (Heichel, 1974).

The Relation Between Commercial Energy and Labour Output

At the end of the 1960s, one dollar's worth of petroleum provided the equivalent in energy terms of 3800 hours of human labour (FAO, 1977)! No wonder that agriculture in the developed world became heavily dependent on fossil fuel.

There is a close relationship between commercial energy input and the productivity of agricultural labour, as shown in Table XXXI.

In the developing countries, there is also a close relationship between the amounts of commercial energy used and yields, as shown by the data in Table XXXII.

Table XXXI Relation between commercial energy input and worker productivity (FAO, 1977, by permission).

Region	Input ($\times 10^9$ joule per worker)	Worker productivity (tonnes of cereal per worker)
North America	556	67.9
Latin America (highest among LDCs)	8.6	1.9
Africa (lowest among LDCs)	0.8	0.5

Table XXXII Relation between commercial energy input and cereal yield (FAO, 1977, by permission).

Region	Input ($\times 10^9$ joule/ha)	Cereal yield (tonnes of grain/ha)
Latin America	4.2	1.4
Africa	0.8	0.8

Energy Requirements of Modern Agriculture

Commodities

Agricultural commodities differ in their energy-input requirements. An example of these differences is provided by data obtained in investigations carried out in France (Table XXXIII).

The differences in *efficiency* in energy use are still more significant. Whilst maize production in modern cropping systems yields 5 calories of digestible energy per calorie of cultural energy, the ratio is 1 or less for rice or peanuts. Oats, soybeans and sorghum require similar investments of cultural energy, but sorghum is twice as efficient in producing digestible energy. Sugar-cane and maize return twice the digestible energy of sugar-beets, rice and peanuts at half the investment of cultural energy (Heichel, 1974).

On a scale of energy requirements from low to very high, crops would generally rank in the following order: small grain, maize, fruits, greenhouse production. High-protein foods from animals, such as eggs, milk and meat require far larger inputs of energy than do plant foods.

In societies with a high standard of living, most of the maize is fed to livestock. When maize is converted into beef, 10 kcal of maize are required to produce 1 kcal of beef and the return of calories per unit caloric input is only 0.30, still further increasing the discrepancy in energy use between modern and traditional farming (Pimental, 1973).

Particularly extravagent in terms of the energy subsidy is feedlot meat production; other high-protein foods such as milk and eggs also have a far poorer energy return than do plant foods. In view of the hopes pinned on the contribution of sea-fishing to the solution of the world's food problem, the high cost of energy involved will have to be taken into account. Soybeans, which possess the best amino-acid balance and protein content of any widely-grown crop are amongst the more economic foods in terms of energy input/output ratio (Steinhart & Steinhart, 1974).

Whilst these figures, obtained in highly mechanized farm systems in

Table XXXIII Energy requirements of selected crops (in thousands of kcal/ha) (Hutter, 1976, by permission).

	Rain-fed	Irrigated (by sprinklers)
Maize	5,730	9,680
Wheat	5,710	—
Sorghum	4,230	7,030
Soya	2,730	4,490
Forage sorgo (ensiled)	8,730	13,030
Forage sorgo (pastured)	6,190	9,640
Lucerne (hay)	5,490	7,930

temperate climates, have no direct application to the labour-intensive agriculture of developing countries, they do illustrate the considerable differences in energy requirements between different crops; choice of crops (and farming systems) may therefore become an important consideration when taking energy inputs into account. The high energy requirements of a forage crop, such as forage sorgo, is an indication of the high cost, in terms of energy, which may be required if forages are produced in order to improve the work output of draught animals, or if domestic animals are stall-fed instead of pastured.

The massive use of energy from non-renewable resources by the developed world has been the major factor in increasing productivity in agriculture, but it has also had negative effects: (a) pollution, due directly to excessive and inefficient combustion of organic fuels and indirectly to the proliferation of industrial products; (b) a depletion of natural resources which are not renewable; and as a result (c) a steep increase in prices. The developing world, as shown by the figures presented above, has had very little share in pollution or depletion, but is all the more disadvantaged by the resultant steep increase in the costs of agricultural production.

Cultural Operations

The energy required for the production of agricultural commodities consists essentially of that expended directly in the forms of labour and mechanical processes (tillage, sowing, harvesting, etc.), and indirectly in the production of inputs (fertilizers, pesticides, etc.), and maintenance of machinery, equipment and infrastructure.

In an investigation in energy consumption of different crops carried out in France, the data of Table XXXIV were obtained for maize (Hutter,

Table XXXIV Direct and indirect energy requirements of cultural operations in maize production (Hutter, 1976, by permission).

	Rain-fed		Irrigated (by sprinklers)	
	× 10^3 kcal/ha	% of total	× 10^3 kcal/ha	% of total
Tillage	1,000	17.5	1,000	10.3
Fertilization	2,690	47.0	4,280	44.2
Sowing	160	2.7	200	2.1
Plant protection	1,000	17.5	1,000	10.3
Irrigation	—	—	2,320	24.0
Harvesting	880	15	880	9.1
Total	5,730	100.0	9,680	100.0
Equivalent tons of fuel oil	0.57		0.97	

Irrigated production as compared to rain-fed (=100): 168.9

Table XXXV Energy equivalents of major inputs in maize production (Hutter, 1976, by permission) (% of total energy requirements of the crop).

	Rain-fed	Irrigation
Fuel	28.5	34.5
Fertilizers	45.0	43.2
Seeds	2.0	1.4
Plant protection chemicals	15.2	9.0
Labour	0.3	0.3
Administration and maintenance	9.0	11.6
Total	100.0	100.0
$\times 10^3$ kcal	5,730	9,680

1976). Together, the direct and indirect requirements for fertilization and plant protection account for 64.5% of energy inputs in rain-fed production of maize, whilst irrigation *per se* adds about 70% energy requirements.

Major Inputs

The energy consumption of the individual inputs are shown in Table XXXV (Hutter, 1976). In mechanized agriculture, the direct labour input in terms of energy is extremely low; more surprising is the fact that direct fuel requirements are remarkably lower than the energy inputs in the form of fertilizers and plant protection chemicals, which account for 52% (irrigated) to 60% (rain-fed) of the total energy requirements of the crop.

This has highly unfavourable implications for agriculture in the developing countries: it is in those inputs which are most essential for increased production and reduced losses that the main savings in energy inputs can be made in labour-intensive farming systems.

Fertilizers

Fertilizers in intensive agriculture account for up to half the energy input required for production. At least 90% of the world supply of fertilizer nitrogen is derived from the fixation of atmospheric nitrogen with hydrogen as anhydrous ammonia (NH_3) (Blouin, 1974). In ammonia plants operated on natural gas, about one-half of the gas is used to provide hydrogen, which combines with nitrogen from the air to produce ammonia. Phosphorus and potash on the other hand, are derived from mineral deposits, and hence fuel is required only for conversion of the raw materials into commercial fertilizers (Bakker-Arkema *et al.*, 1974).

About five times more fuel is required to produce a kilogram of ammonia than to produce a kilogram of phosphorous fertilizer and eight times more than for a kilogram of potash fertilizer. As a result, with the exception of legumes, nitrogen fertilizers account for 70–90% of the total energy input of fertilizers (Hutter, 1976). When ammonia is converted into ammonium nitrate, sulphate of ammonia, urea, etc., still more fuel is needed and the discrepancy with other major plant nutrients is still greater. The direct application of liquid ammonia is therefore the most economical method, both in money and energy cost, of supplying nitrogen to the crops. However, it requires special equipment for its application, but its adoption instead of sulphate of ammonia, urea or other solid fertilizer can result in a significant saving in energy costs.

Insecticides, Fungicides, and Weedicides

Mainly because of the relatively small amounts (as compared to fertilizers) generally applied, the energy input of plant protection chemicals is generally low, amounting, in intensive agriculture, to less than 5% of total crop requirements (Hutter, 1976).

Irrigation

Modern methods of irrigation were developed at a time of relatively low costs of fuel, with the basic objectives of reducing labour costs and improving water use efficiency. As with other factors involved in intensifying agricultural production, the energy costs of different irrigation methods are now receiving increasing attention, especially as modern irrigation methods are highly energy-intensive. In France, sprinkler irrigation resulted in a 70% increase in energy requirements (see Table XXXV); in California, about 68% of all electricity used in the state's agriculture goes into irrigation (Anon, 1971).

It has been estimated that if the entire world's irrigation potential were fully developed, using modern sophisticated irrigation methods, 5–10% of the world's energy expenditure would have to be devoted to irrigation alone (Batty *et al*., 1974).

A study was conducted in Utah on the comparative energy needs of nine different irrigation systems, including surface, sprinkler and drip irrigation methods (Anon, 1974).

A distinction was made between *installation energy requirements* and *energy requirements for pumping:* the ratio between the two varies considerably with irrigation methods.

Assuming zero water lift, i.e. that water is available at the surface of the well, surface irrigation appeared to be far more energy conservative than either sprinkler or trickle systems. However, under conditions in which the latter systems enable marked savings of water—and this is their main justification—and where water lifts are high, sprinkler and trickle irrigation may be more economical in the use of energy than surface irrigation.

Therefore, where water is expensive or relatively scarce, the main consideration will normally be achieving maximum efficiency in water use.

From Farm to Consumer

In modern societies, larger amounts of energy are required for transporting, processing, packaging and marketing food, than for farming it (Pimental, 1973). In the USA, transportation of farm products has grown considerably, hardly any food is eaten as it comes from the fields, and even farmers buy much of their food from markets in town. The food-processing industry is the fourth largest energy consumer among countries. In sugarcane, over 40% of the cultural energy is used to process the crop for human consumption (Heichel, 1974). The manufacturing of packages used for food products consumes a significant proportion of this energy. When the total food chain is considered—production, processing, transport and industries related to agriculture—energy use in the food system is estimated to account for 12.8% of the total US energy use in 1970 (Steinhart, 1973).

A breakdown of the relative energy requirements of the food chain is presented in Table XXXVI. These figures are relevant to developing countries inasfar as modernization of agriculture is basically dependent on the possibility of finding markets, local and foreign, for produce. Road systems and transportation are generally inadequate, so that the energy input in transportation can be far higher than the above figures might suggest. One of the possibilities for industrializing is the processing of high-value agricultural commodities, and here the 'energy crisis' may be a major constraint to development.

The Green Revolution and Energy Requirements

The energy crisis has focused attention on the implications of the green revolution on fossil fuel requirements of the developing countries.

The higher the production potential of plants and animals, the more care and energy inputs they require. Many HYVs require irrigation for full exploitation of their potential, and all require high levels of fertilization

Table XXXVI Food-related energy use (% of total) in the USA 1963 (Bakker–Arkema *et al.*, 1974).

Agricultural production	18.3
Food processing	32.9
Transportation of food	2.4
Wholesale and retail trade	15.9
Home preparation	30.5

and the use of pesticides and herbicides. From the point of view of energy inputs, the green revolution has been dubbed as 'an attempt to export a part of the energy-intensive food system of the highly industrialized countries to non-industrialized countries' (Steinhart, 1973). But most of the developing countries are nations that are poor in fossil fuels and poor in the foreign exchange required to purchase them. Their problems have been intensified by the steep increase in costs of energy in the world market.

The use of improved varieties *per se*, has no significant effect on energy requirements—and even increases efficiency in energy-use when expressed per unit of commodity produced. However, of the three basic factors required to realize the potential of the new HYVs, irrigation and fertilizers are the most high-energy-requiring inputs, accounting for over 60% of the total energy input in modern mechanized agriculture. In a labour-intensive agriculture the *relative* fossil fuel requirements for irrigation and fertilizers are even higher.

IMPLICATIONS OF THE 'ENERGY CRISIS' FOR MODERNIZING AGRICULTURE

The problems faced by countries that wish to modernize their agricultures can be exemplified by a study of energy use in rural India by Revelle (1976). In order to make Indian agriculture competitive and capable of meeting the food requirements of the population in the future, agricultural production must be increased by applying more fertilizers, developing irrigation and wider use of double-cropping. The energy required for pumping, once the full potential for irrigation is developed, would increase threefold, and the additional nitrogen fertilizer required annually for double-cropping 100 million hectares would have an energy requirement of about 1.4 times the total energy use of agriculture in India in 1971.

Most of the projected increase in energy requirements would have to come from fossil fuels. However, despite the high cost of these fuels, according to Revelle's analysis, their use would still be profitable if combined with good management practices. A tripling of fuel energy inputs would cost an additional \$3.2 billion at 1976 world prices, whilst the additional agricultural output that could be achieved would have a value of \$35 billion at 1976 world prices.

Mechanization

The 'energy crisis' should give an additional impetus to labour-intensive agriculture as against mechanization. Where the latter is essential, it should be highly selective. Where mechanization has been adopted, minimum tillage can result in reduction of energy requirements for land preparation by 60–70%. However, additional amounts of energy are required for producing herbicides. There is also a yield reduction of 5–15% to be considered.

Choice of Crops

Most of the modern agricultural systems that are relatively efficient in the use of cultural energy incorporate a photosynthetically efficient crop (Heichel, 1974). Improving the photosynthetic efficiency of crops by breeding and management could therefore achieve significant increases in the efficiency of cultural energy-use.

At first sight the present situation has improved the relative advantages of leguminous crops, and appears to provide an additional impetus to invest in breeding programmes and research on cultural methods for these crops. Among tropical grain legumes suited to the lowland humid tropics mentioned as promising are: cowpeas (*Vigna inguiculata)*, lima beans (*Phaseolus lungatus*), winged bean (*Psophocarpus tetragonolobus*) and others, most of which can be grown mixed with maize (Greenland, 1975).

Considerable optimism has been expressed on the potential benefits of the greater use of nitrogen fixation by microorganisms. Several promising lines of research are being pursued, such as: improving the efficiency of the symbiotic relationship between leguminous crops and microorganisms (in particular breeding for more effective, adapted strains of *Rhizobium*); inducing a similar symbiotic relationship in cereals; genetic transfer of the capability for fixing nitrogen from bacteria to cultivated plants. However, it is generally not appreciated that biological nitrogen fixation is very energy-demanding.

The bacterial process of nitrogen fixation from the air requires approximately the same amount of energy per unit of fixed nitrogen produced (±15,000 kcal/kg) as the chemical manufacturing processes (Revelle, 1976). In the symbiotic processes, the energy required is supplied by the photosynthates produced by the host plant, and therefore competes with other physiological activities in the plant for the products of photosynthesis.

This may explain why the yields of legumes are relatively low and have shown greater resistance to breeding efforts than the grain cereals.

About one-third of the total carbon fixed by photosynthesis is moved to the nodules and less than half of this is returned to the aerial portions of the plant (Hardy & Havelka, 1975).

For every ton of soybeans, with a protein content of 30–40%, 100 kg of nitrogen must be fixed, consuming energy from 400 kg of carbohydrates. By comparison, the natural gas and haphtha commonly used in industrial nitrogen fixation have a high energy content, and consume only $1\frac{1}{2}$ tons of fuel for every ton of nitrogen fertilizer. Even at 1976 oil prices, the cost of the fuel amounted only to 10% of the additional crop yields due to fertilization (Revelle, 1976).

Not only are yields of grain legumes low, in comparison to cereals grown under similar conditions, but the yields of cereals grown after the grain legumes are generally depressed. Already in 1936, Lyon showed that after grain legumes, the soil nitrogen content was markedly reduced and the yields of cereals depressed by over 50%.

The foregoing does not apply to forage legumes harvested before seeds are produced. Therefore, the value of a forage legume, included in the rotation as a provider of atmospheric nitrogen, making possible a considerable saving in energy inputs for the following crops, is an important consideration in appreciating the energy costs of the forage.

However, in tropical regions, legumes appear to be less capable of fixing nitrogen than in temperate and subtropical regions, and do not always carry bacterial nodules. In most parts of Africa, legumes have not proved to be better than grasses or weeds in restoring or maintaining soil fertility (Webb, 1960). These findings were confirmed by Moore (1962), who found that the amounts of nitrogen were rarely significant, and then only after mineral fertilization.

Great hopes are placed on research aimed at inducing microorganisms to perform a symbiotic relationship with cereals, similar to that with legumes, or even to transfer the genetic capability for nitrogen fixation directly from bacteria to plants. It is estimated that the potential benefit of such research, if successful, would make it possible for cereals to fix biologically an amount of nitrogen more than five times greater than current consumption of nitrogen fertilizers by the developing countries. 'Extrapolating from current yield data in the developing countries, this amount of biologically fixed nitrogen could produce 200 million tons of grain, more than half the current output of the developing countries (excluding China) (NAS, 1976). However, these figures do not take into account the probable yield reduction due to energy demands on the plant for biological nitrogen fixation.

Varieties

In reaction to the energy crisis, some agricultural research institutions in developing countries have adopted as a basic objective in their plant breeding programmes, the creation of varieties capable of producing high yields with low levels of fertilizer applications. Borlaug (1976) reacts to these proposals by stating: 'we will succeed in producing such varieties of crop plants about six months after utopian political leaders, sociologists and economists produce a new race of man who needs no food in order to grow strong bodies, maintain health, work effectively, enjoy life and speak eloquently'. Even if varieties are obtained that can make more effective use of the nutrient reserves in the soil, these will have to be replenished if high yields are to be maintained. It therefore appears reasonable to continue to breed varieties with high yield potentials under conditions of favourable moisture and nutrient supply, with high protein yields and with maximum possible resistance to disease and pests.

Manures and Fertilizers

Wherever possible, an adequate level of soil fertility should be maintained by using farm manures, composts, green manures and the inclusion of forage legumes in the crop rotation.

Transporting and applying manure mechanically requires less than one-third as much energy as compared to chemical fertilizers providing similar nutritional levels (Pimental *et al.*, 1973). The energy costs of sowing legumes for green manure are relatively low; however, this requires land, and without phosphoric and/or potassic fertilizers, the production of organic matter is likely to be low.

It should also be mentioned that green manures have given very inconsistent results; in particular, in tropical soils their effect has often been insignificant (Meicklejohn, 1955).

A rough estimate of the total plant, human and animal wastes available in developing countries indicates that if these were used as organic manures, they could supply about 100 million tons of nutrients (consisting of 47% N, 38% K, and 15% P) equivalent to eight times the total fertilizer consumption of these countries in 1970–1 (FAO, 1977). However, there are enormous difficulties involved in mobilizing and using these wastes. Much of the animal husbandry in developing countries, for example, is completely separate from arable cropping, and cannot therefore be used as a source of manure for the latter. Much of the dung is also used as fuel (see below) and is lost as a source for plant nutrients. Even if every effort is made to collect organic wastes and use them efficiently, the use of fertilizers to make up the shortfall will remain indispensable.

Hygiene and pollution hazards need to be taken into account. Run-off and drainage water leaving the farm may carry with it considerable amounts of nitrogen and phosphorus from the manure heaps, and pollute drinking water.

With the use of sewage sludge, account must be taken of the content of heavy metals, which, in the long run, can accumulate to toxic levels.

More attention should be given to applying fertilizers at rates that give optimum returns per unit energy rather than maximum yields. These rates are likely to be the same as those giving the major economic returns.

It is important to remember that the greatest economic return to fertilizers is at rates generally *below* those required for maximum yields. Because of the law of diminishing returns the last increments of fertilizers are rarely justified. Experience has shown that at a rough approximation the first 15–30 kg of nitrogen per hectare increases yields of cereals by 10–15 kg grain/kg N. Beyond these rates, the response declines gradually (FAO, 1977).

In Fig. 61 an example is given of the relationships between yields and price of rice and the economically optimum levels of nitrogen to apply to the crop (Barker, 1970).

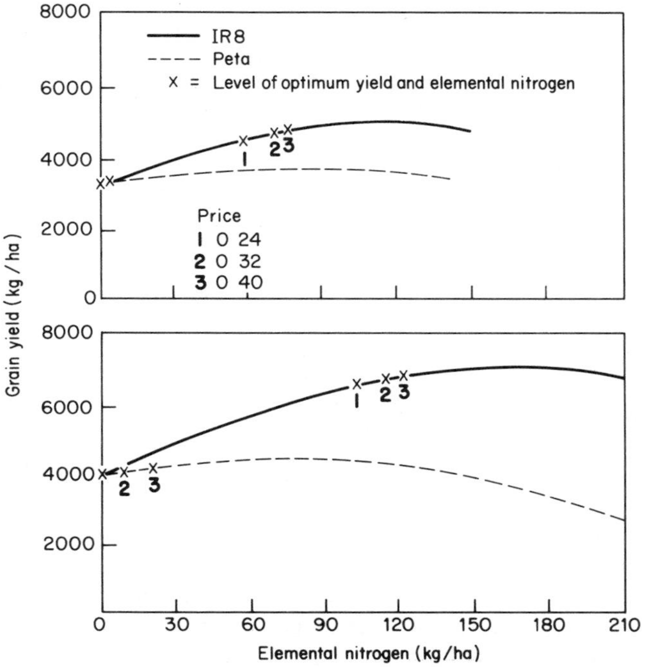

Fig. 61 Optimum levels of nitrogen application to rice varieties for a $2\frac{1}{2}$:1 benefit–cost ratio and yield level: (a) wet season; (b) dry season. Note that for IR8 optimum levels of nitrogen were 60–80 kg/ha in the wet season and 100–120 kg/ha in the dry season. For the traditional variety Peta, small amounts of nitrogen were justified only in the dry season.

It will be observed that with the traditional variety of rice—the tall growing Peta—nitrogen use was not justified at all in the wet season and only at very low rates in the dry season. By contrast, under the same ecological conditions, it was profitable to use relatively high levels of nitrogen at a fairly wide range of fertilizer prices—more in the dry season than in the wet season, for reasons already explained (cf. pp. 311–312). Even if the price of paddy falls from the official support price level of \$100/ton to \$60/ton, the use of fertilizers still remains profitable at the fertilizer costs prevailing at the time. However, in all cases, the amounts of nitrogen that gave the highest economic return were markedly below the rates required for maximum yields.

By foregoing the marginal increase in yields due to the last increments of fertilizers, considerable savings in energy requirements can be achieved with relatively small sacrifices in yields. Therefore, the need to reduce pollution dangers, to limit energy-consuming inputs and purely economic considerations coincide in determining optimum levels of fertilizers use.

Table XXXVII Yield increases and economic returns from input packages applied at several levels to irrigated and rain-fed rice (IRRI, 1978). Comparison with farmer's treatment.

Level of treatments	Yield increase (% of high level)	Net profit	Return per $ invested
High	100%	Lowest	Lowest
Medium	75–80%	Maximum	Medium
Low	50%	Medium	Maximum

Plant Protection

Chemicals required for plant protection are an essential element in the modernization of agriculture. Though they represent, as we have seen, only a small fraction of the total energy inputs in crop production, they are the most energy-intensive of all inputs, requiring 2.3 kg of petroleum equivalent per kg of pesticide (FAO, 1977). Continued growth in use is inevitable, and though the amounts used per unit area are small, economies should and can be made.

These conclusions are born out by recent trials on the economics of irrigated and rain-fed rice, in which management packages of practices including several levels of fertilizer, insecticide and herbicide applications were compared (IRRI, 1978).

Yield increases, net profit and return per dollar invested were calculated. The results are summarized in Table XXXVII.

Energy problems therefore only strengthen the economic logic of not aiming for maximum yields, by using the highest levels of inputs. Medium rates will generally give maximum net returns to farmers, and for those who have difficulties in obtaining cash, low levels provide the highest return per dollar invested. These results confirm that no fertilizer, insect- or weed-control experiment should be considered as adequate, unless the results are submitted to economic analysis. This is, unfortunately, rarely done.

Increased use of *integrated insect control methods* which require only a fraction of the energy used for pesticide manufacture and application could reduce requirements considerably. Hand application of insecticides by knapsack sprayer can reduce energy for application from 45,000 to 750 kcal/ha. (Pimental *et al.*, 1973).

The control of weeds by mechanical cultivation or a combination of hand-weeding, cultivation and herbicides may require slightly less energy than the application of weedicides alone.

Irrigation

The replacement of the Persian wheel and other primitive devices by modern diesel and electric pumps has led to such a reduction in costs of

Fig. 62 IRRI foot-operated diaphragm pump, suited for pumping water from irrigation ditches, open channels, river banks, and shallow wells, can handle muddy water. The entire unit, which weighs 28 kg, can be easily transported. It can be easily manufactured by small machine shops, and is easy to maintain and repair. By courtesy of the International Rice Research Institute, Los Baños.

water than even subsistence farmers have adopted large-scale mechanized pumping.

The main energy-saving methods that can be adopted are in the area of more efficient water-use: (a) by improving irrigation methods; and (b) by increasing yields.

Improved water application involves a more careful preparation of the land so as to enable more uniform spreading of the water; applying the right amounts of water at correct intervals (the general tendency is to over-irrigate); avoiding salinization, etc. More attention could also be paid to developing improved methods for pumping and irrigation suitable for small farmers with limited capital (Fig. 62).

It is estimated that of the total commercial energy used in the world for the manufacture of irrigation equipment and its operation, the developing countries accounted for 59% in 1972–73; their share is expected to rise to 65% by 1985–86 (FAO, 1977), a clear indication of the increasing importance of irrigation in the developing countries.

Alternative Sources of Energy

Considerable technical difficulties will have to be resolved before efficient and economic systems for the use of renewable resources can be developed. Hoffman (1980) states that 'the sheer size of the rural sector and the complex factors that affect the diffusion of new ideas and techniques make it *à priori* unlikely that renewable resources will make much of a contribution in the short term'.

However, because Third World countries are well endowed with renewable energy resources, in particular solar, wind, and biomass, they are eminently suited to be the first countries to 'enter the solar age' and to derive a major portion of their energy needs from renewable sources (Hayes, 1977). The alternative energy technologies are generally small scale and low cost, and therefore most appropriate to the rural areas which suffer the most from energy shortages (Hoffman, 1980).

Sun

In the present state of the art, solar energy is most applicable to uses where low-grade heat is required, such as greenhouses, artificial crop-drying, and for water heating for domestic use. For the latter purpose, efficient flat glass collectors have been developed, but their cost is still relatively high.

The solar photovoltaic cell, which requires little maintenance and is reasonably durable, appears to be well suited for the rural areas in developing countries. A research and development programme on a major scale has been undertaken by many private companies to develop photovoltaic power, and prototypes are already being tested. It is estimated that it will be possible to reduce the cost of the energy produced from the present US $ 1–2/peak watt to 10–30 US cents in the course of the next decade (Hoffman, 1980).

A French state-supported solar energy company—SOFRETES—has concentrated since 1962 on developing autonomous solar pumps, primarily for the developing countries. Since 1962, this company has marketed power stations of 1 kW up to 70 kW to developing countries in Asia, Africa, and Latin America (Hoffman, 1980).

The potential market for the small pumps is estimated to run into thousands for Mexico alone (Clemot, 1978). Similar alternative energy programmes, in collaboration with developing countries, have been undertaken by West Germany and the USA (Hoffman, 1980).

Considerable progress has been made in new processes for producing silicon solar cells, but the direct conversion of solar energy into mechanical or electric power is not yet economical (Makhijani & Poole, 1975).

Wind

Windmills as a source of energy for pumping have been used since time immemorial, though modern forms have been evolved, they have gradually been replaced by fuel-activated pumps. A new look at this source of

Fig. 63 IRRI designed Savonius windmill–piston pump: (a) the combination windmill–piston pump; and (b) detail of pump. By courtesy of the Agricultural Engineering Department, International Rice Research Institute, Los Baños, Philippines.

energy is now justified, especially for water supply on small family farms. The International Rice Research Institute has developed a vertical-axis windmill/piston pump combination, that can provide low-cost water pumping in areas where windspeeds average at least 16 km/h, and the water level is not more than 4.6 m below ground level. It can be built by most metal workshops and its simple construction makes it easy to maintain (Fig. 63).

Windmill power generating electricity is still expensive, mainly because of the need for storage batteries.

Water

Energy generation by large hydro-electric installations is resorted to in many developing countries. However, only about 12% of the world's potential waterpower capacity has been harnessed so far. For the developing countries, the figures are still lower: 5.1% in South America, 4.5% in Asia, and 1.7% in Africa (Usmani, 1976).

For several years, local irrigation and construction authorities in China have been operating under instructions to conceive small dams in such a way that it is subsequently feasible to install power-generating equipment at the outflow. Older dams are systematically modified to add power generation. The Chinese soon perceived the possibilities of rural electrification inherent in the small generating gear which many commune and county plants now manufacture in large quantities in order to make use of even the smallest streams with sufficient water pressure (FAO, 1979).

Fuel from Wood, Biomass, Crop Residues, and Manures

Direct Combustion

About 90% of the world's *fuelwood* consumption is in the developing countries, and demand is increasing by 1–2% annually. It is the dominant source of fuel for low-income families (FAO, 1977). Firewood and charcoal account for 96% of all energy consumed in Tanzania, 91% in Nigeria and 90% in Uganda (Hoffman, 1980).

Though rural populations account for the bulk of fuelwood consumption in developing countries, urban consumption is also very large. Because urban use is very concentrated, this has resulted in large de-forested zones surrounding the towns.

Woodfuel is in principle a renewable resource; in practice, uncontrolled cutting has frequently resulted in the complete destruction of former sources, in particular in parkland savanna areas. World Bank estimates indicate that much of the forests in the developing countries will practically disappear in as little as 40 years (World Bank, 1978).

Destructive cutting of forests can be halted and a sustained production of fuelwood built up by appropriate measures. Large and successful self-help schemes have been initiated in China, Republic of Korea and India (FAO, 1977). In all these cases, the schemes are based on control systems implemented with the active involvement of the people concerned.

Another potential source of fuelwood could be plantations of suitable trees in areas in which the original sources of fuelwood have been destroyed or are no longer sufficient. Though the potential of this solution is very large, it has many limitations in developing countries, mainly the need for fairly heavy investments, and for land that can be made available for this purpose.

Economies in use are also possible: in a study in Indonesia, it was found that the consumption of fuelwood for cooking could be reduced by about 70% as a result of air-drying the wood before use, simple improvements in stove design and a new type of cooking pot (FAO, 1977).

The energy potential of *crop residues* could be exploited for industrial use, as is already done in the case of the Hawaiian sugar-cane industry that uses bagasse for fuel.

The net *oil* potential from one ton of dry organic wastes is estimated at 1.4 barrels (Anderson, 1972). However, the economics and technology for using wood and crop residues directly to raise steam for power production is not very favourable. Capital costs are high and efficiencies are low (about 10%). Another possibility, the anaerobic distillation of wood (pyrolisis), requires plants that are too large for village-scale use (Makhijani & Poole, 1975).

The use of dried *cattle dung* as a source of fuel is a widespread practice in Asia and parts of Africa, and involves an enormous loss of plant nutrients.

In India alone, 68 million tons of dried cow dung, with an energy content of 790×10^{15} joule, were used as fuel in a single year. Together with 39 million tons of *vegetable wastes*, these sources accounted for one-third of non-commercial energy consumption in India (FAO, 1977).

Bio Conversion

Agriculture can produce energy, either by growing crops with a high energy output per hectare or by using crop residues. Sugar-cane can produce 35–90 tons of dry matter/ha/year which can be used as a source of ethanol and methanol.

For every ton of cereal grain, there are 1–2 tons of residues. The energy content of these residues is generally equal to, and sometimes greater than, that of the grain. If only one-half could be recovered by fermentation processes, energy requirements for agriculture, including the production of fertilizers, could be satisfied (Revelle, 1976).

The basic raw materials for producing energy actually increase as agriculture becomes more productive.

Methane Generation from Human, Animal and Agricultural Wastes (NAS, 1977)

The rapid rise in the price of fossil fuels has sparked renewed interest in the generation of methane gas by the decomposition of organic matter under anaerobic conditions. Methods have been developed that are suitable for rural conditions in developing countries, for individual households and for village communities.

In India, the Gobar Gas Research Station, established in Uttar Pradesh in 1971, has published a number of designs for gas plants. Many thousands of plants have been built in rural areas, serving one or more families. More attention is also being given to the production of gas from garbage and night soils in urban areas (Dasgupta, 1977).

In Taiwan, several thousand plants have been built with government aid to generate gas from pig manure on small and medium-sized farms. In the Peoples' Republic of China, tens of thousands of methane generators are operating on night soil and other manures. Other countries that are using biogas plants on a significant scale are the Philippines, Thailand and Taiwan.

A number of firms in the UK and Taiwan are currently marketing a biogas digester made completely of neoprene plastic, with a capital cost significantly below that of the traditional model made of concrete with galvanized iron covering.

FIAT has recently developed a high-efficiency electrical generator fuelled by methane, as a source of household energy. The generator is reported to use methane at a 90% efficiency rate, generating 15 kWh of electricity as well as heat for warming water (Barnett *et al*., 1978).

The use of rural wastes for methane production, instead of for direct use as a fuel or manure, has the following advantages: (a) it produces an energy source that can be stored and used efficiently; (b) the remaining sludge retains the fertilizer value of the original material; (c) it provides a sanitary way to dispose of human and animal wastes and reduces the public health hazards of fecal pathogenes; Weed seeds and plant pathogenes are destroyed; and (d) it reduces the time spent on collecting fuel and cooking.

The non-toxic gas can be used in modified gas-burning appliances for cooking, heating water, lighting, etc. Internal combustion engines can easily be converted to run on methane gas and can be used for water-pumping or traction.

Methane plants are most efficient at relatively high ambient temperatures and are relatively labour-intensive, so they are particularly suited to the conditions prevailing in developing countries. Most cost studies have indicated that economic evaluations are favourable. The gas produced consists of 55–65% by volume of methane, the remainder being CO_2 and traces of H_2S. If the gas is purified, the energy content is similar to that of pipeline quality natural gas (Makhijani & Poole, 1975).

Summary

The energy-crisis with its step increase in the cost of fossil fuels, has compounded the problems encountered by developing countries in the process of modernizing their agricultures.

Import bills of the most seriously affected countries rose by 560% for fertilizers, 338% for fuel and 140% for food (FAO, 1977). The situation created re-inforces the need for a labour-intensive type of production—multiple-cropping and selective mechanization—during the transitional stages from subsistence to market-oriented agriculture. However, the amount of fossil energy used by the developing countries is already very small and cannot be arbitrarily reduced. Whilst agricultural production in the developed countries took 2.9% of the total world consumption of fossil fuels, in developing countries agriculture used only 0.6%. The disparities are even greater if one considers the whole food-chain system (FAO, 1977).

Because the needs of agriculture are relatively minor, there is little scope for reduction in commercial energy consumption, especially as it is needed in the form of the most essential inputs for increasing productivity and reducing losses in agriculture, namely fertilizers, irrigation and plant protection chemicals. These are the ones which require the greatest inputs in terms of fossil energy. Any reduction in their use will inevitably cause corresponding declines in agricultural production.

A further point to remember is that when high yield levels are achieved, far less land is required to produce a given amount of food, and therefore the amount of energy required per unit of produce—for tillage, fertilizers, biocides, etc.—is far lower than at lower yield levels.

In the transition from subsistence to modern farming, the effective use of a modest amount of commercial energy can achieve a doubling of yields.

The input of commercial energy in rice production in the USA is ten times greater than that used for the same crop in modernized agriculture in the Philippines. It is therefore clear that proposals aimed at reducing energy inputs in the course of transforming traditional agriculture (apart from avoiding indiscriminate mechanization), that arbitrarily limit the inputs required for improving production, are discriminatory and counter-productive.

The major components of a policy aimed at reducing fossil energy inputs consist, in essence, of: (a) careful management practices aimed at avoiding waste; (b) increasing production per unit area by the use of modern inputs up to the point of maximum economic benefit, thereby sacrificing the last increments required for achieving maximum yields; and (c) making use of alternative sources of energy with the maximum possible efficiency.

Although the direct and indirect fuel needs of the agriculture of the developing countries represent only a small fraction of world consumption, they are crucial for the modernization of agriculture, justifying a high priority in any rational allocation of world fuel resources (FAO, 1976).

REFERENCES

Aggarwal, P. C. (1973). Green revolution and employment in Ludhiana, Punjab, India. Pp. 40–63 in *Rural Development and Employment* (Ed. C. Gotsch). Ford Foundation Seminar, Ibadan, Nigeria: 774 pp.

Anderson, L. L. (1972). *Energy Potential from Organic Wastes: a Review of the Quantities and Sources*. US Department of the Interior, *Circ*. 8549: 16 pp.

Anon (1974). *Energy Requirements for Agriculture in California*. Joint study prepared by the California Department of Food and Agriculture and the Agricultural Engineering Department, University of California, Davis.

Arnon, I. (1969). *Transition from Extensive to Intensive Agriculture in Israel with Fertilizers* pp. 13–24 in Proceedings of the VIIth Colloquium of the International Potash Institute, Berne. 282 pp. illustr.

Arnon, I., Molcho, S. & Raviv, M. (1975). *Change in Arab Villages in Israel, the West Bank and the Golan*, Settlement Study Centre, Rehovot (in Hebrew), 441 pp., illustr.

Bakker-Arkema, F. W., Birkert, W. G., Boyd, J. S., Doss, H. J., Gerrison, J. B., Hansen, C. M., Heldman, D. R., Loudon, T. L., Maddex, R. L., Stout, B. A., Surbrook, T. C. & White, R. G. (1974). *Energy in Michigan Agriculture*. Michigan State University, East Lansing, Mich.: 45 pp., illustr.

Baldwin, K. D. S. (1957). *The Niger Agricultural Project*. Harvard University Press, Cambridge, Mass.: xvi + 221 pp., illustr.

Balis, J. S. (1968). *An Analysis of Performance and Costs of Irrigation Pumps Utilizing Manual, Animal and Engine Power*. Agency for International Development, New Delhi, India.

Barker, R. (1970). Economic aspects of new high-yielding varieties of rice, IRRI Report. Pp. 29–53 in *Agricultural Revolution in South-east Asia*, Vol. I (Ed. A. Russell). The Asia Society, New York.

Barker, R. & Anden, T. (1975). Factors influencing the use of modern rice technology in the study areas. Pp. 17–40 in *Changes in Rice Farming in Selected Areas of Asia*. International Rice Research Institute, Los Baños: 377 pp., illustr.

Barker, R., Johnson, S. S., Alviar, N. & Orcino, W. (1969). Comparative economic analysis of farm data in the use of carabao and tractors in lowland rice farming. Paper presented at *Farm Management Sem. with Focus on Mechanization, Manila*.

Barnett, A. L., Pyle, L. & Subramanian, S. K. (1978). *Biogas Technology in the Third World: A Multidisciplinary Review*, IDRC, Ottawa.

Batty, J. C., Hamad, S. W. & Keller, J. (1974). *Energy Inputs for Irrigation*. Utah State University, Logan, Utah: iv + 25 pp., illustr.

Black, J. W. (1971). Energy relations in crop production. *Ann. Appl. Biol.*, **67**, 272–8.

Blouin, B. M. (1974). *Effects of Increased Energy Costs on Fertilizer Production Costs and Technology*. Tennessee Valley Authority, National Fertilizer Development Center, Muscle Shoals, Alabama: 33 pp.

Borlaug, N. E. (1976). The green revolution: can we make it meet expectations? *Proc. Am. Phytopathol. Soc.*, **3**, 6–21.

Boshoff, W. H. & Minto, S. I. (1975). Energy requirements and labour bottlenecks and their influence on the choice of improved equipment. Pp. 195–9 in *Farming Equipment Innovation for Agricultural Development and Rural Industrialisation* (Ed. S. B. Westley & B. F. Johnston). Institute for Development Studies, University of Nairobi, Kenya: 238 pp.

Bradfield, R. (1971). *Mechanical Maximum Cropping Systems for the Small Farms of the Rice Belt of Tropical Asia.* Farm Machinery Industrial Research Corporation, Tokyo, Japan.

Brown, L. R. (1970). The green revolution, rural employment and the urban crisis. Paper presented at the *Columbia Conf. on International Economic Development, New York*.

Carr, Marylin (1975). *Some Social and Economic Aspects of Tractorization in Ceylon* (mimeo) cited by Pearse, 1980.

Castillo, Gelia T. (1975). Diversity in unity: the social component of changes in rice farming in Asian villages. Pp. 348–60 in *Changes in Rice Farming in Selected Areas of Asia*. International Rice Research Institute, Los Baños: 377 pp., illustr.

Clemont, M. G. (1978). Contribution of solar energy to the development of arid zones. *ISES (UK) Solar Energy for Industry*.

Collinson, M. P. (1965). *Economics of Block Cultivation Schemes* (mimeogr.). Western Region Research Centre, Ukiriguru.

Crosson, P. R. & Frederick, K. D. (1977). *The World Food Situation*. Resources for the Future, Washington, DC: v + 230 pp., illustr.

Dalrymple, D. G. (1969). *Technological Change in Agriculture, Effects and Implications for the Developing Nations*. USDA, Foreign Agricultural Service, Washington, DC: v + 82 pp.

Dandekar, V. M. (1969). Questions of economic analysis and the consequences of population growth. Pp. 366–75 in *Subsistence Agriculture in Economic Development* (Ed. C. R. Wharton, Jr). Aldine Publishing Co., Chicago, Ill.: 481 pp.

Darmstadter, J. (1972). Energy in the world economy. *Development Digest*, **10** (3), 3–16.

Dasgupta, B. (1977). *Agrarian Change and the New Technology in India*. UNRISD, United Nations, Geneva: xxvii + 408 pp.

Eicher, C. K., Zalla, T., Kocher, J. & Winch, F. (1970). *Employment Generation in African Agriculture*. Institute of International Agriculture, College of Agriculture and Natural Resources, Michigan State University, East Lansing, Mich.: 66 pp.

FAO (1966). *Agricultural Development in Nigeria 1965—1980*. Food and Agriculture Organization of the UN, Rome: 512 pp., illustr.

FAO (1968). Raising agricultural productivity in developing countries through technological improvement. Pp. 73–113 in *The State of Food and Agriculture*. Food and Agriculture Organization of the UN, Rome: viii + 205 pp.

FAO (1969). *Tentative Indicative World Plan*. Food and Agriculture Organization of the UN, Rome.

FAO (1976). *The State of Food and Agriculture 1975*. Food and Agriculture Organization of the UN, Rome: vi + 150 pp.

FAO (1977). Energy in agriculture. Pp. 81–111 in *The State of Food and Agriculture 1976*. Food and Agriculture Organization of the UN, Rome: vi + 157 pp.

FAO (1979). *The State of Food and Agriculture, 1978: Agriculture Series No*. 9. Food and Agriculture Organization of the UN, Rome.

Giglioli, E. G. (1965). Mechanical cultivation of rice on the Mwea irrigation settlement. *E. Afr. Agr. & Forest J.*, **30**, 177–81.

Giles, G. W. (1963). *Opportunities for Advancing Agricultural Mechanization in India and South-East Asia: ASAE Paper No*. 63–154. American Society of Agricultural Engineers.

Giles, G. W. (1967). Agricultural power and equipment. Pp. 175–216 in *The World Food Problems*, Vol. III. A Report of the President's Advisory Committee. Washington, DC: xxii + 332 pp., illustr.

Giles, G. W. (1975). *The Reorientation of Agricultural Mechanization for the Developing Countries*. Shin-Norinsha Co., Tokyo, Japan.

Greenland, D. J. (1975). Bringing the green revolution to the shifting cultivator. *Science*, **190**, 841–4.

Hardy, R. W. F. & Havelka, U. D. (1975). Nitrogen fixation research: a key to world food? Pp. 178–85 in *Food: Politics, Economics, Nutrition and Research* (Ed. P. H. Abelson). American Association for the Advancement of Science, Washington, DC: v + 202 pp., illustr.

Haswell, M. R. (1953). Economics of agriculture in a savannah village. *Col. Office Res. Studies No.* 8, 1–142.

Hayes, D. I. (1977). *Energy for Development: Third World Options*. World Watch Institute, Paper No. 15. Washington DC.

Heady, E. O. (1962). Research and economic development—needs, opportunities and problems. Pp. 1–31 in *Food: One Tool in International Economic Development* (Ed. E. O. Haroldsen). University Press, Ames, Iowa: x + 419 pp

Heichel, G. H. (1974). Comparative efficiency of energy use in crop production. *Bull. Conn. Agr. Exp. Sta.*, **739**, 26.

Hoffman, K. (1980). Alternative energy technologies and Third World rural energy needs. *Development and Change*, **11**, 335–65.

Hülst, H. von (1975). The role of farm mechanisation in developing countries. Pp. 56–67 in *Farming Equipment Innovation for Agricultural Development and Rural Industrialization* (Ed. S. B. Westley & B. F. Johnston). Institute for Development Studies, University of Nairobi, Kenya: 238 pp.

Hutter, W. (1976). Energie consommée pour la production de quelques cultures. *Compt. Rend. Acad. Agric. de France*, **62**, 297–307.

IRRI (1978). *Economic Consequences of the New Rice Technology*. International Rice Research Institute, Los Baños, Philippines: v + 402 pp., illustr.

Kawamoto, A. (1972). Socio-cultural adjustments of farm families and rural communities in the process of mechanisation. Pp. 331–49 in *Farm Mechanization in East Asia* (Ed. H. Southworth). The Agricultural Development Council, New York: 433 pp., illustr.

Khan, A. U. (1974). *Mechanization Technology for Tropical Agriculture: Paper* 74–01. The International Rice Research Institute, Los Baños, Laguna, The Philippines.

Kim, Sung-Lo (1972). A socio-economic analysis of farm mechanisation in Asiatic paddy farming societies with special reference to Korea's and Japanese cases. Pp. 304–30 in *Farm Mechanization in East Asia* (Ed. H. Southworth). The Agricultural Development Council, New York: 433 pp., illustr.

Kline, C. K., Green, D. A G., Donahue, R. L. & Stout, B. A. (1969). *Agriculture Mechanization in Equatorial Africa*. Institute of International Agriculture, Michigan State University, East Lansing, Mich.: 593 pp., illustr.

Lai, Weng-Choch (1972). Current problems of farm management on mechanized farms. Pp. 162–90 in *Farm Mechanization in East Asia* (Ed. H. Southworth). The Agricultural Development Council, New York: 433 pp., illustr.

Leach, G. (1976). *Energy and Food Production*, 2nd edn. IPC Science and Technology Press, Guildford.

Lyon, T. L. (1936). The residual effects of some leguminous crops. *Bull. Cornell Univ. Exp. Sta.*, **645**: 17 pp., illustr.

Makhijani, A. & Poole, A. (1975). *Energy and Agriculture in the Third World*. Ballinger Publ. Co., Cambridge, Mass.: 168 pp.

McPherson, W. W. & Johnston, B. F. (1967). Distinctive features of agricultural development in the tropics. Pp. 184–230 in *Agricultural Development and Economic Growth* (Ed. H. M. Southworth & B. F. Johnston). Cornell University Press, Ithaca, New York: xv + 608 pp., illustr.

Meiklejohn, J. (1955). Nitrogen problems in tropical soils. *Soils and Fertilizers*, **18**, 459–63.

Mettrick, N. (1968). Mechanization of peasant agriculture in East Africa, Pp. 555–66 in Bunting, A. (1970). *Int. Sem. on Change in Agriculture*. Duckworth & Co., London: xiv + 313 pp., illustr.

Moore, A. W. (1962). The influence of a legume on soil fertility under a grazed tropical pasture. *Em. J. Exp. Agric.*, **30**, 239–48.

Morita, Y. (1968). Agricultural mechanization in Japan. Paper presented at *Expert Group Meeting on Agricultural Mechanization, Asian Productivity Organization, Tokyo*.

NAS (1976). *Energy for Rural Development: Renewable Resources and Alternative Technologies for Developing Countries*. National Academy of Sciences, Washington, DC: 306 pp., illustr.

NAS (1977). *Methane Generation from Human, Animal and Agricultural Wastes*: National Academy of Sciences, Washington, DC: xiii + 131 pp., illustr.

Nelson, M. (1973). *The Development of Tropical Lands: Policy Issues in Latin America*. Johns Hopkins University Press, Baltimore, Md.: xvii + 306 pp., illustr.

Norman, D. W. (1970). Traditional agricultural systems and their improvement. Paper presented at a *Sem. in Agronomic Research in West Africa, University of Ibadan, Nigeria*.

Otieno, J., Muchiri, G. & Johnston, B. F. (1975). The implications of Kenya's present economic structure and the rates of growth of the population and labor force. Pp. 90–103 in *Farming Equipment Innovation for Agricultural Development and Rural Industrialization* (ed. S. B. Westley & B. F. Johnston). Institute for Development Studies, University of Nairobi, Kenya: 238 pp.

Pasthasarathy, F. (1975). West Godavari, Andhra Pradesh. Pp. 43–70 in *Changes in Rice Farming on Selected Areas of Asia*. International Rice Research Institute, Los Baños: 377 pp., illustr.

Patil, N. P. (1963). *Economics of Drill Sowing over Broadcasting*. Agric. Sta., Bangalore, India.

Peacock, J. M. (1967). *The Report of the Gambia Ox-Plough Survey, 1966*. Wye College, London.

Pearse, A. (1980). *Seeds of Plenty, Seeds of Want. Social and Economic Implications of the Green Revolution*. Clarendon Press, Oxford. xi + 262 pp.

Pimental, D. (1973). Energy crisis and crop production. Pp. 65–79 in *Energy and Agriculture—Research Implications* (Ed. L. Fischer & A. Biere). North Central Regional Strategy Committee on Natural Resources Development, Rep. No. 2: xxii + 98 pp., illustr.

Pimental, D., Hurd, L. E., Bellotti, A. C., Forster, N. J., Oka, I. N., Sholes, O. D. & Whitman, R. J. (1973). Food production and the energy crisis. *Science*, **182** (4111), 443–9.

Rask, N. (1969). *Analysis of Capital Formation and Utilization in Less Developed Countries: Occasional Paper No.* 4. The Department of Agricultural Economics and Rural Sociology, Ohio State University, Columbus, Ohio.

Revelle, R. (1976). The resources available for agriculture. *Sci. Am.*, **235** (3), 165–80.

Sanchez, P. A. & Buol, S. W. (1975). Soils of the tropics and the world food crisis. Pp. 115–20 in *Food: Politics, Economics, Nutrition and Research* (Ed. P. H. Abelson). American Association for the Advancement of Science, Washington, DC: v + 202 pp., illustr.

Shaeffer-Kehnert, W. (1973). Farm mechanization in the developing countries. Pp. 26–49 in *Applied Sciences and Development*, Vol. 2. Institute for Scientific Cooperation, Tübingen, FRG: 162 pp.

Steinhart, J. S. (1973). Energy use in the U.S. food system. Pp. 41–64 in *Energy and Agriculture, Research Implications* (Ed. L. Fischer & A. Biere). North Central Regional Strategy Committee on Natural Resource Development, Rep. No. 2: xxii + 98 pp., illustr.

Steinhart, J. S. & Steinhart, C. E. (1974). Energy use in the U.S. food system. *Science*, **184**, 307–16.

Tamin, M. bin., & Mustapha, N. H. (1975). Kelawan, West Malaysia, Pp. 201–23 in *Changes in Rice Farming in Selected Areas of Asia.* International Rice Research Institute, Los Baños: 377 pp., illustr.

Tschiersch, J. E. (1978). *Appropriate Mechanization for Small Farmers in Developing Countries*. Publications of the Research Centre for International Agrarian Development. Verlag Breitenbach, Saarbrücken, Germany, iii + 106 pp.

Usmani, F. H. (1976). *Review of the Impact of Production and Use of Energy on the Environment and the Role of UNEP*. UNEP, Nairobi, Kenya.

Webb, R. A. (1960). Problems of fertilizer use in tropical agriculture. *Outlook on Agr*., **2**, 103–13.

Westley, S. B. & Johnston, B. F. (1975) (Ed.). *Farming Equipment Innovation for Agricultural Development and Rural Industrialization*. Institute for Development Studies, University of Nairobi, Kenya.

Wilde, J. C. de & McLoughlin, P. F. M. (1967). *Experiences with Agricultural Development in Tropical Africa*. Johns Hopkins University Press, Baltimore, Md: Vol. I, xi + 254 pp.

World Bank (1978). *Forestry*: Sector Policy Paper, Washington DC.

Wu, Carson Kung-Hsien (1972). Analysis of machinery–labor relationship in farm mechanisation. Pp. 70–92 in *Farm Mechanization in East Asia*, (Ed. H. Southworth). The Agricultural Development Council, New York: 433 pp., illustr.

Yudelman, M., Butler, G., & Banerji, R. (1971). *Technological Change in Agriculture and Employment in Developing Countries: Development Centre Studies, Employment Ser. No.* 4. Development Centre of OECD, Paris: 204 pp.

CHAPTER X

Provision of Essential Conditions for Modernization

SOCIO-CULTURAL AND BEHAVIOURAL FACTORS

The development of agriculture is not a purely technological or economic problem. Its success is frequently dependent on an understanding of the society in which it is to take place, a knowledge of the social and cultural factors that condition farmers' responsiveness to technological change, and the ability to obtain willing cooperation of the people involved.

Importance

There are two opposing schools of thought regarding the motivations, attitudes, and values of the traditional subsistence farmer, or the so-called peasant. One school considers the peasant as 'a miniature economic man' acting within the bounds of economic rationality* and striving to improve his economic position (Schultz, 1964). According to this view, socio-cultural factors only have a marginal role; subsistence farmers live and work in a physical, economic, and cultural environment that is relatively static. Therefore, they have, through a process of trial and error, gradually arrived at an optimal solution to the allocation of the limited resources available to them (Mellor, 1969).

This equilibrium implies that the agricultural sector of the economy has exhausted all the profitable production and investment opportunities inherent in the existing state of the arts. In other words, this is a state of 'technological stagnation' (Dandekar, 1969), which is not due to subjective attitudes of the farmer, but to an economic environment over which he has no control. According to this school of thought, traditional farmers will respond quickly, normally and efficiently to economic incentives in adopting new techniques (Behrman, 1969).

A second school considers that in subsistence agriculture, non-economic forces generally outweigh purely economic forces, leading to behaviour that is *not* within the bounds of economic rationality (Balogh, 1966).

It appears reasonable to assume that both extreme points of view do not

*Economic rationality: behaviour that seeks to maximize economic returns or income.

reflect the real situation, and that individual farmers and groups fall on different points of a continuous scale.

There are certain communities and groups that do not require much persuasion to adopt new techniques, and in which the innovator enjoys respect and prestige, while there are others, and they probably form the majority, in which tradition and established customs are conducive to resistance to change, in spite of all manner of persuasions and inducements, refusing to increase their production and income even if opportunities are available.

Penny (1969), in a study of eight villages in north Sumatra, found that despite the fact that all farmers operated in the same economic environment, their response to opportunities to increase their incomes was extremely diverse. In the course of 50 years some showed an amazing capacity for making rapid changes and became commercial farmers. Others remained tradition-bound, and made no changes in their farming methods or way of life, while yet others were in a transition from subsistence to commercial farming.

The factors influencing farmers' responsiveness to technological change can be due to: (a) cultural values, such as concepts of right and wrong, beliefs and rules of behaviour, attitudes to agriculture, etc.; (b) the social organization of the community: prevailing family and clan relationships, class systems and castes, gerontocracy, attitudes to women; and (c) psychological, such as fatalism.

Beliefs and Rules of Behaviour

Social and cultural factors are characteristic of the society to which the farmer belongs and dominate human behaviour. These include beliefs and rules of behaviour which may constitute formidable barriers to technological innovations, and cannot be easily changed by exhortations, regulations or logic.

Magical and Religious Beliefs

Probably the best known are concepts concerning livestock, leading to non-commercial attitudes to this potentially important branch of production. These attitudes are based on religious beliefs, or on social values attached to the accumulation of cattle as a source of prestige, etc.

In certain areas in China, the introduction of white leghorns, which were undoubtedly an improvement over the local breed, failed in the past because of a taboo against the raising and eating of white birds. A different improved breed, such as Rhode Island red would have had a greater prospect of success (Yang Hsin-Pao, 1949).

Other religious beliefs, such as respect for scattered burial plots, may seriously hamper efforts to consolidate land holdings.

A sizeable portion of the re-settlement budget for the Volta dam was allocated for food and drink to pacify the spirits of shrines which had to be re-located. A fear that handicapped another re-settlement programme arose from the belief of the Abakalti that women who live in tightly settled, integrated areas will become barren (Smock, 1969). In the Philippines, rituals, prayers, and other ceremonies are performed to ward off evil spirits from rice fields, to hasten the growth of the rice and to ensure a good harvest (Tagampay-Casillo, 1970).

Are Religious Beliefs a Serious Obstacle to Modernization?

Lutfiyya (1966), a Palestinian-born Arab anthropologist, in a study of the social institutions and social change in his native village, writes: 'Islam sanctions traditional behaviour and gives it precedence over innovations. This view finds legal support in the shariza doctrine that declares, "al-quadim yabqa zala qidamih", i.e. anything of the past has precedence (over innovations).' Further, 'Innovators are always the objects of shame and ridicule (in the Moslem villages). Invariably there is an outright rejection of anything new that appears to conflict with tradition.'

There is no reason to doubt that the attitudes described by Lutfiyya are authentic, yet it can be stated that all the evidence shows that when favourable conditions occurred for modernization of Arab villages in Israel, and somewhat later in the West Bank, that the process was not hampered by tradition based on religion.

In a study of the transformation of Arab agriculture in Israel and the West Bank, no significant differences in the level and rate of modernization of agriculture were found, between the exclusively Moslem villages in the coastal plain and the mixed Christian, Druse and Moslem villages in Galilee; or even between farmers of different religions within the same village (Arnon *et al.*, 1975).

Even the concept of the 'holy cow' appears to be weakening as modernization reaches the Indian village. In Uttar Pradesh, the slaughter of 'unproductive' bulls and oxen is permitted. Large-scale slaughter of cattle is carried out in large town on the justification that sick animals may be eliminated. The problem is also by-passed by replacing dairy cows by buffaloes; the latter are not included in the religious law against slaughter (Schiller, 1964).

Fatalism

Agriculture in general is subject to forces over which the individual has no control. In traditional societies, floods, droughts, locust invasions, diseases, and epidemics are 'visitations from gods or evil spirits whom man can propitiate, but not control' (Foster, 1962). A 'poverty trap' prevents the majority of villagers from improving their lot. Belief that everyting is 'God's will' with its concomitant fatalistic attitude, is therefore an adjustment to a situation from which the individual has not been able to extricate himself for untold generations.

Attitudes to Agriculture

Unfortunately, agriculture is held in very low esteem in most developing countries. Lutfiyya (1966), describing social attitudes in an Arab village writes: 'Paradoxically, even though there is prestige associated with the ownership of land, the cultivation of land is regarded as a degrading occupation. Consequently, many villagers abandon farming and seek other occupations.'

Pastoral tribes, such as the Masai in Kenya, the Watutsi in Burundi, and the Bedouin in Arab countries despise manual work in general and agriculture in particular; they consider hoeing the soil as a servile and degrading occupation (Dumont, 1966).

It is not uncommon in countries in which 70% of the people are engaged in rural occupations, for only 5% of the university students to elect agriculture as a field of study—and many of them students enroll in agriculture only after having been turned down for study in more prestigious fields. Owing to the fact, perhaps, that agriculture is largely an exercise of laymen, together with the probability that it has not achieved complete professional prestige in almost any developing country, students tend to elect what they consider to be the more certain paths to professional status.

Developed countries which extend help to developing countries, often contribute to a perpetuation of this state of affairs. For example, it is one of the anachronisms of foreign student populations in the United States that, despite the predominantly agrarian nature of their home countries, only 3.6% elect agricultural fields of study. Even these generally do not receive training that improves their suitability for work in their home countries.

In a study of farmers' education aspirations and expectations carried out in seven villages in Laguna district (Philippines), it was found that in spite of improvement in their standard of living following the adoption of HYVs, the dreams and expectations of peasants for their children were definitely away from farming. Even the modernization of farming did not appear to make it attractive enough for them to want their children to stay in it. The college education to which they aspired was not associated with farming, but was considered 'a passport to better jobs, better life and better social standing'. The farmers 'could not understand why persons with college degrees, such as agricultural extension workers, would work in the village and get their feet muddy' (Castillo, 1975).

In a random sample survey of over 700 pupils of the final form of 45 secondary modern schools in western Nigeria, only 6.2% wanted to become farmers (Roider, 1971).

There are exceptions but these are unfortunately rare. For example, the

Sikh religion, a sect of Hinduism practised by the majority of the inhabitants of the Punjab, promotes the dignity of labour and has contributed to the partial breaking down of the castes in that state, which agriculturally is also one of the most progressive in India (Dasgupta, 1977).

The Indian sociologist Nair (1961), found that the tacit assumption made in planning, 'that given equal opportunity, financial incentive and resources, *all* communities will respond similarly, in their productive attitude to work, aspirations and habits' is not valid, She gives numerous examples to show that the social factor ultimately determines whether resources are properly used or not for increasing production. 'A community's attitude to work can be a more decisive determinant for raising productivity in Indian agriculture, than material resources, or for that matter even technology'.

The need to instill the concept of the dignity of farming first in leaders, extension workers and teachers, and through them, into the community as a whole, cannot be overstressed. Respect for work and workers, and in particular for farming and farmers, should become part and parcel of their way of thinking and of feeling. This will be difficult as long as agriculture is primitive and consists mainly of drudgery and back-breaking work. However, as has has been pointed out in Chapter IX, with increasing sophistication of agricultural production and a reduction in the amounts of manual work involved, changes in attitudes to farming can be achieved.

Social Organization

Rural society can be socially organized in a number of ways and these will, in turn, be related to other aspects of the social system such as caste, kinship, gerontocracy, status of women, etc.

Relationships

In tropical Africa, for example, the traditional organization of rural society is based on the primary importance of kinship relationships, gerontocracy, and a strong identification of the individual with the group, frequently associated with distrust of or even hostility towards strangers (Jones, 1965).

These mutual relationships will vary from one society to another. The persistence of certain social and cultural traditions may seriously hamper the adoption of new technology or increased efficiency in agricultural production.

Dunbar & Stephens (1970) describe some of the difficulties faced by a farmer in Uganda, who would like to adopt new practices: he must first obtain the approval of the members of his family before making any change; if he increases his income, he must share his relative affluence with impecunious relatives; his success may provoke the enmity of his neighbours, and he may be exposed to armed robbery.

Traditional systems may be valued and loved, as exemplified by a Gezira settler's statement: 'We hate these straight lines, we would rather be hungry every few years, with freedom to range with our cattle unconfined, than have full bellies and be fined if we stray outside these horrid little squares' (Gaitskell, 1959).

A tradition of cooperation is an integral part of village cultures in many parts of the world (Foster, 1962). Neighbours or family-groups often join in reciprocal activities, such as reclaiming land, harvesting, house-building, etc., often to the sound of music or singing. Foster stresses that this type of reciprocal behaviour is a social imperative, and that these mutual obligations take the place of social security and welfare, cooperative and credit facilities, etc., which are characteristic of more developed societies.

A social tradition, such as the extended family, that ensures that all members of the family survive, however meagerly, has been described as 'a form of social insurance against bad times' (Davy, 1974).

The need to share whatever meagre harvest is available amongst all relatives, is exemplified by the Javanese custom of using a small knife, instead of a sickle, for harvesting rice. This makes it possible for more labour to share in the harvest.

However, the 'pressure to share poverty' in Javanese villages is beginning to weaken, as reflected by the gradual spread of the *tebasan* system, under which the rice crop is sold long before it is harvested. The buyer is responsible for harvesting the crop and is relatively free of the usual cultural obligations to allow as many harvesters as possible, on a privileged sharing basis (Castillo, 1975).

Some hill people of Assam regard it as shameful to sell food; food should be shared freely in times of shortage with those who would otherwise go hungry. The result is that they resist technological innovations aimed at producing and selling surplus food (Hunter & Jiggins, 1978). It is, however, a tragic irony that such ethical traditions should be a major disincentive to economic progress. By implying that any increases in production and hence in income have to be widely shared with people who have made no contribution to the increased production, may therefore be a serious disincentive to increasing productivity.

These traditional forms of cooperation and mutual help have frequently been considered as evidence that the introduction of modern forms of cooperation would be easy because they are in tune with existing cultural patterns. This mistaken assumption has been the cause of many disappointments experienced by well-meaning innovators, who have attempted to introduce cooperatives, for example, in rural areas as an instrument for strengthening their economies.

In most rural societies, side by side with the social forms of traditional cooperation, which are generally based on family ties, and confined to specific traditional activities, exist major divisions, due to mutually antagonistic clans or other types of power blocks, or on a horizontal stratification of the community. These conflicts can be no less traditional and ingrained than the reciprocal self-help activities mentioned above. Factionalism was a major problem which community development programmes had to face in India (Foster, 1962).

Another almost universal trait, is reciprocal mistrust between individual villagers, the result of poverty and of limited opportunities, whereby 'if someone is seen to get ahead, logically it can only be at the expense of others in the village' (Foster, 1962). It finds its most common expression in the bitter litigations and feuds that arise after the death of parents when the inheritance has to be divided amongst siblings.

The pervading fear and distrust is poignantly expressed by a Hindu villager: 'Our lives are oppressed by many fears. We fear the rent collector, we fear the police watchman, we fear everyman who looks as though he might claim some authority over us, we fear our creditors, we fear our patrons, we fear too much rain, we fear locusts, we fear thieves, we fear the evil spirits which threaten our children and our animals and we fear the strength of our neighbour' (Wiser & Wiser, 1951).

It is the factionalisms between groups, and the deep mistrust between individuals, that have proved a major barrier to the institutionalizing of economic activities in villages the world over.

In a study carried out in Brazil, Galjart (1971) found that 'agricultural development was hindered not so much by failure of farmers to adopt technological innovations as by their lack of cooperativeness and solidarity, their refusal to accept responsibility for the facilities and services shared with others, and their ideal of becoming big landed proprietors who employ others to do the manual labour and get rich by producing and marketing a single cash crop.'

The typical attitudes described above should not be construed as an indication that institution building in rural socieities is *a priori* doomed to failure. A clear realization of the difficulties involved is essential in order to avoid facile optimism and to devise an appropriate strategy for replacing the forces that divide people within the village by forces that link them together to their mutual benefit. Of these, the most potent factor may well be the fact that 'a sense of personal identification with small groups seems to be necessary for most people, to provide psychological security and satisfaction in their daily work' (Foster, 1962). The author concludes that unless effective small groups with common binding interests can be created, promotional efforts may be doomed to failure. The ideal of family unity and cooperation may be eventually extended to a larger unit, such as the village.

Class Systems and Castes

Many societies in developing countries have rigid classes of society which constitute barriers to social, economic and agricultural change.

In most backward economies, there exists a 'master–servant' situation, whereby a landed elite dominates the peasants who are passive, dependent and subservient towards those above them. For both groups, this situation discourages initiative and reduces motivation to achievement (Brewster, 1967). Typical for countries with a rigid social order, is the overwhelming concern of individuals within the higher strata for the re-inforcement of their own status.

This attitude entails a reluctance to transfer knowledge to those they consider their inferiors. They, in turn, have been conditioned to believe that 'attempts to raise their own status by acquiring higher knowledge are impermissible' (Brasseur, 1976). These attitudes, which prevail in conservative rural societies, consistently hamper overall change.

In villages in which the caste system dominates, and where it re-inforces the inegalitarian land structure, new technology cannot be adopted equally by everyone, especially if the dominant caste also controls the local institutions, such as cooperatives.

A study on the adoption of new practices in rice cultivation was carried out in the village of Pedapullera (West Godavari district, Andhra Pradesh state, India). This village is fairly representative of an area in which many farmers are not only progressive but also rather prosperous. The rice fields are well drained and the irrigation system is dependable. It has a good all-weather road with easy access to trading and administrative centres. In the vicinity of the village is a seed-farm and a soil-testing laboratory. A village-level extension worker lives in the village. There is also a cooperative credit society with a long and sound history, which supplies fertilizers and insecticides, maintains a rice-mill and purchases the crop from the villagers. And yet, seven years after the introduction of HYVs into the village, of 185 farm households, nearly a third, belonging to the lowest caste, have never grown modern varieties. It was found that a conjunction of small farm size, tenancy, illiteracy and dependence on others for credit and crucial farm assets was the reason that a whole social group in the village was almost completely by-passed by the changes in farm technology taking place in the village around them. The public institutions made no attempt to reduce the impact of these adverse factors. The farmers of this group essentially had no access to cheap institutional credit from the village cooperative and when there were shortages of inputs, they went without (Parathasarathy, 1975).

Gerontocracy

A tradition that is still fairly prevalent in tropical Africa is that of farming by the extended family, under the direction of its head, giving the main economic power to the older people. Dumont (1966) stresses the evil influence of this gerontocracy on the villages. By making the strongest economic power the preserve of the chiefs with their large families, many primitive societies entrust the levers of progress to the oldest, who are often the least receptive to modern technques.

In most underdeveloped and developing countries, the adults are emotionally committed to primitive agricultural techniques and to tribal customs to a degree which seriously hampers or even blocks progress and offers a serious danger of 'counter education' (Chatelain, 1963).

Attitudes to Women

Reliance on, and strengthening of, the existing social patterns will also, in most cases, mean a perpetuation of the degraded role played by women in primitive societies. In most underdeveloped and developing countries, girls and women are an essential part of the farm labour force (cf. pp. 154–156). Where wives do not share either directly or indirectly in cash-crop profits, they may resent increased work-loads. This will inevitably lead to household tensions and hinder farm development (Scudder, 1964). It is not suggested that women should not participate in work on the family farm, but that the traditional forms of participation often need changing. The vocational training of women is usually neglected, as a result of established social patterns and prejudices.

The status of women is far from identical in different countries, and, to quote the recommendation of FAO in this connection: 'Traditional ways of thought must be taken into consideration and can only be modified with caution. But modified they must be, for sooner or later, the day will come when the peasant woman will become conscious of her slavery and of her exhausting and unattractive life; then she will flee to the city and the men will follow suit'.

In certain societies, the wife already has a major influence on the farmer's decison-making, as shown by the data presented in Table XXXVIII, based on a survey of 395 Filipino farmers. In this society, the wife plays a dominant role in decisions affecting all aspects of farming, even the most technical. In certain vital aspects, such as engaging in a new enterprise or buying a water buffalo, the wife is almost the only individual from whom advice is sought.

Table XXXVIII Individuals exerting influence on the farmer's decision-making (Lu, 1968, by permission).

Type of decision	Wife	Landlord	Extension worker	Others
Buying problems	62	52	37	5
Where to sell produce	69	4	—	—
Engaging in a new enterprise	78	2	2	—
Buying a water buffalo	83	7	0.2	—
Buying farm tools and equipment	75	6	6	—
Where to borrow money	84	25	12	—
Changing new varieties	60	52	40	3
Changing cultural practices	58	50	40	8

The ideal of 'dual responsibility for the homestead', or husband–wife teams, as a factor for progress, should be strengthened where it already exists and encouraged where it is not yet the accepted norm. Extending advisory work to women can be a major factor in this process.

Pride and Dignity

A characteristic that contrasts with the poverty and degradation of the majority of villagers in backward areas is an innate feeling of dignity and pride in their traditions. Foster (1962) stresses that 'many technically well designed aid programmes have run into trouble, because culturally forms of pride and "face" which express these strong feelings about role, have not been recognized'.

Effects on Modernization

There is little doubt that the traditional customs and cultural elements which prevail in most underdeveloped rural regions can constitute serious obstacles to rational economic decisions and hence to the modernization of agriculture.

There are limits to the type and volume of new techniques that a given culture can accommodate without losing its validity. The simpler the culture, the narrower are these limits, and the less amenable is the culture to change (Smith, 1966).

However, the resistance to change generally attributed to subsistence farmers should not be exaggerated. Given the proper incentives and supporting services and using appropriate extension methods, farmers adopt new practices that provide greater returns at a low level of risk more easily than is generally assumed.

In a survey which included 200 cultivators in West Pakistan, it was found that in spite of the resistance to change that is often attributed to farmers in developing countries, over a third of the farmers made changes that were recommended by the extension service; in cropping patterns (34%); varieties planted (29%); use of fertilizer (20%); kinds of farm implements used (2%); and use of livestock (15%). These were all changes that were considered to be within the means and ability of the average farmer. Most of the changes that took place required little capital expenditure but some changes were not made even if they required no cash expenditure, when the farmers were not motivated to make changes. The primary motivation of cultivators for making changes was found to be to have more food to eat. Little or no interest was shown in better housing, household ornaments or entertainment, or in investing in land improvement or equipment. These became factors of motivation only after the primary need for sufficient food was satisfied (Sturt, 1965).

Changing Socio-cultural Systems

This brings us to the question of how the general problem should be tackled in order to bring about changes in the habits and mental horizons of rural people, which also predicate changes in socio-cultural systems.

In order to transform a traditional agriculture, dominated by socio-cultural constraints, into a modern agriculture, based essentially on economic decisions, a weakening and possibly even the elimination of traditional constraints is necessary. There is no easy solution to this problem. On one hand, one must attempt to preserve respect and adherence to cultural heritage, essential if one is to ensure social stability; on the other hand, one must attempt social change, in order to enable the concomitant process of economic, technological and political change.

An accepted premise is that it is necessary to adapt to the social framework of farmers, without forcing changes upon them, and that any attempt to replace traditional leaders, or to approach farmers directly, rather than through traditional leaders, is doomed to failure (Admoni, 1963). On the other hand, undue emphasis on working with traditional leaders may be counterproductive, because some villagers may construe this as an attempt by the government to maintain the status quo, by supporting the domination of the wealthier people (Dube, 1956).

There is no doubt that the readiness of farmers to adopt new ideas and practices is influenced by the nature of the leadership and the degree of control it exercises over their community.

Respect for elders and the acceptance of their counsel contribute to social stability. Elders could, and occasionally do, play a key role in encouraging new practices, if they are approached in a way that ensures their cooperation, rather than obstruction.

Reliance on 'traditional leaders' may therefore be justified as a short-term, tactical approach; but normally, a strengthening of their traditional stranglehold—economic, moral and social—on their people will be self-defeating in that it will re-inforce the conservative, anti-change tendencies. These are very strong in any case, and may stifle the initiative of the more progressively minded members of the clan and village. It is therefore essential to develop gradually a local leadership that is committed to progress, and this may require replacing the local power structure in due course (Millikan, 1962).

INFRASTRUCTURE

The transition from subsistence farming to commercial agriculture requires massive investments in the development of an appropriate and efficient infrastructure capable of providing the facilities needed to produce and distribute the inputs required for modernizing production, as well as to handle the increased production.

Generally, in most developing countries, infrastructure is provided by the governmental and subgovernmental sector. When certain facilities are provided by the private sector, there is generally public control or regulation to prevent exploitation. On the other hand, political factors may tend to favour certain regions at the expense of others, though the latter may be the most backward and the most in need of an appropriate infrastructure. It is desirable that most investment decisions for agriculture should be made at regional levels where localized needs and opportunities are best appreciated.

Major Components

The major components of agricultural infrastructures have been classified and listed by Wharton (1967) as follows:

(a) *Capital-intensive infrastructure*
Irrigation and public water facilities: dams, canals, distributaries, drainage system.
Markets.
Transport facilities: roads, railroads, bridges, boats, airplanes, ports, docks, harbours.
Storage facilities: silos, warehouses.
Processing facilities: machinery equipment, buildings.
Utilities: electricity and power, drinking water systems, gas.

(b) *Capital-extensive infrastructure*
Extension education services, statistical reporting services.
Agricultural research and experiment facilities: laboratories, experiment stations.
Crop and animal protection, control, and grading services.
Soil conservation services.
Credit and financial institutions.
Education and health facilities: schools, hospitals.

(c) *Institutional infrastructure*
Formal and informal institutions of a legal, political, and socio-cultural nature.

It is not within the scope of this book to discuss in detail all the components of the agricultural infrastructure listed above. Several have already been treated in previous chapters (extension services, agricultural research facilities, etc.).

In the past, most of the attention of policy-makers and planners in developing countries has been focused on production; whilst the problems of the so-called 'agro-distribution system' (Harrison & Victorisz, 1971) consisting mainly of marketing outlets, a transport network, and storage and processing facilities have been largely neglected.

Markets

Traditional agriculture is not exclusively 'subsistence' agriculture, in the sense that all production is for own consumption. A certain proportion of the production may be available for sale*. However, in a subsistence economy, marketing is a relatively simple structure, requiring little organization and capital investment.

Local fairs and markets, held at regular intervals, are an important element in the economies of quasi-subsistence farmers in developing countries. They serve as the points of initial sale of most farm products which enter the market streams, and also serve as the final point of sale of a small variety of manufactured goods. Total turnover is small, but quite a wide range of commodities is handled. This marketing system is adequate as long as agriculture remains near subsistence level and depends on local exchanges. However, when the stage of commercial farming is reached, a proper market economy must be developed.

Transition from Subsistence to Commercial Production

With modernization of agriculture and rural development, the market processes increase in importance and complexity with a concomitant increase in the expense involved in marketing operations. An efficient marketing system becomes a prerequisite for an efficient production system. The marketing system is no longer concerned only with the disposal of farm products, but must also ensure that the farmer is able to purchase the inputs he requires when and where he needs them.

For example, it has been estimated that the increased use of fertilizers envisaged in modernizing agriculture, especially in 'package programmes' involving high-yielding varieties, will require a five-to-tenfold increase in the quantities of fertilizers to be stored, packed and distributed–implying a considerable development of the marketing process (FAO, 1970).

As the urban population increases, and as crop production becomes more specialized, the commerce of agricultural products from rural to urban areas, and between rural regions, increases in scope. Movement of technological inputs and of manufactured goods from manufacturing centres and from ports of entry, right up to the individual farm, also increases.

The sale and distribution of agricultural machines and equipment and of agrochemicals, will require the estabilishment of specialized agencies capable of providing maintenance and repair services, and salesmen who can also give technical advice.

The developing countries are generally characterized by the inefficiency of their marketing system; as a result they are frequently faced with a vicious circle: the farmer not being able to rely on an economic return from the sale

*Subsistence farming has been defined (in connection with marketing) as one in which the farmer and his family consume more than 75% of the farm's production (Myren, 1970a).

of his surplus production, tends to produce for his own needs only; on the other hand, in the absence of marketable products it is extremely difficult to develop an efficient marketing system.

A typical example of the interrelation between production and marketing is the problem of economic utilization of the large herds of livestock maintained in extensive areas of grazing country in developing countries. Many cattle-owning people are influenced more by tradition and social considerations than by economic motives. Stock is generally sold only when forage is short, and the animals are in poor condition and difficult to market. The need to drive stock over large distances to market generally reduces their value considerably and increases mortality. Therefore lack of access to suitable marketing facilities is undoubtedly one of the main reasons why economic motivation cannot overcome traditional attitudes and why improved rearing, feeding and disease-control practices are not adopted.

Generally, marketing is best organized and most efficient for large-volume export crops such as coffee in tropical countries or grains in the Argentine. It is more efficient for non-perishable articles than for perishable commodities, and is most expensive in percentage of final price for such products as fruits and vegetables. In addition, marketing costs vary more or less inversely with size of farm unit and are high where farms are small and dealers obtain only a small amount of produce from each farmer (Hopkins, 1969).

For perishable produce such as fruits and vegetables, heavy losses are often incurred for three reasons: (1) bad roads sometimes make it impossible to get the produce to market or cause damage on the way (see 'transport'); (2) the expense of local collection (especially among small farmers) is likely to be high because on each trip the local dealer can obtain only a small amount of produce from each farmer; and (3) when such produce reaches the consuming centre the selling expense is further increased because the produce may pass through the hands of wholesalers, jobbers, and retailers—including street vendors—before it gets to the consumer.

As the importance of marketing increases there is a corresponding increase in expenditure in the services provided by the marketing sector in relation to farm production costs. Total marketing costs may equal and even surpass the production costs of a given commodity; they may however be partly offset by improved quality and reduced losses of commodities as a result of improved methods of handling, transport, and storage.

Government Involvement in Marketing

Food-crops

In the transition from subsistence to commercial economies, the capital investment required to enable the marketing system to cope efficiently with the increasing flow of goods may be considerable, in particular in relation to perishables such as fresh meat, milk, eggs, fruit, and vegetables which require special transport equipment, stores, refrigeration units, processing plants, etc. Private enterprise is generally unable or unwilling to make the large-scale investments needed; therefore, unless government takes the initiative, a major bottleneck to agricultural development may ensue. On the other hand, centralized organization of the marketing of food-crops requires management and administrative abilities that are generally beyond the means of public institutions in most developing countries.

In some countries in which there have been acute food shortages, many government departments have often become involved in various aspects of marketing. Responsibilities are often divided among ministries or departments within ministries, whilst cooperation between them is generally lacking (Spinks, 1970).

One solution is for government to foster cooperatives of producers, providing support in various forms such as loans, tax privileges, advice, facilities, and even monopoly rights. For a number of reasons, but mainly because of lack of experience, limited resources and shortage of trained personnel, performance of these cooperatives has often been disappointing (FAO, 1970).

Export-crops

In many developing countries, it is common for governments to attempt to establish adequate control over their country's foreign trade. They assume direct responsibility for the marketing of the main export crops, which are usually the principal sources of foreign currency. Statutory marketing boards–one each for the main commodities—have been established for this purpose, with monopoly control over exports of such commodities as cocoa, palm produce, coffee, bananas, sugar, etc. The boards fix the prices paid for production and marketing services.

In the French-speaking countries of Africa, export marketing is generally left in the hands of private enterprise but is strictly regulated by government through officially fixed prices based on controlled margins and profits. This method obviates the need for an elaborate official organization, but control over the marketing process and its efficiency is minimal.

Road and Transport System

Most roads in developing countries are simply dry-weather tracks. During the rainy season, they may become completely unusable for varying periods of time; the surfaces are generally rough, rutted and dotted with large potholes; severe wash-outs are frequent.

Railways do not generally penetrate far into subsistence areas. In many areas animal or even human transport is still used. In many cases, the overall potential for commercial farming is greater than the possibilities provided by the existing transport and storage infrastructure.

Whilst subsistence farming is practically independent of transport systems, an efficient marketing system essential for commercial farming is not possible without a good transport system.

In India, for example, only one-quarter of villages is within two kilometres of some reasonably good road, and most farmers who have adopted new, improved technologies, are within these villages. The majority of villages are practically isolated from the rest of the economy and their modernization will probably not be possible until their communications with the rest of the country are improved (Owen, 1966).

Abbott (1966) cites a number of examples to show how the lack of means of transport is often responsible both for high marketing costs and for the continuance of subsistence farming. In 1957 it was estimated that in the Syrian region of the United Arab Republic, transport costs for many farmers amounted to about 40% of the export price of wheat and 50% of that for barley. Production is restricted to the needs of local village markets until low cost transport facilities make other outlets accessible.

In Afghanistan, half of all trade, until recently, moved on the backs of men, camels, and donkeys. Poor transport has increased the price of goods to as much as five times their original costs. The movement of freight from coastal areas to interior points of Brazil may take six months by river, and a month to a month and a half by truck. Paraguay's trade involves journeys of a thousand miles by river to the Atlantic Ocean. 90% of farm-to-market transportation is confined to ox-carts on primitive trails (Owen, 1966).

A study of the marketing cost structure in Zaire indicates that transportation costs of farm products account for between 15 and 27% of their final retail price (Mwamufiya & Fitch, 1976).

In most African countries, because of the lack of feeder roads, women carry farm produce to market by the headload. The maximum load an individual woman can carry is about 35 kg (Anschel, 1969), though actual amounts may be much smaller. Trips to the market must therefore be frequent; in parts of western Nigeria women are already going to market almost every day. This mode of transport may become a bottleneck to expanding trade.

Transportation by animals is not necessarily cheap. Bullock transport, for example, is extremely slow, moving loads at speeds of about 2.4 km/h with loads of 1.5–3 tons (Owen, 1966). In India, where much produce is transported to market in bullock carts, the cost per ton per mile may be three to five times as high as transport by truck on drivable roads (Owen, 1970).

Lack of a satisfactory transportation system may have far-reaching effects on the success or otherwise of land development schemes. One of the earlier settlement schemes in southern Brazil was a marked failure because of communication problems (James, 1959). Towards the end of the 19th century, over 50,000 Polish settlers, were sent to a new pioneer zone being established on the northern slopes of the Iguacu valley. All the factors needed for success, excepting one, were present: a favourable climate and a fertile soil, hardworking settlers with a suitable agricultural background, government encouragement and backing. After a few years of producing a variety of amazingly good crops and finding that it was not possible to bring these to market, they abandoned their progressive farming methods and adopted the traditional shifting cultivation, making clearings in the forest, planting maize, permitting hogs to do their own harvesting, and then abandoning the clearings. (The hogs were driven to market over trails impossible for wheeled vehicles.) The final result of the lack of foresight in planning a settlement without access to a market was the destruction, by a small number of people, of a considerable area of the forests of western Parana and the subsequent loss of soil by erosion. By contrast, other colonies in the same general region, with a fair access to markets, have proved to be remarkably successful.

In parts of the Philippines, subsistence crops like rice and maize are often raised instead of potentially more profitable crops, such as Manila hemp, because of the difficulty of bringing the produce to market. Many important agricultural areas are linked with the outside world only during the dry season of the year and valuable fruit is often wasted as a result of sudden rains. In Thailand, the average distance from a farm holding to a railway station is 70 km; to a navigable waterway, 30 km; to a road usuable during most of the year, 10 km.

Large areas of cultivable land may lie largely idle because of the lack of access roads. 4800 km of farm-to-market roads built in Mexico since 1950, out of a planned total of 128,000, have already brought about a striking increase in grain and fruit production in some areas (FAO, 1970).

Poor transport is not only the direct cause of high marketing costs, damage and even loss of perishable food supplies; the supply of essential inputs is also affected, and fertilizers, for example, may reach the farmer too late in the season, causing a drastic reduction in the yields that could be achieved.

Good roads and transportation are, therefore, one of the first priorities for developing agriculture. Without these, agricultural commodities either cannot be marketed at all, or lose a large part of their value by the time they reach market. The construction of a tarred road to the village of Wasirpur in Haryana, India, made it possible for the villagers to grow vegetables instead of grain crops, and transport their daily output to urban markets, obtaining thereby a regular inflow of cash and higher income per hectare. Milk production was also expanded: even a milk collector on a bicycle can carry twice as much milk on a fair road as he can when access to farmers is difficult. Fertilizers became less expensive and deliveries were on time. High-school classes could be conducted in the village because teachers came by bicycle from the nearest town. A dispensary was also established; doctors and veterinarians visit whenever needed. House construction has increased because the transport of building materials by truck was one-tenth the cost of transportation by camel. These changes have been duplicated in many parts of India as a result of the building of feeder roads to isolated villages (Owen, 1970).

In Uganda, the construction of 100 km of roads in one district (an increase of 114% in the area's road mileage) resulted in a 400% increase in cotton production (Whitton, 1966).

In many cases in Latin America, particularly in Colombia and Brazil, it has been found that settlement follows new roads, and that these are therefore a major factor in opening up new areas for agricultrual production. Road-making may also come in response to pressures resulting from increases in the volume or value of agricultural production.

The importance of a satisfactory system of transportation in stimulating agricultural production and marketing is now being increasingly recognized by developing countries, and large sums are frequently earmarked for road construction and transportation. It is important to obtain the greatest possible economic and social benefits from the expenditures.

Air transport is an alternative to highways in solving the access problem to remote areas in the humid tropics; when there is no hope of developing the space between them and settled areas; also for areas that have no hope of highway access within the near future or for transporting equipment for land-clearing and highway construction. Air transport may also prove to be economical for specialized production, when roads would have to be constructed through mountainous terrain, in which the costs of construction and maintenance and vehicle operation are high, and the distance by road may be two to three times the air distance (OAS, 1964).

Storage

In developing countries, fluctuations in yields have had disastrous consequences in the past; these could be overcome, at least partially, by adequate storage facilities. The availability of these facilities is not only an essential precondition to change in agriculture, but also provides a stimulus to farmers to effect such change.

Where no suitable storage is available, the farmer may be better off selling his produce immediately after harvest at a relatively low price and buying some of it back later at a much higher price, than attempting to keep his produce without adequate storage facilities (Fig. 64). In Ghana, for example, because of the losses incurred during long periods of storage, 45% of the crop is sold at harvest time, 30% is stored for household consumption, and only 25% is stored until prices are favourable. This marketing pattern results in large price fluctuations during the year, with depressed farm prices immediately after harvest, when the farmer must sell most of the crop available for marketing (Low, 1975).

In the spring of 1966, scores of village schools had to be closed in northern India, in order to provide storage space for a bumper harvest that was 35% greater than the previous record crop. Even so, a large proportion of the harvest had to be 'stored' in the open with resultant losses (Brown, 1970).

Increases in rice harvest in the Philippines, as a result of the sowing of improved varieties made available to farmers, has created another problem.

Fig. 64 Typical African grain crib. The grain remains easily accessible to rodent and insect attack. (Drawn by D. Arnon)

Not only is the crop considerably larger, but the new, early-maturing varieties have to be harvested during the monsoon period. The grain can no longer be dried by the traditional method of exposing it to the sun and mechanical grain driers are essential if the crop is to be saved.

A modern silo, with driers and elevators, can handle large quantities of produce efficiently and can practically eliminate wastage. It requires a high investment in capital and the mechanized equipment is expensive to operate, though it requires little labour. In the early stages of development, more simple and labour-intensive facilities are generally more justified.

In some countries in Latin America, storage service is provided to the farmer by commercial banks which own an extensive network of warehouses that are located in the principal producing and consuming centres of the country. These banks finance agricultural activities and supply credit for the purchase of agricultural inputs. Grain and cotton are stored by the banks and secured loans are given (Salazar, 1968).

Processing

For many crops, such as rice, tea, coffee, oil palm, fibre, and sugar crops, commercial processing is an essential part in the marketing sequence. For others, such as fruits and vegetables, processing may enable the establishment of important rural industries both for export and for the home market. Processing may also make it possible to improve the nutritional quality of diets on a national scale, by enabling quality control. An additional possibility is the enrichment of certain foods by the use of additives, such as a small percentage of soybean flour added to wheat flour for bread-making, and correcting vitamin deficiencies in certain products, such as margarine, etc. (FAO, 1970).

As long as the marketing of agricultural commodities is fragmented and irregular, processing is bound to be limited to primitive and inefficient methods. Conversely, the development of modern processing plants is only possible after supplies become well-organized and regular distribution costs reasonable and quality adequate. For example, in West Africa, a large proportion of the oil palm crop is processed on or near the farm by primitive methods which extract only 40–65% of the total oil content of the fruit and given an inferior product. Modern mills can extract up to 98% of the oil and provide a high-quality product. As long as collection and transport of the produce cannot be properly organized, the use of modern equipment will not be economically justified (FAO, 1970).

ECONOMIC FACTORS

Aversion to Risk and/or Restricted Access to Resources

Even if farmers do 'act within the bounds of economic rationality' certain economic constraints may effectively prevent the adoption of new technology. In certain cases, such as plantation crops, the considerable time required between planting and the first returns may be too long for farmers to bridge. The possibility of a too low rate of net return may be another disincentive. However, the major constraint, apart from lack of money wherewith to buy the necessary inputs, is the aversion to risk shown by many subsistence farmers.

On the whole, it appears fairly certain that traditional farmers have been conditioned by the experience of generations to a sense of fatalism and to be suspicious of change.

The *per caput* annual income of these farmers is commonly between US$80 and $150. The soil they farm is usually exhausted and the land resources of individual farmers are limited. Crop failures do not simply mean a reduced income, but may result in outright starvation. Traditional methods of farming have at least proved over the centuries that they are fairly well adapted to local conditions. The fear that new and untried practices may fail to produce as much as the old practices, and thereby endanger the farmers's very survival, is not irrational; on the contrary, it is a legitimate economic consideration.

The degree of aversion to risk is re-inforced by the following factors: (a) when the farm is almost entirely devoted to food-crops, alternative sources of food are restricted, there is no opportunity for work outside the farm, credit is scarce or expensive; and (b) when the value of the family's minimum subsistence level is close to the family's assets (Wharton, 1969).

Though most authors are agreed that risk aversion is a major factor preventing the adoption of new practices by small farmers, Griffin (1972) does not accept the plausible argument that small farmers are less able to bear risks than large landowners, and are therefore unwilling to do so. He is of the opinion that high risk will deter investment in innovations in general, but it will not necessarily deter one group more than another: 'If many peasants fail to benefit from the process of technical change, one should seek for the cause not in their attitudes, but in their restricted access to resources.'

Whether farmers do not adopt improved practices because of risk aversion or because they lack the means to do so, or for both reasons, is not of overriding importance. It is an indisputable fact that the underprivileged rural sector cannot afford modern inputs and in the vast majority of cases, as we shall see below, does not have equal access to institutional credit or other incentives.

Experience in developing countries has shown that improved practices are adopted, and increased agricultural production can be obtained within a relatively short period, if the right incentives are provided and the necessary resources are made equally available to all sectors.

New HYVs of rice were first adopted by rice farmers in the Philippines in 1966; by 1970 they were sown on 57% of the lowland rice area (Castillo, 1975). This can be compared with the eight years it took for 59% of Iowa farmers to adopt hybrid maize (Rogers, 1962).

INCENTIVES

The principal incentives are insurance for innovators, prices and subsidies, production credit, tax structure and policy, and land tenure arrangements.

Insurance for Innovators

An extremely important proposal to reduce at least part of the risk involved in adopting new practices, has been suggested by Marglin (1965). The kernel of this proposal is insurance against the failure of a recommended innovation to produce a sufficient, agreed increment in output over that obtained by traditional methods. The insurance can be offered as an incentive to farmers participating in schemes of supervised credit linked to the purchase of inputs, such as fertilizers, improved seeds, etc.

Only 'packages' with a high cost/benefit ratio, not less than 1:3, should carry this insurance. The cost of the inputs would be the limit of the liability. The only condition is that the farmer must actually implement the innovation with a minimum of competence.

Prices

In subsistence farming, only a small fraction of the production occasionally reaches the market, so that the prices of the farm commodities have almost no impact on crop production. Any surplus marketed is generally pledged to his creditors, who usually dictate the price he will receive.

A system of efficient pricing is one of the basic economic requirements for the development of agriculture. Prices can be used to change cropping patterns, first by encouraging a shift from low to higher yields, and subsequently from low-value to high-value crops. The unfortunate tendency in most developing countries, especially in those with large urban populations, is to base pricing more on the need for low food costs in cities than on the production incentives needed by farmers, thereby impairing the effectiveness of agriculture in contributing to the overall development of the economy.

There are three sets of prices to be considered: of *farm products*, of *agricultural inputs*, and of *consumer goods* and *services* that farmers have to buy.

Farm Products

Because agriculture is a biological process, production is seasonal and output irregular, resulting in the inherent instability of prices of agricultural commodities. The factors involved are:

(a) the seasonal concentration of output;
(b) the great fluctuations in output because of unpredictable weather conditions, the prevalence of diseases and pests, etc.;
(c) the relatively low price elasticities of many basic farm products.

These price fluctuations cause the greatest hardships to farmers in developing countries, because it is they who are generally obliged to sell immediately after harvest and are at the mercy of merchants and moneylenders. In Cambodia, for instance, prices of paddy in January and February, just after harvest, are said generally to be barely half those prevailing in July and August, while in Colombia the immediate post-harvest price of potatoes is often no more than one-third of that realized later in the season (Abbott, 1966).

In India, grain prices fell so steeply after the good harvest of 1958 that the total returns to farmers for the larger output were appreciably smaller than their returns for the poor harvests of the two or three preceding years. 'After such an experience, not many farmers are likely to listen attentively to extension men telling them how to get higher yields. Nor are the likely to buy more fertilizers for their grain crops' (Barker, 1970).

The less developed the marketing system of the country, the greater the seasonal variation in prices of agricultural commodities tends to be.

Examples have been quoted from countries as far apart as Colombia and Somaliland, where prices of staple foods are five or six times as high before the harvest as after, though these are no doubt extreme cases. Governments attempt to overcome or at least mitigate these problems by official price policies.

The objectives of a pricing policy may be either complementary or conflicting. They may aim at providing an incentive to farmers to increase production of a certain commodity important to the national economy, and of stabilizing prices. The main objective, however, may be to reduce the cost of living of the urban population, and thereby keep costs of production low in industry. This would in effect signify a subsidy provided by the agricultural sector to industrial development.

There are two main dangers inherent in price stabilization schemes: price policies may be uneconomic, or government decisions may be arbitrary and inconsistent. This has led to the unsatisfactory record of many national price stabilization agencies (FAO, 1970). Further, official edicts on price levels are generally ineffective if the means for effective control or regulation are absent as is usually the case. The operating costs involved in price stabilization schemes, which require storage for more or less extended periods, and transport from surplus to deficit supply areas, may be excessive. A study in Chad indicated that if government was to purchase millet in surplus growing areas in order to re-sell six months later in deficit areas, the cost of the millet would have to be doubled in order to break even (FAO, 1970).

Increasing the price of essential commodities may lead to an undesirable sequence of events. The cost of living will increase, and this may force a demand for higher wages. As a result, the costs of production of farm commodities and manufactured goods increase, reducing the purchasing ability of the local population and the ability to compete on foreign markets. The result may be inflationary pressure on the economy (Barker, 1970).

Finally, there are problems of quality control and incentives. If there is no provision for official grading and inspection of produce, supported by price differentials according to quality, price controls can lead to a deterioration in the quality of the products marketed. Traders can arbitrarily reduce official prices by claiming that the produce is substandard; producers can react by marketing ungraded produce or even by adding various inert materials.

The rate of response of small farmers to changes in the relative prices of agricultural products and farm input costs have generally been found to be positive and as high as for farmers in developed nations (Stevens, 1977).

Government price policies were found to be the most important factor influencing the adoption of HYVs in the Punjab (Pakistan). In this country the objectives of government is to miximize foreign exchange earnings and support prices were established in order to maintain the production of the high-quality Basmati variety for export, thereby slowing down the adoption of modern varieties (Chaudhari *et al.*, 1975).

Stability of prices cannot be achieved by legal means only; these will be realistic only if there exists an adequate transport system and storage facilities. Nor can stability of prices be absolute. A certain amount of fluctuation must be allowed according to the success of the harvest and the time that has elapsed since the last harvest. If prices remain stable throughout the year, much greater storage capacity will have to be provided. The stability of prices should be tempered by flexibility between a minimum and a ceiling price. Prices of agrochemicals are frequently unreasonably increased by taxes, by wide price margins taken by dealers, and by expensive methods of handling.

An efficient agrodistribution system, through better marketing and cheaper transport, extensive storage facilities, preservation and processing of perishables, can make important contributions to price stability, without involving official regulation.

Wilde & McLoughlin (1967) point out that cattle in a traditonal society, because of certain social values, do not have the character of a cash-crop and their marketing is therefore only slightly, if at all, influenced by prices. When grazing is ample, few cattle will be sold even if prices are very favourable; conversely, during years of drought, sales will expand regardless of price levels, as the death of the animals is the only alternative to their sale.

Inputs

For modernizing production, farmers must buy: fertilisers; chemicals for the control of pests, diseases and weeds; tools, equipment, machines and fuel. In most developing countries the prices of these inputs are excessively high. A *balance* must be maintained between the prices the farmer obtains for his products and the cost of the inputs he is being encouraged to make. A favourable relationship between the cost of fertilizer and other inputs and the price the farmer receives for his product is considered as probably the most important factor in making possible the extraordinary achievement of the green revolution (Wellhausen, 1970).

In Thailand and Pakistan, government price policies have resulted in high prices for fertilizers, in relation to the price of rice. In 1971, the price ratio of a kilogram of nitrogen to one kilogram of paddy-rice was about 7 to 1, compared with a ratio of 3 to 1 or less in other parts of Asia (Castillo, 1975). Evidently this affects the numbers of users of fertilizers and the rates at which these are applied.

Consumer Goods and Services

The prices of consumer goods and services that farmers buy are the key to the purchasing power of the net income earned by farmers, hence their importance. In most underdeveloped and developing countries these prices have been rising relative to the prices farmers obtain for their products. The price relationship between what the farmer gets for his produce and what he has to pay for inputs and consumer goods is further distorted by the special encouragement provided in many countries in the form of high levels of protection to local industries. Whilst it is reasonable that agriculture should contribute to other sectors of the economy, this must be held within reasonable limits, so as not to keep agriculture itself stagnant.

In marked contrast, is the policy followed by the Chinese government—aimed at raising living standards in rural areas. By the end of the 1950s, a given amount of agricultural produce was purchasing about 35% more industrial goods than at the beginning of the decade. By the end of the 1960s, this had almost doubled to 67%. In order to maintain prices to the urban consumer at a reasonable level, the government is purchasing grain at higher prices than that at which it is sold to city dwellers (Gurley, 1973).

Subsidies

Subsidies of agricultural prices or inputs are generally justified in the early stages of development to stimulate the introduction of new crops or the adoption of new techniques that have proven to be beneficial. The subsidies must be large enough to make the desired adoption sufficiently attractive to the farmer and reduce the element of risk involved; however, it is usually desirable that a reasonable part of the cost should be borne by the farmer. Once innovations are accepted, and the farmer knows the benefits that accrue to him, subsidies can be gradually reduced and finally discontinued.

The great advantage of subsidizing inputs such as fertilizers, pesticides, equipment, etc., rather than increasing prices of commodities produced is that the subsidies directly encourage the use of inputs that increase productivity. By contrast, if the price of the product is increased, the profit accrues to both progressive and backward farmers. At the same time, the political and economic consequences of increasing the price of food and fibres to the non-farming population are involved. However, the two methods should not be considered as mutually exclusive, and a judicious policy of prices and subsidies has to be established in accordance with each specific situation.

Credit

Importance

The use of modern agricultural inputs generally causes a considerable increase in the cash costs of farming (cf. p. 294).

Traditionally, credit in rural areas is obtained from money-lenders, who provide credit in the preharvest months against the security of the farmer's growing crops. In an area study in Iran, it was found that almost all vegetable growers received only 12% of the retail value of their crops, because they had accepted advanced credit and pledged their crops (FAO, 1970). It is the necessary outcome of the need to re-pay previous debts to stave off disaster, to cope with a crisis such as disease of a family member, or to finance the marriage of son or daughter, according to whether the prevailing norms are bride-price or dowry.

The high cost of credit from traditional sources is one of the main reasons why subsistence farmers do not adopt more rapidly improved practices which require cash outlays.

In a survey carried out in the CENTO countries, it was found that the main reason given by farmers for not using fertilizers was lack of money. They had no money of their own to invest and if they borrowed money they had to pay double the amount after the crop was harvested. Interest rates range from 60–250% (Central Treaty Organisation, 1962). At such high rates, the use of fertilizers will generally not be economical.

In all fairness, it must be pointed out that interest on agricultural loans is bound to be high, because capital is scarce, farm loans are costly to administer, the demand for credit is seasonal, and because the uncertainties of agriculture result in considerable loss through default (Long, 1968).

Experience in developing countries has shown the importance of a good system of agricultural credit in promoting agricultural development and the necessity of providing farmers with short-term credit at reasonable cost for recurrent seasonal inputs such as fertilizers, insecticides, pesticides, etc., even when the prices of the inputs are subsidized. The credit must be available at the time when the inputs are purchased and must be extended until the crop is marketed.

The need for credit to enable the adoption of a 'package' of high-yielding varieties and improved practices is exemplified by the following examples: it is estimated that whereas the total cash costs of production for the average Filipino rice-farmer, using traditional varieties and methods, is about $20 per ha, the cost rises to $220 when the high-yielding IR8 is grown. Therefore, although the yield may increase threefold, leading to a net return four times greater than with traditonal varieties, the farmer must have access to greater credit, at reasonable cost, to finance improved production methods (Wharton, 1969). Also needed are longer-term loans for the establishment of new crops, in particular perennial crops, purchase of new equipment, and the construction of new buildings. For example, in southern Brazil it has been shown that the changeover from extensive ranching to intensive arable cropping has caused a 30-fold increase on a per hectare basis in credit requirements of the farmers involved (Adams, 1970).

Credit Institutions

Agricultural credit institutions are usually grouped into three types (Wharton, 1967): (a) private informal; (b) private formal; and (c) public. The less developed the agriculture of a country, the greater the relative weight of the informal private sources: relatives, neighbours, local moneylenders, and merchants. It is exactly in these areas that the importance of institutional public credit facilities is greatest, in order to prevent exploitation of the farmer.

Credit requirements for modernizing agriculture on a national scale are so considerable that only government initiative can provide the necessary funds and promote the organization required for providing credit to farmers.

In view of its importance, agricultural credit is the second largest category of World Bank lending to the developing countries. Experience has shown that lending to credit institutions, to be effective, must be accompanied by extensive technical assistance in order to develop sound management and operating procedures. The effective use of credit is also dependent on sound technical advice to farmers in guiding their operations. Centralized agricultural credit is usually provided in developing countries by agricultural banks, and cooperative credit societies, generally sponsored by governments, the latter providing most of the capital. To be productive, it is desirable that credit be linked with an undertaking by the borrower to adopt certain improved practices, in which case the bank operates in close collaboration with the extension service. Such collaboration will ensure that the advice proffered by the extension service on improved methods of production would be backed by the means to obtain the required inputs. Once credit is given to the farmer, the extension worker can supervise the new cultivation practices.

For the de-centralization of agricultural credit, the most effective instrument is the village-type cooperative, provided the farmer can be educated to accept and support this type of institution. Cooperative purchase of inputs will reduce their cost; while cooperative sale of farm produce will protect the farmer from pressures to sell at low prices, especially if his products are perishable. Quality control and improved packing made possible by cooperative marketing will also have a favourable effect on his income. When such cooperatives also handle credit, they have satisfactory guarantees that the credit will be repaid.

Reasons for Success and Failure

Despite the proliferation in developing countries of government-sponsored rural banks, farmers' associations, cooperatives, etc., to supply credit, the traditional sources of credit to the small farmer, such as landlords, private money-lenders and relatives, frequently remain important (IRRI, 1975).

Credit cooperatives have generally had a high rate of failure in developing countries. One of the many causes has frequently been the high costs and risks in administering small loans to numerous farmers in low-income areas.

Many programmes of credit administration involve complicated procedures and the processing of numerous forms. As a result, they are highly demanding of scarce trained manpower. In Africa, for example, these administrative systems and the manpower they require appear to have been a far greater constraint to the provision of credit to the small farmer than were finances (Lele, 1976). Efforts to reduce risks generally result in drastic reductions in the numbers of farmers eligible for loans—generally those who are in greatest need of credit.

Some other causes of failure have been: the lack of trained management personnel; the use of cooperatives as political tools to implement various government programmes unrelated to credit; and conflicts between managers and strong interest groups within cooperatives.

If governments are really interested in developing cooperative credit systems in those areas in which lack of credit is a major obstacle to modernization and development, they 'must be prepared to bear, as social overhead, the cost of developing such alien institutions, until management can be trained and the institutions gain sufficient experience to be autonomous' (Johnson, 1971).

The success of any programme of rural credit in increasing production depends to a large degree on a number of factors, apart from the availability of money. There must be an opportunity for farmers to utilize additional capital properly. This signifies first of all that new technical inputs must be available, the usefulness of which has been proven in the field; there must be markets that can absorb increased production and an effective repayment of loans must be ensured. In the absence of new techniques, loans will be mainly used for non-productive expenditures (Long, 1973). Often credit is tied to improved practices, but if no attempt is made to ensure that the inputs are actually aquired, or properly used, it should not be surprising if difficulties are encountered in getting farmers to repay loans.

An interesting side-light on the reasons why institutional credits to subsistence farmers have frequently proved disappointing is provided by the results of a sample field survey of Colombian farmers. Over 50% of those who had borrowed money stated that they preferred money-lenders to banking institutions as a source of loanable funds, even though the former charged 24–96% annually, as against 8–12% charged by the banks (Nisbet, 1971). The main shortcomings of the institutional credit market, from the point of view of subsistence farmers, as they emerge from this study, are the following:

(a) Applicants for bank loans were generally required to anticipate the date of loan disbursement by as much as six months, whilst money-lenders are prepared to provide loans on short notice.
(b) Application procedures for bank loans are frequently complicated; long forms have to be filled out which the farmers cannot read or understand. By contrast, the borrower simply has to talk to the money-lender about his financial needs; there are no application or financial statements to fill out, no land titles to present and generally no references are required. The money-lender accepts or rejects the request immediately.

(c) Many credit programmes only grant credits for required inputs, such as seeds and fertilizers, to be used under the direction and supervision of extension workers acting as agents of the lending institution. The farmer may not wish to use inputs with which he is not familiar, or distrust the young, inexperienced urban-raised agents frequently employed in the field; or he may require the loan for consumption purposes. By contrast, the money-lender shows no interest in the ultimate use of the funds borrowed.

(d) In may cases, money-lenders are also the buyers of produce. When the small farmer applies for a crop loan to a bank, he loses his accustomed market outlet for his produce.

Training of Staff

A major factor in ensuring the success of agricultural credit schemes on a broad scale is the training of the staff who man the branches serving farmers. This training should include, apart from banking procedures, knowledge of basic agricultural technology, input–output relationships, the agricultural marketing structure, input supply sources, etc.

Lack of Equity

Traditionally, the small farmer has always been more dependent on credit than has the large farmer who frequently serves as a money-lender whilst having the best access to institutional credit. Larger farmers, when they borrow, do so mainly for buying farm assets so that credit serves to increase their productive capacity; whilst small farmers use about half the credit they obtain for consumption and social ceremonies, thereby further impairing their economic situation (Dasgupta, 1977).

One major negative aspect of many public credit schemes is that most of the available capital goes to the larger farmers. In India, for example, following the green revolution, larger farmers have generally increased their borrowing in order to finance the higher costs of production of HYVs, whilst the amounts borrowed per hectare by small farmers have remained practically unchanged, as shown by the example given in Table XXXIX. It should be pointed out that the credit system in the Punjab is less conservative and less prejudiced against the small farmer than in most other areas in India (Dasgupta, 1977).

Even programmes with the specific objective of improving production on small farms make few loans to the bottom 40% of farmers. In Pakistan, the smaller farmers, who constituted 60% of farmers, received 3% of institutional credit; in Bangladesh, the minority of farmers holding more than three acres received 80% of the loans from the Agricultural Development Bank (Long, 1974).

A surprisingly similar development in Mexico is described by Alcantara

Table XXXIX Amount borrowed (rupees per ha) by farm size in Ferrozepur district, Punjab (Kahlon & Singh, 1973).

Farm size (ha)	1967–68	1971–72	1971–72 in % of 1967–68
Small (below 6 ha)	252	245	97
Medium (6–14 ha)	86	305	355
Large (over 14 ha)	87	423	+486

(1976). In the 1920s the National Agricultural Credit Bank (BNCA) was established for the express purpose of providing low-interest public loans to authentically small farmers, many of whom had received their land as a result of the first land reform decrees following the revolution. With the initiation of all-out government support for the modernization of agriculture in the irrigable areas, the emphasis of the BNCA shifted to financing capital improvements on large commercial farms. Instead of its former policy of dealing only with smallholders, funds were increasingly channelled through credit unions and other banks. In the critical years immediately preceding the green revolution large sums of long-term credit were made available to large landowners.

Cooperatives, which are supposedly democratic institutions, also favour large farmers in the supply of credit, as shown in the following example in Table XL from Ferrozepur district in the Indian Punjab.

The credit institutions are under political, economic and administrative pressures which discriminate against the smaller farmer. Loans to small farmers are more costly, more difficult to administer and often have higher default rates. If one includes administrative cost, supervising cost, default, etc., costs may run as high as 100% in some programmes, and public credit is generally as expensive as private loans (Long, 1973). Crop insurance

Table XL Total borrowing in rupees per holding from various sources in Ferrozepur district, Punjab (Kahlon & Singh, 1973).

	Size of farms		
Source	Small	Medium	Large
Cooperatives	417	658	3064
Land mortgage bank	—	1545	673
Government	—	23	—
Others	533	385	4727
Total	950	2611	8464
in %	100	275	891

agencies give preference to large farmers and price-support agencies buy their produce. Those who do not get credit, do not generally participate in the two other programmes. 'The result is a type of circular causation that benefits the present modern sector but is of limited use to the sectors which we want to bring into the market economy.' (Myren, 1970b).

Repayment of Debts

In many countries the repayment of credit has been a serious problem, and if neglected, organized credit schemes may be discredited. In many developing countries, there are either no effective sanctions against default, or sanctions are simply not applied.

The poor repayment of credit has been variously attributed to the following reasons (Lele, 1975): (a) poor quality of extension; (b) low returns from the technology packages for which credit was obtained; (c) unfavourable weather conditions resulting in low yield; (d) inadequate efforts at collection by the credit-supplying authorities.

Analysis of the repayment figures in three Ethiopian credit programmes showed that a higher percentage of defaulters was found among the larger farmers than among the small-scale farmers. One reason given for this surprising result is that larger farmers, being politically more powerful, feel they can renege on their debts with impunity (Lele, 1975).

Amongst the means proposed to improve loan recoveries, are group responsibility, linking the provision of credit with marketing, or with the provision of storage facilities. If the farmer's produce can be stored in public warehouses, not only can they get better prices for it, but recovery of loans is easy to achieve, so that both sides derive an advantage.

Corporate responsibility has been applied as a means of reducing debt defaults; those cooperatives which have loans in default and do not make good their defaults cannot borrow on behalf of any of their members. Whilst this system may ensure a high rate of loan recovery, it deprives non-defaulting members of defaulting cooperatives of the right to credit. Experience has shown that farmers may even withhold repayment of loans if they feel they will in any case be deprived of future credit because the cooperative to which they belong will default (Sanderatne, 1970).

Crop Insurance

The first government-operated multi-risk crop-insurance schemes in the world were created in 1938 in the United States and Japan, and have since served as models for many other countries (Oury, 1970).

Crop insurance can be either voluntary or compulsory. It is a means of sharing the burden of losses occurring in agriculture as a result of disasters, whether natural (drought, hail, frost, etc) or man-made (fires). In essence, it may have some of the elements of a minimum income guarantee scheme. In many of the developing countries it would strengthen the position of tenants vis-à-vis landlords, merchants, and money-lenders. By enabling farmers to repay their debts in case of failure due to factors beyond their control, it would improve their credit worthiness for funds required for development. 'Crop insurance and credit schemes are therefore mutually reinforcing' (Oury, 1970).

An all-risk crop-insurance scheme provides essentially a specified yield guarantee, which generally does not exceed 75% of the average yield of five previous years, for example. The amount of indemnity is generally linked to the actual costs of producing the insured crop.

Under compulsory crop insurance, which is the form most suited to developing countries, premiums can be relatively low, because the risks are widely spread. Under certain circumstances, government is the insurer.

In Israel, for example, farmers who grow wheat—a crop extremely dependent on the vagaries of rainfall—receive a government guarantee against crop failure, which covers the costs of production. However, this insurance holds only for wheat sown to the north of a line, called the 'drought line', below which average rainfall is less than 250 mm.

Even in cases of mutual compulsory insurance, government will have to establish a reserve fund to serve as a cushion for spreading losses over several years.

Before instituting an all-risk crop-insurance scheme in a developing country a pilot scheme, on a small but appropriate scale, is required. It should be operated on a limited number of crops, in a limited area, for a sufficient number of years. It would enable the development of an appropriate administration for the scheme, provide essential data for building an actuarial basis for crop insurance, and train a nucleus of staff to be involved in the subsequent generalization of the scheme.

Capital

Capital Requirements for Developing Agriculture

The transformation of subsistence agriculture, traditionally based on the exploitation of biological and land resources, into modern agriculture using improved technology, implies the investment of appreciable capital resources.

Present projections suggest a rate of growth of the labour force resulting in its doubling in the course of 30 years. As a result, the ratio of population to land in most developing countries will double by the end of this century, and the need for capital growth as a means of increasing the productivity of land and of creating jobs can be seen as enormous (Edwards, 1973).

At National Level

Capital is needed at the national level for the financing of infrastructure projects, such as electric power, transportation, irrigation and flood control projects, which do not as a rule attract private investors.

The World Bank, in twenty years of operation, has provided large amounts of capital for loans and credits in indirect support of the agriculture of developing countries. However, only in recent years has the bank become aware that its earlier efforts did not take sufficient account of the numerous human and institutional problems that impeded the effective use of capital resources. Experience has shown that capital can be applied rapidly and effectively only when the other elements of the agrarian structure are adequate, in particular, extensive land reform and efficient agricultural administration. For this reason many countries at the present time are not able to absorb much capital for agricultural development (Burke-Knapp, 1966).

At Farm Level

Capital is also needed at farm level because agricultural techniques are typically primitive in subsistence agriculture and productivity is low. The developing countries have an enormous potential to increase agricultural production by relatively moderate injections of capital into small-scale operating units.

In traditional agriculture the main inputs are labour and land. Even other inputs, such as seeds, manure, tools and draught animals, are frequently farm-produced.

With modern agriculture, at the farm level, capital substitutes in large part for labour and land: two-thirds of the total inputs are represented by capital. The chemicals, machines, improved seeds, fuel, etc., required for increasing productivity and efficiency are purchased from sources outside the farming sector.

Capital in most developing countries is extremely scarce and relatively expensive; hence, the need to utilize it in a way that ensures maximum increases in production potential.

It is generally agreed that where capital is limited and labour plentiful, it is desirable to use capital in order to promote labour-intensive production methods, thereby covering a greater labour force. This will result in an increase of purchasing power and its wider distribution.

The more intensive use of capital for replacing labour in agriculture may be uneconomic and anti-social as long as labour is underemployed, the cost of labour is low, and alternative means of employment are insufficient.

This is not always recognized, and premature capital investment in certain technologies may be counter-productive. For example, at a certain stage of development, mechanization leads to a conflict of interests between individual large-scale farmers and the interests of the community. Certain farmers save labour by investing capital, without concern for the fact that the economy cannot usefully absorb the labour thereby made redundant.

The incentives to replace labour by machinery is greater in plantation and estate farming than on family farms, not only because of economy of scale or for economic reasons, but frequently because owners or operators resent and fear labour's increasing political awareness and capacity to organize. The resulting displacement of labour further contributes to unsettled social and political conditions.

There are many cases in which government policy encourages capital investment in mechanization, even in areas characterized by underemployment in agriculture (and its concomitant low wages) by unjustified price, exchange rate and credit policies which distort the relative costs of mechanical power relative to labour or animal-draught. There are many cases in Latin America, for example, of marked distortions inducing premature mechanization (Ruttan & Hayami, 1972).

However, as general economical development and industrialization of a country proceeds the labour force available to agriculture decreases. Agricultural production can still increase substantially, even with an ever-decreasing labour force, provided more capital is invested and the level of knowledge improves (OECD, 1967).

An illustration is provided by the change of techniques adopted in Mexican agriculture since 1945, which has resulted in the use of less land and labour, and more capital equipment and supplies. Agricultural crop output in 1945–49 was about 60% greater than in 1925–29. In producing this 60% greater output, only 16% more labour and 23% more land were used, but over 300% more capital, equipment and supplies were used (More, 1955).

Agriculture as a Source of Capital

The capital requirements for the transformation of agriculture are considerable. In developing countries, agriculture is frequently the only major potential source of this capital. Success depends on agriculture becoming capable of making capital investments out of its own revenues for sustained growth and development; then, capital justifiably becomes a substitute for labour.

Economists, in the past, tended to take an oversimplified view of development as an accumulation of a stock of physical assets such as equipment, inputs, irrigation layouts, etc. It is now generally recognized that the efficiency with which capital is used is at least as important as the quantity and variety of physical forms available. This efficiency will be dependent on scientific and technical knowledge, and the know-how to use this knowledge. Therefore, investment in human skills and managerial capacity, as well as in research, is essential (Johnston & Kilby, 1973).

This can be illustrated by a few examples: in Japan, during the period 1950–60, there was almost no increase in the three traditional basic agricultural resources: land, labour, and capital. The increase in cultivated areas devoted to rice was only 2%, and yet a 20% increase in rice production was registered. Mexico, during the period 1945–60, increased investment in the physical factors of production by 22%, but more than doubled agricultural output (Adiseshiah, 1970). In Israel, during the same period, a 10% increase in investment in the three factors (mainly for irrigation) resulted in a 60% increase in agricultural production. These results were achieved mainly through improved use of the resources available to agriculture.

A major factor in creating new capital from agriculture itself is therefore the rational adoption of new technologies, a subject already discussed in detail in Chapter IX.

The Institutional Framework

Societies not only have characteristic patterns of values, they also have different modes of organization.

Inadequacies

Many of developing nations, even when they are rich in natural resources, remain poor because their social institutions are either inadequate or not oriented to meet new economic and social needs (Deutsch, 1971). Certain authors argue that a major constraint to modernization of agriculture is inadequate service from the institutions responsible for supporting agriculture.

In a study on the factors impeding the diffusion of new technology in the development plan for the Pueblo valley (Mexico), Diaz (1974) states that 'the very roots of backwardness are found in the dysfunctional institutional structure which serves agriculture in Mexico'. Farmers who wished to acquire credit from Banco Agricola and Banco Ejidal encountered cumbersome and time-consuming procedures for obtaining (and even for repaying!) loans; the promised inputs arrived late and they received little technical assistance from the banks in their use. This meant that obtaining credit was too costly in terms of time required, returns to inputs were too low because of late delivery and/or inadequate advice.

Efforts were made to improve the service of the banks and of the commercial supplier of fertilizers, by speeding the processing of loans and repayments, ensuring timely delivery of inputs and improving extension services provided by the technical staff of the bank, counselling farmers on the use of modern inputs. One measure of the success of these institutional reforms was that the rate of repayment of loans increased from 40% to 90%!

However, the major shortcoming of the institutions providing public services in the developing countries is that access to their services is markedly biased in favour of larger and prosperous farmers.

In as study prepared within the framework of the World Employment Programme, Harvey *et al.* (1979) state: the operation of service institutions in peasant communities continues to exhibit the regular features of differentiation, inequality and exclusion. The authors reach the pessimistic conclusion that 'rural planning should assume that services will benefit mainly the better off, even when the intention is to benefit mainly the poorer rural groups. What is more, planners should assume that even those programmes aimed specifically at poverty groups will partially miss the target'.

We have encountered this bias in several problem areas such as extension, services, credit, etc., covered in previous chapters.

In the next chapter, the implications of this apparently ubiquitous situation and possible solutions will be discussed in more detail.

Bureaucracy

Bureaucracy in developing countries can be a major factor in impeding progress in general, and agricultural development in particular. The main characteristics of such a bureaucracy are: authoritarian attitudes coupled with a sense of insecurity leading to an unwillingness to make decisions involving responsibility; a rigid adherence to rules and regulations which can be overcome, however, by bribery; and barriers of communication between the heads of the hierarchy (Axinn & Thorat, 1972). Of these, probably the single most important factor preventing progress is the *corruption* prevailing in many developing countries, from the ruling elite right down to the lowest bureaucrat. By the time each one has taken his 'bite', very little is left for productive investment.

The services rendered by officials are 'used to reward those that are obedient, compel wanted patterns of behaviour, and restrain unwanted initiative' (Owens & Shaw, 1972).

Owens and Shaw state that the gathering of tribute by officials should not be confused with what is called 'corruption' in developed societies; and that 'these practices are not considered corrupt in traditional communities'. They are, in fact, the discipline of such a society, the principal way in which those who rule keep the mass of people under their control. Corruption decreases respect for government authority and endangers further political stability. When corruption is widespread, the development processs is slowed down by inertia and inefficiency, 'speed money' becomes necessary, not to get anything unlawfully, but simply to overcome red tape.

In most developing countries, government services are highly centralized. The younger, less experienced and lower-paid officials are posted to the rural areas. As a result, implementation of agricultural policies and programmes is frequently ineffectual.

The weakness of the bureaucracy is compounded by the fact that most officials are, by education and provenance, urban oriented, and have little empathy with, and understanding for, the rural people they are supposed to serve.

Examples of how bureaucracy can hamstring development are provided by the large irrigation projects. The reason for the disappointing results of these projects are many and complex. Crosson (1975) ascribes the major reason for the inefficiency encountered in these projects as follows: 'the management of large irrigation projects is generally in the hands of public officials who are too far removed from the on farm situation to know the conditions of efficient water use, who lack economic incentives to achieve it even if they knew how, and who typically are bound by inflexible operating rules of water allocation impeding their response to economic incentives, even if they had them. The inflexibility of operating rules is the most obvious of these limitations and in itself would be sufficient to explain inefficient use of water.'

Even in countries that have adopted a policy based on promoting the development of the peasantry, bureaucracy can be an impediment to progress. Pearse (1980) notes that when the State sets up a system to modernize agriculture that provides inputs and organizes marketing, the administration of these services creates a powerful rural bureaucracy with its hands on the flow of wealth and goods. This monopoly can lead to new types of abuse and exploitation.

Social Services

One of the reasons for migrating from rural to urban areas is the considerable difference in the level and quantity of service amenities available in cities as compared to villages. Education, health services, water supply, communications, etc., are all biased in favour of urban areas. This discrepancy not only encourages abandonment of the villages, usually by the younger people with initiative and/or a better education, but also makes it almost impossible to attract qualified people, civil servants, teachers, physicians, agronomists, etc., who could make a considerable contribution to rural development to settle in rural areas.

The provision of social services is an essential concomitant of the modernization process. Apart from the humanitarian aspect, improvements in health, water supply, and education are also economically justified, because they contribute to increases in agricultural productivity as a result of the increased availablity and improved quality of labour.

However, considerable problems arise because of the conflicting demands for scarce financial, manpower, and organizational resources, between the provision of social services and the need to invest in technical development programmes and infrastructure.

Hence, the importance of developing low-cost social service systems adapted to the limited resources available. The willingness of villagers to contribute their labour to building health clinics, schools, etc. and other forms of self-help and participation should be mobilized. The public health programmes organized in Niger provide an example of such adaptation. Volunteers from each village are trained at the nearest dispensary in hygiene, preventive medicine and first aid, and supplied with a village medical kit. The villagers repay their medic in kind or by helping him with his farm work. Periodical visits by qualified medical personnel from the district dispensary make follow-up visits to check on the village health worker and to replenish his supplies (Lele, 1975).

The Need for Institutional Change

Ruttan (1977) graphically describes the difficulties faced by developing countries: 'The developing world is still trying to cope with the debris of nonviable institutional innovations; with extension services with no capacity to extend knowledge, or little knowledge to extend; cooperatives that serve to channel resources to village elites; price stabilization policies that have the effect of amplifying commodity price fluctuations; and rural development programmes that are incapable of expanding the resources available to rural people.'

'. . . unless social science research can generate new knowledge leading to viable institutional innovation and more effective institutional performance, the potential productivity growth made possible by scientific and technical innovation will be underutilized' (Ruttan, 1977).

It is now generally accepted that technological change cannot be adequately undertaken in isolation from the institutional framework within which it is to take place. All economic activities involve the mutual interaction of individuals whose rights and obligations are socially defined.

The adoption of technological change required for modernizing agriculture, will therefore not be effective unless many of the basic institutional systems are revised and adapted to development needs. Simpler and more precise rules for political and administrative procedures, as well as closer supervision of implementation are needed. The remuneration of low-paid civil servants should be improved. Punitive action against corrupt officials and the givers of bribes should be speedy and effective.

Overstaffed government services may constitute a heavy financial burden for developing countries, without a concomitant contribution to development. The pressure of numerous people of different educational levels who cannot find productive employment has frequently led to overstaffing of government services. Abdul Hameed (1977) describes the situation in Sri Lanka, which is fairly typical.

As a result of the increasing involvement of government in providing farm inputs, credit and marketing services and the need to implement certain institutional reforms, 'the number of civil servants working in the rural areas of the country has increased enormously. A sizeable segment of these rural officials has been drawn from the most vociferous section of the non-agricultural and university-educated population. However, many of these officials do not seem to have undergone adequate training in extension work, credit administration, rural living, etc., and do not appear to participate with the same degree of involvement in improving the country's agriculture in contrast to their counterparts in some other Asian countries.' The author further notes that besides the multiplicty of offices and officials the work is compartmentalized and no efforts are made to solve the problems of agriculture by an integrated approach.

A proliferation of rules and regulations to justify the existence of an overblown bureaucracy may be a serious obstacle even in the absence of corruption and even when special incentives are made available to farmers by government (credit, subsidies, etc.). Bureaucratic procedures involving much travelling and office-sitting, may be so time- and money-consuming in relation to the possible benefits, that many small farmers will not even make the attempt to obtain them.

Rural Organizations

'Frequently, the major obstacles to progress lie with the underprivileged and uninvolved rural people themselves, and with the deficiencies within their own and government organization, on which reliance must of necessity be placed for the support needed' (FAO, 1969).

In most developing countries, power is monopolized by the large landowners and by those who control access to the services and opportunities provided by Government agencies. These include: 'those who manage Government agencies connected with law and order, health, public works, agriculture and development programmes; and those who become the recognized political chiefs and are responsible for arousing and maintaining support for their parties among the local population, and canalizing those favours that the parties, in or out of power, can pass down; and those who control communications and transport. All these elements are to be found in the élites who handle agricultural development at the local level in a variety of alliances and compacts' (Pearse, 1980).

The underprivileged sectors have little hope of achieving progress unless a change in the distribution of power within the rural community occurs. Hence the importance of creating an institutional framework which involves the rural population in planning their own future (Owens & Shaw, 1972).

Local informal groups and more formal organizations are necessary to make the wishes and needs of the rural sector known to the local authorities for community-level planning, to mobilize local resources, to perform certain services for themselves and to serve as an effective framework for the transfer of certain government services to their members.

Rural people's organizations have been defined by FAO as 'bodies which are run and controlled by their members to a large extent'. Members decide upon and engage in socio-economic programmes, as well as in bargaining and claim-making activities (Heck, 1979). However, in most existing rural organizations the members play a possive role whilst elite leaders make the decisions. These leaders are generally local chiefs or politicians of a higher socio-economic background than most of the members, so that the organization perpetuates the local power structure. The lack of qualified staff frequently leads to the appointment of government officials as managers. Governments often consider their organizations mainly as tools for implementing their policies.

Popular participation in rural organizations depends on three prerequisites: participation by members in decision-making, effective contribution by members to development efforts, and equitable sharing of the benefits of development.

Unfortunately, in most developing countries, the farmers' capacity to organize themselves in order to protect their interests, solve their problems, and to deal with state agencies has been neglected.

Most rural organizations are founded by a government agency with a top–bottom approach to development and are generally elite-oriented and dominated. Decision-making is the prerogative of governments and old and new elites are actually aided in re-inforcing their dominance of the disadvantaged sector of the rural population (Heck, 1979).

A major part of the success achieved by the agricultural sector in Taiwan is attributed to the organization of the farmers into associations. A federated four-tiered system of multi-purpose organization was set up under the aegis of an autonomous central development agency (Joint Commission on Rural Construction). At the base are small agricultural units, each consisting of several families; at the next level are township farmers' associations; the national organizations is at the apex. The results achieved have been spectacular: an average yearly increase in output of 5%, during the period 1950–70, with the greatest increase obtained by small farmers with less than one hectare of land, whose income exceeded $300 *per caput* in 1970 (World Bank, 1975).

The FAO is making a major effort in a number of countries to help small farmers and landless workers to organize themselves into informal self-help groups for various income-improving activities. These 'small farmers and peasant production groups' also serve as an effective recipient framework for government services and facilities (Heck, 1979).

These activities are initiated by a full-time *group organizer* who lives with the people and guides them in problem-identification and in the planning and implementation of income-increasing activities. The group members select the leaders, establish rules and define the activities of the group.

After several groups are operating in an area, they are federated in an association that concerns itself with common problems such as adoption of new technology, inputs, credit, processing, marketing, etc.

External initiatives and funds are essential to start these groups and associations, in particular to train group organizers and pay their wages until the groups become self-sustaining.

An example of the successful implementation of FAO's 'small farmers' field action projects' (FAPS) are the three FAPs operating in Bangladesh since 1976, in three separate zones covering eight villages with 155 groups, including three women's subgroups. By 1978, of 2712 eligible families,

1633, nearly all illiterate, had joined the groups. Each group elected a chairman and a secretary–treasurer. Most of the chosen leaders were between 20 and 40 years old. The groups engaged in beef fattening, rice growing, potato growing, goat and dairy-cow rearing, fish culture, silkworm rearing, etc. With increased cooperation between groups through their associations, contacts with government agencies increased. The groups received loans totalling over $140,000; the repayment rate was 94%, a very impressive achievement.

FAPs were also initiated in Nepal and the Philippines, and were rather successful in all three countries. Other examples of successful participation of small farmers in rural organizations that meet communal needs are the 'Harrambee self-help movement' in Kenya, the 'Nboa system' in Ghana, and the 'groupings of producers' in Senegal.

All three have in common that all members participate in decision-making and contribute labour and cash to the projects; they are built upon traditional forms of material aid; they re-inforce social cohesiveness and they foster local leadership in youth.

Experience has shown that if a peasant group functions satisfactorily, people in neighbouring villages become interested and ask for help from local leaders from the successful village (Heck, 1979).

Since the disadvantaged sectors perceive themselves as helpers and are mostly illiterate and resigned to their marginal position it would be illusory to expect them to take the initiative and organize themselves successfully for group actions.

If the fostering of popular participation in rural organizations is accepted as the most effective tool for promoting the interests of the disadvantaged sectors of the rural population, active support from government is essential. Funds for mobilizing and training people whose role is to initiate group activities in villages, for training the leaders subsequently elected by these groups, to provide credit for group activities, etc., are a first prerequisite. Government must also provide political support to counteract the influence of those who fear a threat to their vested interests.

The major problem encountered in these endeavours is to provide guidance and assistance to these groups whilst at the same time strengthening their ability to decide and act for themselves.

Agrarian Reform

Implications

It is not within the scope of this book to deal in detail with the problem of land reform. Here, we are concerned only with some of the implications of land reform for the modernization of agriculture.

Most developing countries have problems due to traditional land ownership and tenture patterns, such as absentee landlords, short-term leases and share-cropping on unfavourable terms to the farmer. A characteristic pattern that has evolved in many developing countries is one in which the land-owning class finds it far more profitable to invest capital in acquiring land for renting out, and allied activities of money-lending and trading, rather than in farming proper. Whether the latter is based on exorbitant rents or on share-cropping, the tenants, in order to subsist, must exploit the land to the maximum without returning anything to it.

In Latin America, the traditional and dominant forces of settlement have been the large estates—latifundia—on one hand and extrmely small holdings—minifundia—on the other. Also, the minifundia usually occupy the poorest and least accessible land.

These traditional systems, based on old property laws and practices, such as *repartimiento* and *encomienda*, entrusted the indigenous populations and their labour to the owners of land (Moore, 1961). The most important effect of this system was the establishment of a powerful elite dedicated to maintaining the existing order.

By contrast, in tropical Africa, the land tenure problem does not, in general, consist of the concentration of land ownership in the hands of the few, and the resulting relationship of landlords to tenant farmers. Much of the land is held to belong to ancestors and to unborn children in a sense of tribal trust.

In the words of a Nigerian chieftain, 'Land belongs to a vast family of which many are dead, few are living and countless numbers are still unborn' (Meek, 1949).

Within this concept an individual can claim a holding for his subsistence according to his need, that is to say the number of wives and children he has, but not for aggregation. When the need decreases by death, the holding reverts to the community for re-allocation by the chief to others in need (Gaitskell, 1968).

The problem here is generally one of defining individual rights to the land as against traditional communal rights, and it becomes more acute as land becomes scarcer (Wilde & McLoughlin, 1967).

In Kenya, for example, the greater part of the land (about 73%) is regarded as the exclusive domain of particular tribes. Each tribe is unwilling to allow members of another clan to establish rights on what they consider their exclusive domain, thereby denying access to good agricultural land to people who could develop and use it.

Certain areas of high agricultural potential in Kenya, because of the jealously guarded tribal grazing rights, support populations of seven pastoralists per square kilometre as comapred with 150 to 200 persons and even more in other areas with similar potential, and 54 persons per square kilometre in districts with far lower potential (FAO, 1974).

In Japan and Taiwan, the available land was traditionally fairly distributed among farm households. Therefore, in these two Asian countries, land reforms were mainly aimed at ensuring tenant's rights and the institution of rent controls (Yudelman *et al.*, 1971).

Objectives

The necessary reforms may be concerned with problems of land tenure and ownership *per se* or with the excessive fragmentation and dispersal of holdings. In many traditional societies, land ownership confers social prestige, wealth, control of labour, and political power. As development progresses, the pressures increase to change the distribution of power, status, wealth and income resulting from owning of land.

Land reforms are, therefore, primarily undertaken for social reasons. However, it has been proven that the concentration of land ownership and the proliferation of minifundia is wasteful of labour and land and militates against the introduction of improved techniques and efficient land-use (Prebisch, 1971).

In a study sponsored by CIDA in 1963 on land tenure problems in the Argentine, Brazil, Chile, Colombia, Ecuador, Guatemala and Peru, it was found that not only 'were traditional land tenure systems to blame for the disquiet in rural areas and the perpetuation of social injustice, but, in addition, they contributed significantly to the economic backwardness of Latin America and prevented the introduction of modern agricultural technology' (CIDA, 1966).

The latifundia–minifundia system discourages the adoption of new technologies in agriculture for the following reasons (Crosson, 1970):

(a) Smallholders do not have the means and the knowledge to adopt new practices.
(b) Sharecroppers working on the latifundias do not have the necessary incentives, as they can be dismissed from the land at any time by the landowner.
(c) The latifundista, with a high and assured income, usually independently of agriculture, is under no economic impulsion to improve farming techniques.

Whilst there is a wide variety of possible reforms of agrarian systems, minimum objectives would appear to be: (a) greater security of tenure for farmer-tenants; (B) an equitable distribution of agricultural output between landlords and shareholders or tenants; (c) an upper limit in the size of holding per family; (d) consolidation of small, fragmented holdings; (e) prevention of fragmentation by sale or succession; and (f) the abolition of absentee land ownership.

Limitations

It has become increasingly clear that the adoption of improved techniques is not possible unless the farmer operates land which he owns or holds securely, and from which he obtains an equitable share of the produce. If he is a tenant, who retains only a small proportion of his yield and all investments conducive to higher yields have to be made at his own expense, he has no incentive to adopt improved practices. An equitable re-distribution of the land is therefore an essential precondition to agricultural progress.

Land reform, in itself, does not however necessarily ensure economic development, and is therefore no guarantee of rapid progress. To be successful, it must be accompanied by appropriate planning and followed up by a whole series of institutional innovations in the fields of credit, marketing, processing, and prices, in agricultural research and extension.

For example, land reform in Mexico suffered from the lack of proper land surveys and insufficient land classification studies, which could show the distribution of soil, slope and water supply. As a result, the land allocated to many 'ejidos' was not suitable for crop production at all, and they had no source of water allocated to them. Of the more than 24 million hectares of ejido lands in 1960, only 7 million hectares could be used to raise crops and only 2 million could be irrigated (James, 1959). Further, rapid population growth and lack of credit have kept most of the ejidos in stagnant subsistence.

Similarly, while the Bolivian land reform, at least partly, liquidated social injustice, it did not result in a marked increase in the land cultivated by the peasant or the yield from the land, because it was not followed by necessary institutional reforms.

In adopting new forms of land tenure, one may be faced by the need to choose between solutions that are effective from the standpoint of economics, but undesirable from the social viewpoint on one hand, and solutions that are socially desirable but are less productive economically, on the other hand.

Whilst a more equal distribution of incomes as a result of land reform is a laudable and socially justified objective, it may, at least in the short run, reduce savings and investment. For example, where ownership passes from large landowners to potential farmers who do not have the necessary managerial and farming skills, or the funds needed to develop their newly acquired land, overall productivity may be reduced.

In many cases, where political and social considerations have been dominant, productivity and resource allocation have been virtually ignored, or considered only passively, by exempting efficient estates for subdivision. Grave difficulties of implementation are commonly encountered and these may be complicated by reaction of one sort or another; consequently, the effect of land reform on productivitiy has often been at least limited and at worst negative.

Organization into cooperatives is usually seen as the answer to this problem but in many developing countries such organisations do not as yet have the managerial skills or the resources to operate effectively.

The problem of land reform is particularly acute when introducing irrigation in semi-arid regions, where rain-fed agriculture is traditionally practised and pressure on the land is extreme. The existing land ownership pattern may then be the greatest obstacle to irrigation development. If the land is divided into very small holdings, and the parcels of land of the individual farmer are widely scattered, no modern irrigation farming is possible before these holdings are consolidated. On the other hand, the promise of development made possible by irrigation may be the lever to achieve the necessary land reform.

In contrast, one advantage of a desert that is to be developed by irrigation, is that the present population is sparse, land as a distinct from water rights has little value and problems of tenure are not critical. In areas of limited rainfall, land reform is almost meaningless unless it also gives title to the use of the available irrigation water.

Size of Farms

A problem closely related to that of land tenure is the size of the individual farm. In most developing countries, part of the land is either excessively split up into very small producing units that are difficult to farm economically and part is concentrated in the hands of a few large landowners.

In regional development planning, the question of farm size assumes considerable importance. The land available for settlement is generally limited and the number of potential settlers great. Hence considerable political pressure may be exerted to adopt farm units of a size that will not be able to provide minimum adequate incomes.

Farm size *per se*, however, has little significance. It must be considered within the context of the norms of the community, the fertility of the land, the effectiveness of the infrastructure, the services available, the intensity of land-use, population pressure, the tenure system, the availability of capital and know-how, the diligence of the farmers, etc.

Minimum farm size depends on many factors. The concept of 'livelihood threshold' relates to a farm size which makes possible the production of family needs in calories plus a further 50% to be used or sold to purchase supplementary foods or other essentials (Pearse, 1977).

Large Farms

From the point of view of propensity to adopt new technologies, there are at least four main types of large farms: the large commercial farms or plantations, state farms, farms based on various degrees of collective ownership or cooperation, and latifundias.

Large Commercial Farms

Large farm enterprises are characterized by 'professionally managed operations with a high degree of division of labour and specialisation, the reliance on an entirely hired, non-family labour force, the routine utilisation of advanced technologies, a high degree of mechanisation, and an overriding objective: the maximum return on invested finance capital' (Powell, 1972). Capitalistic large plantations and ranches, etc., belong to this category.

Plantations

Plantations are large production units devoted to a single crop. Typically they produce a tropical commodity of high value in the markets of the developed countries. Processing of the products, such as cotton, sugar, coffee, cocoa, rubber, palm oil, etc., is generally an integral part of the plantation's activity.

Plantations generally have a bad record: they have been associated with colonialism, foreign appropriation of the country's assets, exploitation of labour, dependence on a single crop, etc. However, these are not necessarily built-in defects of the system. Plantations can play a useful role in the economy of developing countries if they provide a source of employment paying fair wages and adopt profit-sharing incentives. They can even be planned to become worker-managed or to dissolve into smallholdings at a later date (Harvey *et al.*, 1979). They can also provide services and guidance to small-holders in their vicinity (see below).

State Farms

In many developing countries, governments are interested in setting up large state farms, with the most modern equipment, for reasons of prestige, political conviction or as a short-cut from subsistence agriculture to modern farming. The expectation is that such farms will serve as models of high productivity, provide food for the cities and commodities for export, and employ large numbers of workers.

The first systematic approach to modernize agriculture adopted in Ghana after independence was the establishment of fully mechanized state farms. This approach proved to be a costly failure: of the 123 state farms established between 1960 and 1965, only 3 remained in production in North Ghana by 1979. While all the state farms cultivated 1.4% of the total agricultural land, they contributed less than 0.5% of the total agricultural production.

Apart from corruption and mismanagement, the main reasons for the failure of the state farms are ascribed to 'maladjustment of the model to the cultural background of the native peoples and to the constraints dictated by the natural environment' (Bennett & Schork, 1979).

On the whole, experience with state farms has generally not been encouraging. Rigid bureaucratic control, lack of adaptability and rigid organizational forms generally characterize this type of farming unit. The difficulty of finding people with the necessary managerial abilities and technical knowledge, the lack of cost consciousness, the low output of labour on state-owned enterprises, are all handicaps to success. Ghana is an example of a country in which enormous funds were invested in state farms as a short-cut to modernization and in which the results have long been disappointing (Wilde & McLoughlin, 1967). This does not imply that all state farms are bound to fail, but merely to stress the factors that militate against success, and the limitations that have to be overcome.

Large-scale state farms, may be justified in the early stages of development as a means for rapidly developing an area for intensive production (irrigation projects, large-scale land reclamation projects, etc.), as a means of training future farmers in modern production methods, providing an assured income during the period until the land becomes fully productive, etc. After these objectives are achieved, the land is divided into family-sized farms and handed over to prospective farmers. This approach, to be successful, requires: (a) physical planning at the outset based on the final objective of splitting up the large estates after a number of years into viable farm units; (b) qualified managers and technical personnel to run the farm efficiently, and educate future farmers; and (c) guidance and provision of basic services to the farmers after their assuming responsibility for their individual holdings.

*Collective Farms**

Communes, kolkhozy, sovkhozy in the USSR: Though Soviet Russia cannot be considered a developing country, the elimination of family farms in the 1920s, and their replacement by huge farming units occurred under conditions of subsistence farming.

Historically, large-scale farming was not established in the USSR as a means of modernizing agriculture, reducing costs of production, or improving the income of the peasants. The dominant motive was to overcome the difficulty of organizing 'procurement'. Collectivization consisted of agglomerating the dwarf peasant farms into relatively small collective farms, and at a later stage, increasing the size of the collective farms through amalgamation and by the conversion of part of the collectives into even larger state farms (Straus, 1969).

Three types of large farms were originally established: communes, which were based on communal working and living, these were unsuccessful and short-lived; the kolkhozy and the sovkhozy.

The kolkhozy (or artels), are collective farms in which the land is worked cooperatively, whilst family smallholdings provide food for own consumption and subsidiary production. The farm is managed by a chairman (formally elected by the members of the collective) assisted by a management

*The communes of the People's Republic of China and the Ujama villages of Tanzania are described and discussed in Chapter XI.

board and a number of specialists. 'Brigadiers' are in charge of special work units, each assigned to the production of a specific commodity. A certain amount of practical initiative is allowed the 'chairman' of the kolkhoz, but he has to account to the authorities for the performance of the farm. The collective farmer is considered as a partner in a cooperative enterprise and is not entitled to a firm wage but to a proportion of the net income produced by the collective farm. In practice, this means the residue remaining after the State has claimed its share in various forms: income tax, allocation to the State of the farm produce at dictated prices; sale to the farm of inputs and commodities provided by industry, also at dictated prices. Therefore revenue accruing to farmers depended only partly on the efficiency with which the farm was managed, or on labour productivity and yields, and to a larger extent on government price policy for products and inputs.

The residual income is distributed among members on the basis of the number of 'labour-days' they have worked; 'labour-days' being calculated according to a scale in which, for example, a day of the lowest type of work is valued as half a 'labour-day' and the most difficult mechanical jobs are rated as two 'labour-days'.

In order to provide an incentive for increased production, substantial premium payments have been provided for production above government-set norms, and the kolkhoz members are allowed to keep part of the excess production in kind. To placate the farmers, the possession of individual family plots was legalized. Houses, small vegetable gardens and orchards, and a certain proportion of the livestock and poultry are not socialized. The size of these allotments is generally between one-quarter and one-half a hectare.

In 1962, these family mini-farms accounted for 45% of total milk and meat production, 76% of eggs, 22% of wool, 70% of potatoes, 42% of vegetables, and 66% of fruits. The share of the private sector in total agricultural production has been estimated by a Soviet source at more than 36% (Karcz, 1967). The kolkhoz is therefore essentially a bimodal structure of farm size consisting of a large cooperatively managed farm together with tiny family farms.

The tendency has been to increase the size of the collectives. In 1950, a collective farm, on the average, had 165 families and had a sown area of less than 960 ha; after the merger campaign, by 1963 the average number of families had increased to 411 and the cultivated area to 2860 ha (Karcz, 1967).

Initially, the kolkhozy tended to continue the typical subsistence-type farming of the peasant economy on which they had been imposed by force. Modernization and specialization were difficult because of the lack of a supporting system providing the inputs necessary for modernization. A solid infrastructure is still lacking in wide rural areas of the Soviet Union. It is only in the 1960s that the 'need for making good much of the arrears in agricultural investment, which have accumulated over decades' became government policy (Straus, 1969).

Sovkhozy or state farms were intended to be 'specialized agricultural factories' under direct government control and were to be established especially for rapidly expanding production of essential commodities on land previously unused either because it was marginal or in areas where climatic hazards had previously prevented permanent agricultural settlement. The farm staff consists of salaried workers. In the 1960s, there were over 8000 state farms with an average sown area of about 10,000 ha each. Some state farms had an area exceeding 70,000 ha of agricultural land (Straus, 1969). The state farms have a much broader top managerial and technical administration than the collective farms. The director is assisted on the average by three divisional managers, three to four chief specialists, each of whom has a staff of several skilled experts. The need for such very large specialist staff at farm level is dictated not only by the size of the farms but also by the enormous amount of paperwork required from them, and the weakness or lack of supporting institutions in the supply of essential services and in the marketing of produce.

In Summary

The collective and state farms of the USSR are not large farms in the accepted sense, but are super-large: 'huge in area and Leviathan in administration' (Laird, 1953). Beyond a certain size of farm, there are no further economies of scale, but organization and administration becomes extremely cumbersome. Even in the modern agriculture of the USA, experience has shown that 'not only do costs seem to rise after a certain size, but at least in some cases yields begin to drop off, because of lack of timeliness in operations resulting in declines in yields (Brewster & Wunderlich, 1961). However, the low productivity of labour and low yields per unit of land of the kolkhoz and the sovkhoz cannot be imputed to their super-large sizes alone. Government policies have resulted in the undermechanization of production, backwardness of the whole infrastructure of the agricultural economy, lack of amenities for the rural population, and the oversized peasant population (Swearer, 1967). All these have compounded the effects of hugeness on productivity of land and labour.

Collective and Cooperative Farms in Non-communist Countries

Whilst at first glance the communal and cooperative forms of farming in the communist and non-communist countries may appear very similar, there are two basic differences of considerable importance: in the communist regimes coercion has been used in the establishment of the farm units and planning is unilateral and from above. In the non-communist countries, joining communal or cooperative forms of settlement is a matter of individual choice among alternatives, and the planning and management (apart from guidance and provision of Governmental incentives) are the prerogative of the members of each group.

The 'Kibbutz'

The kibbutz is a unique form of settlement that originated in Palestine at the beginning of the century by Jewish immigrants, who not only believed that a return to the land was essential for the redemption of their people, but were also imbued with social ideals. Whilst the amount of land available per family is the same as that allocated to family farms (see the moshav, pp. 460–465). All available resources—land, water, capital, and human—are managed and operated as a single unit. The management is elected, and is responsible for planning and implementation; individuals are responsible for different branches of production as well as for the multiple services required by a large collective. Elected committees orient and aid management, but matters of major importance are submitted to the general meeting of all members.

No salaries are paid. Whilst each family has its own housing unit, the main meals are served in a communal dining hall, whilst the children generally live together in separate buildings.

The kibbutz has all the advantages of scale of the large farming unit; individual members specialize in various forms of production and can therefore achieve a high level of competence in their respective fields; because of the special way in which the farms are managed it is fairly easy to release kibbutz members for training courses, workshops, lectures, and visits to other settlements.

The kibbutz is therefore able to diversify and specialize at the same time. For this reason it is also able to undertake the risks involved in experimenting with new forms of production, and as a result, the kibbutz has been the spearhead of agricultural progress in Israel and has achieved levels of production unsurpassed elsewhere. In recent years, major investments have been made in suitable industries, in order to make full use of the available human resources and provide employment for the population increase that has occurred since their establishment.

The kibbutz structure originated in response to a hostile environment; climate was harsh, the land, generally marginal, had to be reclaimed; disease was endemic and attacks by hostile neighbours frequent. The communal form of settlement made possible collective responsibility for the individual; farming could be continued, even when a large proportion of the members were sick; guard duties could be rotated; the children were well cared for whatever the workload of the parents; and—as mentioned above—individuals could specialize and achieve proficiency in various branches of production.

However, the kibbutz has proven itself stable and enduring even after the conditions which predicated its establishment have disappeared. Membership now includes third and fourth generations. Productivity is very high and the kibbutz is prosperous even by urban standards.

Small wonder that many visitors from foreign countries, impressed by the economic and social achievements of the kibbutz, have thought it to be an appropriate model for establishing settlements elsewhere in areas to be developed under difficult environmental conditions. Several attempts have been made in developing countries to establish Kibbutz-type settlements (Ghana, Uganda) none of which has been successful.*

The driving force that has made the Kibbutz possible, and the cement that has held it together in times of adversity and prosperity, is motivation and ideals that have been forged under very special and unique conditions. Without these, it appears that human nature does not readily adjust to the many constraints imposed by communal living, especially in the absence of any relation between individual effort and personal gain.

Collective Moshav (Moshav Shitufi)

The 'collective moshav' is an intermediary form of settlement between 'kibbutz' and 'moshav' (see below).

The basic concept is that of communal ownership and management of all means of production, whilst retaining intact the personal life of the individual (Desroche & Gat, 1973).

As in the Kibbutz, production is communal; a committee assigns work to each member; the community is responsible for the economic, social and cultural requirements of its members. Education of the children is also a joint responsibility.

Different from the kibbutz with its communal dining hall and childrens' houses, the life of the individual is entirely confined to his family. Each family has its house, owns its furnishings, prepares its meals and raises its children. A monthly budget is allocated to each family, in accordance with its size and disregarding completely the number of days worked or the nature of the work provided by the members of the family. The family is free to spend its budget in any way desired. No private plots or individual economic enterprise of any kind is allowed.

The women's main task is considered to be centred on her household, but she is called upon to provide a certain number of hours daily for communal work, as requested by the works committee. This includes house-keeping for a mother who has fallen sick.

In brief, the 'moshav shitufi' combines the main characteristics and advantages of the 'kibbutz': efficient organization of production and services, collective solidarity, and communal ownership of means of production, whilst maintaining the main elements of family life.

The moshav shitufi, wherever established in Israel, has generally been very successful both economically and socially. And yet, notwithstanding its manifest advantages, it has not been widely adopted. The very fact that it is

*In Japan, there are a number of reportedly successful kibbutz-type collectives (Tezuka & Kusakari, 1969).

a compromise between kibbutz and moshav, is probably the reason why most prospective settlers chose the one or the other, according to their personal inclinations, and not the hybrid form.

As to the applicability of the 'moshav shitufi' concept to other developing countries, the same constraints preventing the adoption of the kibbutz are also relevant to that of the 'moshav shitufi'. *Farmers' Groups in Ghana* are reportedly operating satisfactorily in Ghana, where they have replaced the largely defunct State Farms as a model for modernizing agriculture. In these groups, large areas of land on which a single specialized crop such as cotton, kenaf, rice, maize, or groundnuts is grown, are cooperatively farmed. This is in addition to the normal occupation of the participants. Farmers Groups are actively supported by Government which provides the following incentives: fixed market prices; subsidized inputs; complete input packages, including seed of improved varieties; cheap credit and the hire of machinery from state-run machine stations.

The main drawback of the system is that its advantages accrue to a relatively small group of people, who have the capital and/or access to the credit required; the majority of members of the Farmers' Groups are not farmers but merchants, contractors, civil servants, and army officers (Benett & Shcork, 1979).

The Ghanaian experience should however not be considered as an indictment of the concept of 'Farmers' Groups'—on the contrary, it should serve as an object lesson that the success of a basically sound concept depends on the way it is implemented. Some form of communal or cooperative production unit is probably the only viable solution for the multitudes of small farmers, share-holders, tenants, and farm labourers who form the bulk of the rural population.

Appraisal of Group Farming

Group farming can take a number of different forms and can be established in countries with different political regimes.

In the communist countries they are characterized by coercion and lack of alternatives, planning and direction from above, and extremely large size. Political and macro-economic considerations originally outweigh the adoption of measures for increased productivity and satisfaction of individual aspirations; however, failures resulting from this approach have generally caused a retreat from policies dictated by political ideology.

In the non-communist countries group farming is voluntary; the units are sufficiently large to offer the advantages of scale but small enough to be managed efficiently by their members. Various degrees of communal ownership and various hybrid forms have evolved.

Notwithstanding their manifest advantages, the rate of adoption of group farming in its different forms in developing countries has been minimal. And yet, group farming has a great potential as a form of settlement for the most disadvantaged sectors of the rural population. These possibilities will be discussed in more detail in Chapter XI.

Latifundias

The latifundias are a type of large farm with entirely different characteristics to those of former types. They have the following features: reliance on traditional, often extensive, technologies; the binding of the labour force by various means other than monetary contractual inducements; and a low degree of mechanization. Land ownership in this case is important primarily for reasons of prestige and as a hedge against inflation; because of the size of the property, income is ample even with primitive production techniques. Therefore, as opposed to 'capitalist enterprises, the latifundia does not tend to invest financially, but uses its control over land resources to exploit its labour force' (Powell, 1972).

Under the prevailing social and economic conditions of many of the developing countries, latifundias may be a factor retarding modernization and intensification. If large landowners lack interest or initiative, ability or capital to develop their land, large units are a handicap to progress, and acute problems of productivity arise. The owners of latifundias tend to transfer the income they extract to urban areas, either for industrial investment or to finance the consumption of luxuries. This is especially true of absentee landlords (Yudelman *et al.*, 1971).

Small Farms

In most developing countries the proportion of small farms ranges from 51 to 100%. Those smaller than one hectare make up 46% of the total in India, 52% in Liberia, 57% in the Philippines and 67% in the Republic of Korea (FAO, 1977). In Taiwan, 66.4% of farms are less than one hectare, and in Indonesia 70.1% (Griffin, 1972); in India 7.7% of rural families own over half the cultivated area.

Even in regions in which population densities are not particularly high in relation to land resources, as in North Africa, the Middle East, the Philippines and Latin America, the great majority of farmers have small farms, because most of the land is owned by relatively few.

The number of small farms has increased steeply in recent years as a result of population increase. During the period 1950–70 the number of farms smaller than 5 ha has doubled in India, trebled in Brazil and quintupled in Iraq (FAO, 1977).

Part-time Peasant Farms and Minifundias

The process of subdivision of the land eventually reaches a point at which holdings are too small to support a household even at minimal sub-

Fig. 65 The 'crazy-quilt' pattern of tiny smallholdings of the Pueblo subsistence farmers is typical of minifundias in other parts of Latin America. By courtesy of the Rockefeller Foundation.

sistence level. Some holdings may be so small that they deserve the name micro-lots (Fig. 65).

Of 172 million rural families in 19 Latin American countries, 5–7 million subsist on minifundias (Bachman & Christensen, 1967). The farmer must then choose one of the following alternatives:

(a) To lease land to supplement his holding.

(b) To lease out his land and search for employment elsewhere.
(c) To supplement his income from outside work with that from his land; in Japan, for example, over 50% of the farm family's income is obtained from off-farm sources (Schickele, 1971).
(d) To sell his land.

The forced liquidation of small family farms and the resultant consolidation of holdings is, in purely economic terms, 'an efficient re-allocation of resources'. This is only true if there exist alternative sources of employment for former farmers. In most developing countries, we have seen that such sources of remunerative employment are generally non-existent, and the social cost of the 'efficient re-allocation of resources' is high (Yudelman *et al.*, 1971).

Another solution that is frequently proposed for small uneconomic holdings is to organize them into cooperatives. The major weakness of this solution is that the uneconomic size of the holding is due to population pressure; the amount of excess labour remaining unchanged, this problem is not solved by consolidation into larger units. If alternative employment is found for redundant labour, it is just as relevant to the individual holding as it is to the cooperative; the mini-farm can then become an auxiliary source of income to the family.

Family Farms

Adequate-sized family farms are enterprises which control sufficient land resources to provide full-time employment to a farm family at a productive level, and which underwrite their basic life requirements. Such families do not rely on the labour market for employment, nor do they usually employ hired labour within the enterprise (Powell, 1972). The head of the family is both operator and worker.

Family farms can vary considerably in size; the acreage required to generate the same level of income depends on the work potential of the family, the level of capital investment, and, of course, the ecological conditions.

A recent survey in Kenya, for example, indicated that for rainfed agriculture, the farm size needed to produce approximately $40 *per caput* per year, increased progressively from 2.6 ha up to 16.4 ha as one moved to less and less favourable ecological zones. In order to generate the same level of income in range areas bordering the Sahel, from 90 to 135 ha are needed (World Bank, 1975).

Within certain limits, rice-farm size tends to be inversely correlated with intensity of cultivation. In Kahuman village, central Java, a 0.6 ha rice farm is the major means of support for a family of six to seven persons; this contrasts with a 6 ha farm in Rai Rot village (Suphan Bas, Thailand) for the same size family.

The Kahuman farmer obtains from his tiny, garden-like weedless plots a 5 t/ha yields twice a year as compared to 2 t/ha obtained once a year on the 6 ha in Rai Rot. To save time, seedlings for the second crop are prepared before the old crop is harvested. The difference in annual yield between 'large' and 'small' family farms is 5 to 1 (Barker & Anden, 1975).

Between the family-type farm and large farms, there is a whole spectrum of medium-sized farms, on which, in addition to the farm family, there is regular utilization of hired labour. These farms differ in the amount of land, capital goods and financial credit incorporated in their operation (Powell, 1972).

The Moshav

The moshav is a type of village that has evolved in Israel. In principle, it is a cooperative of freeholders' family farms, comprising from 60 to 120 units. Whilst each family constitutes an independent unit, many operations and services are on a cooperative basis. The village cooperative purchases all farm inputs and other supplies, and organizes the handling and sale of produce. Institution credit is made available to the individual farmer exclusively through the cooperative, so that responsibility for repayment of credit is corporate. The cooperative is also responsible for municipal and cultural services.

Government planning units, established in each region, have been responsible for planning the physical layout of each village and its agricultural production. The basic guidelines on which planning is based, are that production should be based on family labour and the income of the farming unit should be equivalent to that of a small urban entrepreneur. According to these plans, each village specializes generally in two fields of production, which are considered as most appropriate to its specific ecological conditions, and provide a well-balanced programme of employment throughout the year. Examples are villages planned to consist of mixed dairy–vegetable farms, citrus and vegetable growers, industrial crops and poultry, etc. Credit and subsidies for the infrastructure required are supplied by the authorities in accordance with these plans. It is of interest to note that after 10–15 years, generally no more than one-third of the farms are still operating according to the original plan; the others have embarked on forms of production completely different to those planned—an indication that initiative has not been stifled by planning and the constraints imposed by the cooperative form of production.

The basic principles of the moshav can be summarized as follows:

(a) The moshav consists of 60–120 family farms, organized for self-government and cooperation in agricultural production, as well as social, cultural, and municipal services.

(b) All farm units have equal allocations of land, water, capital and equal access to credit.
(c) These resources are calculated to enable a family, using its own labour, to achieve an income approximately equivalent to that of a small urban entrepreneur.
(d) The land is held in freehold, under a 49-year lease; tenancy rights can be inherited, but the farm unit cannot be divided among heirs.
(e) Whilst each family constitutes an independent unit, as many operations and services as possible are carried out cooperatively.

Applicability of the Moshav Concept in Developing Countries

The moshav concept as a 'method of integrating individualism in an accepted cooperative community pattern' (Klayman, 1970) can be a useful model for adoption by developing countries, after it is adapted to the particular social and cultural characteristics as well as the economic level of the communities involved. Its advantages are:

(a) The possibility of using a transitional framework whilst introducing new settlers gradually to modern farming, self-government, and cooperative enterprise,
(b) its flexibility, making it adaptable to different economic, social, and cultural levels.
(c) It conforms to the aspirations of the majority of farmers, who are individually minded and prefer their own family farm to other forms of settlement, and
(d) its cooperative framework (marketing, supply, credit, and community services) make possible economies of scale and no less important, mutual aid.

Moshav-type villages have been established with varying degrees of success as model villages, among others in the Central African Republic, Senegal and Dahomey, and on a larger scale in Venezuela and Iran. Whilst, in general, it is still too early to judge the merits of these schemes, a detailed study of 'settlement schemes in western Nigeria' by Roider (1971), exemplifies the considerable difficulties and problems encountered in the execution of such an attempt because of 'the lack of a cultural base, the low levels of agricultural technology and project administration' (Eicher, 1967).

Prerequisites for success in the adoption of the moshav concept are (Frank, 1968):

1. Careful selection of the candidates for settlement, according to predetermined criteria adjusted to each situation.

2. The farm unit must be economically viable and enable the achievement of a standard of living at least equal to that of an urban qualified worker. Size of the farm unit will therefore depend on the production potential of the area and the capital available for investment.
3. Production must be planned so that it corresponds to the work potential of the family and the need for hired labour should be confined to periods of peak requirements or sickness.
4. Public facilities, for education, health, recreation and cultural purposes, must be planned and provided for.
5. Both fragmentation of holdings and accumulation of physical resources (land and water) by individuals must be prevented.
6. Adequate guidance and training in production methods and community management must be provided.
7. Until the settlers become self-supporting, additional sources of income, mainly from public works (land development, infrastructure, etc.) must be provided.
8. Detailed statutes have to be drawn up, detailing the rights and duties of the individual as well as details of the investment programme and conditions of repayment of credits provided.
9. A probationary period of 3–5 years, before the settlers are given land title or a final lease contract. During this period the authorities are entitled to remove settlers who neglect their lands, or do not fulfil other basic obligations.

The Nigerian Experience

In 1959, at the initiative of the Minister of Agriculture and Natural Resources, a decision was taken to start a settlement scheme in Nigeria, based on the Israeli Moshav model (Roider, 1971). According to this model, the settlement was to be a collective community of settlers who, after having benefited from initial guidance and financial aid from government, and after gaining sufficient experience and self-reliance, would undertake the operation themselves, as free enterpreneurs within a central cooperative.

The Plan

The build-up of the settlements was to be accompanied simultaneously by the establishment of five 'farm institutes' in which prospective settlers would receive two years training. At first, the following ministries were involved in the settlement schemes:

(a) Agriculture and Natural Resources—for planning and execution;
(b) Health and Social Welfare—for problems of community development;

(c) Land and Labour—for planning of the layout of the villages;
(d) Works and Transport—for building of roads, water supply, construction of houses and other buildings; and
(e) Trade and Industry—for organization of all cooperative aspects.

To ensure cooperation, a working party comprising all the ministries involved was established, with six committees, each representing a special field of competence. At a later stage, recognizing the insufficiencies of this set-up, a special department for settlement planning and administration was proposed.

The interval between the decision to start the scheme and the beginning of field work was very short, so that the planning staff were only able to outline a rough framework of the scheme.

All the settlements had the following common characteristics:

(a) In the village centre were grouped all the communal services: school, medical station, church, mosque, shop, market stalls, community centre, etc.
(b) Around the village centre, 'home plots' of 0.2–0.4 ha, containing house, garden and eventually livestock buildings. Behind the home plot, an arable plot of about 0.8–1.0 ha for the production of subsistence food-crops.
(c) The remaining areas were grouped around the village site, into blocks, according to the type of crop grown.

According to the ecological conditions of the region, two basic types of settlements were planned: type A, in the rain and dry forest areas in which tree-crops dominated; and type B in the northern savanna areas, in which farm-crops were emphasized. Cattle-keeping was considered only for the type B settlements, as the type A could not provide an acceptable nutritional basis for the livestock.

Special importance was assigned to mechanization as a means to expanding the area cultivated per family unit, and for psychological reasons, mainly as evidence of the move away from primitiveness and drudgery.

The agricultural machinery was to be operated initially by government on a hire basis, and subsequently to be handed over to the central cooperatives of the settlements.

After a breaking-in period, the farm settlements were to govern themselves on the moshav pattern, each forming a large multi-purpose cooperative for the purchase of production inputs, sale and processing of produce, operation of central services, and eventually, assumption of responsibility for certain production branches in which economy of scale was important.

The estimated costs per settler were £2000 of which £300 was an outright government grant and the remainder a loan to the settlers with easy conditions of repayment. It was calculated that the scheme would bring an internal rate of return of more than 6%, and could be considered an economic venture.

Implementation

By 1966, 20 settlements (out of a planned 35 settlements) with approximately 1200 settlers were operating, of which 18 were oriented primarily towards the production of tree-crops (type A) and two settlements producing annual and perennial crops (type B).

Only very few communal amenities were in existence. Almost all had a central water supply, and a few also had a community centre and cooperative shop. No other social amenities (school, hospital, etc.) were completed, mainly because of the heavy financial burden created by directly productive activities.

As to production itself, by 1966 only 39% of the planned tree-crops had been planted; these areas had, as yet, scarcely come into bearing, as only 12% of the plantations were more than four years old, and thus able to bring the first small yields. Most of the plantations were in an unsatisfactory state of cultivation; some areas were even in need of re-planting.

On type B farms, the main crop grown was maize, which accounted for 80% of the sown area. Actual yields of the annual and perennial crops were lower than planned, though an observable trend towards improvement occurred in the last years. Mechanization proved to be very expensive, and the settlers were convinced that income from a non-mechanized area would have been higher than from the larger mechanized area.

Poultry was intended to compensate for the low income in the early years of establishment of the tree-crops. However, by 1966, only 44% of the planned poultry housing had been built, and of the existing poultry houses, only a quarter was fully used. Production was also relatively low.

Cattle-raising had been introduced on a small scale, and hand not proved to be profitable.

The settlers were still under the authoritative direction of the administrative and had a voice only in minor decisions. They were however relatively satisfied and accepted the lack of responsibility resulting from this situation. They continued to expect free services from government*.

In 1973, Oni & Olayemi of the University of Ibadan wrote: ‘After thirteen years of existence, the hope for an agrarian revolution based on the farm settlement scheme has dwindled considerably. It has now become an established fact that the project has not succeeded in terms of achieving the set-out objectives. For instance, several researchers have shown, that the profitability in terms of rates of return in investment is too low for the

*There have been isolated cases of exceptional success, with earnings of up to £1380 for a production season, and 5% of the settlers earning over £500 (net income) in a given year.

average farmer to endure; its demonstration effects on the neighbouring farmers have been very peripheral; and its impact in alleviating the unemployment problems of the primary school leavers has been virtually insignificant.'

How far can these shortcomings of the settlement scheme be imputed to the moshav system on which it was to be modelled?

The Nigerian example highlights the difficulties of judging whether the model of the Israeli moshav, with its combination of collectivity and individuality, is successful or not in other developing countries.

The shortcomings in execution and achievement described above, do not reflect any deficiencies in the moshav concept. These have occurred in the initial period during which the settlements were under the tutelage of government, and the moshav concepts had not yet been applied. Hence, whatever shortcomings were experienced, these were due to administrative failures and inadequate management by the authorities involved, and not to the failure of the model *per se.*. Probably the main indictment against the scheme, as it stood some 6–7 years after initiation, was the fact that the settlers were satisfied with their passive role.

However, it should be remembered that in Israel too, in the case of settlers with low levels of education and technical ability, and who were not imbued *a priori* with the ideology of cooperation and other principles underlying the moshav concept, it generally took up to a decade to transform them into independent farmers, and still longer before their villages functioned according to moshav principles. Hence, any conclusions regarding the applicability of these principles to other situations, can only be made on a long-term perspective.

Relative Advantages and Disadvantages of Farm Size Types

Evidence from many countries indicates that economic progress in agriculture is possible under a great variety of farm size conditions and that size of farms *per se* is neither a prerequisite nor an obstacle to agricultural development. 'A large farm unit is not in itself a guarantee of productivity' (Bachman & Christensen, 1967).

In research on the relation of farm size to economic development, Kanel (1967) has investigated the unit costs of production on family and larger farms, the problems of providing basic services, such as extension, credit, and marketing facilities and the differences in entrepreneurial attitudes. He found that in terms of economies of scale, there is no overwhelming advantage to any particular size of farm in developing countries, and that in most types of farming, the advantage was more likely with the family than with the large farm.

Technical know-how and managerial ability are very scarce in developing countries, and this is one reason why family farms, under certain circumstances, may be more efficient than large ones. Family farms also favour labour-intensive forms of production, and use less scarce capital. This relationship has been tested and verified in Asia, Africa and Latin America (Dorner, 1973).

All other conditions being equal, family farms may be more productive than large farms, because the small farmer will devote more care in preparing his plot, weeding, irrigating, etc., than the larger farmer who is dependent on hired labour. In India, average yields of farms of less than two hectares were found to be nearly 50% greater than on farms of more than 20 hectares; in Taiwan, farms with less than one hectare have far higher yields than do those with more than two hectares (Grant, 1973).

Ghose (1979) makes the interesting comment that these facts do not necessarily reflect an intrinsic superiority of family production over wage-labour based production, as is often supposed. He argues that the observed superiority is intimately linked with primitive technology and insufficient development of markets. The factors involved in the inverse relationship between farm size and land productivity are: (a) an overwhelming importance of human labour in the production process; (b) dependence on farmyard manure for the maintenance of soil fertility; (c) primitive, labour-intensive, methods of irrigation; (d) a virtual absence of markets in some inputs, such as farmyard manure, and (e) imperfections in the wage-labour market. Ghose argues that with the introduction of chemical fertilizers, labour-saving mechanization and modern irrigation equipment, the superiority of small-scale production is likely to be eroded, even though it may remain more labour-intensive than the large farms.

An example of this process—though it does not constitute conclusive evidence—is provided by a decade of considerable technological process in Ferozepur (India), during which the relationship between farm size and land productivity changed from negative to positive, whilst that between farm size and intensity of labour use continues to be negative.

The views expressed by Ghose may explain the apparently conflicting results obtained from recent studies in India on the relationship between yield per unit area and size of holding. In nine studies in areas that had adopted HYVs, and were producing rice, wheat, and maize respectively, yields were found to be higher on larger farms than on smaller ones in five areas; the opposite was found to be true in three areas, and in two areas, there was no significant correlation between farm size and yields, (Dasgupta, 1977). These apparently inconsistent results could be explained by the differences between individual farms, irrespective of size, in the use of inputs and cultural practices accompanying the adoption of the HYVs.

In the production of certain crops, family farms have proven to be able to compete effectively with large plantations or estates. For example, cocoa production in Ghana and Nigeria, which dominate the world cocoa market, is entirely in the hands of smallholders.

Farm size tends to grow with the decline in labour supply relative to available land and capital.

Large farms and estates have the advantages of being better able to utilize capital, to hire managerial staff, and to assure a large volume production to the market or a sizeable flow of raw materials to local industries. The latter advantage is of particularly importance in crops that require fairly elaborate processing immediately after harvest. Where such estates are efficiently managed and modern methods are used, no problems of productivity arise.

The application of improved technology makes the use of large units of machinery and equipment profitable. On an average, each person in the farm population of developed countries will have about 20 times as much land as those in developing countries. For the Far East alone, it would be 50 times as much (OECD, 1967). In certain countries, the investment in pumping equipment, soil-levelling and conditioning, recurrent investment in improved farm equipment, make large units a precondition for economic operation (Gregor, 1959). Four-hundred hectares are considered the minimum area required for economic operation of irrigated farms in California.

For tubewells in Asia, the minimum command area below which the costs of water rise sharply is estimated to vary from 10 to 20 hectares (Shaw, 1971). The calculated minimum size of holding, using bullock power, beyond which tubewell equipment is no longer profitable for the Punjab, was found to be 6 hectares (US Aid, 1969).

Incentives to achieve careful management, intensive use of scarce land resources, and of family labour can be most effective in family farms. Because labour is abundant, capital scarce and managerial talents scarcer still in most developing countries, the relative advantage of the family farm are enhanced.

Adoption of New Technology

Though, in principle, most biological innovations in agriculture are highly divisible, larger farmers have easier access to information and inputs, face lower transaction costs per hectare and can better afford to take risks than smaller farmers. For these reasons they are more likely to adopt new practices at an earlier stage than smaller farmers.

There is also a natural tendency for government planning and implementation of development policies to rely mainly on the large-scale farmers for spearheading progress. Credit, tax relief, import licenses and other concessions are channelled to this sector, easily leading to the neglect of the small farmers in need of modernizing their production methods. However, most studies indicate that as the new practices prove themselves, the differences in the percentages of adopters between larger and smaller farmers tend to disappear, *provided the latter can afford to buy the required inputs.* In four out of six studies in different parts of the world on factors influencing the rate of adoption of new practices, the availability or use of credit was significantly related to the adoption of high-yielding varieties

Table XLI Percentage of rice-farmers using HYVs and fertilizers, in relation to farm size (Barker & Anden, 1975).

Practice	Farm size		
	Less than 1 ha	1–3 ha	Over 3 ha
Modern varieties			
Wet season	84	86	93
Dry season	89	91	89
Fertilizers			
Wet season	76	75	82
Dry season	84	83	85

(Perrin & Winkelmann, 1976). For example, in the low elevation areas of Colombia, only 19% of small farmers had adopted new varieties of maize whilst at the same time 65% of larger farmers were already using them. If, credit and extension contacts were considered, however, the multi-variate analysis indicated that the residual effect of farm size was minor, with each hectare increase in size being associated with only a 0.1% increase in the probability of adoption. Similarly, in the valleys of El Salvador, where the respective adoption rates for new varieties for small and larger farmers were respectively 34 and 71%, the multi-variate analysis of hybrid maize adoption failed to show any effects of farm size *per se* (Perrin & Winkelmann, 1976).

In a survey of 32 villages, carried out in a broad geographic area representative of a wide variety of rice-growing conditions in Asia with high production potentials, relations between farm size and the adoption of new practices were found as shown in Table XLI. The great majority of the farmers in villages surveyed have adopted modern varieties and use fertilizers. Farm size had no marked or consistent effect in this respect. However, it must be stressed that the survey was carried out in *selected* villages with a high potential for production and where the majority of farmers were willing and *able* to adopt the new practices. Where these conditions do not pertain, farm size is bound to be more critical, not because of economies of scale, but because of the relationship between levels of income and farm size.

Fragmentation of Small Farms

A major problem of small farms, usually due to traditional inheritance procedures, is excessive fragmentation that may even defeat the purpose of land reform. In Taiwan, for example, though the average size of the farm is only 1.03 ha, more than half the farms have at least three plots of land each, and at least 75% of these plots are not on farm roads (Wu, 1972).

Consolidation of fragmented holdings is a prerequisite to endeavours aimed at increasing labour productivity on the farm, and in certain cases to development. In many Arab villages in Israel, the main constraint to the introduction of irrigation, where water was available, was the fragmentation of the farmsteads into small and scattered parcels of land.

Land consolidation should be planned to improve the usage of fields, re-arrange farm roads, and make possible efficient irrigation and drainage schemes.

Minimum Size of Farms

It has been stressed that relatively small family farms can also be efficient and productive. However, farm units must be large enough, not only to ensure efficient production, but to provide an acceptable standard of living.

Practically all the recent reforms have subdivided the land into many units too small to be economically viable or to yield more than a meagre living. The Mexican reform, for example, increased the proportion of landowners from 3% to 50% of the rural population, but nearly 85% of the holdings, occupying 74% of the land, are less than five hectares. Most of the occupants of such holdings called 'minifundistas'* are in dire need, at income levels hardly sufficient for subsistence. By their efforts they can do little or nothing to improve productivity. Such is the situation in Bolivia and other states of Latin America where land reform has not yet had any positive economic effect.

After the Egyptian reforms, 95% of owners had less than two hectares, and half of these had less than half a hectare.

Social, Political, and Security Aspects of Farm Size

Farm size has social, political, and security aspects in addition to technical and economic considerations.

Efficient estates in the developing countries are, or were, giving rise to local hostility. Ownership has been frequently transferred either voluntarily as in Malaysia and Guatemala, or by expropriation as in Cuba. Such transfers serve legitimate national, social, and political ends but often result in decreased productivity in comparison to the well-managed large estates.

Land concentration in the hands of a minority effects the distribution of political power and enables the monopolization of the benefits of technological change. For example, in Pakistan, 69% of tubewells were sunk on farms larger than 10 ha and only 4% on smaller farms. Thereby, large landowners obtained even more control over the scarce resource water (Griffin, 1973).

It can also result in subjection of the small farmers by permanent manipulated debt.

*Minifundistas are 'peasants who do not have secure access to sufficient land resource to absorb the family labour force so as to supply the basic living requirements of the family' (Powell, 1972).

By contrast, where land is fairly evenly distributed, as in Taiwan for example, smallness does not necessarily entail domination or discrimination (Pearse, 1980).

Very strong emotional and social problems militate against the abolition of small farms, and the problem therefore arises of making small farms viable in the age of mechanization. Many improved cultural practices that can transform traditional agriculture, such as better-yielding varieties, use of fertilizers, improved water management, etc., do not involve economies of scale and can be used equally rationally on small and large farms. For the relatively few functions that do require large-scale investment and organization, the problems can be overcome by the provision of cooperative specialized services and/or government or private contractual services.

The examples of Japan and Taiwan show that as labour becomes scarcer and more expensive, small farms can also be mechanized and use labour efficiently. Many farmers in these countries have moved out from subsistence to commercial farming without any appreciable increase in the size of the farms.

Where farms are too small, the only way to provide an acceptable standard of living for the family is to find employment outside the farm. In Japan, the high farm household income of small-size farms is mostly due to non-farm income. Part-time farming comprises 84% of all farm households in Japan (Kim, 1972).

A different approach that has been proposed is to consolidate a large number of small holdings by forming large agricultural corporations in which the small landowners are represented by stocks (Ozal, 1963).

'Mixed' Structures of Large and Small Farms

There is much evidence to show that excessive emphasis on size and the exclusive use of large units also has drawbacks and dangers, particularly if such units are centrally directed. A mixed structure of units of various sizes appears to be the most promising in the developmental stages of agricultural expansion. Large and small farms are not mutually exclusive, and can well exist in the same region, and even in proximity to each other. A formula that has many merits is that of a large state- or corporation-owned farm, surrounded by relatively small producting units. Among farmers who still require a long period of education towards cooperation, a neighbouring large farm can play an important role. The fully equipped and mechanized large units can help smaller farms and also provide facilities for processing, grading, packing and marketing their produce.

More important still: the large farm should be the first to adopt improved cultural practices and can then serve as a guide and demonstration area, and thereby help in influencing the small farmer to accept the same methods.

An example of the 'spread-effects' of commercial plantations is provided by the smallholder tea production scheme in Tanzania (Sabry, 1967). In this country, tea was always produced on large commercial estates. It was generally assumed that growing tea on smallholdings would not ensure an adequate supply of the required quality to the processing factories. However, many farmers holding land in the vicinty of the estates, who had gained experience in the cultivation of tea by working as labourers on these estates, wished to grow the crop themselves. The Ministry of Agriculture decided to help them establish smallholder plantations near existing commercial estates, provided the latter were prepared to assist in the scheme by providing planting material, technical assistance in the field, and processing of the leaves. The area to be planted by each farmer in stages over a period of three to four years was limited to one acre (0.4 ha). It was estimated that with the help of his family, it would be possible for the small farmer to tend his one acre of tea without curtailing the production of food-crops.

Financial assistance and technical supervision were provided by government for the establishment of smallholder plantations and their maintenance during the first five years.

The success of the scheme is shown by the progress of the smallholder's plantings (Fig. 66).

An earlier combination of large-scale and family farming is the Gezira scheme in Sudan, which involves over 25,000 tenant farmers on an irrigated area of about a quarter million hectares. Cotton is produced under central management and supervision, on large-scale farming conditions, whilst other crops and animal husbandry are reserved for family-type farming conditions (Gaitskell, 1959).

At first sight, the situation described for the kolkhozy and sovkhozy (cf.

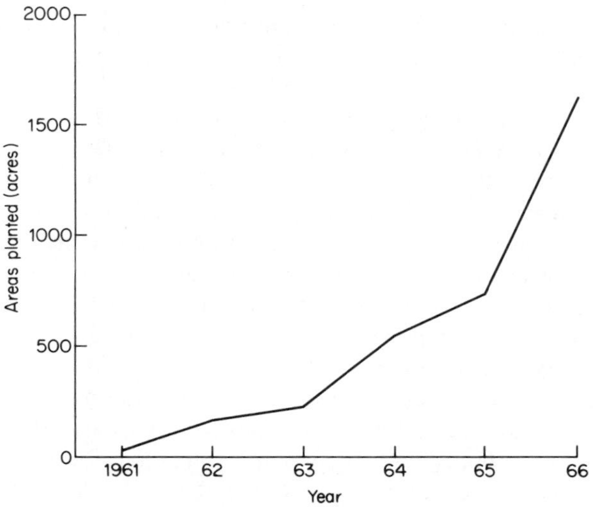

Fig. 66 Areas planted with tea by smallholders in Tanzania from 1961 to 1966 (based on Sabry, 1967).

pp. 451–453), whereby the main assets in land, equipment, livestock, etc., are collectively managed on large units whilst each family has its own mini-farm, would appear to conform to the 'mixed' structure of large and small farms described above, and therefore the advantages outlined for this model should apply to the Soviet system, too. However, large and small farms of mixed structure will be mutually supporting and advantageous only if small farms enjoy at least equal backing and incentives from the authorities as large units. In the Soviet Union, family mini-farms were established only as the result of a whiplash to the coercion employed in establishing collective farms. They are considered politically undesirable, and are, at best, tolerated by the authorities. The resemblance to the model described above is therefore more apparent than real.

Summary

The need to reckon with socio-cultural and behavioural factors when attempting to modernize agriculture is becoming increasingly recognized. Magical and religious beliefs, negative attitudes to agriculture, certain aspects of the social system, such as caste, kinship, gerontocracy, the low status of women, mutual distrust, etc., can be serious obstacles to rational-economic decisions, and hence to modernization of agriculture. It is therefore essential to develop gradually a local leadership that is committed to progress.

In the past, whatever attention was given to agriculture in developing countries was generally focused on production, whilst the problems of the agro-distribution system (marketing outlets, transport network, storage and processing facilities), were largely neglected. It is now recognized that the transition from subsistence farming to commercial agriculture requires massive investments in the development of an appropriate and efficient *infrastructure*.

Subsistence farmers are generally averse to taking risks and do not have the means to aquire modern inputs. Hence, a number of *economic constraints* may prevent the adoption of new technology and economic incentives are necessary to promote change. Insurance for innovators, a system of fair pricing of farm products, agricultural inputs and consumer goods and services are essential; subsidies for inputs are generally justified in the early stages of development; credit needs to be supplied for investment and the purchase of inputs. A system of crop insurance must be organized.

Inadequate service from the *institutions* responsible for supporting agriculture may be a major constraint to modernization. Amongst the main institutional weaknesses are: an ineffective bureaucracy, excessive centralization of authority, lack of empathy with the small farmer and a low level of municipal services. It is now generally accepted that technological change cannot be undertaken unless the basis institutional systems are revised and adapted to development needs. Of particular importance is the fostering of popular participation in rural organizations, in order to enable the farmers to protect their interests and cooperate in the solution of their problems.

Most developing countries have problems due to traditional land ownership and tenure patterns; under these conditions, *agrarian reform* is an essential prerequisite to agricultural development. The problems themselves differ from country to country; they may concern land ownership and tenure *per se*, or the excessive fragmentation and dispersal of holdings.

Evidence from many countries indicates that economic progress in agriculture is possible under a great variety of *farm size* conditions and that size of farms, *per se*, is neither a prerequisite or an obstacle to agricultural development. Whilst there is generally no overwhelming advantage to any particular size of farm in developing countries, in most types of farming the advantage is more likely to be with the family than with the large farm. A fact that must be given due consideration is that farm size has social and political implications, in addition to technical and economic aspects. Because very strong emotional and social problems militate against the abolition of small farms, the problem of making these viable in an age of mechanization, assumes great urgency.

REFERENCES

Abbott, J. C. (1966). The role of marketing in the growth of agricultural production and trade in less developed countries. Pp. 166–78 in *Getting Agriculture Moving* (Ed. R. E. Borton), Vol. 1. Agricultural Development Council, New York: 526 pp.

Abdul Hameed, N. D. (1977). *Rice Revolution in Sri Lanka.* United Nations Research Institute for Social Development, Geneva: x + 282 pp.

Adams, D. W. (1970). Institutional reform: the conflict between equity and productivity (Discussion). *Am. J. Agr. Econ.,* **52**, 613.

Adiseshiah, M. S. (1970). *Let My Country Awake; the Human Role in Development.* UNESCO, Paris: 375 pp.

Admoni, Y. (1963). Adapting extension methods to different social groups in Israel. Paper presented at *the Rehovot Conf. on Comprehensive Planning of Agriculture in Developing Countries, Rehovot, Israel:* 11 pp. (mimeogr.).

Alcantara, Cynthia H. de (1976). *Modernizing Mexican Agriculture; Socio-economic Implications of Technological Change 1940–1970.* UN Research Institute for Social Development, Geneva: xii + 350 pp.

Anschel, K. R. (1969). Agricultural marketing in the former British West Africa. Pp. 140–56 in *Agricultural Cooperatives and Markets in Developing Countries* (Ed. K. R. Anschel, R. H. Brannon, & E. D. Smith). Praeger, New York: xxiii + 373 pp.

Arnon, I., Molcho, S., & Raviv, M. (1975). *Change in Arab Villages in Israel, the West Bank and the Golan.* Settlement Study Centre, Rehovot: (in Hebrew): 441 pp. (mimeogr.).

Axinn, G. H. & Thorat, S. T. (1972). *Modernizing Agriculture; A Comparative Study of Agricultural Extension Education Systems.* Praeger, New York: xv + 216 pp., illustr.

Bachman, K. L. & Christensen, R. P. (1967). The economics of farm size. Pp. 234–54 in *Agricultural Development and Economic Growth* (Ed. H. M. Southworth & B. F. Johnston). Cornell University Press, Ithaca, NY: xv + 608 pp.

Balogh, T. (1966). *The Economics of Poverty.* Weidenfeld & Nicolson, London.

Barker, R. (1970). Economic aspects of new high-yielding varieties of rice; IRRI report. Pp. 29–53 in *Agricultural Revolution in Southeast Asia* (Ed. A. Russell), Vol. I. The Asia Society, New York: 376 pp.

Barker, R. & Anden, T. (1975). Factors influencing the use of modern rice technology in the study areas. Pp. 17–40 in *Changes in Rice Farming in Selected Areas of Asia.* International Rice Research Institute, Los Baños: 377 pp., illustr.

Barter, P. G. H. (1966). Special problems of agricultural planning. Pp. 471–84 in *Getting Agriculture Moving* (Ed. R. E. Borton) Vol. I. Agricultural Development Council, New York: x + 526 pp.

Behrman, J. R. (1969). Supply response and the modernization of peasant agriculture. Pp. 232–42 in *Subsistence Agriculture in Economic Development* (Ed. C. R. Wharton, Jr). Aldine Publishing Co., Chicago: 481 pp.

Bennett, Alice & Schork, W. (1979). *Studies toward a Sustainable Agriculture in Northern Ghana.* Research Centre for International Agrarian Development, Heidelberg. 125 pp. + 12 appendices, illustr.

Beteille, A. (1971). The social framework of change. Pp. 114–64 in *Regional Development—Experiences and Prospects in South and Southeast Asia* (Ed. L. Lefeber & M. Datta Chaudhuri). Mouton, Paris.

Brasseur, R. E. (1976). Constraints in the transfer of knowledge. *Focus.* **18** (3), 12–19.

Brewster, J. J. & Wunderlich, G. (1961). Farm size, capital, and tenure requirements. Pp. 207–212 in *Adjustments in Agriculture: A National Basebook.* Iowa State University Press, Ames, Iowa.

Brewster, J. M. (1967). Traditional social structures as barriers to change. Pp. 66–98 in *Agricultural Development and Economic Growth* (Ed. H. M. Southworth & B. F. Johnston). Cornell University Press, Ithaca, New York: xv + 608 pp.

Brown, L. R. (1970). *Seeds of Change*, Praeger, New York. xv + 205 pp., illustr.

Burke-Knapp, J. (1966). The role of international agencies in aiding in world food production. Pp. 11–17 in *Proc. Symp. on World Food Supply.* National Academy of Sciences, Washington, DC: 84 pp., illustr.

Castillo, Gelia, T. (1975). *All in a Grain of Rice.* Southeast Asian Regional Centre for Graduate Study and Research in Agriculture: xii + 410 pp.

Central Treaty Organization (1962). *Travelling Seminar for Increased Agricultural Production.* Report of a regional tour, Office of the US Coordinator for CENTO Affairs, Ankara: 160 pp.

Chatelain, R. (1963). Main problems of agricultural training at village level. Paper presented at *Rehovot Conf. on Comprehensive Planning of Agriculture in Developing Countries*, Rehovot, Israel: (mimeogr.), 10 pp.

Chaudhari, H. A., Rashid, A. & Mohy-Ud-Din, Q. (1975). Gujranwala, Punjab. Pp. 225–41 in *Changes in Rice Farming in Selected Areas of Asia.* International Rice Research Institute, Los Baños: 377 pp. illustr.

CIDA (1966). *Land Tenure Conditions and Socio-Economic Development of the Agricultural Sector Brazil.* Organization of American States, Washington, DC.

Crosson, P. R. (1970). *Agricultural Development and Productivity*, Johns Hopkins University Press, Baltimore, Md: xvi + 198 pp., illustr.

Crosson, P. R. (1975). Institutional obstacles to expansion of world food production. Pp. 17–22 in *Food: Politics, Economics, Nutrition and Research* (Ed. P. H. Abelson). American Association for the Advancement of Science: v + 202 pp., illustr.

Dandekar, V. M. (1969). Questions of economic analysis and the consequences of population growth. Pp. 336–75 in *Subsistence Agriculture in Economic Development* (Ed. C. R. Wharton, Jr). Aldine Publishing Co., CHicago: 481 pp.

Dasgupta, B. (1977). *Agrarian Change and the New Technology in India.* UNRISD, United Nations, Geneva: xxvii + 408 pp.

Davy, E. G. (1974). Drought in West Africa. *WMO Bull.* 23, Geneva.

Desroche, H. & Gat, Z. (1973). *Opération Mochav*, Edition Cujas, Paris, 429 pp.

Deutsch, K. W. (1971). Developmental change: some political aspects. Pp. 27–50 in *Behavioral Change in Agriculture* (Ed. P. Leagans & C. P. Loomis). Cornell University Press, Ithaca, New York: xii + 506 pp., illustr.

Diaz, H. (1974). An institutional analysis of a rural development project: the case of the Pueblo project in Mexico *Ph.D. Thesis*, University of Wisconsin.

Dorner, F. (1973). Land reform, technology, income distribution and employment. Pp. 574–617 in *Rural Development and Employment* (Ed. C. Goetsch). Ford Foundation Seminar, Ibada, Nigeria: 774 pp.

Dube, S. C. (1956). Cultural factors in rural community development. *J. Asian Studies,* **16**, 19–30.

Dumont, R. (1966). *African Agricultural Development.* Food and Agriculture Organization of the UN, Rome.

Dunbar, A. R. & Stephens, D. (1970). Social background. Pp. 98–108 in *Agriculture in Uganda* (Ed. J. D. Jameson), 2nd ed. Oxford University Press, London.

Edwards, E. O. (1973). *Employment in Developing Countries.* Ford Foundation, New York: 104 pp., (mimeogr.).

Eicher, C. K. (1967). Israeli innovations. Paper presented at *Sem. on Adapting Agricultural Cooperatives and Quasi-Cooperatives to the Market Structure and Conditions of Underdeveloped Areas, University of Kentucky, Lexington, Ky.*

FAO (1969). *Mobilization of Human Resources for Rural Development.* Food and Agriculture Organization of the UN, Rome: 14 pp.

FAO (1970). *Marketing a Dynamic Force in Agriculture: World Food Problems, No.* 10. Food and Agriculture Organization of the UN, Rome.

FAO (1974). *Kenya Case Study in Improving Productivity in Low Rainfall Areas,* COAG/74/4. Food and Agriculture Organization of the UN, Rome.

FAO (1977). *The State of Food and Agriculture, 1976 FAO Agricultural Series No.* 4. Food and Agriculture Organization of the UN, Rome: vi + 157 pp.

Foster, G. M. (1962). *Traditional Cultures: and the Impact of Technological Change.* Harper & Row, New York: xiii + 292 pp.

Frank, M. (1968). *Cooperative Land Settlements in Israel and their Relevance to African Countries.* Veröffentlichung der List Gesellschaft E. V. Band 53 Kyklos Verlag, Basel; J. C. B. Mohr (Paul Siebeck), Tübingen. xii + 168 pp.

Gaitskell, A. (1959). *Gezira a Story of Development in the Sudan.* Faber & Faber, London: 372 pp., illustr.

Gaitskell, A. (1968). Problems of policy in planning agricultural development in Africa south of the Sahara. Pp. 214–238 in *Economic Development of Tropical Agriculture* (Ed. W. W. McPherson). University of Florida Press, Gainesville, Florida: 684 pp.

Galjart, B. (1971). Rural development and sociological concepts, a critique. *Rural Sociology,* **36**, 31–41.

Ghose, A. K. (1979). Farm size and land productivity in Indian agriculture: a reappraisal. *J. Development Studies,* **16**, 27–49.

Grant, J. P. (1973). *Growth from Below—A People-Oriented Development Strategy.* Overseas Development Council, Dev. Pap. No. 16: 32 pp.

Gregor, H. F. (1959). Push to the desert. *Science,* **129**, 1329–1339.

Griffin, K. (1972) *The Green Revolution: an Economic Analysis.* UN Research Institute for Social Development, Geneva: xii + 153 pp., illustr.

Griffin, K. (1973). Policy options for rural development. Pp. 618–79 in *Rural Development and Employment* (Ed. C. Gotsch). Ford Foundation Seminar, Ibadan, Nigeria: 774 pp.

Gurley, J. G. (1973). Rural development in China, 1949–1972, and the lessons to be learnt from it. Pp. 306–56 in *Rural Development and Employment* (Ed. C. Gotsch). Ford Foundation Seminar, Ibadan, Nigeria: 774 pp.

Harrison, B. & Victorisz, T. (1971). Agro-distribution systems for Asian countries. Pp. 97–113 in *Regional Development—Experiences and Prospects in South and Southeast Asia* (Ed. L. Lefebre & M. Datta-Chaudhuri). Mouton, Paris: 278 pp.

Harvey, C., Jacobs, J., Lamb, G. & Schaffer, B. (1979). *Rural Employment and Administration in the Third World.* The International Labour Office, Teakfield Ltd., Westmead, England. xi + 111 pp.

Heck, B. van (1979). *Participation of the Poor in Rural Organizations*. Food and Agriculture Organization of the UN, Rome: xiii + 98 pp.

Hopkins, J. (1969). *The Latin American Farmer,* US Department of Agriculture, Washington, DC: 137 pp.

Hunter, G. & Jiggins, Janice (1978). *Farmer and Community Groups.* Agricultural Administration Unit, Overseas Development Institute, London: 21 pp. (mimeogr.).

IRRI (1975). *Changes in Rice Farming in Selected Areas of Asia.* International Rice Research Institute, Los Baños, Philippines: 377 pp., illustr.

James, P. E. (1959). *Latin America,* 3rd edn. Odyssey Press, Indianapolis, New York: xiii + 942 pp., illustr.

Johnson, W. F. (1971). Agricultural credit in Southeast Asia. *Development Digest*, **9** (2), 55–60.

Johnston, B. & Kilby, (1973). *Agricultural Strategies, Rural—Urban Interactions and the Expansion of Income Opportunities.* OECD Development Centre, Paris: v + 279 pp., illustr.

Jones, O. (1965). Increasing agricultural productivity in tropical Africa. Pp. 19–50 in *Economic Development in Africa* (Ed. E. F. Jackson). Basil Blackwell, Oxford: vii + 368 pp.

Kanel, D. (1967). Size of farm and economic development. *Indian J. Agric. Econ.,* **22**, 26–44.

Kahlon, A. S. & Singh, G. (1973). *Social and Economic Implications of Large Scale Introduction of High-Yielding Varieties of Wheat in the Punjab with Special Reference to the Ferozopur District.* Punjab Agriculture University, Department of Economics and Sociology, Ludhiana (India).

Karcz, J. F. (Ed.) (1967). *Soviet and East European Agriculture.* University of California Press, Berkeley, Calif.: xxv + 445 pp., illustr.

Kim, Sung-Lo (1972). A socio-economic analysis of farm mechanization in Asiatic paddy farming societies with special reference to Korea's and Japanese cases. Pp. 304–30 in *Farm Mechanization in East Asia* (Ed. H. Southworth). The Agricultural Development Council, New York: 433 pp., illustr.

Klayman, M. I. (1970). *The Moshav in Israel.* Praeger, New York: 371 pp.

Laird, L. D. (Ed.) (1953) *Soviet Agricultural and Peasant Affairs*. Univ. of Kansas Press, pp. xi + 333, illustr.

Lele, Uma, J. (1975). *The Design of Rural Development—Lessons from Africa.* The Johns Hopkins Press, Baltimore and London. xiii + 246 pp.

Lele, Uma J. (1976). Designing Rural Development Programs: Lessons from Past Experience in Africa. *Econ. Dev. Cult. Change,* **24**, 287–308.

Long, M. (1968). *Interest Rates and the Structure of Agricultural Credit Markets.* Oxford Economic Papers, 20.

Long, M. (1973). Conditions for success of public credit programs for small farmers. Pp. 680–712 in *Rural Development and Employment* (Ed. C. Gotsch). Ford Foundation Seminar Ibadan, Nigeria.

Long, M. (1974). Conditions for success in small farmer credit programs. *Development Digest,* **12** (1), 47–53.

Low, A. R. C. (1975). Small farm improvement strategies. *Oxford Agric. Stud.,* **4** (1), 3–19.

Lu, H. (1968). Some socio-economic factors affecting the implementation at the farm level of a rice production program in the Philippines. *Ph.D. Thesis*, University of the Philippines.

Lutfiyya, A. M. (1966). *Baytin, A Jordanian Village: A Study of Social Institutions and the Social Change in a Folk Community.* Mouton, The Hague: 202 pp.

Marglin, S. A. (1965). Insurance for innovators. Pp. 257–60 in *Policies for Promoting Agricultural Development* (Ed. D. Hapgood). Massachusetts Institute of Technology, Cambridge, Mass.: 321 pp.

Meek, C. M. (1949). *Land, Law and Custom in the Colonies.* Oxford University Press, London.

Mellor, J. W. (1969). The subsistence farmer in traditional economics. Pp. 209–26 in *Subsistence Agriculture in Economic Development* (Ed. C. R. Wharton, Jr). Aldine Publishing Co., Chicago: xiii + 481 pp.

Millikan, M. F. (1962). *Education for Innovation, Restless Nations: A Study of World Tensions and Development.* Agricultural Development Council, New York.

Moore, W. E. (1961). The social framework of economic development. Pp. 57–82 in *Tradition, Values and Socio-Economic Development.* Duke University Press, Durham, NC.

More, C. A. (1955). Agricultural development in Mexico. *J. Farm. Econ.,* **37**, 72–80.

Mwamufiya, M. & Fitch, J. B. (1976). *Maize Marketing and Distribution in Southern Zaire.* CIMMYT, Mexico: 11 pp.

Myren, D. T. (1970a). Integrating the rural market into the national economy of Mexico. *Development Digest,* **8** (4), 65–70.

Myren, D. T. (Ed.) (1970b). *Strategies for Increasing Agricultural Production on Small Holdings.* CIMMYT, Mexico: 86 pp., illustr.

Nair, Kusum (1961). *Blossoms in the Dust.* Duckworth, London: 200 pp., illustr.

Nisbet, C. T. (1971). Moneylending in rural areas of Latin America: Some examples from Colombia. *Am. J. Econ. Sociol.,* **30**, 71–4, 80–4.

O.A.S. (1964). *El Papel de la Aviacion Civil en los Proyectos de Colonizacion.* Doc. No., UP/G. 36/17, Conferencia de Expertos en Aviacion Civil, Santiago.

O.E.C.D. (1967). *The Food Problem of Developing Countries.* OECD, Paris: 114 pp.

Oni, S. A. & Olayemi, J. K. (1973). Determinants of Settlers Success under the Western Nigerian Farm Settlement Scheme. *The Nigerian Agr. J.,* **10**, 228–39.

Oury, B. (1970). Crop insurance, credit worthiness, and development, *Finance and Development,* **7** (3), 37–42.

Owen, W. (1966). Immobility and poverty. Pp. 268–71 in *Getting Agriculture Moving* (Ed. R. H. Borton), Vol. 1. Agricultural Development Council, New York: x + 526 pp.

Owen, W. (1970). Transport and agriculture in India. *Development Digest,* **8** (1), 22–6.

Owens, E. & Shaw, R. d'A. (1972) *Development Reconsidered*. Lexington Books, D. C. Heath & Com. Lexington Mass. xix + 190 pp.

Ozal, K. (1963). A review of irrigation development in Turkey. Paper C/261, *UN Conf. on Application of Science and Technology for the Benefit of the Less Developed Areas.* United Nations, Geneva: Vol. 3, viii + 309 pp.

Parathasarathy, G. (1975). West Godavari, Andhra Pradesh. Pp. 43–70 in *Changes in Rice Farming in Selected Areas of Asia.* International Rice Research Institute, Los Baños: 377 pp., illustr.

Pearse, A. (1977). Technology and peasant production: Reflections on a global study. *Development & Change,* **8**, 125–59.

Pearse, A. (1980). *Seeds of Plenty, Seeds of Want. Social and Economic Implications of the Green Revolution.* Clarendon Press, Oxford. xi + 262 pp.

Penny, D. H. (1969). Growth of economic-mindedness among small farmers in North Sumatra, Indonesia. Pp. 152–61 in *Subsistence Agriculture and Economic Development* (Ed. C. R. Wharton, Jr). Aldine Publishing Co., Chicago: 481 pp.

Perrin, R. & Winkelmann, D. (1976). Impediments to Technical Progress on Small Versus Large Farms. Paper presented to *Meeting of the American Agricultural Economics Association, CIMMYT, Mexico City:* 11 pp., (mimeogr.).

Powell, J. D. (1972). Agricultural enterprise and peasant political behaviour. Paper presented at the *Purdue Workshop on Empirical Studies of Small-Farm Agriculture in Developing Nations, Purdue University, Lafayette.*

Prebisch, R. (1971). *Change and Development: Latin America's Great Task.* Praeger, New York: xxi + 293 pp., illustr.

Rogers, E. M. (1962). *Diffusion of Innovations.* The Free Press, New York: xiii + 367 pp., illustr.

Roider, W. (1971). *Farm Settlements for Socio-Economic Development.* Weltforum Verlag, Munich, FRG.

Ruttan, V. W. (1977). Induced innovation and agricultural development. *Food Policy*, **2**, 196–216.

Ruttan, V. W. & Hayami, Y. (1972). Strategies for agricultural development. *Food Res. Inst. Studies in Agric. Economics Trade and Development,* **11**, 129–48.

Sabry, O. A. (1967). Smallholder tea production in Tanzania. *Development Digest,* **4** (4), 40–9.

Salazar, J. M. (1968). Latin America: The need for increased agricultural production. Pp. 51–61 in *Strategy for the Conquest of Hunger.* The Rockefeller Foundation, New York: 131 pp.

Sanderatne, N. (1970). Agricultural credit: Ceylon's experience. *South Asian Rev.,* **3** (3), 215–25.

Schickele, R. (1971). National policies for rural development in developing countries. Pp. 57–72 in *Rural Development in a Changing World* (Ed. R. Weitz). MIT Press, Cambridge, Mass.: 587 pp.

Schiller, O. (1964). *Agrarstruktur und Agrarreform in den Ländern Süd-und Südostasiens.* Verlag Paul Parey, Hamburg & Berlin: 128 pp.

Schultz, T. W. (1964). *Transforming Traditional Agriculture.* Yale University Press, New Haven: xiv + 212 pp.

Scudder, T. (1964). Relocation, agricultural intensification, and anthropological research. Pp. 31–39 in *The Anthropology of Development in Sub-Saharan Africa* (Ed. D. Brokensha & M. Pearsall). University of Kentucky Press, Lexington, Ky: 100 pp., illustr.

Shaw, R. d'A. (1971). *Jobs and Agricultural Development—A Study of the Effects of a New Agricultural Technology on Employment in Poor Nations.* Overseas Development Council, Washington, DC. 74 pp.

Smith, M. G. (1966). The communication of new techniques and ideas: Some cultural and psychological factors. Pp. 121–30 in: *Social Research and Rural Life in Central America, Mexico and the Caribbean Region* (Ed. E. de Vries), UNESCO, Paris, 257 pp.

Smock, D. R. (1969). The role of anthropology in a Western Nigerian resettlement project. Pp. 40–7 in *The Anthropology of Development in Sub-Sahara Africa* (Ed. D. Brokensha & M. Pearsall). The Society for Applied Anthropology, Monogr. No. 10, University of Kentucky Press, Lexington, Ky: 100 pp.

Spinks, G. R. (1970). Attitudes toward agricultural marketing in Asia and the Far East. *Monthly Bull. Agric. Econ. Stat.* FAO, Rome.

Stevens, R. D. (1977). *Tradition and Dynamics in Small Farm Agriculture* The Iowa State University Press. Ames, Iowa: xiii + 266 pp., illustr.

Straus, E. (1969). *Soviet Agriculture in Perspective.* Praeger, New York: 328 pp.

Sturt, D. W. (1965). Producer response to technological change in W. Pakistan. *J. Farm Econ.*, **47**, 625–33.
Swearer, H. R. (1967). Kruschev's reforms—comments. Pp. 51–6 in *Soviet and East European Agriculture* (Ed. J. F. Karcz). University of California Press, Berkeley, Calif.: xxv + 445 pp., illustr.
Tagampay-Castillo, G. (1970). Impact of agricultural innovation on patterns of rural life (focus on the Philippines). Pp. 13–52 in *Agricultural Revolution in Southeast Asia*, Vol. II (Ed. A. Russell). The Report of the Second SEADAC International Conference on Development in Southeast Asia, The Asia Society, New York.
Tezuka, N. & Kusakari, Z. (1969). *Collectives in Japan.* The Japan Kibbutz Association, Tokyo: 82 pp., illustr.
US Aid (1969). *Country Field Submission: India.* Washington, DC.
Wellhausen, E. J. (1970). The urgency of accelerating production on small farms. Pp. 5–9 in *Strategies for Increasing Agricultural Production in Small Holdings* (Ed. D. T. Myren). CIMMYT, Mexico: 86 pp., illustr.
Wharton, C. R. Jr. (1967). The infrastructure for agricultural growth. Pp. 107–42 in *Agricultural Development and Economic Growth* (Ed. H. M. Southworth & B. F. Johnston). Cornell University Press, Ithaca, NY: xv + 608 pp., illustr.
Wharton, C. R. Jr. (1969). Risk, uncertainty and the subsistence farmer. *Development Digest,* **7** (2), 3–10.
Whitton, R. M. (1966). Highways and development. Pp. 266–67 in *Getting Agriculture Moving*, Vol. 1 (Ed. R. E. Borton). Agricultural Development Council, New York: x + 526 pp.
Wilde de, J. C. & McLoughlin, P. E. M. (1967). *Experiences With Agricultural Development in Tropical Africa.* Johns Hopkins University Press, Baltimore, Maryland: Vol. 2. xii + 446 pp., illustr.
Wiser, Charlotte & Wiser, W. H. (1951). *Behind Mud Walls.* Agricultural Missions, Inc., New York.
World Bank (1975). *The Assault on World Poverty.* Johns Hopkins University Press, Baltimore, Maryland: xi + 425 pp., illustr.
Wu, Torng-Chuang (1972). Government policies promoting farm mechanization. Pp. 75–115 in *Farm Mechanization in East Asia* (Ed. H. Southworth). The Agricultural Development Council, New York: 433 pp., illustr.
Yang Hsin Pao (1949). Planning and implementing rural welfare programmes. *Human Organization,* **813**, 17–21.
Yudelman, M., Butler, G., & Banerji, R. (1971). *Technological Change in Agriculture and Employment in Development Countries: Development Centre Studies, Employment Series, No.* 4. Development Centre, OECD, Paris: 204 pp.

CHAPTER XI

Development Planning and Strategy

Planning the Transition from Traditional to Modern Agriculture

Objectives of Rural Development

The modernization of agriculture is not a purely technological process; to be effective it must occur within the overall framework of rural development.

It is now generally accepted that the objective of rural development is the improvement of the standard of living of the rural population as a whole, not only through increased agricultural production and the resulting rise in incomes, but also through improvement in the 'quality of life'—as in the fields of employment, health, education, culture and active participation in the development process. Basic to this approach is the concept of social equity, whereby the low-income rural population shares equitably in the benefits of development.

Whilst it is difficult to envisage rural development without increases in output and productivity, modernization of agriculture does not in itself ensure an improvement in the welfare of the bulk of the rural population. On the contrary, experience has shown that it has frequently resulted in a widening of the gap between the advanced and the disadvantaged sectors of the rural population.

For these reasons an understanding of some of the basic approaches to rural planning and the strategy of development is required by those charged with introducing technological change in agriculture.

Levels of Planning

Development planning can be carried out at three levels: the national level (macro-planning), the regional level and that of the individual villages and farms (micro-planning).

National Planning

This is essential in order to define overall goals, to allocate national resources to different sectors of the economy and regions of the country.

The macro-economic approach which is characteristic of the national level has the advantage of an overall perspective of the national economic potentials and needs, and of intersectoral relationships, but lacks the precise and detailed knowledge of the technological, economic and socio-cultural relationships effective at village level (Nicholls, 1964). It is the exclusive preoccupation with planning at the national level which has been the main weakness of most development plans in the past.

Regional Planning

Agricultural development depends on encouraging a large number of small producers to modernize their farming practices, under widely differing ecological, social, and economic conditions. Therefore, a de-centralized method of planning and implementing agricultural development is essential (Schickele, 1971).

Whilst national planning is based on statistical aggregates and averages, and micro-planning is linked to human behaviour and social environment, the regional level is essential in order to integrate macro- and micro-planning. Regional planning is therefore carried out at the level at which;

(a) a direct contact between the planner and the people involved in the development process becomes possible;
(b) the elements of institution building needed for a specific area can be defined; and
(c) a feasible strategy for achieving the necessary institutional changes can be formulated (Weitz, 1971).

Regional planning is therefore the level at which national development policies are translated into concrete projects reaching down to the smallest units of production. A *region** for development purposes can be one of the following units (Weitz, 1971):

(1) A *physically defined area*, of which a watershed basin is probably the most logical unit for development purposes. The Tennessee Valley Authority (TVA) is the first, and probably best-known example of comprehensive integrated regional development.
(2) *Social factors*: a region may be characterized by its backwardness in relation to other parts of the country, and therefore be singled out for intensive development. Examples are the Mezzogiorno region in southern Italy and the Nordeste region in Brazil.

*Much confusion results from the use of region for entirely different concepts: (a) an area comprising a number of different countries with certain common ecological features, such as the Sahel region in Africa; (b) the division of a country into smaller units, on the basis of the factors mentioned above.

(3) *Geo-political factors*: Thinly populated border areas may be slated for regional development to prevent intrusion of foreign elements or encroachment by neighbouring countries.
(4) *Administrative factors*: A country may be divided into regions for development purposes along purely administrative lines. For example, Jamaica has been divided into three regions for development: the western, southern and northern regions respectively with hardly any ecological factors being involved. The borders of the three regions follow those of the existing administrative units, the parishes. The first region slated for development is the western region.

Comprehensive Integrated Regional Planning

Regional planning, as it is widely accepted today is both comprehensive and integrated.

Comprehensive planning signifies that it is equally concerned with all the economic, social, and institutional aspects of planning at the regional, village and farm levels. It embraces production programmes, and controls the allocation of land and water, the supply of all services and inputs required by the farmers, the provision of credit and other incentives, the development of infrastructure, institutions, etc.

The micro-planning at farm and village level must be in harmony with the macro-planning at regional and national levels.

Integrated planning signifies that it is concerned with the simultaneous development and coordination of the different economic sectors: agriculture, industry and services, in accordance with the resources of the region and its economic development. Such coordination is essential because of the interdependence of development of agriculture and of the other sectors of the economy.

The essential role of agriculture in providing the means for a take-off into self-sustained growth has been discussed in Chapter I. Agriculture cannot, however, develop in isolation; its own further development is entirely dependent on the expansion of non-farming activities.

Development of agriculture consists in the gradual transformation of subsistence agriculture to production for the market and depends on the purchase of production inputs from other sectors of the economy. A precondition is the existence of a non-agricultural population with a reasonably high standard of living. For this reason, industrial development has a profound effect on agriculture. By making possible a rapid increase in income of the non-agricultural sector, it creates expanded markets for agricultural produce. It also causes changes in the social structure and norms, which in turn influence the values and behaviour of farmers that have an impact on farm technology.

Modern farming is also dependent on a supply of industrial products: machinery, chemicals, packing materials, etc. Formerly, the farmer himself provided the seeds he required, maintained soil fertility (by composts, manure, cover crops and night soil), and controlled weeds, pests and diseases as best he could; he also used home-made simple tools or implements. It is in the essence of modernization of agriculture that the farmer should use 'improved' varieties produced by plant breeders, and purchase fertilizers, herbicides, chemicals for pest and disease control as well as implements and machinery—all produced by industry (Jones, 1971). Agriculture is most successful in regions in which industry and other resources, such as mining, tourism, etc., promote the development of an adequate infrastructure of roads, commerce, social services and amenities, and share in their costs. Developments in agriculture and other economic sectors are therefore highly interdependent on, and complementary to each other, and this must be reflected in the integrated planning of the region.

Industries that process agricultural products (such as oil expressing and refining in the production of vegetable oils and soaps, milling of cereals and pulses, textile production, sugar and starch production, food preservation, etc.) make possible the diversification of agricultural production. Modern food processing also makes a considerable contribution to the economy of the country by reducing waste and preventing seasonal gluts.

The integration of industries in rural areas has additional benefits. It reduces the mass emigration of the rural population to the towns and, as a corollary, improves the standard of services provided to the rural population in health, education, culture, etc. Local markets for agricultural products increase profitability to the farmer; the production of raw materials for local industries makes possible diversified farming; seasonal work in industry may balance seasonal work in agriculture thereby providing fuller employment throughout the year. And, possibly most important of all, industry can absorb redundant labour from agriculture; the resultant scarcity of labour and the higher wages practically force farmers to improve the productivity of agriculture.

Taiwan provides a highly interesting example of the interaction between agriculture and other economic sectors. First, in the 1960s a high priority was given to agricultural development; this resulted in a potential demand for manufactured agricultural inputs. Concurrently, an equitable distribution of land was achieved by implementing an agrarian reform; this resulted in a fairer distribution of income and created a market for consumer goods in the rural areas. Farmers' organizations were established and strengthened, making possible the efficient purchase of farm inputs, and sale of farm products. Within industry, firms which used labour-intensive techniques were favoured, creating rapidly expanding opportunities for employment of surplus rural labour. As a result, real wages rose rapidly in both the agricultural and industrial sectors and the urban–rural wage differential practically vanished (Griffin, 1973).

One example of the opposite approach, on a massive scale, is the policy first adopted by the revolutionary regime of the People's Republic of

China. The *first* objective in economic planning was industrialization; agriculture was relegated to second place and the few programmes adopted for its development were generally piecemeal and of a local nature. It was only after serious setbacks in food production had occurred, that the policy of 'developing agriculture as the foundation of the national economy and industry as the leading factor' was adopted. The most important role assigned to industry was to provide the materials and machinery required to modernize agriculture (Kuo, 1972).

In brief, the limiting factor in the process of economic development may be either an inadequate increase in agricultural production or an inadequate growth of a non-agricultural sector engaged in industry and services.

Agricultural and industrial development are therefore interdependent, and neither can be neglected if development is to proceed at an acceptable rate. Planning should therefore be concerned with ensuring a complementary advance of both sectors, with agriculture producing food for the local population and for export, as well as raw materials for local industry whilst the latter produces the inputs for a modern agriculture and consumer goods for the population as a whole.

Dynamic Planning

Effective regional planning cannot be a 'one-time act'; 'it must be a continuing and dynamic process closely interwoven with implementation' (Weitz, 1971). A feedback mechanism, based on observing the results of implementation, makes it possible to adjust the plan continuously to changing circumstances and needs. This may involve modification of the original plan or even changes in its objectives.

According to these concepts, regional planning involves the following stages:

(1) The determination and formulation of the goals of the development project.
(2) The identification of the economic, social, political constraints to implementation.
(3) The collection of basic data required for planning.
(4) Assignment of administrative responsibilities for implementation.
(5) Planning the implementation.
(6) Observation and evaluation of the implementation process and re-assessment of the original plan.

Effectiveness of Implementation

Experience has shown that the planning procedures briefly described are no guarantee in themselves that agriculture will actually move forward. Erroneous strategies such as the exclusive concentration of the investment effort in industry to the neglect of agriculture; non-productive investments in prestige projects such as monumental buildings and developing military potential; poor government, as reflected by an overblown, inefficient and dishonest bureaucracy; faulty planning, as evidenced by the lack of success of many spectacular irrigation schemes; uncontrolled population increases and lack of employment policies; these are some of the main obstacles to the modernization of agriculture in many developing countries, even when significant sources of development capital become available to the national economy.

Strategies of Implementation

Even in the best of circumstances, governments of developing countries will be faced with a number of basic dilemmas resulting from the constraints imposed by scarce economic and human resources; hence the need for defining priorities in the mobilization and allocation of resources between social and economic objectives, between various economic sectors, such as industry and agriculture, between sectors of the population, between regions, etc.—in brief, the need for a strategy of development. 'A strategy is a general plan of how to use power and resources to attain specified objectives' (Barraclough, 1971).

National Strategies

Any strategy of development must work through national power structures.

In most developing societies, it is only government which has the power and the means to create the climate and the infrastructure favourable to agricultural development. The state will play entirely different roles in agricultural development according to the political philosophy which dominates its actions. This may or may not result in drastic changes in society. Modernization of agriculture can therefore be achieved by different social systems, using different strategies and pursuing different socio-economic aims.

Griffin (1973) distinguishes between three distinct rural development strategies which he calls respectively, the *technocratic strategy*, the *reformist strategy*, and the *radical strategy*. This classification is based on social and political considerations regarding the intended beneficiaries of agrarian policy. The strategies differ in their objectives, ideology, the dominant forms of land tenure, and the distribution of the benefits of the economic system. Few countries can however be placed unequivocally into one category or another, and most are situated somewhere along a continuum between the two extreme strategies.

Technocratic Strategy

The technocratic, productivity-oriented strategy for rural development is the one pursued by most developing countries. Also called *modernization strategy* by Barraclough (1973), it assumes that: 'Rural development can be achieved by adopting the technologies of the developed countries without simultaneously and profoundly reforming social structure. Existing power relationships, traditional land tenure system and class structures are accepted as the starting point.' The main economic objective in this case is to increase agricultural output, relying on a liberal capitalist ideology, with emphasis on competition, free markets and private property. Land ownership is highly concentrated in latifundia, plantations, and large corporate farms. By concentrating income and wealth, it is expected that a large proportion of this capital will be devoted to investment and growth, thereby achieving increased output.

To achieve development goals, bureaucracy is considered the most effective instrument of action. The main beneficiaries of this strategy are the bigger landowners who have the most access to government-initiated facilities (Heck, 1979).

Traditionally, this was the strategy generally followed in the poor agrarian countries of Latin America. Reliance was placed mainly on large landowners and on domestic and foreign investors to increase food production, mobilize labour and accumulate capital. Representative countries that have followed this approach are the Philippines, Brazil and the Ivory Coast (Griffin, 1973), Indonesia, Thailand, Malaysia and South Korea (Heck, 1979), and Tunisia (Pearse, 1980).

Reformist Strategy

The reformist (or 'solidarity-oriented') strategy of rural development aims at increasing agricultural productivity, community solidarity and changes in the attitudes of the rural population. Though the bureaucracy is the main instrument of action, in certain cases attempts were made to ensure popular involvement in rural development through political parties, local governments, and cooperatives.

Changes in the rural power structure are marginal. The main beneficiaries of these policies are the big and medium-size landowners (Heck, 1979).

The reformist strategy is basically one of compromise between the technocratic and the radical approaches and is based largely on a nationalist or populist ideology. It places priority on re-distributing income to some sections of the rural community (in particular the middle peasantry) and accords lower priority to increasing agricultural output than does the technocratic approach. Reforms of agrarian institutions are generally partial, fragmented and incomplete and concentrated in certain regions to the exclusion of others, thereby creating a 'dual economy' system (cf. pp. 510–511).

The land tenure is mixed: medium family farms tend to predominate, coexisting with minifundia, small cooperatives and large capitalist farms or neo-latifundia. There is a tendency for agricultural entrepreneurs to replace the traditional landlords. Among the owners of large 'progressive' farms are to be found many of urban origin, such as retired army officers, civil servants or politicians.

The re-distribution of income is generally from upper to middle income groups; the lowest levels may benefit somewhat by greater employment opportunities. However, the majority of peasants are left on subsistence plots or as partially-employed landless workers (Barraclough, 1973). Mexico is a fairly representative example of reformist strategy. The Mexican strategy consisted of the application of highly capital-intensive techniques on a small fraction of the country's land in the arid North Pacific and northern regions.

Large public investments in irrigation were made, permitting a considerable increase in crop area, the speedy adoption of high-yielding varieties, and the increased use of chemical fertilizers and mechanical equipment. The main reason for this policy was the existence of large amounts of unutilized and sparsely populated land in the coastal and northern regions. With irrigation, a rapid increase in agricultural production was possible, and this could be done most efficiently on large mechanized farms.

This policy has resulted in the creation of an enclave of large, modernized commercial farms within a national agrarian structure in which the predominantly near-subsistence sector remained practically unchanged (Alcantara, 1976).

As a result, in spite of the exceptionally vigorous growth of output and employment on a national scale, about 80% of the rural population have derived little benefit from development achieved outside agriculture or within the modern subsector of agriculture (Johnston & Kilby, 1973). In fact, 5% of the farmers produce 32% of the total national agricultural product, mainly on large market-oriented farms. Half the total number of farmers—in the main, subsistence farmers—produce no more than 4% of the total agricultural product (Sanchez, 1970).

In the subsistence sector holdings are still very small, techniques remain primitive, and as a result, yields and incomes are very low; the opportunity for gainful employment is confined to about 150 days in the year. Thus, the remarkable success of a strategy of chanelling investments into those sectors which can yield the surest, largest, and quickest economic return, has had the adverse side-effect of accentuating the differences between the relatively few 'haves' in the market-oriented farming communities and the great multitude of 'have-nots' in subsistence farming (Gaitskell, 1968).

Most Asian countries have chosen the reformist strategy. These include India, Sri-Lanka and Pakistan (Heck, 1979).

Radical Strategy

The radical or egalitarian strategy based on communist ideology, has as its main objective to achieve rapid social change and a re-distribution of political power. Political theory has a greater influence on policies than economic or technological considerations and these have lower priority than social objectives. In regard to agriculture, the main tool employed is the enforced establishment of large collective or 'state' farms.

Countries following the radical strategy of modernization of their agriculture are faced with the dilemma of deciding whether agriculture should be collectivized or modernized first. One school assumes that prior collectivization provides a more rational and efficient organization of labour and use of land, and creates suitable conditions for the rapid large-scale adoption of modern methods; another school advocates the postponement of collectivization until the productivity of agriculture can be increased through modern inputs, and this in turn implies a sufficient degree of industrialization able to supply the chemicals, machines and other inputs required for modernization of agriculture. Both the USSR and at a later date the People's Republic of China decided on collectivization first, professedly because it was supposed to facilitate the rapid transition of millions of peasants from traditional to modern agriculture, but actually in order to overcome the otherwise almost unsuperable difficulties of procurement at dictated prices of most of the agricultural produce from millions of recalcitrant peasants. The success of the plans for industry, commerce, and consumer goods depended on the ability of the authorities to extract from agriculture the required commodities and investment funds. Hence the importance attached to collectivization of agriculture. These policies were carried out despite the clear evidence that they were seriously affecting production.

The principal means adopted for increasing agricultural production was the mobilization of human labour, by extending the number of days worked, and increasing the intensity and efficiency of labour. Rough equality was achieved by abolishing private property in land, and establishing large production units: collectives, communes, and state farms. These are required to deliver specific quantities of commodities at prices fixed by the planning authorities.

The major beneficiaries of the radical strategy were the small peasant and landless labourers, but because of lack of emphasis on modern technology, the overall level achieved was very low.

One of the most striking and recent examples of radical strategy is that adopted by the People's Republic of China. Starting from a feudal system of land ownership where 80% of the land was owned by rich peasants and landlords who constituted 5% of the population, considerable material progress appears to have been achieved within a relatively short period.

In 1950, the land of large landowners was confiscated and distributed between 400 million tenants, in plots of 0.5 to 1.0 hectares. These small peasant farms were replaced in turn by mutual aid teams, subsequently by agricultural producers' cooperatives, and, finally, by people's communes. Each stage brought more socialization of the factors of production and increasingly larger farming units (Walker, 1965). Resistance to collectivization was violent, production fell drastically, and faced with famine in the cities, the government was obliged to purchase major quantities of wheat in the United States. It was only gradually that the system was made to work (Dauphin-Meunier, 1976).

According to Gurley (1973), the overall rural development strategy adopted appears to have consisted of the following steps:

(a) Destruction of the feudal-landlord–bureaucrat class structure, and the re-distribution of land, other assets, and power to peasant workers.
(b) Establishment of production on socialist principles and educating peasants and workers in socialist values and ideals. Goals were to be provided that inspired people to work hard, and basic decision-making at the lowest possible level was to be encouraged.
(c) Nationalization of industry as soon as feasible. Emphasis to be given to industries directly linked to agriculture.
(d) Agriculture was to be labour-intensive before mechanization was attempted in order to avoid creating a 'dual economy' in which part of the rural sector employed machinery and the remainder had to continue relying on manual labour.
(e) Replacement of market-price determined allocation of resources and distribution of incomes by a full planning mechanism.
(f) Encouragement of savings at all levels and their use for self-financed investment, especially in small-scale industries using indigenous methods.
(g) Provision of health-care facilities and general education.
(h) On-going revolution at all levels of society was to be maintained.

Structure

The rural population has been organized into communes—'multi-purpose political and organizational bodies covering the full range of economic, social and administrative activities necessary and feasible in a rural community' (Azis, 1973). Essentially, the commune is a unit of labour rather than of land.

The basic unit of the commune is the *work team*, consisting of 50 to 200 workers (Dauphin-Meunier, 1976). The work team is responsible for organizing its production so as to meet its own needs and supply the quotas allotted to it by the commune.

A number of work teams are grouped together into a *production brigade*, consisting of 1000–2000 workers, which serves as a link between the commune and its member work teams. The brigade operates certain production facilities, such as workshops and agricultural machinery stations and performs certain planning and management functions.

Operation

All available land is collectively owned by the communes; it is allocated to production brigades, and in turn subdivided between production teams. Each family also disposes of an individual plot; the produce of which (pork, poultry, vegetables) may be sold on the free market. These plots account for about 5% of the total area of the commune. Because the family plots are not in accordance with the accepted ideology, two attempts were made by the authorities, within a period of six years (1956–61), to eliminate these plots. Each was restored under pressure of discontent and lowered production (Walker, 1965).

The members of the commune elect a 'People's Council' of about 100 representatives, who in turn elect a 'Revolutionary Committee' of 100–25 members, responsible for the management of the commune. The chairman of the committee is the chief executive of the commune, whilst other members supervise the departments concerned with various production and cultural activities of the commune. A similar organization supervises the production brigade (Gurley, 1973). Each commune has considerable freedom in planning its activities and investing its revenues (Dauphin-Meunier, 1976). Because of the scarcity of capital, the main resource used is human labour, which substitutes not only for machinery but also for animal power.

In the early stages, all labour not directly required for agricultural production was mobilized in public works aimed at improving the land. Dykes and dams were built, irrigation systems laid out and roads constructed. In agriculture itself, efforts were devoted to improving production by labour-intensive methods.

In the next stages, the communes diversified their activities and achieved a large degree of self-sufficiency: first in the agricultural sector; in forestry, fisheries, and animal husbandry, and then developed small industries, mainly those using raw materials produced by agriculture or those providing inputs for agriculture. These activities gradually absorbed the internal increase of the labour force.

The mobilization of labour for agricultural development continues but it has shown diminishing returns over time; also the need to free labour for additional activities (see below) has resulted, since the 1960s, in increasing importance being accorded to the adoption of modern inputs and methods (Barnett, 1979).

Distribution of income to individuals follows the principle 'From each according to his ability, to each according to his work'. The basic accounting unit is the production brigade. After deducting expenses (agricultural tax, production expenses, investment fund, welfare fund and overhead) from the gross revenue, the balance accrues to the members.

The members income, which generally amounts to 50–55% of the gross revenue, is distributed amongst individuals according to a work point system, based on kind of work and number of hours worked.

Most recently, several important new initiatives have been reported in agricultural planning and policy-making in China (FAO, 1979); however, agriculture remains the foundation of the entire economy. Its capability for production growth, labour absorption and the supply of food and materials for industrial processing continues to govern the rate of expansion of the other sectors.

During 1977 and 1978, farmland improvement, land reclamation, irrigation works, expansion of chemical fertilizer supply, the breeding and spread of improved varieties, increased provision of farm equipment and tractors, have intensified.

China's planners are now basing their food strategy on the selection of a number of base areas for grain production, on which all available modern inputs will be concentrated. The objective is to enable these 'surplus output bases' to double or treble their marketed grain by 1985. Other selected provinces are to establish base areas for the intensified production of cash crops, such as cotton, oilseed, and sugar crops. Similarly, greater specialization is being promoted in the grazing areas of the north and northwest, where livestock has traditionally played a dominant role (FAO, 1979). Development administration has been decentralized to the lowest level. Whilst there is a unified centrally directed national programme, local units have considerable leeway to plan and manage within the overall framework. Local problem-solving is encouraged, making possible pragmatic, realistic and flexible microplanning.

Achievements

In recent years, many individuals, as well as missions consisting of scientists and development specialists from Western countries, international organizations and institutions, have visited the People's Republic of China. Judging by their published accounts, impressive material results have been achieved in the relatively short period since the communes have been established.

An FAO study mission to China, several of whose members have long been involved in promoting rural development, described the achievements of China's agriculture as 'breathtaking'. Starting from a situation of endemic famines, China had achieved by the early nineteen seventies the second highest production of food staples *per capita* in Asia. Grain output, in normal years, now exceeds current requirements and has made possible significant stockpiling of reserves (FAO Study Mission, 1978).

Though the Chinese farmer remains poor, hunger has been eliminated, and despite periodic set-backs, adequate food reserves have been created. Whilst China's average agricultural production is still half that of Japan, by 1973 the average farm income was two- to three-times higher than that in India and Pakistan, and 50 to 100% higher than that in Thailand, Indonesia, and the Philippines, in terms of real purchasing power (Aziz, 1973).

China has probably been the most successful country in reducing agricultural and rural underemployment in recent years (FAO, 1976). Whilst other developing countries with similar demographic problems have massive underemployment and unemployment, China has been able, mainly through a tremendous rural development programme 'to provide productive, remunerative, and socially satisfying employment' either in production or services, for the entire potential labour force (FAO Study Mission, 1978).

A major achievement has been the fairly egalitarian distribution of income and hence the prevention of significant differences in wealth leading to a dual economy and the elimination of the old forms of exploitation by landlords and money lenders (Barnett, 1979).

China has also made enormous progress in the emancipation of women. The rights of women are protected by law and a policy of translating these rights into practice is actively pursued.

Significant advances have been made in medical and health work in the rural areas by using unorthodox methods of which the best-known is 'the barefoot doctor'. On the other hand, the population explosion that plagues other developing countries has been successfully brought under control (FAO Study Mission, 1978).

Relevance of the Chinese Experience to Other Developing Countries

Apart from sheer scale, many of the problems of the rural areas faced by China were similar to those of other developing countries: recurring famines and widespread poverty, masses of illiterate peasants, a population explosion with concurrent lack of productive employment sources, migration from the rural areas to the towns, obsolete agrarian structures, social inequity, low status of women, etc.

In view of the remarkable achievements of Chinese agriculture in recent years in all the above mentioned problem areas, the possibility of transferring or adapting the Chinese development model to other developing countries is of considerable importance.*

The major difficulty in analysing the Chinese achievements and assessing their relevance to other countries, is in separating the effects of various components of policy, strategy, and implementation on the development process, such as the communist ideology; the cultural characteristics of the Chinese people, their traditions and historical background, and the actual methods of development that were applied.

An answer to the question of whether similar results can be achieved under different political and ideological regimes is bound to be largely speculative. The Chinese understandably stress that their achievements are

*The factual information on which this section is based, is mainly derived from the FAO Mission Report (1978).

inextricably bound up with China's political, economic, and social system, and cannot be transferred in isolation from it.

Many components of the ideology and strategy of the Chinese Revolution are indeed inseparable from the Chinese communist ideology. But the converse is not necessarily true: many of the elements that have contributed to the Chinese success story are not necessarily dependent on communist ideology and can be adopted within the framework of ideological contexts that are different to that of the Chinese, provided that they strive for development and social equity. Even in China, the principles of Communism have been diluted by the acceptance, albeit reluctantly, of personal incentives alongside communal interests, and of private enterprise in combination with communal ownership.

It is, however, certain that attitudinal, ideological, and organizational changes in many non-communist countries are an essential prerequisite for the adoption of some of the major successful components of the Chinese experience, without however predicating acceptance of communist ideology.

The required changes involve, amongst others, the recognition of the fundamental role of agriculture in economic development and the adoption of a national strategy to promote rural development; the transfer of responsibility to farmers for their affairs through viable farmers' organizations, to which access of the disadvantaged rural sectors is assured; stimulating people's participation in the planning process; increasing concern for the small farmer and the adoption of new forms of group farming.

These changes may be difficult, but not impossible to achieve.

The areas in which the Chinese experience can be of relevance to other developing countries are in the fields of technology, policy, planning and implementation, attitudes and ideology.

Technology

Agricultural progress has been more difficult to achieve in China than in other developing countries, for the paradoxical reason that their traditional agriculture was far more advanced than that of the others. Before World War II, yields in China were amongst the highest of all countries in which traditional, pre-scientific agriculture predominated. In the 1930s, yields of many major crops were higher in China even than in the USA! (Barnett, 1979). Chinese traditional agriculture was characterized by the intensive application of labour and use of organic manures. In contrast to other developing countries, 'there were no obvious and gross inefficiencies in Chinese farming that could be quickly overcome' (Barnett, 1979).

The remarkable increases in agricultural production of recent years have been due:

(a) to the development of irrigation, drainage, and other water control projects, which have reduced the impact of droughts and floods. At present, over 40% of all cultivated land is irrigated, enabling multiple-cropping and intensive intercropping over large areas. The Chinese average cropping-index of 140 is amongst the highest in the world (Barnett, 1979);

(b) reclamation through terracing, soil transfer, removal of rocks, drainage, etc. of large areas of marginal soils.

(c) Full use of all material resources.

The FAO Study Mission that recently visited China (1978) was particularly impressed by the way the Chinese conserved and used every resource and avoided waste. Every possible effort is invested in reclaiming land and every fragment of available land is used intensively; enormous amounts of compost, town rubbish, and night soil are collected, processed, and applied to the land.

All bodies of water are controlled, conserved, and utilized. Rivers and irrigation canals are 'harvested' systematically by producing food (fish, ducks, and geese), producing forage (water hyacinths) and organic matter (water weeds and river silt). Organized drives are undertaken against food waste, pests (mass anti-fly and anti-rodent campaigns, for example) (FAO Study Mission, 1978).

(d) Intensive use of labour in agricultural operations: the Chinese continue to apply the traditional, intensive use of labour, especially in hand-transplanting, fertilization, weeding and pest control. This meticulous care of the fields ensures maximum returns from the improved practices, such as HYVs, fertilizers, insecticides, etc.

The marginal productivity of labour in agriculture is already reaching a very low level. More priority is therefore being accorded to mechanization in order to increase productivity, reduce the drudgery involved in many farm operations, and release labour for rural development projects, mainly infrastructure and industry (FAO Study Mission, 1978).

(e) Adoption of modern practices: new practices are introduced in the wake of an intensive research programme, that is oriented towards the solution of the practical problems encountered in the field. The farmers are involved in the planning and implementation of the research programme, which involves large numbers of experimental plots in the communal farms.

Adoption of new findings and techniques by farmers is rapid and general (FAO Study Mission, 1978).

Whilst the methods used to achieve technological progress have undoubtedly been the result of the policies adopted by the communes, there is no evidence that this form of organization is essential in order to achieve similar results elsewhere.

Policy

Though the economy of the Chinese People's Republic is centrally planned, and firmly rooted in communist ideology, the leadership has shown a considerable degree of non-dogmatic pragmatism, which can be an object lesson to politicians and planners in other developing countries.

Primacy of agriculture: After attempting to follow the Russian example, by concentrating on the development of heavy industry, the Chinese have reassessed their priorities and now consider agriculture as the foundation of their economy.

Unremitting propaganda, sometimes draconian measures, mass mobilization of human resources, as well as many positive incentives, were expressions of the firm and single-minded commitment of the Chinese leadership to the development of the rural sector.

Of these measures, the incentives—material and non-material—were the components of greatest direct relevance to other countries.

Amongst the means used to promote agriculture are fiscal and pricing policies:

The tax on agriculture is low and does not exceed 4–5% of gross income. Because the tax is fitted in absolute terms, it diminishes gradually in relation to total income as production increases. As a result, the flow of money to agriculture exceeds that obtained from agriculture by 23%, ensuring a more rapid rise of the standard of living in the rural as compared to the urban sector (FAO Study Mission, 1978). Price policies are designed so as to ensure farmers a fair and stable return, to encourage production of socially and politically desirable commodities; to avoid inflation and to provide the State with income.

Availability of credit for productive purposes at low interest rates and with a minimum of bureaucratic red tape not only promotes development, but also enables Government to influence the direction of rural development. There are no longer landlords, middle-men, and money-lenders to exact tribute from the small farmer.

These measures have resulted in a profound change in the terms of trade between agricultural and industrial products, ensuring an increase in the share of the rural population of total national output (FAO Study Mission, 1978).

These measures should be compared to the policies prevalent in most developing countries, namely to keep food prices low, and thereby reduce the cost of living in the cities and subsidizing industrial development; unfavourable relationships between the prices the farmers receive for their produce and those they pay for essential inputs, services, and consumer goods (see Chapter X).

The entire educational system, in both rural and urban areas, has a marked bias towards agriculture and related activities.

By contrast with the situation in most developing countires, agricultural education is oriented to the needs of the majority of farmers. Teachers are encouraged to work with and learn from farmers. Outstanding farmers are recruited as teachers for agricultural schools. Only students with practical farming experience are admitted to the agricultural schools and colleges. Three times a year (at sowing, weeding and harvesting time) the students return to work on the farms, thereby keeping up to date with actual problems. On completion of their studies, they are required to return to their original place of work (FAO Study Mission, 1978).

Full mobilization of labour: Whilst most developing countries are beset by problems of unemployment and underemployment, in China the mass mobilization of labour has been the main moving force of rural capital formation.

Everybody is a member of a work unit, which provides employment and food. In addition to routine agricultural production, manpower has been used in large productive, labour-intensive projects, such as irrigation and drainage works, land reclamation and improvement, etc., which have thereby created durable capital assets. It is fair to assume that the organization of the peasant masses into communes has made these achievements possible. And yet, certain strategies used by the Chinese could be usefully adopted by other countries. These aspects have already been discussed in Chapter V.

On the other hand, vigorous and successful birth control measures have prevented the possible recrudescence of problems of unemployment or underemployment in the future.

Retention of people in the rural areas: The Chinese have been successful in preventing the exodus from the rural areas, and even in reversing the process, by a policy aimed at increasing the relative attraction of the countryside. Amongst the means employed to this end are: the provision of better housing, improved public facilities, the narrowing of income differentials; reduced drudgery of farm operations by selective and opportune mechanization; emphasis on cultural activities and opportunities for travel outside the home village; availability of a reasonable selection of consumer goods (FAO Study Mission, 1978).

All these measures need not be a monopoly or a radical strategy; they can just as well be adopted in any developing country desirous of stemming the rural migration to the towns.

The commune structure has provided the basis for the growth of farm employment and the general modernization of rural life by assuring that as far as possible each administrative county has its own industries for the production of iron, steel, fertilizers, machinery and energy. Within these industries, appropriate activities are carried out at county, commune and production brigade levels (Harvey *et al*., 1979).

Integration of agriculture, industries and services in the rural areas, is a basic tenet of 'integrated rural development' and can be applied irrespective of the ideology of the political leadership.

Rewarding individual effort: Though the communist ideology is basically egalitarian, the realities of human nature have forced certain changes, which have frequently been accepted under duress. The slogan is no longer: 'from each according to his ability, to each according to his needs' but '. . . to each according to his work'. This change in policy must inevitably lead to cumulative differences in individual income levels, even within each commune. The range between the lowest and the highest earners in a commune is already from 1:3 to 1:4 (FAO Study Mission, 1978).

It has been recognized that individual effort generally needs to be stimulated by material rewards and that remuneration has to be proportional to the effort invested by the individual if overall productivity is to be achieved.

Even the concession to human nature has not been sufficient. After several attempts to abolish private plots had resulted in a drastic drop in production, the communist leadership had to authorize the *de facto* ownership and management of individual households, including private plots and their produce, small livestock and simple tools. The care lavished on these private plots produces a disproportionate share of the family income, leading inevitably to a certain mount of tension between collective and private interests (FAO Study Mission, 1978).

The relevance of this experience to other situations will become evident in the discussion of communal production units in non-communist countries in the following section.

Planning and Implementation

The Chinese agricultural planning methods are decentralized, flexible and based on active participation by farmers. As a result, 'agricultural plans can breathe'; they are realistic and relevant to local needs and enjoy the support of the farmers. The targets set are generally below expectations. The production teams can therefore meet their obligations to the State even if environmental conditions are unfavourable. The teams are encouraged to produce beyond the target, because they are assured of a 30% premium above the basic price.

Supply of inputs is planned in cooperation with the supply and marketing cooperatives of the communes so that purchased inputs are available when required.

The main policies are expressed in simple and easily remembered slogans and the mass media are harnessed to their propagation. About two-thirds of the space or time in the media are devoted to agricultural development messages (FAO Study Mission, 1978).

The Chinese have retreated from their first attempts, based on the Russian model, of trying to replace peasant production by agricultural production along the lines of factory organization. The deployment of vast work units proved to be impossible to organize and manage (Harvey *et al.*, 1979). What could be considered a political retreat, namely the adoption of the three-tier organization of commune, production brigade and work team—has actually been the key institution responsible for translating Chinese political concepts and policies into actual practice.

The basic unit—the work team—consists of some twenty to thirty households, and therefore approximates in size a small village. It is large enough to enjoy the advantages of economy of scale, but small enough to preserve individual interest and initiative. Though it fulfils cooperative functions, it is basically different to the majority of cooperatives in the developing countries, which aim at servicing individual production, whilst the work team promotes collective production (FAO Study Mission, 1978).

A question of major importance is whether the latter type of cooperation could be a solution to the problems of non-viable small farming units, sharecroppers and landless peasants in non-communist countries.

The following statement by Robert McNamara, the former President of the World Bank, who can hardly be suspected to communist leanings, gives an unqualified affirmative answer:

> 'Obviously, it is not possible for governments to deal directly with over 100 million farm families. What is required is the organization of local farm groups, which will service millions of farmers at low cost, and the creation of intermediate institutions through which governments and commercial institutions can provide the necessary technical assistance and financial resources for them.
>
> Such institutions and organisations can take any number of forms: smallholder associations, county or district level cooperatives, various types of communes. . . . What is imperative is that at each organisational level financial discipline be rigorously required, and that the entire structure be oriented towards initiative and selfreliance. Experience shows that there is a greater chance of success if the institutions provide for popular participation, local leadership, and decentralisation of authority' (McNamara, 1973).

In brief, production cooperatives of small farmers, in which sharing of income is based on the labour invested by an individual and not on the size of his former holdings, need not be limited to communist regimes.

Admittedly, it may be difficult to establish the essential preconditions for their adoption in most countries.* Fundamental political and social changes may be essential before and during their establishment. These have been outlined above.

*The negative results of various experiments in group farming and the lessons to be learnt from them, should not be forgotten.

The large-scale adoption of communal production units is possible where spare land exists, enabling resettlement (mainly in Africa and Latin America). Elsewhere, land reform will be an essential prelude to the initiation of group farming.

As McNamara correctly points out, appropriate types of group farming do not have to be exclusive as in China, but can be confined to solving the problems of the most disadvantaged rural sector, leaving scope and initiative to other forms of settlement from viable-sized family farms to largescale commercial plantations and state farms. The possible symbiosis between these different forms of production have been discussed in the previous chapter.

The group farms themselves can allow a limited amount of private enterprise on the individual households, thereby enabling additional income resulting from individual diligence and initiative.

Effective Participation of the Farmers

The Chinese have avoided the dichotomy between planning and implementation that is so characteristic of development programmes in most developing countries. Planning and implementation are the responsibility of the same groups; at each level those that are to be responsible for implementation are directly involved in plan formulation.

This approach has ensured that planning has been realistic and adaptable to changing conditions, an important object lesson for other countries.

Attitudes and Ideology

The ideological teachings of the Chinese Communist Party have many admirable ethical precepts which are fully relevant to other developing countries.

The dignity of physical labour in general, and of farming in particular, is an inherent component of the Chinese work ethic, which stands in stark contrast to the low status accorded to physical work and agriculture in many other rural societies.

Whilst the Chinese peasant has traditionally been hard-working and frugal, the status and recognition accorded to work, the fostering of the ideal of service to others, combined with appropriate material incentives, have undoubtedly played an important role in improving the productivity of labour and its social status. Conern for the individual is expressed in a number of ways: upwards mobility, through election to increasingly responsible posts—not necessarily within Party cadres—is within the reach of every individual who makes the necessary effort and proves his ability; Individual talents are given scope for development by access to the necessary training facilities (FAO Study Mission, 1978). And last but not least, the leaders generally attempt to set a good example: lack of ostentation, a reputation for integrity, accessibility, familiarity with farming problems, and commitment to their improvement.

The leadership submits to continuous, organized criticism, aimed at preventing alienation from the people, vested interests, and the corruption inherent in power (FAO Study Mission, 1978). In all probability, here lies the main key to the extraordinary changes, reported by unbiased observers, that have occurred in China in the wake of their revolution.

It can be stated unequivocally, that the first, and most essential prerequisite for progress in developing countries, is a leadership of integrity, that is entirely committed to the welfare of their people, a precondition that is not dependent on the acceptance of communist ideology.

Intermediate Strategies

We have already mentioned that few countries can be placed unequivocally into one category or another, and that most are generally distributed along a continuum between the two extreme strategies.

In East Africa, for example, governments have opted for socialism and accept the need for the state to intervene in development strategy. However, none of these states—Kenya, Tanzania and Uganda—had the capital or the personnel needed to assume ownership of all means of production; hence, certain sectors of the economy had to be left in private, or joint state–private hands. Of the three, Tanzania is the country that has moved nearest to the 'radical' strategy. It is making a considerable effort to build an agrarian, socialist society that is equity based. It is, however, different from the Chinese prototype of the radical strategy model, in that it places a high value on individual liberties (Blue & Weaver, 1977).

Basic to the ujama concept is the belief that the traditional African extended family is an indigenous form of socialism, and that this unique social and cultural heritage provides a framework for development along socialist lines. Hence, the ujama was conceived as an up to date and larger version of the traditional African family (Nyerere, 1967).

The strategy adopted in the second five-year plan was based on the following principles (Seidman, 1972):

(1) *social equality*, to be achieved by spreading the benefits of development widely throughout society, in particular the rural areas;
(2) *ujama villages*, in which private ownership and control of agricultural production was to be replaced by collective production, thereby avoiding wide differences of wealth and income;
(3) *self-reliance* resulting from maximum mobilization of domestic resources for development, particularly through mobilization of the people;
(4) *economic and social transformation*—the plan urged rapid expansion of productive capacity to create the basis for future economic and social transformation;

(5) *African economic integration* based on economic cooperation with other African States.

Tanzania is a very large country, approaching Nigeria in size. Until independence in 1961, the majority of the people lived in isolated homesteads, on individual smallholdings. This diffuse and scattered population made it extremely difficult to initiate a programme of modernization and intensification of agriculture.

The Tanzanian government saw in the existing settlement pattern of widely scattered individual households a major constraint to the modernization of agriculture, to cooperation, and to the provision of social amenities, as well as to the creation of a sense of national unity. 'Villagization' was to overcome these problems. It has taken place through a mixture of social and economic incentives, and in some cases, physical force has been used (Blue & Weaver, 1977).

The ujama village passes through three stages. In the first stage, a group of 50 to 100 farmers decides to form a village and where to locate it, the type of production to be adopted and the initial infrastructure that needs to be established. They elect their leaders and begin to shape their organization.

The second stage consists of the implementation of the decisions taken: construction of buildings and infrastructure and beginning of farming. During this period, the villagers may request technical advice from government and financial and material help on a limited scale. In the final stage, the village becomes a fully viable economic unit, in which all activities—from production to marketing—are carried out cooperatively, and the proceeds are distributed according to the work contributed by each member.

The intention was that initially the ujama villages would have private and communal plots, the former to be gradually phased out as the ability of the ujama to handle cooperation operations improved (Lele, 1975).

Frequently, political initiatives or directives 'were either side-tracked or watered-down by bureaucratic elements that were luke-warm, if not hostile, towards the ujama village programme (Lele, 1975).

At present, the villages are the primary legal entities of the Tanzanian administrative and legal structure. Through village cooperatives, villagers can obtain credit, set up banking accounts, receive limited fees on agricultural sales, and receive small-scale assistance grants. The village as a whole can initiate construction projects and mobilize the necessary labour (Blue & Weaver, 1977).

By the end of 1976, 85% of Tanzania's 15.4 million people were reported to be living in 7684 ujama villages, which vary considerably in the level of their organization and functioning (FAO, 1977).

The Tanzanian government has made considerable progress in achieving some of its objectives, such as moving the country towards greater equity, participation and self-reliance. On the other hand, food production has stagnated, an unfavourable atmosphere for industrial investment has been created, and the goal of communal agriculture appears to have been shelved.

The Tanzanian model is increasingly falling into discredit. The reasons adduced for this situation are (Verhagen, 1980): (a) The Party and Government have appropriated such a degree of authority that very little participation in policy-making at the base remains; (b) there is virtually no protection against misuse of power from above; (c) lack of loyal and equitable implementation at all levels of the bureaucracy; (d) mismanagement of the cooperatives and frequent embezzlement of goods and cash; (e) limited control over marketing by the farmers and ineffective marketing practices; (f) lack of identification of the Party with farmers' interests; (g) the members of the village councils are inadequately prepared for their role, whether politically, morally or technically. As a result they are often ignorant about their responsibilities, embark on misguided projects or abuse their power.

As a remedial measure, Government has decided to appoint trained 'Village Managers' to each village and the World Bank is providing financial assistance to train 1500 technicians for this work.

The stated goals of Kenya's 'development plan for 1970–74' were essentially similar to those of Tanzania. The main difference was in the manner of implementation. In Tanzania, it was hoped to achieve a more balanced, internally integrated economy by State action in vertical areas of the economy, to re-direct the pattern of productive activity, including the establishment of poles of growth in each region. By contrast, in Kenya, the approach adopted was to expand government expenditures for social and economic infrastructure throughout the country, in the expectation that private enterprise would eventually find it profitable to invest in new areas (Blue & Weaver, 1977).

Agents of Implementation

A problem frequently encountered when attempting to define the kind of strategy adopted in a given country is that the declared aims and objectives may be in accordance with one type of strategy, while implementation appears to be on entirely different lines. Where these contradictions between declared national policy and the situation in reality are encountered, they can be ascribed to the fact that the political leadership is in the hands of those who are opposed to reform.

It is no longer in keeping with the times to fight against progressive legislation; therefore the strategy adopted in these cases is to pay lip service to reforms and to apply political power to block their implementation. Mathur (1965) states that in India 'organisation and institutional supports built up to implement agricultural policy have been rendered ineffective largely due to the socio-political forces at work'. The same is said in other words by Barraclough (1973): 'a policy is as much a matter of role application and adjudication as of legislation'; in Latin America, too, 'minimum wage laws, land or income taxes and social welfare programmes or land tenure reforms may be legislated only to become completely ineffective in practice'.

In addition to the large landowners, the traditional holders of power in the rural areas, an emergent rural middle class consisting of agricultural entrepreneurs, petty landlords and the new bureaucracies, views agrarian reforms as 'a threat to their cheap and subjected labour force, their rents, their interest and their petty monopolies' (Pearse, 1980). They have therefore every interest in attempting to block land reforms that favour the underprivileged rural sectors, by a variety of devices—such as distribution of the land among members of the landowner's family, changing tenants annually, and others.

In brief, strategies aimed at achieving rural development based on social equity will be effective only if implementation is the responsibility of forces genuinely devoted to achieving these aims.

Investment Strategies

Whatever the ideological approach, the problem remains of having to establish priorities in investment of capital and efforts. Rural development is to be viewed as 'a continuous, dynamic process, requiring a sequential approach in planning and implementing a rural development strategy, involving the establishment of priorities and time phasing of activities' (Lele, 1976).

Two alternative investment strategies for developing countries have been proposed (Hirschman, 1958):

(a) To invest mainly in *social overhead capital*, on the assumption that by creating conditions essential to profitability of productive investment by the private sector, the latter will be stimulated.
(b) To emphasize initially *directly productive investment* which in turn will create a demand for social overhead development.

The elements of social overhead capital include investments in physical facilities, such as roads and storage facilities, and institutional investments, such as research and extension services, a credit system, cooperative societies, etc.

Schutjer & Coward (1971) suggest that the more logical sequential priority in development planning is to emphasize initially directly productive investment. The reasons given for this choice are: (a) an initial increase in productivity will bring in its wake an automatic adjustment of the social and economic structure, and should the change not be automatic, it will be more easily stimulated; (b) even within the existing social and economic structures of the country, it is possible to bring about an initial increase in agricultural productivity through improved technology; and (c) sustained change will eventually require a revision of the social and economic structures of the nation.

There are considerable differences between developing countries in regard to their institutional structures, and any general formula will certainly not be applicable to all cases. Certain existing institutional patterns, or the absence of others, may prove to be barriers to development and hence the existing ones will have to be changed and those lacking will need to be established as a prerequisite to technological advance; others may be less restrictive and their reform can therefore be delayed. In other cases, large-scale market-oriented farmers may have access to the resources required for technological innovation within the existing institutional structures, whilst the ability of small farmers with limited resources to adopt new practices will depend on appropriate institutional change or adjustment.

In brief, decision-makers will have to envisage a mix of simultaneous investments in social, overhead and in directly productive activities that is adapted to each particular set of circumstances.

After a policy of investment has been adopted, there follows the need to establish priorities in the implementation of technological innovations.

PRIORITIES IN THE ADOPTION OF TECHNOLOGICAL INNOVATIONS

Interrelationships Between Factors for Development

The complementary nature of agricultural techniques and the futility of attempting to introduce them one at a time has been discussed previously (cf. pp. 262–265). The adoption of technological improvements is also completely dependent on many social, economic, and political factors.

Relatively simple changes 'may lead to events which wrench social, economic and political relationships lying deep in the structure of society'. For example, they may project millions of people out of subsistence living into a monetarized economy for the first time (Krebs, 1964). A single innovation may require a fundamental transformation. For instance, if animal- or tractor-drawn equipment is introduced to overcome a labour bottleneck, a whole train of innovations may become necessary. The use of implements may make necessary the complete clearing of land, removing all stones and tree stumps. This makes the traditional farming methods uneconomical; other methods such as crop rotation, manuring, and fertilization become essential. If livestock are maintained for traction and production of manure, a forage crop has to be introduced into the crop rotation, etc. (Wilde & McLoughlin, 1967).

Because of these highly important interrelationships among various factors, the effectiveness of a development programme will depend on the ability to find an appropriate combination of simultaneous changes.

It is characteristic of developing countries 'that nearly everything seems to need doing at once' (Kellogg, 1962). Because of the interlocking links between production factors, the need for a package programme, covering all aspects of a development plan cannot be disputed. However, equally characteristic of developing countries are the limitations in human resources and capital, which simply make it impossible to tackle everything at the same time. Indeed, it is the inherent complementarity of agricultural factors that makes the planning of a logical, coordinated action programme, each stage of which is dependent upon the others, so extremely difficult. This is probably the main cause for the poor performance of most agricultural development plans.

'Minimum Essential Needs' Strategy

Between the extremes of a complete 'package programme', highly desirable but not always feasible, on one hand, and the concentration on a single factor, usually ineffective, on the other, is a policy called 'minimum essential needs strategy' (Gaitskell, 1968). The complementarities described above are not necessarily rigid and it is therefore possible to a certain degree, to make the inputs of the package programme sequential. Priority is given to a minimum number of factors that must be applied simultaneously, in order to interact concurrently if they are to have an appreciable impact on development, whilst others have to be added in sequence as the means therefore become available.

The 'minimum essential needs' strategy must be evolved specifically for each particular situation and set of circumstances. The first step is to decide on the objectives to be achieved; the second is to examine available resources in relation to requirements; and the third is to decide on priorities in implementation.

While satisfactory agricultural progress cannot be piecemeal, it must perforce proceed in carefully planned stages that take into account potential resources. Hence, agricultural development first and foremost requires a determination of priorities. No hard and fast rules can be laid down in regard to priorities that are applicable to all situations. Development planners are faced with such an enormous variety of environmental, economic, and social conditions in different combinations, that strategy and a programme of priorities must be based on a detailed diagnosis of each particular situation, and adapted to each set of circumstances.

In view of the difficulties involved in modernizing agriculture on a national scale, two alternate approaches for accelerating agricultural production have been adopted: national commodity programmes and defined-area projects (Wortman & Cummings, 1978).

Several countries have been reportedly successful in increasing production of food crops through national commodity programmes. Wortman & Cummings (1978) cite as instructive the experiences of Kenya with maize, Turkey with wheat, and Colombia and the Philippines with rice. However, a closer look at these programmes indicates that, though labelled 'national' programmes, their implementation was limited to well-defined, and generally most-favoured regions.

In the Philippines, the rice campaign was launched following a critical rice shortage in 1973, under the slogan *masagna 99*—masagna meaning abundance, and 99 referring to the largest yield of 99 cavans per hectare (4356 kg/ha).

The major components of the campaign were (Drilon, 1976):

(a) A package of improved practices developed for rain-fed production in lowland farms by IRRI in collaboration with the National Department of Agriculture and Natural Resources and first tested in a pilot programme with about 100 farmers.
(b) Government promoted the adoption of recommended HYVs, and private dealers handled the distribution.
(c) Fertilizers were provided to participating farmers at a subsidized price.
(d) A campaign to control pests and diseases was organized by the Pesticide Institute of the Philippines and the chemicals were made available through private dealers.
(e) Credit was made available through a number of institutions; they were guaranteed 85% of possible losses by the Governmental Agricultural Loan Fund. The borrowers were organized into small groups with collective responsibility; as a result, an impressive 91% loan repayment average was achieved.
(f) Adequate water throughout the growing season was made available through irrigation improvement and pump distribution programmes; making double-cropping possible.
(g) Over 2000 technicians were trained and provided with motorcycles.
(h) All available media, i.e. radio, newspapers and posters, were used in an educational drive.
(i) Price support, procurement, and storage was organized.
(j) A management system was organized to plan, implement, and monitor the programme, operating at all levels of government. At national level, a policy-making council and a working group to handle operations were established; the directives passed through the regional, provincial, and municipal levels to production group leaders and farmers.

Though a number of problems arose during implementation, rice production was increased, though rather modestly (25% in 1974 over 1973—a year of depressed production; 1% in 1975—a year of numerous typhoons; and 10% more in 1976).

However, among the indirect benefits of *masagna 99*, the experience gained in organizing institutional credit, the timely supply of inputs, and a delivery system, will certainly have an impact on the further development of agriculture (Wortman & Cummings, 1978).

Defined-area Projects

The Pueblo project was initiated in 1967 in order to devise techniques for an efficient transfer of research results from the experiment stations to the small farmers. The area chosen for the project, in the valley of Pueblo, comprises over 100,000 hectares of rain-fed cropland, about three-quarters of which was in maize, with 43,000 farm families. Though defined as an 'area' project, its basic concern was with a single crop—maize.

Overall control and coordination of the project were vested in CIMMYT and the National Agrarian University at Chapingo. A team for planning and implementation was established, consisting of a project coordinator, four agronomic research workers, an evaluator, and five technical assistants. Twenty-five local farmers were selected and trained to assist the professional team.

The first step was to survey the resources and activities of the people in the project area and to define the existing constraints. Field trials were established in 26 locations, in order to provide information on maize varieties, fertilizer rates and times of application, plant populations, dates of planting, etc.

Experimental work on farmers' fields was a part of the plan from the outset. The staff also studied the organization of the communities, the power structure, and the informal systems of communication. Working through farmers' groups was initiated in 1969 (Winkelmann, 1976).

During numerous visits to the villages, the project was discussed with the participants. As a result, farmers felt involved, leaders were identified and the main problems to overcome were recognized.

On the basis of these findings, the following steps were taken. The project area was divided into five ecological zones. A technical assistance agent, assisted by one or two locally trained farmers, was assigned to each zone. The participating farmers (over 2500) were organized into groups.

The nearby offices of government agencies for credit, fertilizer supply, crop insurance, and price supports, who previously had not served small farmers, originally balked at becoming involved in the project, but finally participated under pressure from farmers' groups.

The combination of activities—experimenting, demonstrating, promoting the plan and facilitating the access to inputs—led to a rapid expansion in the number of farmers associated with the plan. By 1973, an estimated 43,300 farmers were involved. From 1967 to 1972, the average increases in the use of fertilizers were 129% for nitrogen and 95% for phosphorus. The average yield of maize increased by 88% (however, if the influence of more favourable precipitations in 1972 is discounted, the increase due to improved technology is estimated at about 30%). The number of man-days required to produce one hectare of maize increased from 40.6 man-days to 52.7 man-days, when a change was made from the traditional to new technology (CIMMYT, 1974).

If these achievements appear to be relatively modest, it must be remembered that the project was oriented towards improving the production of a single rain-fed crop (maize), grown in an area with relatively unreliable precipitation. This limited the choice of improved technologies, the levels at which they could be applied, and the potential benefits that could be achieved. On the other hand, the greater risks involved increased the resistance of farmers to change their methods of production.

Pearse (1980) does not accept this explanation, but ascribes the discrepancy between the expectations of the initiators of the Pueblo Project and the actual results to the fact that the participating farmers 'felt they were objects rather than subjects of the plan'. In contradiction to the stated aims of the project, as described above, the farmers who took part in the plan 'had little voice in planning the campaigns, in field trials and in the evaluation of the results of packages'. The participation of farmers in field trials was generally limited to the loan of some land, and farmers remained ignorant of the results of field research. As a result, the personnel of the Plan and participating farmers have become alienated from each other to the detriment of the Plan.

Stages of Development

The basic objective of the first stage is to enable the subsistence sector to produce a marketable surplus of food, which can be exchanged for producer and consumer goods and services from other sectors of the economy. A start will thereby have been made in the development process. Modern farm inputs are generally not yet available; credit and technical assistance are still rudimentary; extension services are still inadequate; and there are few urban jobs yet available at this stage (Barraclough, 1973).

First, the preconditions for development have to be provided; of these, changes in the land tenure system need to receive high priority. Tenure conditions need to be reformed to encourage more intensive land-use, and a more equitable distribution of income. At the same time, existing land reserves should be developed. If the area to be developed is arid, water resources must be developed and the whole infrastructure providing water to the individual farm must be established. Improved transportation facilities may make it possible to bring more land under cultivation and irrigation.

Surplus rural labour should be mobilized to aid in productive public works.

At the farm level itself, priority should be given to methods that do not require heavy cash investments, but that give substantial returns by increasing the efficiency and productivity of the existing labour-intensive agriculture. Efforts will aim at introducing high-yielding and disease-resistant varieties, at providing high-quality seed, at improving seedbed preparation and sowing techniques, at proposing more rational crop sequences and diversification of crops, at providing improved tools and equipment, and at recommending adequate levels of fertilization.

These inputs have the following characteristics: (a) they come mainly from outside traditional farming; (b) they depend on government initiative and help; and (c) the complementary nature of the various factors requires the simultaneous adoption of a number of techniques (Mellor, 1962).

Although the cash outlays required by the individual farmer are relatively modest at this stage, he is generally incapable of providing them himself; therefore, these inputs have to be financed by credit advances to the farmers.

This credit can be tied to improved practices. This makes it easier for extension agents to get the attention and participation of farmers who might otherwise be uninterested in methods or projects proposed by the agricultural ministries.

Where resources are limited, and this is generally the case, it is better to concentrate at this stage successively on relatively limited areas at a time; applying simultaneously an appropriate combination of changes, rather than dealing nationwide with a single variable, such as an improved variety or the application of fertilizers.

An example of such a comprehensive approach is shown by Takahashi (1970). In a study of a Philippine village in central Luzon, in 1961, he found that 'the peasants did not have any will at all to increase the level of productivity of the land they cultivate'. This attitude he attributed to the land tenure system in force at the time, whereby only the landlords prospered from increased output from the land. A shift from share-tenancy to leasehold occurred in 1968, as part of a comprehensive land reform. The new rice varieties were introduced with intensive extension work by an adequate number of experts. Irrigation facilities were improved by the National Irrigation Administration and the Asian Development Bank; a farmers' cooperative marketing association was activated under the land reform plan. As a result of these activities, by 1970, yields of rice had increased two and a half times over 1966, in a village in which the peasants had previously demonstrated their apathy towards the adoption of new practices.

Market-oriented Agriculture

As a result of the factors brought into play in the first stage, production increases, total productivity increases ranging from fourfold to eightfold are possible in favoured areas, while from twofold to fourfold increases averaged over large areas are considered conservative (Mellor, 1962).

The use of machinery and chemicals, the complexity of farm organization increase considerably, with a concomitant increase in the importance of education, extension, and research. An increasingly larger proportion of farm output is sold, and agriculture becomes capable of making a positive contribution to the overall economic development of the country. It is during this stage that conditions are most favourable for the promotion of cooperative frameworks. This should include government aid to cooperative marketing services.

Special attention needs to be given to improving farming systems, through: diversification, more effective crop rotations, integration of animal husbandry and arable cropping, soil conservation and selective mechanization.

The connecting link between the various stages of development is comprehensive land-use planning. Extensive data-gathering is required on ecological conditions, yields, farming systems, soil fertility and erosion problems, demographic and socio-cultural factors, opportunities for expansion of cultivation, required investments in water development and infrastructure, etc. (Lele, 1976).

Industrialized Agriculture

As a result of industrial development, the non-farm population that is dependent on purchased food becomes bigger, and the demand for agricultural products rises rapidly. Labour becomes gradually more scarce and costly; by contrast, capital becomes more abundant. As a result, agriculture becomes more and more mechanized; specialization becomes prevalent and farm size generally increases as holdings become consolidated.

As a result of specialization, farmers become willing to buy a substantial part of their own food requirements, thereby contributing to a rapid increase in effective market demand for food and other products.

For example, it was found that in parts of western Nigeria that had reached the third stage, rural families in seven villages in the cocoa growing region purchased food that supplied from 46 to 75% of their caloric requirements. By contrast, in a region adjacent to the cocoa belt, the specialized production of staple foodstuffs for sale to the cocoa region has become prevalent (Eicher, 1970).

As farming becomes more modernized, the level of rural education must be raised accordingly. The absorption of excess rural labour in productive public works, community development and rural industries needs to be expanded.

Strategy and the Dual Economy

Agriculture will remain the dominant economic sector in developing countries for a long time to come. The strategies adopted in implementing rural development will have a considerable influence on the kind of society that will emerge.

One is accustomed to thinking of the enormous gap between developing and underdeveloped countries, and between urban and rural sectors. There is another equally disquieting imbalance *within* the rural population. In the agriculture of many developing countries exists what Watters (1971) calls a 'dual economy'. This consists of the existence of a commercialized, capitalistic farming sector on the one hand and of a peasant, traditional sector on the other.*

Examples of the dual economy can be found in many countries, including Brazil, Colombia, Indonesia, Ivory Coast, Mexico, Morocco, and Sri Lanka (FAO, 1973).

Where resources are limited, a particularly embarrassing problem with which most governments are faced is the need to make a choice between two generally conflicting alternatives: to apply the scarce resources in such a way as to (a) achieve the maximum economic returns on a national scale, or (b) to achieve 'social equity' by improving the situation of the most underprivileged sectors of the society.

In terms of agricultural development this signifies either investing in the 'leading sectors', e.g. in the most promising regions and the most progressive farmers, or concentrating all resources on the improvement of the most backward regions and of subsistence farmers. These two conflicting approaches have been called respectively 'crash modernization strategy' and 'progressive modernization strategy' (Johnston & Kilby, 1971).

India, in the first stages of its development plan, adopted the progressive modernization strategy. Subsequently, policy-makers came to the conclusion that this approach had led to failure. The strategy of focusing on increasing agricultural production in the areas with the best conditions for economic progress was then adopted (Mathur, 1965).

If the premise is accepted that the objective of agricultural development is not only to increase agricultural production but also to achieve 'social equity' in terms of ensuring employment opportunities and improving the standard of living of the entire rural population, the question of how development investment and the rewards from increased agricultural production are distributed among the different strata of the population is of paramount importance.

'Crash Modernization' Strategy

In view of the 'time is running out' consideration that is always present, Hill (1966) argues that at the outset efforts should be concentrated in

*Pearse (1980) objects to the term 'dualistic' because 'it is misleading and mystifying, since it has been allowed to suggest a society with two levels of "civilization": a "modern" enlightened elite struggling for progress, but held back by an ignorant peasantry.' He suggests instead the term 'bi-modal'. We have continued to use the term 'dual-economy', in exactly the same sense envisaged by Pearse for the term bi-modal, namely: 'agrarian structures in which agricultural farming units divide into two clearly distinguished strata in respect of the magnitude of the farms as economic undertakings, the lines of cropping, their market orientation, and the socio-political situation of the cultivators who control them'.

those geographic areas and on those sectors where promise of quick success is greatest. An important objective at the outset should be to demonstrate, if one can, that progress is possible. If this can be demonstrated, time will be gained to work with the 'slow adopters'.

It has been stated by economists that the deliberate creation of imbalances may catalyse the whole development process. By concentrating investment in a strategic region or sector, economies of scale can be achieved, leading to breakthroughs. From these zones, development can spread to other areas, setting off a phase of national economic progress. This concept has been called the 'spearhead' approach to development (Ojala, 1967).

There is no doubt that it is usually far easier to introduce modern methods of production, irrigation, fertilizers, mechanization, etc., to the capitalistic sector than to that of the peasant farmer. From a purely economic standpoint, concentrating investments and extension efforts in the more favourable areas with the better farmers may also return the largest short-term dividents. Policy-makers tend to take the easy way and as a result government as well as private investment are mainly channelled into the capitalistic farming sector. The same holds true for extension efforts. Price supports and subsidies mainly benefit the same sector, as the subsistence farmer in any case has very little to sell and no easy access to markets. Most of the subsidized inputs are used by the larger farmers. It has been calculated that for India and Pakistan, of every $10 transferred via the price-support system, only about $1 goes to small farmers (Falcon, 1970).

Where statistics indicate that overall agricultural output has increased markedly in developing countries, they frequently reveal that most of the increased output may be due to a small percentage of the agricultural holdings, whilst the majority remains stagnant. An illustration of this approach is the 'intensive agricultural district programme' established by the Indian government in cooperation with the Ford Foundation. A few districts were chosen with good prospects for a dramatic agricultural breakthrough; these had an assured water supply by irrigation and farmers reasonably open to innovations. A project officer was appointed for each district to coordinate the programme; the extension staff was doubled; transport and equipment were provided. Within each district an intensive effort was made to provide farmers on a priority basis with a full 'package' of inputs and services—from improved seeds to short-term credit (Hunter, 1970).

Not only were the most favoured districts chosen, but the large farmers predominated in the new programmes. In a report of the Ministry of Food and Agriculture in eight districts, it is stated that the average holding of participants was double that of non-participants.

In Kenya, institutions and services are focused primarily on cash-crops and are largely limited to the most favoured areas—with good level soils and a reliable rainfall of over 750 mm. These areas account for only 7% of the country. Subsistence farmers living in the more marginal areas are limited not only by poor natural resources, but by lack of services and absence of high value crops that could be a source of steady cash income (Gerhart, 1973).

Even in egalitarian China, a major effort in development is concentrated on a number of base areas, significantly called 'areas of high and stable yields'—namely areas in which land and water development have been completed and water supply is controlled (Pearse, 1980).

In Tanzania, efforts were concentrated on areas and crops where they are most likely to be effective, such as cotton-growing in the Sukuma area. This programme was called the 'focal point approach' (Dumont, 1966).

Seidman (1972) describes the gradual creation of two groups of farmers and widening of the gap between them in Tanzania as a result of this policy. 'Thousands of poorer peasants, seeking to escape the poverty of the subsistence sector, unable to benefit from or to compete with the transformation schemes, found themselves confined to less desirable land, increasingly unable to earn enough from their own farms to support their families as producer prices fell. More and more of them were joining the growing numbers of hired labourers working for those who had benefited from historical accident and/or government assistance in producing large amounts of relatively profitable cash crops on large areas of good land. Thousands began to join those looking for better paying jobs in urban areas. There appeared to be emerging an entrenched group of better-off private commercial farmers who could be expected to envisage their futures as linked with the *status quo*—not with essential structural change.'

It were these developments that led to the Tanzanian government to adopt a complete change in strategy resulting in the concept of ujama villages described above.

Everywhere, larger farmers have built-in advantages over small farmers, and this tends to increase the already considerable gap between them. Institutional credit systems are generally geared to lending against land as security, and therefore the dominant proportion of available credit goes to large farmers.

The powerful agricultural entrepreneurs enter into various forms of collaboration with financial groups, storage agencies, agro-industries and centres that supply modern technology (FAO, 1979).

Even cooperatives can be an instrument for discrimination against the small farmer. In a tightly hierarchical society, with a gencrally low level of education and of commercial skill, the cooperative will generally be dominated by the large landowners, providing benefits to themselves and frequently mismanaging the cooperative financially (Hunter, 1970).

One of the means to exploit the poorer members of the cooperative is the practice of paying for capital investment, such as coffee processing machinery, by an equal deduction from each member's proceeds. In a typical Kenyan coffee cooperative with 800 members, the income of the top

72 members from the cooperative was ten times as great as that of the remainder (Harvey *et al.*, 1979). The larger farmers also control, organizationally, cooperatives that have been set up to break the hold of moneylenders, and it has been shown in India that a higher proportion of credit is supplied to large farmers than to the small ones.

Harvey *et al.* (1979) state that 'there is little evidence to suggest that marketing and service cooperatives are likely to effect any significant improvement in the productive potential of poor peasants'. The authors conclude that, on the contrary, cooperatives and farmers' associations tend to reinforce the comparative advantages of the more prosperous farmers and in many cases to strengthen the hegemony of the dominant rural classes.

The so-called 'small farmer credit programme' in Kenya, financed by the International Bank of Reconstruction and Development, has been effectively limited to at most 3% of peasant farmers (Heyer, 1973).

The same holds true for other scarce resources, such as water. Influential farmers have first call on water supplies from publicly owned irrigation systems, and large landowners with greater capital assets and better access to credit, are more able to establish and effectively use tubewells, purchase machinery, establish storage facilities, etc. (Shourie, 1973).

The economic weakness of the small subsistence farmer, his indebtedness, his lack of social and political power limit his access to government officials, extension officers, irrigation controllers, etc., and prevent his competing on equal footing with large farmers for factors of production and essential services, or to obtain a fair price for his product (Pearse, 1977).

A study of the effects of various programmes of rural development in four villages in the state of Uttar Pradesh (India) showed that the class/caste hierarchy in these villages enabled the village elites to corner all government facilities which were supposedly equally available to all; the lower strata were also excluded from places where public broadcasts were relayed. Because of their illiteracy, they remained unaware of essential information on welfare and development programmes, facilities for obtaining credit and inputs, several years after their introduction (Pearse, 1977).

The 'Green Revolution' and the Dual Economy

It has already been pointed out that the 'green revolution' favours the stronger elements of the rural economy and thus tends to accelerate the increase in the gap between the privileged and the underprivileged sectors.

Alcantara (1976) describes how, with increasing modernization of agriculture, wealth was increasingly concentrated in the hands of a small fraction of the rural population of Mexico, whilst an absolute worsening occurred in the standard of living of the lowest income groups in the country; other consequences were: the declining relative production and productivity of non-irrigated land; the stagnation of the yields of maize and beans, which are the staple diet of the majority; and the persistence of widespread hunger in the birthplace of the green revolution. Notwithstanding the investment of billions of pesos of the national budget into agriculture, 83% of all farmers could only maintain their families at subsistence or near-subsistence levels.

Larger landowners generally have the technical knowledge and the incentives for innovation which the small, poor peasants lack, so that the former are in a far better position to grasp the opportunities provided by the green revolution.

The extremely unequal distribution of income also limits access to education, even at primary level, of a large proportion of the rural population, thereby ensuring a continuation of the vicious circle.

In brief, the small farmer is discriminated against in every area. 'Extension agents concentrate on the large farmers, credit agencies favour the low-risk borrowers, fertiliser and pesticide salesmen prefer the large buyers' (Griffin, 1973).

As a result of these factors, the seed–fertilizer revolution is actually very reactionary. It has been mainly confined to large market-oriented farms in the northern part of Mexico; in Pakistan it has occurred mainly in the prosperous areas of the Punjab, and in other irrigated areas of the Indus basin. In India, new wheat varieties have been adopted most widely in the most prosperous districts and states (Punjab and Haryana) where capitalist farming is the most advanced, and far less in the less prosperous areas such as Uttar Pradesh and Bihar. In the Philippines, the so-called 'rice and corn self-sufficiency programme' was given priority in central Luzon, the most advanced and prosperous agricultural region of the country (Griffin, 1973).

Under the slogan of 'sound farm management' the larger commercially managed farms are encouraged to combine the introduction of the new varieties with that of labour-saving equipment. As a result, productive employment in agriculture may actually diminish. Thus, Griffin (1973) points out that 'in the three Asian countries which are in the vanguard of the green revolution (Philippines, India, and Pakistan), the most prosperous people in rural areas are subsidised but not taxed, and policies have been introduced which ensure that they receive the benefits of agrarian change while those who are less prosperous receive its costs. The consequences are greater inequality and, in some instances, greater misery.'

There is also increasing evidence that as farming opportunities become more profitable as a result of the new technology, owners take over the land for cultivation on their own account, thereby transforming tenants into landless labourers.

Finally, it should be pointed out that the seed–fertilizer revolution, because of the increase in production, frequently results in lowered prices of the product; the small farmer, therefore, has not only been unable to increase his production, but may see his income from sales actually diminish.

This was the case in Turkey, for example, where wheat is the principal crop. The adoption of high-yielding varieties has been almost entirely confined to the high-rainfall coastal lowlands. Production in these regions doubled, causing a decline in prices, so that the farmers in the drier Anatolian plateau are worse off than they were before (Brown, 1970).

Taiwan is one of the exceptional countries in which the general discrimination against small farmers does not occur. The adoption of improved practices in the agriculture of Taiwan is widespread, and Griffin (1972) ascribes the reason for this to the fact that, as a result of land reform, landownership is fairly equally distributed in small farm units; in addition, water, fertilizers, technical knowledge and credit are equally accessible to all farmers. As a result, profitable technologies can be adopted by all farmers, and in consequence, innovation is rapid and general.

Implications

It is clear from the foregoing that problems of rural development cannot be considered exclusively within an overall national framework, but that increased attention must be given to the specific regional and sectoral problems within countries. We have seen that innovations, such as the green revolution, the expansion of irrigation facilities, the introduction of multiple-cropping, the cultivation of more profitable commercial crops and mechanization have usually accentuated existing trends towards inter-regional inequality and greater social differentiation; they tend to be adopted in regions with favourable conditions or where large-scale agriculture is found; and, therefore, from the socio-economic point of view their main effect has been to strengthen market-oriented agriculture, and discriminate against traditional farming.

Economic Effects

A policy discriminating against the backward sector, though it may possibly be based on short-term economic considerations, will in the long run, lead to a widening of the economic and technological gap between the two sectors.

Overall national development cannot be sustained without a more equitable distribution of income which makes possible a wider and more effective market. This is particularly true in countries in which the rural poor form a majority of the population. In 1970, 48% of the population of the 18 tropical countries of Central and South America lived in rural areas directly dependent on agriculture and forestry. 70% of the active rural labour force of 35 million is at the subsistence level with a gross annual output of less than \$200 for workers. 70–90% of what this group produces is either consumed directly by the family, in the case of agricultural products, or is spent on subsistence-level food and shelter. Under these conditions, rural demand for industrial consumer goods and for manufactured production inputs must perforce be very low and cannot provide an impetus for industry (Nelson, 1973).

In a predominantly agrarian economy, it is more likely that the overall level of saving and investment will be higher if the productivity of land and labour is increased by the widespread adoption of new technology on a large number of family farms, than by concentrating resources on a relatively restricted scale in an advanced subsector of farming. It is also probable that the latter will, in any case, adopt new practices even when most government efforts and incentives are concentrated on the more backward sectors of the rural economy. In other words, there does not have to be a conflict between equity and productivity.

Further, the concentration of income due to the quasi-exclusive development of a limited sector of large farms results in demands for luxury goods which are generally not suited for efficient local production and therefore constitute a drain on foreign currency. By contrast, modernizing the traditional sector increases the productivity and buying capacity of a larger section of the rural population in agriculture, and thereby creates a greater market for locally produced goods, incidentally increasing employment opportunities. Also, as the result of the improvement in the quality

Table XLII Comparison of economic and social indicators in countries favouring different roads to growth and development (Grant, 1973).

	Countries favouring labour-intensive policies and small producers		Countries favouring capital-intensive policies and large producers		
	Korea	Taiwan	Philippines	Mexico	Brazil
GNP growth rates in the 1960s (%)	9	10	6	7	6
Investment costs of increasing GNP by $1.00 in the 1960s ($)	1.70	2.10	3.50	3.10	2.80
Agricultural working population per 100 ha	197	195	71	35	43
Ratio of income controlled by top 20% to bottom 20%	5.1	5.1	16.1	16.1	25.1
Income improvement of poorest 20% over last 20 years (%)	over 100	200	negligible	negligible	negligible

and quantity of the foods consumed, it makes possible a significant improvement in health and productive capacity of the rural population (Johnston & Kilby, 1973).

A comparison of economic and social indicators in countries favouring different approaches to growth and development, confirms that 'social efficiency' can in the long run also be the most economically efficient approach (Table XLII).

Equitable income distribution due to increased productivity of small farms also favours the growth of small-scale agro-industries in the rural regions, which can further improve income distribution (Schutjer & Coward, 1971).

Social and Political Implications

Even if the economic situation of the most depressed sectors of the rural population is somewhat improved as a result of the modernization of agriculture, it is the *disparities* between sectors which is politically and socially most dangerous. It is well known that inequalities in relative wealth can be more conducive to social tensions than general outright poverty (French & Lyman, 1969). 'An agricultural revolution geared only to new techniques and higher productivity', and ignoring the social implications, 'will unwittingly cause festering existing tensions to grow apace' (Ladejinsky, 1970).

Regional socio-economic disparities are already the source of considerable friction in Pakistan, India, Nigeria and other countries (French & Lyman, 1969). Modernization will generally accentuate these disparities; unless government policies favouring the weaker sectors are adopted.

We have mentioned the situation in Mexico a number of times as an example of a country in which government policies have encouraged an increased disparity between large and small farmers as a result of the green revolution. It is only fair to mention that it is now becoming increasingly obvious to the Mexican government that agricultural production cannot continue indefinitely to be based on production by a small fraction of the farmers, and that an increase in the purchasing power of subsistence farmers is a prerequisite towards widening internal markets for domestic industry. A new emphasis is being given to raising productivity in the non-irrigated areas, in re-asserting the rights of land reform beneficiaries and helping them to recover their land from renters; in strengthening local peasant organizations and once again promoting cooperatives, so as to lessen the competitive disadvantage in buying inputs and selling farm produce. A larger volume of credit, both public and private, is being devoted to small farmers in the relatively remote areas. Efforts are being made to remove corrupt employees. Agricultural insurance is being made available to more small farmers. The extension programme has been extended towards working with ejidatarios and minifundistas. Research is increasingly concerned with rain-fed crops. These new programmes have been included into a comprehensive framework of national and regional planning (Alcantara, 1976).

Diaz (1974), who describes the problems encountered in the diffusion of the new technology in the Pueblo valley, also states that efforts have been made to improve the services of the banks by speeding the processing of loans and repayments, and of the commercial suppliers of fertilizers by ensuring timely delivery of inputs. The extension services provided by the technical staff of the bank, counselling farmers on the use of modern inputs has also been improved. One measure of the success of these institutional reforms was that the rate of repayment of loans increased from 40% to 90%! In Indonesia and Malaysia too, the governments are attempting to redress the balance in favour of the underprivileged farming sector by means of farmers' associations, mobile bank units, and input coupons for individual farmers (Palmer, 1976).

Summary

Rural development planning concepts have progressed considerably in recent years and have become the concern of specialized institutions of research and education. Whilst the planning and implementation are carried out by interdisciplinary teams of development specialists, the process remains basically politically oriented, and must work through national power structures.

National political philosophy determines the choice of the type of development strategy to be adopted: technocratic, reformist or radical—from this choice follow implications regarding the main intended beneficiaries of the policy adopted.

Whatever the strategy chosen, the governments of developing nations are faced with a number of basic dilemmas resulting from the constraints imposed by scarce economic and human resources.

Of these dilemmas the most embarrassing and difficult to resolve are decisions on priorities in the mobilization and allocation of resources between two generally conflicting alternatives: to apply the scarce resources in such a way as to achieve the maximum economic returns on a national scale; or to achieve 'social efficiency' by improving the situation of the underprivileged sectors of rural society.

Many developing countries have opted for the first approach, investing available resources in the most favoured regions and promoting the interests of the advanced rural sector in order to achieve the most rapid increase possible in agricultural production. As a result, the already great disparities between the regions and sectors have been widened, a process further accelerated where agriculture has been modernized following the introduction of the green revolution.

There is no general consensus on whether the strategy to be adopted in agricultural development should be based on a policy of efficiency or of equity. A rigid adherence to one or the other policy is probably not tenable. Efficiency alone cannot provide a satisfactory guideline because the increased disparities between sectors is politically and socially most dangerous, and in the long run economically counterproductive. Complete commitment to equity is not realistic in view of the scarcity of available resources.

Equality is probably impossible to achieve or maintain in the long term, excepting possibly in highly motivated and ideologically strong societies, such as the Kibbuz. The abilities—physical and mental—of the individual, his initiative and drive, are bound to create differences in income levels, unless repressed by authoritarian measures. These, in turn, are generally self-defeating, because they eliminate major incentives for individual productivity.

Also, a certain widening of interregional and interfarm inequalities is probably an inevitable concomitant of the progressive modernization of agriculture, whatever the strategy adopted. There will always be regions that are ecologically more favoured, and farmers with more than average ability or greater resources than others.

The potential inequalities resulting from the adoption of new technologies *can be reduced* by designing agricultural programmes and by developing infrastructures with preferential attention to the small farmer; by safeguarding the land rights of tenants and small farmers; by gearing credit programmes to the needs of small farmers; by adopting appropriate mechanization and irrigation programmes; and probably most important of all, promoting the ability of the small farmers to defend their own interests.

What is essential and possible to achieve is (a) provide the minimum physical and economic resources enabling the achievement of an acceptable standard of living by the underprivileged rural sector; (b) prevent accumulation of physical resources, such as land and water rights, beyond a predetermined limit; (c) outlaw various forms of exploitation of the weaker elements of society.

The essential precondition for the realization of such a policy is that the necessary measures are taken by government to prevent the appropriation of the means of production by the stronger elements of the rural economy.

This is not an easy objective to achieve, because the rural elite generally combines status, political influence and economic power. Only governments genuinely devoted to national objectives, and not to the perpetuation of monopoly power structures and sectoral interests, can achieve these aims.

In order to secure the interests of the disadvantaged rural sector, Government benevolence cannot be sufficient. Effective political action is required in order to enhance the power of the small farmers in the market and to secure equal access for them to the benefits available from institutions providing (or controlling) rural services. (Harvey *et al*., 1979).

REFERENCES

Alcantara, de, Cynthia H. (1976). *Modernizing Mexican Agriculture: Socio-economic Implications of Technological Change 1940–1970*. UN Research Institute for Social Development, Geneva: xvii + 350 pp.

Aziz, S. (1973). The Chinese approach to rural development. *Int. Dev. Rev*., **15** (4), 2–7.

Barnett, A. D. (1979). *China and the World Food System*. Overseas Development Council, Washington, DC. x + 115 pp.

Barraclough, S. L. (1971). Rural development strategy and agrarian reform Paper presented at the *Latin American Seminar in Agrarian Reform and Colonization*, FAO, UNDP, Chiclayo, Peru.

Barraclough, S. L. (1973). Rural development and agrarian reform. Pp. 515–48 *Rural Development and Employment* (Ed. C. Gotsch). Ford Foundation Seminar, Ibadan, Nigeria: 774 pp. (mimeogr.).

Blue, R. N. & Weaver, J. H. (1977). *A Critical Assessment of the Tanzanian Model of Development*. The Agricultural Development Council, Inc.: Reprint No. 30, 19 pp.

Brown, L. R. (1970). *Seeds of Change*. Praeger, New York: xv + 205 pp., illustr.

Castillo, Gelia T. (1975). *All in a Grain of Rice*. Southeast Asian Regional Center for Graduate Study and Research in Agriculture: xii + 410 pp.

CIMMYT (1974). *The Pueblo Project: Seven Years of Experience: 1967–1973*, CIMMYT, Mexico: ix + 116 pp., illustr.

Dauphin-Meunier, (1976). Le modele chinois de développement rural. *C.R. Acad. Agric. de France*, **62**, 184–207.

Dawson, O. L. (1970). *Communist China's Agriculture: Its Development and Future Potential*. Praeger, New York: xvii + 826 pp.

Diaz, H. (1974). An institutional analysis of a rural development project: the case of the Pueblo project in Mexico. *Ph.D. Thesis*, University of Wisconsin.

Drilon, J. D., Jr (1976). Masagana 99: an integrated production drive in the Philippines. Paper presented at a *Sem. on Accelerating Agricultural Development and Rural Prosperity, 5–18 September 1976, University of Reading, England:* (mimeogr.).

Dumont, R. (1966). *African Agricultural Development*. Food and Agriculture Organization of the UN, Rome: vi + 243 pp.

Eicher, C. K. (1970). Some problems of agricultural development: a West African case study. Pp. 196–246 in *Africa in the 70's and 80's* (Ed. F. Arkhurst). Praeger, New York: xiii + 405 pp.

Falcoln, W. P. (1970). The green revolution: generations of problems, *Am. J. Agric. Econ*., **52**, 698–710.

FAO (1973). *The State of Food and Agriculture*. Food and Agriculture Organization of the UN, Rome: xii + 222 pp., illustr.

FAO (1976). *The State of Food and Agriculture 1975: Agricultural Series No*. 1. Food and Agriculture Organization of the UN, Rome: vi + 150 pp.

FAO (1977). *The State of Food and Agriculture 1976: Agricultural Series No*. 4, Food and Agriculture Organization of the UN, Rome: vi + 157 pp.

FAO Study Mission (1978). *Learning from China: A Report on Agriculture and the Chinese People's Communes*. Food and Agriculture Organization of the UN, Rome. viii + 112 pp.

FAO (1979). *The State of Food and Agriculture 1978: Agricultural Series No.* 9, Food and Agriculture Organization of the UN, Rome.
French, J. T. & Lyman, P. N. (1969). Social and political implications of the new cereal varieties. *Development Digest,* **7** (4), 111–18.
Gaitskell, A. (1968). Importance of agriculture in economic development. Pp. 46–58 in *Economic Development of Tropical Agriculture* (Ed. W. W. McPherson). University of Florida Press, Gainesville, Florida: xvi + 328 pp., illustr.
Gerhart, J. D. (1973). Management problems and rural development in Kenya. Pp. 196–217 in *Rural Development and Employment* (Ed. C. Gotsch). Ford Foundation Seminar, Ibadan, Nigeria: 774 pp. (mimeogr.)
Grant, J. P. (1973). *Growth from Below—A People-Oriented Development Strategy: Dev. Pap. No.* 16. Overseas Development Council: pp. 32.
Griffen, K. (1972). *The Green Revolution: an Economic Analysis*. UN Research Institute for Social Development, Geneva: xii + 153 pp. illustr.
Griffen, K. (1973). Policy options for rural development. Pp. 618–79 in *Rural Development and Employment* (Ed. C. Gotsch). Ford Foundation Seminar, Ibadan, Nigeria: 774 pp. (mimeogr.).
Gurley, J. G. (1973). Rural development in China, 1949–1972, and the lessons to be learnt from it. Pp. 306–56 in *Rural Development and Employment* (Ed. C. Gotsch). Ford Foundation Seminar, Ibadan, Nigeria: 774 pp. (mimeogr.).
Harvey, C., Jacobs, J., Lamb, G., & Schaffer, B. (1979). *Rural Employment and Administration in the Third World*. The International Labour Office, Teakfield Ltd., Westmead, England. xi + 111 pp.
Heck, B. van (1979). *Participation of the Pool in Rural Organizations.* Food and Agriculture Organization of the UN, Rome: xiii + 98 pp.
Heyer, J. (1973). *Smallholder Credit in Kenya Agriculture*: *IDS Working Pap. No.* 85.
Hill, F. F. (1966). Developing an effective extension service. Pp. 331–5 in *Selected Readings to Accompany Getting Agriculture Moving* Vol. 1 (Ed. R. A. Borton). Agricultural Development Council, New York: 526 pp.
Hirschman, A. O. (1958). *The Strategy of Economic Development*. Yale University Press, New Haven, Conn.: xiii + 217 pp.
Hunter, G. (1970). *The Administration of Agricultural Development: Lessons from India*. Oxford University Press, London: 160 pp.
Johnston, B. & Kilby, F. (1973). *Agricultural Strategies, Rural–Urban Interactions and the Expansion of Income Opportunities*. OECD Development Centre, Paris: 279 pp., illustr.
Jones, G. (1971). *The Role of Science and Technology in Developing Countries.* Oxford University Press, London: xviii + 174 pp., illustr.
Kellogg, C. E. (1962). Interactions in agricultural development. Pp. 12–24 in *The Application of Science and Technology for the Benefit of the Less Developed Areas*, Vol. 3. United Nations, Geneva: viii + 309 pp.
Krebs, W. A. W. (1964). The developing countries—new frontiers for research. Pp. 46–51 in *Proc. 17th Natl. Conf. on Administration in Research, University of Denver, Colorado.*
Kuo, L. T. C. (1972). *The Technical Transformation of Agriculture in Communist China.* Praeger, New York: xx + 261 pp., illustr.
Ladejinsky, W. (1970). Ironies of India's green revolution. *Foreign Affairs*, **38**, 758–68.
Lele, Uma J. (1975). *The Design of Rural Development—Lessons from Africa.* The Johns Hopkins Press, Baltimore and London. xiii + 246 pp.
Lele, Ulma J. (1976). Designing rural development programmes: lessons from past experience in Africa. *Econ. Deve. & Cult. Change*, **24**, 287–308.

McNamara, R. S. (1975). The Nairobi Speech, pp. 90–98 in *The Assault on World Poverty* (The World Bank) The Johns Hopkins University Press, Baltimore & London, xi + 425 pp.
Mathur, G. (1965). *Planning for Steady Growth*. Basil Blackwell, Oxford: xvi + 386 pp., illustr.
Mellor, J. W. (1962). The process of agricultural development in low-income countries. *J. Farm. Econ.*, **44**, 700–16.
Nelson, N. (1973). *The Development of Tropical Lands: Policy Issues in Latin America*. Resources for the Future, Baltimore, Md: xii + 306 pp., illustr.
Nicholls, W. H. (1964). The place of agriculture in economic development. Pp. 11–44 in *Agriculture in Economic Development* (Ed. C. Eicher & L. Witt). McGraw-Hill, New York: vi + 415 pp., illustr.
Nyerere, J. K. (1967). *Socialism and Rural Development*. Government Printers Office, Dar-es-Salaam, Tanzania.
Ojala, E. M. (1967). The programming of agricultural development. Pp. 548–85 in *Agricultural Development and Economic Growth* (Ed. H. M. Southworth & B. F. Johnston). Cornell University Press, Ithaka, NY: xv + 608 pp. (mimeogr.).
Palmer, Ingrid (1976). *The New Rice in Asia: Conclusions from Four Country Studies*. United Nations Research Institute for Social Development, Geneva, vii + 146 pp.
Pearse, A. (1977). Technology and peasant production: Reflections on a global Study. *Dev. & Change*, **8**, 125–59.
Pearse, A. (1980). *Seeds of Plenty, Seeds of Want. Social and Economic Implications of the Green Revolution*. Clarendon Press, Oxford. xi + 262 pp.
Ruttan, W. V. & Hayami, Y. (1973). *Technology Transfer and Agricultural Development: Staff Pap.* 73–1. Agricultural Development Council, New York: 32 pp.
Sanchez, L. J. (1970). The Pueblo project. Pp. 11–18 in *Strategies for Increasing Agricultural Production on Small-Holdings: CYMMIT Int. Conf., Pueblo, Mexico* (Ed. D. T. Myren). CYMMIT, Mexico: 86 pp., illustr.
Schickele, R. (1971). National policies for rural development in developing countires. Pp. 57–72 in *Rural Development in a Changing World* (R. Weitz). MIT Press, Cambridge, Mass.: 587 pp.
Schutjer, W. A. & Coward, E. W. Jr (1971). Planning agricultural development: the matter of priorities. *J. Dev. Areas*, **6**, 38–39.
Seidman, A. (1972). *Comparative Development Strategies in East Africa*. East African Publishing House, Nairobi: 299 pp.
Shourie, A. (1973). Growth and development. Pp. 387–425 in *Rural Employment and Development* (Ed. C. Gotsch). Ford Foundation Seminar, Ibadan, Nigeria: 774 pp. (mimeogr.).
Takahashi, A. (1970). *Land and Peasants in Central Luzon: Socio-Economic structure of a Philippine Village*. East–West Centre Press, Honolulu: x + 169 pp., illustr.
Verhagen, K. (1980). Changes in Tanzanian rural development policy 1975–78. *Development & Change*, **11**, 285–295.
Walker, K. R. (1965). *Planning in Chinese Agriculture*. Frank Cass & Co., London: xviii + 109 pp.
Watters, R. F. (1971). *Shifting Cultivation in Latin America*. Food and Agriculture Organization of the UN, Rome: 305 pp., illustr.
Weitz, R. (1971). *From Peasant to Farmer: A Revolutionary Strategy for Development*. Columbia University Press, New York: 292 pp.
Wilde, J. C. de & McLoughlin, P. F. M. (1967). *Experiences with Agricultural Development in Tropical Africa*. Johns Hopkins University Press, Baltimore, Md: Vol. 2, xii + 446 pp., illustr.

Winkelmann, D. L. (1976). *The Adoption of New Maize Technology in Plan Pueblo, Mexico*. CIMMYT, Mexico: vi + 24 pp.
Wortman, S. & Cummings, R. W. (1978). *To Feed this World—The Challenge and the Strategy*. Johns Hopkins University Press, Baltimore Md: xiv + 440 pp., illustr.

Author Index

Subject Index